Microplastics Pollution

Vijay Kumar · Pardeep Singh · Shailender Kumar ·
Ankit Kumar Singh
Editors

Microplastics Pollution

A Threat to Biosphere

Springer

Editors
Vijay Kumar
Department of Environmental Studies
Satyawati College
University of Delhi
New Delhi, Delhi, India

Pardeep Singh
Department of Enviornmental Studies
PGDAV College
University of Delhi
New Delhi, Delhi, India

Shailender Kumar
Department of Environmental Studies
and Disaster Mangement
University of Delhi
New Delhi, Delhi, India

Ankit Kumar Singh
Department of Botany
Lalit Narayan Mithila University
Darbhanga, Bihar, India

ISBN 978-3-031-99773-0 ISBN 978-3-031-99774-7 (eBook)
https://doi.org/10.1007/978-3-031-99774-7

This Springer imprint is published by the registered company Springer Nature Switzerland AG
The registered company address is: Gewerbestrasse 11, 6330 Cham, Switzerland

Contents

Chapter 1
Introduction to Microplastics: Origin, Environmental Pathways, and Impacts on Ecosystems and Human Health

Vijay Kumar, Anupam Sunny, Khushaboo Verma, Ambneesh Mishra, Md. Izhar Alam, Hira Lal Yadav, Javed Alam, Sweta Mohan, and Devendra Kumar Singh

Abstract The plastics which were valued for their durability and versatility, have become ubiquitous in modern applications, from packaging to medicine. However, their non-biodegradable nature and poor waste management have led to severe environmental pollution, with over 430 million tons produced annually, which is projected to double by 2040. The improper disposal of such a huge plastic waste results in the release and accumulation of microplastics (MPs, < 5 mm) and nanoplastics (NPs, < 100 nm) in air, water, soil, and biota, which is being produced from both intentional production and degradation of larger plastics. These persistent MPs contaminate the various ecosystems, enter into the food chains through ingestion, inhalation, and dermal contact, and act as vectors for several toxic chemicals like heavy metals and

V. Kumar (✉) · A. Sunny
Department of Environmental Studies, Satyawati College, University of Delhi, New Delhi, India
e-mail: kumarv.rs.apc13@itbhu.ac.in

K. Verma
Faculty of Education, Banaras Hindu University, Varanasi, India

A. Mishra
Department of Environmental Science, Ramanujan College, University of Delhi, New Delhi, India

Md. I. Alam
Department of Environmental Studies, University of Delhi, New Delhi, India

H. L. Yadav · J. Alam
Department of Civil Engineering, G.B. Pant Institute of Engineering and Technology, Pauri Garhwal, Uttarakhand, India

S. Mohan
Department of Management Studies, Madan Mohan Malviya University of Technology, Uttar Pradesh, Gorakhpur, India

D. K. Singh
Pharmacopoeia Commission for Indian Medicine & Homoeopathy (PCIM&H), Ghaziabad, India

V. Kumar
Department of Chemistry, Govt. P. G. College, Amarpatan, M. P, India

V. Kumar et al. (eds.), *Microplastics Pollution*,
https://doi.org/10.1007/978-3-031-99774-7_1

polycyclic aromatic hydrocarbons, and increase their ecological and health impacts. In aquatic systems, MPs cause physical harm through ingestion and entanglement and reduces the growth and survival of organisms. In terrestrial environments, MPs disrupt soil structure, microbial activity, and nutrient cycling and impairs the health of plants. Human exposure to MPs is linked to respiratory, immune, and reproductive health issues, exacerbated by toxic additives like phthalates and flame retardants. The current chapter explores the sources, pathways, and impacts of MPs and NPs, emphasizing the urgent need for improved waste management, biodegradable alternatives, and global cooperation to mitigate plastic pollution and safeguard ecosystems and human health.

Keywords Microplastics · Nanoplastics · Plastic pollution · Environmental contamination · Toxic chemicals · Food chain contamination · Plastic additives · Biodegradable plastics

1.1 Introduction

Plastics were discovered as a most versatile material and started to use widely in packaging, construction, and automobiles due to its durability, impermeability, nonbiodegradability, anticorrosive, insulating, lightweight, and inexpensive properties (Pan et al., 2020). Plastics are also used widely in spacecrafts, buildings, agriculture, medicine, and are a chief ingredient in various materials such as paints, adhesives, and polymer composites. Despite of numerous applications, the mismanagement and haphazard use of plastic items and lack of ethics, carelessness, responsibility the human being have made it ubiquitous in nature and greatest threat to the survival of flora and fauna on planet earth. Since 1950, the total amount of plastic produced is more than 9.2 billion tons, out of which 5.3 billion tons have been discarded, 1 billion tons is incinerated, 0.7 billion tone is recycled, and 2.9 billion tons are still in use. At present, the total amount of plastic generation is more than 430 million tons (Pilapitiya et al., 2024) wherein 227 million tons are mismanaged, 129 million tons are disposed-off, and only 73 million tons are recycled (Fig. 1.1). According to an estimate, the global production of plastic is to be doubled by 2040.

Since, the plastics are non-biodegradable in nature, its mismanagement and improper disposal in environment has created an unavoidable threat to all kinds of lives present on the earth. It has now become one of the greatest pollutions which is being accumulated in the all the segments of the environment, i.e., hydrosphere, atmosphere, lithosphere, and biosphere. After the poor disposal and mismanagement in the environment, the plastic is exposed to various physical, chemical, and biological processes such as mechanical abrasion, hydrolysis, photodegradation, chemical oxidation, and biodegradation. Thus, the plastic gets degraded, fragmented, and disintegrated and releases very fine particles in the environment. These fine particles with size < 5 mm are called as MPs (MPs) and smaller than 100 nm are called nanoplastics (NPs). These MPs and NPs which are generated by the fragmentation

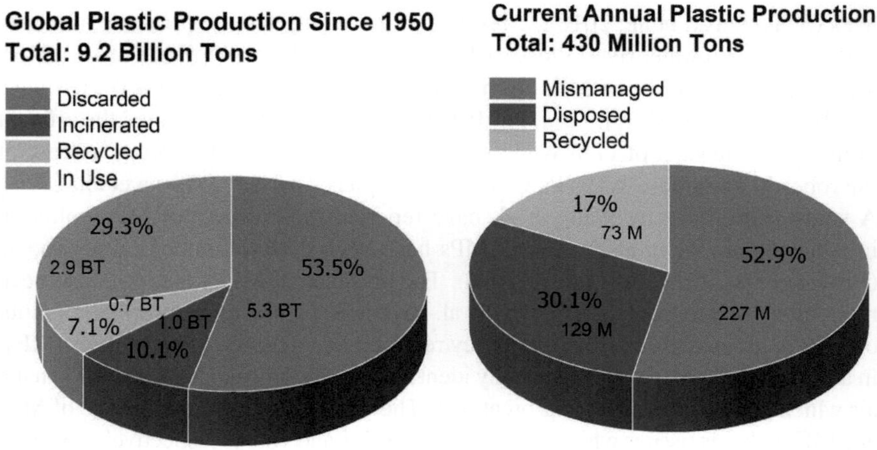

Fig. 1.1 Global plastic production since 1950 and current annual production

and degradation of larger pieces of plastics are regarded as secondary MPs whereas the MPs which are deliberately manufactured for use in various commercial products such as cosmetics, detergents, medicines, insecticides, paints, etc., are referred as primary MPs (Fig. 1.2). The further disintegration of primary and secondary MPs, generates NPs. MPs and NPs are highly stable in the environment due to their nonbiodegradability nature. Hence, they can persist in the air, water, and soil for decades.

The intentional production and unintentional release of MPs in the environment is continuously increasing and contaminating the air, water, and soil. Schymanski et al. have reported the contamination of drinking water and bottle mineral water with

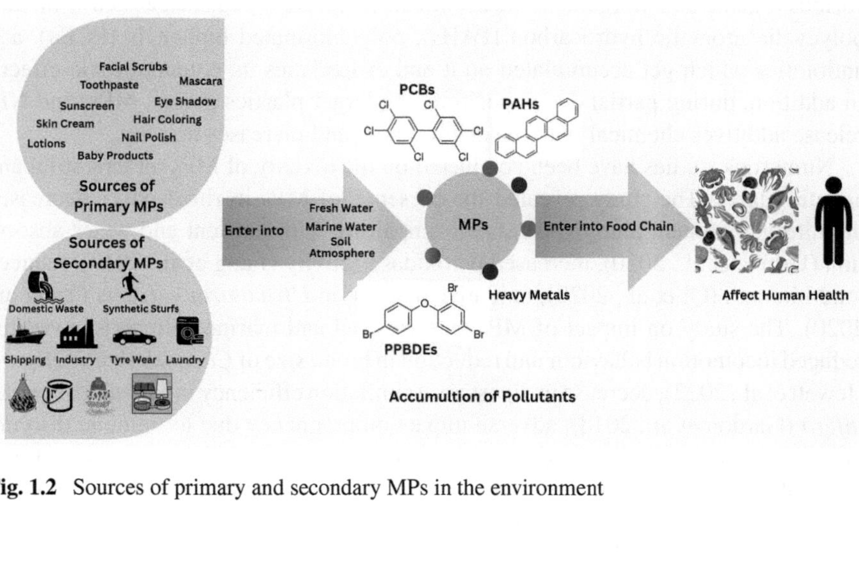

Fig. 1.2 Sources of primary and secondary MPs in the environment

MPs (Schymanski et al., 2018). Eriksen et al. have detected the presence of higher MPs in the fresh water rivers close to densely populated areas (Eriksen et al., 2013). The substantial release of MPs from wastewater treatment plants (WWTPs) have also been reported (Murphy et al., 2016). Mason et al. have investigated the effluent of treated wastewater of 17 WWTPs of USA. They found that the average release of microbeads from these WWTPs was 13 billion pieces per day (Mason et al., 2016). A study carried out by Wang et al. have reported the problem of MPs pollution in aquaculture (Wang et al., 2020). MPs have been detected in coastal sediments (Browne et al., 2011; Frias et al., 2016). The presence of MPs fibre have also been reported by several researchers. Dris et al. have investigated the atmospheric fallout of MPs in urban and suburban site. They reported the presence of MPs fibres (MFs) in air (Dris et al., 2016). A recent study identified the presence of MPs and MFs in the air which can be inhaled during breathing. The study revealed the presence of MPs and MFs in lungs tissues which were less than 5. 5 μ and 16.8 μ respectively (Amato-Lourenço et al., 2020). The MPs and MFs present in air get deposited in water and over the land surfaces and exacerbates the contamination of water and soil. More than 70% of the global plastic waste generated is discarded via landfilling which make the soil as a great sink of MPs. A farmland-based study of Sanghai, China have reported the presence of 62.5 MPs items per Kg in deep soil and 78.0 MPs items in shallow soil (Liu et al., 2018). MPs have also been reported in home garden (Lwanga et al., 2017), floodplain and high mountain areas (Scheurer & Bigalke, 2018), as well as soil of industrial areas (Fuller & Gautam, 2016).

Due to their tiny size, the MPs, and NPs can easily be ingested by various aquatic organisms, absorbed by root of the plants, and inhaled from air during breathing and enter the terrestrial and aquatic food chains. Subsequently, the MPs get bioaccumulated in the various tissues of plants and animals. The bioaccumulation of MPs in gradual higher trophic level poses significant adverse impact on individual organisms as well as whole ecosystem. These MPs and NPs also provide suitable surface for various organic and inorganic toxic chemicals to get adsorbed such as heavy metals, polycyclic aromatic hydrocarbon (PAHs), polychlorinated biphenyls (PCBs), and antibiotics which get accumulated on it and exacerbates its potential toxic effects. In addition, during partial decomposition, the larger plastics pieces, MPs, and NPs release additives chemical in to the environment and increases toxicity.

Numerous studies have been conducted on the toxicity of MPs on terrestrial and aquatic plants. The study revealed the presence of MPs in rhizosphere decreased the nitrogen content and growth due to impairing with nutrient and water absorption (Urbina et al., 2020), increased peroxidase activity (Jiang et al., 2019), reduced total biomass (Qi et al., 2020), oxidative damage in *Utricularia vulgaris* (Yu et al., 2020). The study on impact of MPs on terrestrial and marine animals showed the reduced locomotion behaviour and reduction in brood size of *Caenorhabditis elegans* (Jewett et al., 2022), decrease in digestive assimilation efficiency in *Pinctada margaritifera* (Gardon et al., 2018), adverse impact on pregnancy due to immune disorder in mice (Hu et al., 2021), and several metabolic, immune, and neurotoxic effects (Li et al., 2023). The impact of MPs and microfibre on human health corroborated the drastic health issues including chronic respiratory disease in polymer factories workers (Kremer et al., 1994), interstitial lungs disease (Eschenbacher et al., 1999). Horvatits et al., have detected the higher concentration of MPs in cirrhotic liver tissues (Horvatits et al., 2022). Exposure to MPs resulted into asthma, and allergy in children (Bamai et al., 2014), and shorter pregnancy duration due to release of phthalates (Latini et al., 2003).

1.2 Formation of Plastics

The first step in the production of plastics is the extraction of raw materials such as coal, natural gas, and crude oil. These materials, which are made up of intricate compounds, are processed to yield monomers, which are necessary chemicals that serve as the building blocks for polymers (Lodge et al., 2020). Naphtha, which is made from crude oil, is a crucial component in this process; however gas can also be a source (Speight, 2016). Crude oil is heated during the refining process and then distilled into various fractions, such as naphtha, which is necessary for the synthesis of monomers (Fahim et al., 2009). These monomers, including propylene and ethylene, go through a process called polymerization, in which they combine chemically to create polymers (Makio et al., 2011). This polymerization occurs through two pathways. The first pathway is addition polymerization where monomers link in a chain reaction using catalysts like peroxide which produce plastics like polyethylene and polystyrene (Russell, 2002). The second pathway is condensation polymerization in which different monomers join, releasing small molecules like water, resulting in polymers like polyester and nylon (Sharma & Kundu, 2008).

Finally, polymers are combined with additives during compounding and processing to modify their properties. Pellets generated from the mixture are molded into finished or semi-finished plastic products using extrusion and other techniques. Plastics are ensured to meet specific design, size, form, and property requirements for a range of applications in this way.

1.3 Types of Plastics

Plastics can be categorized according to a number of factors, including their origin, structure, molecular forces, polymerization mechanism, behavior that varies with temperature, and preservation methods.

1.3.1 Based on Origin

On the basis of origin, the plastics are categorized into natural, synthetic and semi-synthetic categories. Natural plastics are made of polymers extracted from natural materials obtained from plants and animals including tortoise shell, amber, shellac, rosin, and horn. The major examples of such polymers are as lignin, cellulose, starch, silk, collagen, keratin, and chitin (Klemm et al., 2005).

Synthetic polymers are derived from petroleum crude oil during its purification process. Such polymers have a wide range of applications due to their durability and versatility. These polymers are advantageous over natural polymers due to their purity and reproducibility (Andrady & Neal, 2009). The examples of such polymers include polyethylene, polystyrene, polyvinyl chloride, polyamide, polyethylene terephthalate, silicone, synthetic rubber, etc.

The natural polymers which are chemically modified are known as semi-synthetic plastics. The major examples of such plastics include, rayon (regenerated cellulose fibers), cellulose acetate (acetylating cellulose), celluloid (derived from or wood pulp), vulcanized rubber (natural rubber modified with sulphur) (Gilbert, 2016).

1.3.2 Based on Structure

There are three categories for the structure-based classification: linear, branched, and cross-linked. In linear plastics, the monomers are connected in a simple straight chain and have high density, and flexibility and higher solubility in solvents (Painter & Coleman, 1997). The examples include, high-density polyethylene polyethylene, polypropylene, polystyrene. The branched polymers are similar to linear polymers with side branches with lower density, and flexibility, and lower solubility in solvents such as low-density polyethylene (Sperling, 2015). In cross-linked polymers, the different polymeric chains are connected in a network-like structure and have extremely high strength, rigidity, melting point, and insolubility. Volcanized rubbers and bakelite are examples of cross-linked plastics (Mark, 2007).

1.3.3 Based on Molecule

The molecular-based classification categorizes plastics into four types: elastomers, fibre, thermoplastic, and thermosetting plastics. The elastomers are plastics with elasticity which can stretch and return back to their original shape. For example; natural rubber, silicon rubber, and polyurethane (Mark, 2004). The polymers which spun into long thin thread and used in textiles are called fibres such as nylon, polyester, acrylic, etc. (Hearle, 2001).

Thermoplastic and thermosetting polymers are separated by their physio-chemical characteristics both before and after heat treatment. Thermoplastics may solidify when cooled and melt into any shape when heated. Thermoplastics can therefore undergo repeated deformation with normal plasticity. About 80% of all plastic is made up of thermoplastics (Nicholson, 2017). Whereas the thermosetting plastics break-down after heating and become permanently hardened during the molding process.

These plastics cannot be remelted and reshaped after molding process (Brydson, 1999).

1.3.4 Based on Biodegradability

Based on biodegradability, the plastics are categorized into biodegradable and non-biodegradable plastics. Biodegradable plastics breakdown naturally by microorganisms in relatively shorter period. The examples of biodegradable plastics include polylactic acid (obtained from fermented starch), polyhydroxyalkanoates (produced by microbes from organic materials) (Vert et al., 2012). However, the non-biodegradable plastics don't decompose naturally and remain in the environment for hundreds of years. Few examples are polyethylene, polypropylene, polyvinyl chloride, polystyrene, etc. (Andrady, 2011).

1.3.5 Based on Life Cycle

On the basis of life cycle, the plastics are classified as single-use plastics and reusable plastics. The plastics which are intended for single use only and are meant to be thrown away after their use are called single-use plastics. Because of their widespread use and short lifespan, it pollutes the environment tremendously due to mismanagement by human (Geyer et al., 2017). The examples of single-use plastics include, plastics bags, plastic straws, plastics cutlery, plastic bottles, plastic wrappers, sachets, containers, etc. For reducing the plastics waste and environmental impacts of plastics, some plastics are designed for their multiple use which are durable, can be cleaned, and reused. For examples, reusable water bottles, plastic utensils, plastic containers, etc. (Hopewell et al., 2009). In addition to this, the Society of the Plastic Industry (SPI) introduced a number coding system in 1988 for the recycling of the plastics which is called SPI codes. The SPI codes are represented in a three recycling arrows in triangular shape which carry a number within it. The number within the recycling arrow represents different types of plastic resin. There are seven different types of plastics classified in this classification system with 1 denoting polyethylene terephthalate (PET), 2 high-density polyethylene (HDPE), 3 polyvinyl chloride (PVC), 4 low-density polyethylene (LDPE), 5 polypropylene (PP), 6 polystyrene (PS), and 7 other types of plastics (Hussein et al., 2023). This categorization makes recycling easier for both recyclers and consumers.

1.4 Types of Additives Used in Plastics

Any ingredient that is intentionally added to plastics during processing to improve its physical or chemical properties is referred as additive. The common additives used in plastics are antioxidants, flame retardants, plasticizers, stabilizers, etc. The kind of additive is determined by the plastic polymer and the end product's specifications. The commonly used additives in plastics are following:

1.4.1 Brominated Flame Retardant

Brominated flame retardants (BFRs) are additives used in plastic products to reduce flammability and are used in electrical/electronic equipment. There are five major class of BFRs used in plastic industry are following:

(i) ***Hexabromocyclododecane (HBCD)***: HBCD is used as thermal insulator in building industry. It is listed as a persistent organic pollutant (POP) under the Stockholm Convention which exists in three stereoisomers namely α-, β-, and γ-HBCD. It is used in expanded and extruded polystyrene (EPS and XPS) at 4–7% concentrations. HBCD has been regulated under the EU's Registration, Evaluation, Authorization and Restriction of Chemicals (REACH) regulation since 2006 and was listed for elimination under Annex A of the Stockholm Convention in 2013, with exemptions for EPS and XPS (Marvin et al., 2011).

(ii) ***Polybrominated Diphenyl Ethers (PBDEs)***: PBDEs are extensively used as flame retardants and commonly used in plastic and electronic casting. These are hydrophobic, persistent, bioaccumulative, and toxic in nature and pose significant environmental and health risks. There are three main commercial formulations of PBDEs; penta, octa, and deca BDEs which have faced increasing regulatory restrictions. The European Union banned penta- and octa-BDEs in 2004 and deca-BDE in 2009 for electronic and electrical applications, with additional restrictions limiting their concentration to 0.1% in mixtures or products. Tetra to hepta BDEs are listed for elimination under the Stockholm Convention (Darnerud, 2003).

(iii) ***Tetrabromobisphenol A (TBBPA)***: TBBPA are produced by brominating bisphenol A and is the most widely produced BFR globally, accounting for 60% of the BFR market. It is primarily used in acrylonitrile butadiene styrene (ABS) plastics and other materials like high-impact polystyrene and phenolic resin (Covaci et al., 2011).

(iv) ***Polybrominated Biphenyls (PBBs)***: PBBs are majorly applied in plastic foams. These are extremely stable and persistent in nature.

The emerging BFRs, such as 1,2-bis (2,4,6-tribromophenoxy) ethane (BTBPE), decabromodiphenylethane (DBDPE), and hexabromobenzene (HBB), have raised concerns due to their presence in the environment, organisms, food, and humans.

Unlike TBBPA, most BFRs are not chemically bound to polymers, allowing them to leach into the environment (Law 2009).

1.4.2 Plasticizers

A plasticizer is a material or substance added to a material (often a plastic or elastomer) to improve its workability, flexibility, or distensibility, according to the Council of the International Union of Pure and Applied Chemistry (IUPAC) (Vert et al., 2012). According to this description, a plasticizer can decrease the product's elastic modulus, melt viscosity, or second-order transition temperature.

Based on chemical composition, plasticizers are often divided into two classes: phthalates and non-phthalates. Phthalates, also known as phthalic acid esters (PAEs), that are frequently employed in the manufacturing of PVC and make about 10–60% of PVC by weight (Schettler, 2006). These are colourless liquid and soluble in hexane and other organic solvents which make them soluble in body fluids like saliva and blood plasma also. The plasticizers diethylhexyl phthalate (DEHP) and diisononyl phthalate (DINP) are two of the most widely utilized for PVC. It is now widely accepted that exposure to phthalates may pose a health risk to human health. Phthalate plasticizers can leach, migrate, or evaporate into indoor air and atmosphere, food, and other items as these are not chemically bound to PVC. Phthalates found in consumer goods can expose people directly through contact and consumption, indirectly through product leaching, or generally through environmental contamination. Phthalates can act as endocrine disruptors.

Benzoates, trimellitic esters, phosphoric esters, citric esters, adipates, or sebacates are examples of non-phthalate plasticizers. Adipates already widely employed in PVC products, while others are only utilized in niche items due to their high cost, which prevents them from being used in common applications. Regarding their possible impacts on the environment and human health, the majority of these plasticizers have not been thoroughly investigated (Bernard et al., 2015).

1.4.3 Antioxidants

Antioxidants are additives used in synthetic polymers, particularly in polyolefins like PE and PP, which make up 60% of the global demand for antioxidant additives. They help prevent plastic aging and delay oxidation. However, like other additives, antioxidants can leach from plastics, potentially migrating into food from packaging and posing food safety risks (Mesquita and Abrantes, 2022). Common antioxidants used in plastics include Irganox 1076, Irganox 1010, and Irgafos 168, all of which belong to the Irganox series.

Bisphenol A (BPA) is a widely produced chemical, with over 3 million tons manufactured annually. It is primarily used as a monomer in polycarbonate plastics

(65%) and epoxy resins (30%), such as those lining aluminum cans. BPA is also used as an antioxidant or plasticizer in other polymers like PP, PE, and PVC. BPA can leach from food and drink packaging, posing exposure risks to humans. Despite being identified as a significant endocrine disruptor, BPA is still permitted for use in food-contact materials in the European Union (Vandenberg et al., 2007).

1.4.4 Nonylphenols (NP)

Nonylphenols (NP) are byproducts of the degradation of nonylphenol ethoxylates (NPE), which are used in a variety of applications such as paints, pesticides, detergents, and personal care products. NP can also serve as antioxidants and plasticizers in plastics production. They can leach from plastic bottles into water and are primarily released into the environment through wastewater treatment plant effluents. NPs are considered endocrine disruptors, and due to their harmful effects on the environment and human health, their use is prohibited in the European Union (Soares et al., 2008).

1.5 Weathering and Degradation of Plastics: A Pathway to Generation of MPs

The various types of plastics with different chemical additives when exposed to physical, chemical, and biological actions break the macroplastics into MPs with the size range from 1 μ to 5 mm and nanoplastics with the size less than 1000 nm. There are two ways of degradation of plastics and formation of MPs.

1.6 Abiotic Degradation

Abiotic degradation is the change in physical and chemical properties of plastics due to abiotic factors like temperature, light, water, air, and mechanical force.

1.7 UV Radiation and Photo-Oxidation

Exposure to ultraviolet (UV) radiation from sunlight is one of the main factors contributing to the breakdown of plastics. UV radiations like UV-B (290-315), and UV-A (315-400) are the main UV radiations due to having sufficient energy to breakdown chemical bonds in polymers and initiate photo-oxidation (Chamas et al., 2020).

The photo-oxidation of plastics break down the polymer chain and lead to the formation of free radicals. These free radicals react with oxygen, causing oxidation, embrittlement, and discoloration. PE has a strong resistance to photodegradation due to lack of chromophores which absorb UV-radiations and initiated degradation process. However, during manufacturing or environmental exposure, the impurities and structural irregularities can act as chromophores which absorb UV-radiations (Chamas et al., 2020). Consequently, free-radical formation occurs and plastic get breakdown via a process known as chain-scission. In contrast, PP is more prone to photodegradation due to presence of tertiary carbon atom structure and chromophores from impurities which make it susceptible to oxidation when exposed to UV-radiation (Ibrahim, 2024). As a result, free-radical formation occurs which accelerate the degradation process through cross-linking and chain-scission reactions.

When exposed to UV radiation, PVC rapidly undergoes dehydrochlorination and result in the formation of short sequences of conjugated unsaturation within the polymer. These C=C are inherently unstable and get degraded under continued UV exposure. Similar to PE and PP, the presence of chromophores in PVC, often introduced by impurities which enables the absorption of UV radiation and initiate free-radicals formation. These radicals promote the formation of hydroperoxides which in-turn break the backbone of the polymer and generate MPs. Recently, a new route of degradation of PVC has identified in environmentally relevant conditions. This degradation pathway starts with the formation of polyene structures which oxidised into ketones and alcohols after interacting with O_2 and $OH^{\cdot}$ radicals (Najafi et al., 2013).

1.8 Thermal Stress

Thermal degradation is the process by which the plastic breakdown due to exposure to high or variable temperatures. At high temperature, the polymers can undergo thermos-oxidative reaction where the absorbed heat provide enough energy to break long chain polymers and generate free-radicals. These radicals interact with oxygen and form hydroperoxides, which then decompose into hydroxy and alkoxy radicals (Aboulkas & El Bouadili, 2010). This decomposition reaction is continued until the energy input is stopped or formation of stable inert product via colliding two radicals. The thermal degradation can lead to both chain-scission reaction which reduces molecular weight and cross-linking reaction which increases molecular complexity.

The temperature at which thermal degradation occurs depends on the thermal properties of plastic and availability of oxygen. There are two key thermal properties which influences the thermal degradation of plastics, glass transition temperature (T_g) and melting temperature (T_m). The T_g is the temperature at which the polymers shift from a rigid glassy state to flexible and rubber like state. Below T_g, the plastics remain hard and brittle due to restricted molecular movement, whereas above T_g, the increased molecular mobility allows greater flexibility. The Tm is the temperature at which a crystalline polymer transitions from a solid to a liquid state (Pielichowski et al., 2022).

Another critical factor in thermal degradation is the activation energy which determines the minimum temperature required to initiate the breakdown of polymers chains. In the presence of oxygen, plastics that contain hydrogen as a primary component undergo exothermic oxidation at temperature around 200 °C, releasing heat and accelerating degradation. However, for polymers with a higher T_m, more heat energy is required to surpass their threshold activation energy. The combination of oxygen availability, polymer composition, and thermal properties influences the rate and extent of thermal degradation, affecting material performance, stability, and longevity (Peterson et al., 2001).

1.9 Moisture and Hydrolysis

Moisture is essential for the breakdown of polymers, especially those containing hydrolytically unstable linkages like amides and esters. When these bonds are broken by water molecules, hydrolysis takes place, which results in chemical degradation (Devi et al., 2019). Plastics in humid or marine environment are especially vulnerable, as moisture accelerates surface erosion and promotes microbial activity and further degrade the material. Over the time, this contributes to the breakdown of plastics into MPs (Chamas et al., 2020).

1.10 Mechanical Forces

Weathering of plastics is also greatly influenced by mechanical forces from wind, waves, and physical abrasion (Chamas et al., 2020). In aquatic environment, the repeated freezing and thawing can accelerate the mechanical breakdown of plastics (Cooper, 2012). The degree of degradation largely depends on the mechanical characteristics of plastic, particularly elongation at break which measures how much a plastic can be stretched before fracturing. It is also referred as fracture strain which varies significantly among different plastic and ranges from 1 to 900% (Chamas et al., 2020). The plastics with lower elongation at break are more susceptible to fragmentation when exposed to external tensile forces. The continuous mechanical stress over time leads to polymer chain scission, further weakening the structure. The synthetic fibers which make up over 60% of the global fiber consumption, are especially prone to mechanical degradation. The common synthetic fibers include polyester, polyolefin, acrylic, and polyamide. Mechanical stress is the key factor in deterioration of these fibers during washing. The combination of shear forces, abrasion, and impact stress during laundering significantly contributes to the shedding of MPs fiber into the air, making airborne pollution just as significant as fiber loss from washing (Akyildiz et al., 2024).

The mechanical forces also play a major role in generating tire wear, brake wear, and road wear particles. The interaction between tires and road surfaces as well as brake pads, and brake disks, produces fictional stress which gradually removes tiny fragments of rubber. As vehicles are in motion, continuous pressure and stretching of tire treads result in material fatigue, ultimately leading to the breakdown of rubber and release of micro-sized particles into the environment (Henry et al., 2018).

1.11 Biotic Degradation

The breakdown of plastic via living organisms is known as biotic degradation. This process can be physical which involve activities such as biting, chewing or fragmentation within the digestive system or biochemical, where microbial enzymes facilitate degradation. Various microorganisms including bacteria and fungi as well as insects significantly contribute to this process (Mohanan et al., 2020). The current research has identified various microbial strains of bacteria and fungi which degrade plastics via their metabolic activities.

Generally, conventional plastics exhibit low bioavailability due to their solid and durable nature, meaning only a small fraction is accessible to microorganisms. In addition, large polymer molecules cannot be directly consumed by microbes, instead the extracellular enzymes secreted by microbes, first breakdown into a smaller and more absorbable compounds for further metabolic activities. The process of abiotic degradation which creates pores and fractures on the polymers surface while generating low molecular weight degradation products can accelerate microbial degradation.

Based on presence or absence of ester or amide, the plastics are of two different categories; hydrolysable or non-hydrolysable which are attacked by various extracellular hydrolases. PE, PP, and PVC are non-hydrolysable polymers which do not degrade easily but due to their structural resemblance with lignin, the lignin degrading enzymes can also degrade (Danso et al, 2019).

1.12 MPs in Air

MPs are ubiquitously present in both outdoor and indoor air which raise health and environment related concerns. The distribution and behaviour of MPs present in air is similar to that of other airborne pollutants. The concentration, transport, dispersion, and removal of MPs depend on meteorological condition, source of emission, and long-range transport.

1.12.1 MPs in Outdoor Air

The presence of MPs in outdoor air results from industrial processes, vehicular emission, and environmental degradation of plastics by wind abrasion, resuspension of tyre wear, mechanical drying of synthetic fibres, sea spray, industrial activities, incineration, and landfill.

1.13 Sea Spray

Studies indicate that over 30% of airborne particulate matter over oceanic regions consists of plastic-based materials. Similarities in polymer composition, concentration, and morphology between atmospheric and surface seawater MPs suggest that the ocean may be a significant source of aerosolized MPs (Trainic et al., 2020).

MPs averaging 20 ± 13 μm can become airborne through bubble bursting or sea spray ejection, contributing up to 11% of atmospheric MPs emissions (Allen et al., 2022; Brahney et al., 2021). Once suspended, they are transported by wind, facilitate particle aggregation, and influence cloud and droplet formation due to their hydrophobicity (Allen et al., 2022). The ocean surface facilitates a cyclical "grasshopper" effect, where MPs are lifted, transported, and redeposited. While marine MPs pollution was previously linked to terrestrial sources (Duis & Coors, 2016, Allen et al., 2022) estimate that 136,000 tonnes of plastic are transported from marine environments to land annually, reintroducing them into terrestrial ecosystems.

Laboratory studies show that nanoplastics, particularly those < 350 nm, can transfer from the air–water interface into the atmosphere via bubble bursting. This process is influenced by factors such as hydrophobicity, UV degradation, polymer aging, and bubble size, which enhance aerosolization. Polystyrene has been observed in airborne environments as individual 600 nm particles or clusters of up to three, often coated with sulfur atoms (Masry et al., 2021). While laboratory experiments confirm the transfer of plastics from water to air, further studies are needed to assess efficiency across different polymer types and verify nano plastics in environmental air samples.

Beyond marine environments, MPs in terrestrial aquatic systems correlate with atmospheric deposition, suggesting that inland water bodies also contribute to airborne MPs pollution via entrainment in moving water (Kallenbach et al., 2022).

1.14 Wind Abrasion

MPs in terrestrial environments degrade into smaller fragments more quickly than those in marine settings due to increased exposure to UV radiation and greater temperature variations (Horton et al., 2017). Wind tunnel experiments have demonstrated that surface MPs can become airborne at rates ranging from 0.08 to 1.48 mg/m^2/min, highlighting wind erosion as a major source of atmospheric MPs pollution (Rezaei et al., 2022).

In outdoor environments, wind abrasion detaches MPs from textiles and resuspends them from landfill sites (Rillig, 2012).

Dust storms also contribute to MPs dispersion. Research in Iran indicates that approximately 2×10^{12} MPs particles can become airborne during a dust storm, with around 1.5×10^{12} settling within the affected area. Globally, an estimated 4×10^{13} MPs particles are lifted from soil sources into the atmosphere, accounting for approximately 5% of total airborne MPs (Abbasi et al., 2022; Brahney et al., 2021).

1.15 Tyre Wear

MPs from roads contribute heavily to air pollution, with tyre wear being a major source. When tyres rub against the road or break down from heat, they release tiny particles (smaller than 2.5–350 µ) made of synthetic rubber and other materials from road surfaces (Abbasi et al., 2019). These particles mix with road dust and get lifted into the air, with tyre wear making up around 5% of airborne MPs (Wik & Dave, 2009). In busy traffic areas, the amount of tyre wear particles in the air is usually between 2 and 5 µg/m^3, making up 1% of PM_{10} pollution, with 15% of that coming from tyres (Panko et al., 2013).

The amount of MPs released from tyres depends on factors like tyre condition and vehicle weight. Studies in several countries estimate that 146,000 tonnes of tyre wear particles enter the air each year, with global emissions reaching 2.9 million tonnes annually. This includes 288,000 tonnes of PM10 and 29,000 tonnes of PM2.5 (Evangeliou et al., 2020). Despite different estimates, tyre wear and road dust are clearly major sources of MPs pollution in the air.

1.16 Industrial Activities

Industrial activities are significant contributors to atmospheric MPs pollution through the production, handling, and processing of synthetic materials. Various industries, including plastic manufacturing, textile production, and construction, release MPs into the air through mechanical abrasion, high-temperature processes, and improper waste management.

Plastic production facilities release MPs into the air through resin pellet spills, cutting, molding, and grinding operations (Prata, 2018). During plastic extrusion and injection molding, airborne MPs can be emitted due to heat-induced degradation and mechanical wear. Improper disposal of plastic waste and mismanagement of industrial effluents also contribute to MPs dispersion in the atmosphere (Kole et al., 2017).

Synthetic textile manufacturing is a major source of airborne MPs, particularly in the form of microfibers released during fabric production, dyeing, and drying (Dris et al., 2017). Studies have shown that workers in textile industries are exposed to high concentrations of airborne synthetic fibers, which can contribute to MPs pollution beyond occupational settings (Zhang et al., 2020). Additionally, garment finishing processes such as sanding and brushing release significant amounts of MPs fibers into the environment (Browne et al., 2011).

The construction industry generates MPs through the degradation of synthetic materials such as insulation foams, paints, and polymer-based coatings (Wright et al., 2020). Sandblasting, drilling, and demolition activities release MPs particles into the air, which can be transported over long distances by wind currents (Liu et al., 2019). Road construction and maintenance, including the use of polymer-modified bitumen in asphalt, also contribute to airborne MPs pollution (Andrady, 2011).

Waste incineration facilities release MPs into the atmosphere due to the incomplete combustion of plastic-containing waste materials (Zhang et al., 2020). Additionally, industrial exhaust from plastic-reliant industries, such as packaging and automotive manufacturing, contains MPs fragments and fibers (Prata et al., 2020). These emissions contribute to the long-range transport of MPs, impacting both urban and remote environments (Allen et al., 2019).

1.16.1 MPs in Indoor Air

The major sources of MPs in indoor air are use of synthetic fibres which releases MPs fibres during cleaning, drying, and wearing (Napper & Thompson, 2016). Opening of plastic packaging and the use of 3D printer also contributes MPs significantly (Stephens et al., 2013).

MPs in indoor environments originate from various sources and their formation mechanisms are critical for analysis. While research on this topic is limited, available studies provide essential insights. One of the primary sources of indoor MPs is synthetic textiles, including clothing, bedding, curtains, carpets, and other fabric-based household items. Materials such as nylon, acrylic, polyester, polyamide, polyolefin, and rayon shed microfibers into the air during regular use and movement (De Falco et al., 2018). This release occurs in both residential and commercial settings, with concentrations influenced by population density and air circulation (Dris et al., 2017).

Another significant contributor is the degradation of household materials and surface finishes. Items such as painted walls and ceilings, PVC flooring, polyurethane coatings, wallpapers, plastic utensils, cleaning tools, and general cleaning products release MPs due to wear and tear (Vianello et al., 2019). Activities like rubbing, cleaning, cutting, and scratching these surfaces accelerate their breakdown, leading to airborne MPs dispersion (Prata et al., 2020). The intensity of this release varies based on usage and maintenance levels, with residential kitchens generating higher

MPs pollution than office spaces. In contrast, offices are more likely to have MPs from electronic equipment, printing, and document shredding (Gasperi et al., 2018).

Additionally, indoor environments are affected by external MPs sources. Industrial and agricultural emissions can introduce MPs into buildings, while particles from vehicle tires contribute to indoor contamination, particularly in areas near busy roads (Evangeliou et al., 2020). These external pollutants enter indoor spaces through windows, air infiltration, or mechanical ventilation, further increasing exposure to airborne MPs (Wright & Kelly, 2017).

1.17 MPs in Fresh Water Bodies

There are several sources of MPs in freshwater bodies which are following:

1.18 Industrial Processes

Industrial operations play a crucial role in MPs pollution, contributing to environmental contamination through multiple pathways. Primary MPs, intentionally produced in the form of plastic pellets (nurdles), synthetic fibers, and cosmetic additives, frequently enter the environment during manufacturing, transport, and processing due to insufficient containment practices (Geyer et al., 2017). Furthermore, secondary MPs arise from industrial processes, including the wear and tear of plastic machinery components, degradation of polymer coatings, and wastewater discharges from synthetic textile production (Ziajahromi et al., 2017). A significant portion of these MPs is released into the environment through industrial effluents, as conventional wastewater treatment facilities are not equipped to completely remove them (Prata, 2018). Research indicates that industrial regions located near water bodies, such as rivers and coastal zones, often exhibit heightened MPs contamination, underscoring the impact of industrial activities on ecosystem health (Browne et al., 2011).

The presence of industrial MPs affects various environmental mediums, including air, soil, and aquatic ecosystems. Emissions from plastic manufacturing facilities contribute to airborne MPs, which can travel over large distances before settling into natural habitats (Leslie et al., 2022). MPs discharged through industrial wastewater enter freshwater and marine environments, where they are ingested by aquatic organisms, leading to bioaccumulation and potential toxicological effects (Katare et al., 2021). Additionally, industrial plastic waste degrades over time, breaking down into MPs that persist in soil and disrupt terrestrial ecosystems. To mitigate this issue, industries must adopt stricter pollution control measures, enhance waste management systems, and invest in sustainable materials to reduce dependence on conventional plastics (de Souza Machado et al., 2018). Advanced filtration technologies in wastewater treatment plants, alongside circular economy initiatives in industrial production, can play a key role in reducing MPs emissions and promoting environmental sustainability (Hale et al., 2020).

1.19 Agricultural System

MPs contamination in agricultural systems has become an increasing environmental concern due to the extensive use of plastic-based materials in farming. Various agricultural practices contribute to MPs pollution, including the use of plastic mulch films, seed coatings, controlled-release fertilizers, irrigation tubing, and greenhouse coverings, all of which degrade over time into smaller plastic particles (Liu et al., 2014). Additionally, the application of biosolids as organic fertilizers introduces MPs into farmlands, as wastewater treatment processes are unable to completely eliminate synthetic fibers and plastic fragments from sewage sludge (Corradini et al., 2019). These MPs accumulate in the soil, altering its physical and chemical characteristics and potentially impacting plant growth and microbial biodiversity (de Souza Machado et al., 2018). Research suggests that agricultural soils may retain higher concentrations of MPs than aquatic systems, as plastic residues continuously enter the soil with limited degradation under terrestrial conditions (Nizzetto et al., 2016).

The movement and effects of MPs in agricultural environments extend beyond soil contamination. Wind and water erosion can disperse MPs into nearby ecosystems, including freshwater sources, worsening environmental risks (Rillig et al., 2019). Moreover, these plastic particles can interact with soil contaminants such as pesticides and heavy metals, potentially altering their bioavailability and toxicity (Wang et al., 2022a, 2022b). Studies indicate that soil organisms, including earthworms, may ingest MPs, which can affect soil health and disrupt natural ecosystem functions (Wang et al., 2022a, 2022b). There is also growing concern that MPs may be absorbed by crops, raising the possibility of their transfer into the food chain. To combat MPs pollution in agriculture, adopting sustainable alternatives such as biodegradable plastics, enhancing waste management strategies, and implementing policies to regulate plastic use in farming are essential steps toward minimizing environmental impact.

1.20 Urban Sewage and Drainage System

Urban sewage and drainage systems serve as significant conduits for MPs pollution, enabling the movement of plastic particles from residential and industrial sources into both aquatic and terrestrial environments. MPs enter wastewater treatment plants (WWTPs) through household activities such as washing synthetic fabrics, using personal care products containing microbeads, and the breakdown of plastic waste

(Murphy et al., 2016). Industrial wastewater further contributes to MPs contamination, particularly from sectors like textile manufacturing, cosmetics production, and plastic processing (Ziajahromi et al., 2017). While advanced wastewater treatment technologies can capture a large portion of MPs, conventional systems still allow substantial amounts to be released into surrounding water bodies. Additionally, sewage sludge, which retains a considerable quantity of MPs particles, is frequently repurposed as agricultural fertilizer, introducing synthetic polymers into soil ecosystems [4]. Stormwater runoff from urban landscapes exacerbates the issue by carrying plastic debris from road surfaces, construction sites, and landfills into drainage systems (Sun et al., 2019).

Urban drainage networks play a critical role in MPs transport, especially in densely populated areas where impermeable surfaces prevent the natural filtration of pollutants. During periods of heavy rainfall, combined sewer overflows (CSOs) release untreated wastewater directly into rivers, lakes, and coastal waters, significantly increasing MPs pollution (Dris et al., 2017). In addition to direct sewage discharge, sources such as tire wear particles, polymer-based road markings, and street dust contribute to the presence of MPs in stormwater, which often enters water bodies without adequate treatment (Kole et al., 2017). Furthermore, airborne MPs settle into urban drainage systems, highlighting atmospheric deposition as an overlooked source of contamination. The persistence of MPs in sewage and drainage systems underscores the need for improved filtration methods, stricter industrial discharge regulations, and the implementation of sustainable urban drainage solutions, such as biofilters and retention ponds, to reduce environmental risks (Iyare et al., 2020).

1.21 Sewage Sludge

Sewage sludge, a by-product of wastewater treatment plants (WWTPs), serves as a major repository for MPs, contributing to their accumulation in both terrestrial and aquatic ecosystems. WWTPs collect MPs from multiple sources, including synthetic fibers from laundry processes, microbeads from personal care products, industrial wastewater, and urban stormwater runoff. While advanced treatment methods can remove a significant portion of MPs from effluent water, these particles often accumulate in sewage sludge, leading to high concentrations in processed biosolids (Murphy et al., 2016). Research indicates that the number of2020 MPs particles in sewage sludge can range from thousands to millions per kilogram of dry weight, depending on wastewater composition and treatment efficiency (Li et al., 2020). In many regions, treated sludge is repurposed as an agricultural soil amendment, unintentionally introducing MPs into farming systems (Corradini et al., 2019). Once in the soil, these particles can persist for long durations, alter soil properties, and interact with microbial communities, potentially influencing plant growth and soil health (de Souza Machado et al., 2018).

The impact of MPs in sewage sludge extends beyond soil contamination, as wind and surface water runoff can redistribute these particles into freshwater and marine environments, intensifying pollution concerns. Additionally, studies suggest that MPs in sludge can adsorb pollutants, including heavy metals and organic contaminants like pesticides, increasing their environmental stability and toxicity (Hodson et al., 2017; Wang et al., 2020). There is also growing concern about the potential uptake of MPs by crops, which could facilitate their entry into the human food chain (Li et al., 2020). To address the environmental risks associated with sewage sludge-derived MPs, sustainable waste management approaches are essential. These include advancements in sludge treatment technologies, regulatory interventions to minimize MPs inputs at the source, and the promotion of biodegradable alternatives to conventional polymers (Hale et al., 2020). Further studies are necessary to evaluate the long-term ecological consequences of MPs introduced through sludge applications and to develop effective mitigation strategies.

1.22 MPs in Marine Water

The major sources of MPs in ocean can be divided into two groups: land based and marine based sources. The land based sources contribute significant amount of MPs to the ocean through surface run-off, wastewater discharge, sewage sludge, agricultural activities, as discussed above.

1.23 Fishing Activities

One of the primary contributors to marine MPs pollution is Abandoned, Lost, or Discarded Fishing Gear (ALDFG). Fishing equipment such as nets, lines, and traps, primarily made from synthetic polymers like polyethylene (PE), polypropylene (PP), and polyamide (PA), undergo degradation due to prolonged exposure to sunlight, mechanical stress, and microbial activity (Richardson et al., 2019). Research suggests that ALDFG accounts for a significant proportion of floating plastic debris in various ocean regions, sometimes contributing up to 70% of macroplastic waste (Lebreton et al., 2018). Over time, these plastic materials break down into MPs, posing threats to marine ecosystems and leading to entanglement and ingestion by marine species.

Fishing operations also contribute to MPs contamination through the wear and tear of synthetic materials, including nylon nets, ropes, and buoys. The repeated use and retrieval of these materials during fishing activities result in the release of MPs fibers into seawater (Napper et al., 2021). Studies have shown that a single fishing expedition can shed thousands of MPs particles, which persist in the marine environment and may be consumed by aquatic organisms, leading to bioaccumulation within the food chain (Cole et al., 2013).

Apart from fishing gear, other plastic waste from fishing vessels, such as discarded food packaging, plastic bait containers, and synthetic gloves, exacerbates oceanic MPs pollution. Although international regulations, such as MARPOL Annex V, prohibit the disposal of plastic waste at sea, non-compliance and improper waste management continue to be major challenges (Vuong, 2023). Large plastic waste items degrade over time, eventually fragmenting into MPs and further contaminating marine ecosystems.

1.24 Maritime Transport

Another significant source of MPs from maritime activities is the use of polymer-based paints and anti-fouling coatings on ships. These coatings, designed to prevent biofouling and corrosion, contain MPs particles that are gradually released into the ocean due to abrasion, weathering, and maintenance procedures (Turner et al., 2021). Studies indicate that ship paint is a major contributor to MPs pollution, particularly in high-traffic shipping corridors and port areas (Song et al., 2015).

Additionally, plastic pellets, commonly referred to as nurdles, serve as raw materials for plastic production and are frequently transported via cargo ships. Accidental spills during their loading, transport, or unloading result in substantial MPs pollution (Karlsson et al., 2018). Once released into the marine environment, these tiny pellets are difficult to retrieve and often resemble food to marine organisms, leading to ingestion and potential ecological harm (Hodson et al., 2017). Despite existing regulations, the illegal disposal of plastic waste by cargo ships, cruise liners, and fishing vessels remains a persistent issue. Plastic debris from maritime transport, including discarded packaging materials, food containers, and synthetic ropes, gradually breaks down into MPs under oceanic conditions (Vuong 2023). Moreover, tire wear particles (TWP) generated by port machinery, tugboats, and cargo-handling vehicles contribute to MPs contamination. These particles are carried into the ocean through surface runoff, increasing MPs concentrations in and around harbors and shipping terminals (Kole et al., 2017).

1.25 MPs Contamination in Food Chain

MPs contamination has become one of the greatest environmental threats to the survival of future life on the earth. These tiny particles accumulated in different aquatic organisms and different parts of vegetation and their products and have contaminated the trophic levels of food chains. These synthetic particles originate from various sources, including agricultural and industrial activities, wastewater discharge, sewage sludge, landfills, fishing, maritime transport, and atmospheric deposition. As a result, MPs have been detected in terrestrial and freshwater, and marine ecosystems, where they are consumed by organisms of different trophic levels, and ultimately reach in human food chain.

In aquatic ecosystems, MPs are ingested by lower trophic organisms such as zooplankton, copepods, fish larvae, and filter feeders like mussels and oysters.

Through trophic transfer, MPs accumulate in larger predators, including fish, seabirds, and marine mammals. Studies have shown that benthic fish reflect sediment contamination, while pelagic species indicate MPs presence in open waters.

Beyond seafood, MPs have also been detected in honey, sugar, beer, and sea salt, as well as in packaged drinking water. Research highlights that bottled water, particularly in plastic containers, contains higher MPs concentrations compared to glass bottles and beverage cartons. Similarly, canned seafood products have been found to contain MPs, increasing the potential for human exposure.

Terrestrial food chains are also affected by MPs pollution. Agricultural soils accumulate MPs from sources such as irrigation with contaminated water, atmospheric deposition, sewage sludge applications, and the degradation of plastic mulch films. These particles impact soil properties, influencing water retention, nutrient cycles, and microbial interactions. More critically, studies have confirmed the uptake of MPs by crops like lettuce, wheat, and cucumbers through root absorption, leading to their accumulation in edible plant parts.

MPs enter the human body primarily through three main pathways: ingestion (via food and water), inhalation (through indoor and outdoor air), and dermal contact (from personal care products, textiles, and dust) as shown in Fig. 1.3 (Prata et al., 2020). The primary route of exposure is through the consumption of food and beverages, including staple items such as fish, meat, fruits, vegetables, cereals, legumes, and water. Common MPs found in food and drinks include polypropylene (PP), polyethylene terephthalate (PET), polyvinyl chloride (PVC), polyethylene (PE), high-density polyethylene (HDPE), low-density polyethylene (LDPE), polyamide (PA), polystyrene (PS), and expanded polystyrene (EPS) (Karami et al., 2017).

Inhalation is another major route, as MPs are released into the air from multiple sources, including textile production, wear and tear of materials like car tires and building structures, and resuspension of MPs from surfaces. These airborne particles

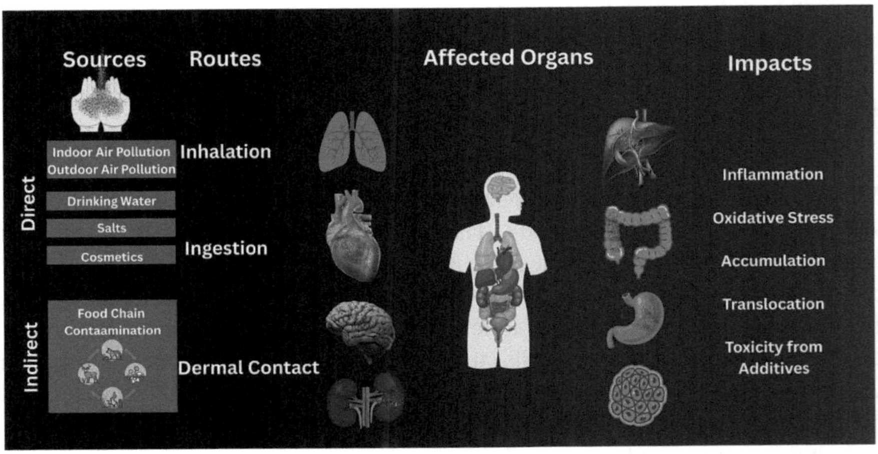

Fig. 1.3 Illustration showing the direct and indirect sources of MPs contamination to human being via inhalation, ingestion, and dermal contact and its possible impacts

primarily consist of synthetic fibers, PP, acrylic, PA, polyester, and PE (Gasperi et al., 2018; Lehner et al., 2019). Skin contact, though considered a less significant exposure route, can still occur through dust, synthetic fibers, and microbeads in cosmetics, which often contain MPs like PE, PP, PET, and nylon (Prata et al., 2020).

1.26 Impact on Aquatic Environment

MPs are widespread pollutants in aquatic environments, impacting both marine and freshwater ecosystems through physical, chemical, and biological processes. Due to their tiny size, durability, and capacity to carry harmful substances, they pose a considerable environmental risk.

1.27 Physical Impact of MPs on Aquatic Organism

MPs pose a physical threat to aquatic organisms through ingestion and entanglement. Various marine species, such as zooplankton, fish, seabirds, and marine mammals, often confuse these tiny particles with food (Pits et al., 2023). When these MPs are ingested, causes digestive blockages, reduce food consumption, and create a false feeling of fullness, ultimately hindering the growth, reproduction, and survival (Wright et al., 2013). For example, studies on European perch (*Perca fluviatilis*) demonstrated that larval fish showed a preference for MPs over their natural prey, leading to slower growth and higher mortality rates (Galafassi et al., 2021). Moreover, smaller aquatic creatures can become trapped in MPs fibers, often derived from synthetic fabrics, limiting their mobility and making them more vulnerable to predators (Fjelstad, 1988).

1.28 Chemical Impact of MPs on Aquatic Organism

MPs are not merely passive particles; they can act as carriers of harmful chemicals. The hydrophobic surfaces of MPs enable them to attract and absorb persistent organic pollutants (POPs), such as PCBs, PAHs, and heavy metals from the surrounding environment easily. When aquatic organisms consume these contaminated particles, the toxins can build up within their bodies and magnify through the food chain, posing threats to top predators, including humans. In addition, MPs often contain chemical additives like plasticizers, flame retardants, and stabilizers, which gradually release into the water. Substances such as BPA and phthalates are known to interfere with hormonal systems and can disrupt growth and reproduction in aquatic species.

1.29 Impacts of MPs on the Terrestrial Environment

It is evident from the literature survey that majority of MPs research are focusing on marine environment. However, the terrestrial environment is more affected than it as 80% of marine MP originates from land sources (Horton et al., 2017). MPs enter the terrestrial environment through various pathways such as plastic mulching, littering, using sewage sludge as manure, atmospheric deposition, and urban run-off (Blasing & Amelung, 2018). These MPs affects the soil in various ways which are following:

1.30 Soil Structure and Function

MPs primarily affect the physical properties of soil by reducing its bulk density, porosity and water retention capacity. de Souza Machado et al. (2018) reported that the presence of polyesters fibre and PE beads decreased the bulk density up to 25%. Zhang et al. (2018) have found that the presence of fibrous MPs can increase the soil bulk density which reduces the aeration and water infiltration, thereby inhibiting root penetration and affect the plant growth. However, Rillig et al. (2019) have identified that the fragmented MPs may create macropores which increases water loss and reduces the stability of soil, particularly in sandy soil. MPs also affect the water retention and hydraulic conductivity. de Souza Machado et al. (2019) reported that the presence of fibrous MPs from textile industries increased the water retention by creating heterogeneous soil matrix whereas the spherical beads enhanced the drainage by disrupting the capillary action. MPs also affect the soil aggregation which is an important determinant of structure and stability of soil. MPs interfere with binding of soil particles by organic matter and clay and thus weaken the stability of aggregate. Zhang et al. (2018) showed that the PE MPs reduced the aggregate stability by up to 15%. This reduced stability increases the susceptibility to erosion and reduces the capacity of soil to support plant growth. In addition, MPs also alter the thermal properties of soil. Qi et al. (2018) reported that plastic sheet increases the retention of heat which affects the germination of seed and microbial processes too.

1.31 Disruption of Soil Microbiome and Nutrient Cycling

Microbes play an important role in degrading organic matter and nutrient cycling. MPs affect soil microbial communities and their enzymatic activity adversely. Fei et al. (2020) reported the change in bacterial diversity and enzymatic activity in soil contaminated with MPs. Leifheit et al., (2021) observed that MPs altered the abundance of arbuscular mycorrhizal fungi (AMF) which is very critical for the uptake of nutrient by plants. The disruption of microbial communities weakens the resilience of plant under stress condition like drought. de Souza Machado et al. (2019) identified that the polyester fibres and PE beads decreased the enzymatic activity by 30%.

The nutrient cycling is closely related to microbial activity which is vulnerable to MPs contamination. MPs disrupt the nutrient cycling like carbon cycle by interfering with organic matter decomposition. A study conducted by Liu et al. (2018) reported that the presence of PP MPs caused physical obstruction for microbial degradation or release of toxic chemical like phthalates, bisphenol A from the plastic additives which reduced the respiration rate by 15–20%. Similarly, MPs also alter the nitrogen cycling by decreasing the microbial population. Wang et al. (2020) found that PS MPs decreased the population of ammonia oxidizing bacteria and reduced the rate of nitrification by 25% which affect the yield.

1.32 Chemical Pollution and Toxicity

MPs act as a potential vector for various toxic chemicals including pesticides, heavy metals, and persistent organic pollutants (POPs). These MPs also act as a source of several chemicals as chemicals such as BPA, DEHP, HBB, etc. are added as additives. These chemical get leached out into the soil over time (Wang et al., 2020). Such chemical contamination affects the soil fertility,, plant health and groundwater quality too. The agricultural lands irrigated with wastewater are at greater risk of MPs contamination and associated toxicity with heavy metal accumulation (Corradini et al., 2019).

1.33 Conclusion

Plastics, valued for their durability, low cost, and versatility, are widely used in packaging, construction, agriculture, and medicine. However, their non-biodegradable nature and poor management have made them a major environmental problem. Over 430 million tons of plastic are produced yearly, with only a small portion recycled, and this amount may double by 2040. The improper disposal leads to plastic accumulation in air, water, soil, and living organisms, causing widespread pollution. When the plastics break down through physical, chemical, and biological processes, they form MPs and NPs, which are tiny particles less than 5 mm and 100 nm, respectively. These particles, whether intentionally made for products like cosmetics or formed from larger plastics, released into the environment and persist for decades. When the MPs enter the ecosystems through various pathways like wastewater, littering, and atmospheric deposition, contaminates the food chains and human health via ingestion, inhalation, and skin contact. These MPs provide the surface for toxic chemicals like heavy metals and polycyclic aromatic hydrocarbons, increases their harmful effects.

In aquatic environments, MPs affects organisms by causing digestive blockages and entanglement, and reduces the growth and survival rates. MPs disrupt soil structure, fertility, and microbial activity, affecting plant growth and nutrient cycles. For humans, MPs in food, water, and air are linked to health issues like respiratory diseases and immune disorders. Additives used in plastics, such as phthalates and flame retardants, further increase toxicity of MPs released into the environment. To tackle this issue, better waste management, reduced use of single-use plastics, and biodegradable alternatives are needed. Improved wastewater treatment and sustainable farming practices can also limit MP spread. Global cooperation and public awareness are crucial to reduce plastic production and pollution. Future studies should explore the long-term effects of MPs and NPs and find ways to clean up contaminated environments.

In summary, plastics, once a remarkable invention, now threaten ecosystems and human health due to mismanagement. With collective action, including better policies and research, we can reduce plastic pollution and protect the environment for future generations.

References

Abbasi, S., Keshavarzi, B., Moore, F., Shojaei, N., Sorooshian, A., Soltani, N., & Delshab, H. (2019). Geochemistry and environmental effects of potentially toxic elements, polycyclic aromatic hydrocarbons and microplastics in coastal sediments of the Persian Gulf. *Environmental Earth Sciences, 78*, 1–15.

Abbasi, S., Rezaei, M., Ahmadi, F., & Turner, A. (2022). Atmospheric transport of microplastics during a dust storm. *Chemosphere, 292*, Article 133456.

Aboulkas, A., & El Bouadili, A. J. E. C. (2010). Thermal degradation behaviors of polyethylene and polypropylene. Part I: Pyrolysis kinetics and mechanisms. *Energy Conversion and Management, 51*(7), 1363–1369.

Ait Bamai, Y., Shibata, E., Saito, I., Araki, A., Kanazawa, A., Morimoto, K., Nakayama, K., Tanaka, M., Takigawa, T., Yoshimura, T., Chikara, H., Saijo, Y., & Kishi, R. (2014). Exposure to house dust phthalates in relation to asthma and allergies in both children and adults. *Science of the Total Environment, 485*, 153–163.

Akyildiz, S. H., Fiore, S., Bruno, M., Sezgin, H., Yalcin-Enis, I., Yalcin, B., & Bellopede, R. (2024). Release of microplastic fibers from synthetic textiles during household washing. *Environmental Pollution, 357*, 124455.

Allen, D., Allen, S., Abbasi, S., Baker, A., Bergmann, M., Brahney, J., Butler, T., Duce, R. A., Eckhardt, S., Evangeliou, N., Jickells, T., Kanakidou, M., Kershaw, P., Laj, P., Levermore, J., Li, D., Liss, P., Liu, K., Mahowald, N., …, Wright, S. (2022). Microplastics and nanoplastics in the marine-atmosphere environment. *Nature Reviews Earth & Environment, 3*(6), 393–405.

Allen, S., Allen, D., Phoenix, V. R., Le Roux, G., Durántez Jiménez, P., Simonneau, A., Binet, S., & Galop, D. (2019). Atmospheric transport and deposition of microplastics in a remote mountain catchment. *Nature Geoscience, 12*(5), 339–344.

Amato-Lourenço, L. F., dos Santos Galvão, L., de Weger, L. A., Hiemstra, P. S., Vijver, M. G., & Mauad, T. (2020). An emerging class of air pollutants: Potential effects of microplastics to respiratory human health? *Science of the Total Environment, 749*, Article 141676.

Andrady, A. L. (2011). Microplastics in the marine environment. *Marine Pollution Bulletin, 62*(8), 1596–1605.

Andrady, A. L., & Neal, M. A. (2009). Applications and societal benefits of plastics. *Philosophical Transactions of the Royal Society B: Biological Sciences, 364*(1526), 1977–1984.

Bernard, L., Cueff, R., Chagnon, M. C., Abdoulouhab, F., Décaudin, B., Breysse, C., Kauffmann, S., Cosserant, B., Souweine, B., Sautou, V., & Armed Study Group. (2015). Migration of plasticizers from PVC medical devices: Development of an infusion model. *International Journal of Pharmaceutics, 494*(1), 136–145.

Bläsing, M., & Amelung, W. (2018). Plastics in soil: Analytical methods and possible sources. *Science of the Total Environment, 612*, 422–435.

Brahney, J., Mahowald, N., Prank, M., Cornwell, G., Klimont, Z., Matsui, H., & Prather, K. A. (2021). Constraining the atmospheric limb of the plastic cycle. *Proceedings of the National Academy of Sciences, 118*(16), Article e2020719118.

Browne, M. A., Crump, P., Niven, S. J., Teuten, E., Tonkin, A., Galloway, T., & Thompson, R. (2011). Accumulation of microplastic on shorelines woldwide: Sources and sinks. *Environmental Science & Technology, 45*(21), 9175–9179.

Brydson, J. A. (1999). *Plastics materials*. Elsevier.

Chamas, A., Moon, H., Zheng, J., Qiu, Y., Tabassum, T., Jang, J. H., Abu-Omar, M., Scott, S. L., & Suh, S. (2020). Degradation rates of plastics in the environment. *ACS Sustainable Chemistry & Engineering, 8*(9), 3494–3511.

Cheng, W., Zhou, Y., Xie, Y., Li, Y., Zhou, R., Wang, H., Feng, Y., & Wang, Y. (2023). Combined effect of polystyrene microplastics and bisphenol A on the human embryonic stem cells-derived liver organoids: The hepatotoxicity and lipid accumulation. *Science of the Total Environment, 854*, 158585.

Cole, M., Lindeque, P., Fileman, E., Halsband, C., Goodhead, R., Moger, J., & Galloway, T. S. (2013). Microplastic ingestion by zooplankton. *Environmental Science & Technology, 47*(12), 6646–6655.

Cooper, D. A. (2012). *Effects of chemical and mechanical weathering processes on the degradation of plastic debris on marine beaches*. The University of Western Ontario (Canada).

Corradini, F., Meza, P., Eguiluz, R., Casado, F., Huerta-Lwanga, E., & Geissen, V. (2019). Evidence of microplastic accumulation in agricultural soils from sewage sludge disposal. *Science of the Total Environment, 671*, 411–420.

Covaci, A., Harrad, S., Abdallah, M.A.-E., Ali, N., Law, R. J., Herzke, D., & de Wit, C. A. (2011). Novel brominated flame retardants: A review of their analysis, environmental fate and behaviour. *Environment International, 37*(2), 532–556.

Danso, D., Chow, J., & Streit, W. R. (2019). Plastics: Environmental and biotechnological perspectives on microbial degradation. *Applied and Environmental Microbiology, 85*(19), e01095-e1119.

Darnerud, P. O. (2003). Toxic effects of brominated flame retardants in man and in wildlife. *Environment International, 29*(6), 841–853.

De Falco, F., Gullo, M. P., Gentile, G., Di Pace, E., Cocca, M., Gelabert, L., Brouta-Agnésa, M., Rovira, A., Escudero, R., Villalba, R., Mossotti, R., Montarsolo, A., Gavignano, S., Tonin, C., & Avella, M. (2018). Evaluation of microplastic release caused by textile washing processes of synthetic fabrics. *Environmental Pollution, 236*, 916–925.

de Souza Machado, A. A., Kloas, W., Zarfl, C., Hempel, S., & Rillig, M. C. (2018). Microplastics as an emerging threat to terrestrial ecosystems. *Global Change Biology, 24*(4), 1405–1416.

de Souza Machado, A. A., Lau, C. W., Kloas, W., Bergmann, J., Bachelier, J. B., Faltin, E., Becker, R., Görlich, A. S., & Rillig, M. C. (2019). Microplastics can change soil properties and affect plant performance. Environmental Science & Technology, 53(10), 6044–6052.

Devi, R. S., Ramya, R., Kannan, K., Antony, A. R., & Kannan, V. R. (2019). Investigation of biodegradation potentials of high density polyethylene degrading marine bacteria isolated from the coastal regions of Tamil Nadu, India. *Marine Pollution Bulletin, 138*, 549–560.

Dris, R., Gasperi, J., Mirande, C., Mandin, C., Guerrouache, M., Langlois, V., & Tassin, B. (2017). A first overview of textile fibers, including microplastics, in indoor and outdoor environments. *Environmental Pollution, 221*, 453–458.

Dris, R., Gasperi, J., Saad, M., Mirande, C., & Tassin, B. (2016). Synthetic fibers in atmospheric fallout: A source of microplastics in the environment? *Marine Pollution Bulletin, 104*(1–2), 290–293.

Duis, K., & Coors, A. (2016). Microplastics in the aquatic and terrestrial environment: Sources (with a specific focus on personal care products), fate and effects. *Environmental Sciences Europe, 28*(1), 2.

Eriksen, M., Mason, S., Wilson, S., Box, C., Zellers, A., Edwards, W., Farley, H., & Amato, S. (2013). Microplastic pollution in the surface waters of the Laurentian Great Lakes. *Marine Pollution Bulletin, 77*(1–2), 177–182.

Eschenbacher, W. L., Kreiss, K., Lougheed, M. D., Pransky, G. S., Day, B., & Castellan, R. M. (1999). Nylon flock–associated interstitial lung disease. *American Journal of Respiratory and Critical Care Medicine, 159*(6), 2003–2008.

Evangeliou, N., Grythe, H., Klimont, Z., Heyes, C., Eckhardt, S., Lopez-Aparicio, S., & Stohl, A. (2020). Atmospheric transport is a major pathway of microplastics to remote regions. *Nature Communications, 11*(1), 3381.

Fahim, M. A., Al-Sahhaf, T. A., & Elkilani, A. (2009). *Fundamentals of petroleum refining*. Elsevier.

Fei, Y., Huang, S., Zhang, H., Tong, Y., Wen, D., Xia, X., Wang, H., Luo, Y., & Barceló, D. (2020). Response of soil enzyme activities and bacterial communities to the accumulation of microplastics in an acid cropped soil. *Science of the Total Environment, 707*, 135634.

Fjelstad, E. J. (1988). The ghosts of fishing nets past: A proposal for regulating derelict synthetic fishing nets. *Washington Law Review, 63*, 677.

Frias, J. P., Gago, J., Otero, V., & Sobral, P. (2016). Microplastics in coastal sediments from Southern Portuguese shelf waters. *Marine Environmental Research, 114*, 24–30.

Fuller, S., & Gautam, A. (2016). A procedure for measuring microplastics using pressurized fluid extraction. *Environmental Science & Technology, 50*(11), 5774–5780.

Galafassi, S., Sighicelli, M., Pusceddu, A., Bettinetti, R., Cau, A., Temperini, M. E., Gillibert, R., Ortolani, M., Pietrelli, L., Zaupa, S., & Volta, P. (2021). Microplastic pollution in perch (Perca fluviatilis, Linnaeus 1758) from Italian south-alpine lakes. *Environmental Pollution, 288*, 117782.

Gardon, T., Reisser, C., Soyez, C., Quillien, V., & Le Moullac, G. (2018). Microplastics affect energy balance and gametogenesis in the pearl oyster Pinctada margaritifera. *Environmental Science & Technology, 52*(9), 5277–5286.

Gasperi, J., Wright, S. L., Dris, R., Collard, F., Mandin, C., Guerrouache, M., Langlois, V., Kelly, F. J., & Tassin, B. (2018). Microplastics in air: are we breathing it in? *Current Opinion in Environmental Science & Health, 1*, 1–5.

Geyer, R., Jambeck, J. R., & Law, K. L. (2017). Production, use, and fate of all plastics ever made. *Science Advances, 3*(7), Article e1700782.

Gilbert, M. (Ed.). (2016). *Brydson's plastics materials*. William Andrew.

Hale, R. C., Seeley, M. E., La Guardia, M. J., Mai, L., & Zeng, E. Y. (2020). A global perspective on microplastics. *Journal of Geophysical Research: Oceans, 125*(1), e2018JC014719.

Hearle, J. W. (Ed.). (2001). *High-performance fibres*. Elsevier.

Henry, B., Laitala, K., & Klepp, I. G. (2018). Microplastic pollution from textiles: A literature review.

Hodson, M. E., Duffus-Hodson, C. A., Clark, A., Prendergast-Miller, M. T., & Thorpe, K. L. (2017). Plastic bag derived-microplastics as a vector for metal exposure in terrestrial invertebrates. *Environmental Science & Technology, 51*(8), 4714–4721.

Hopewell, J., Dvorak, R., & Kosior, E. (2009). Plastics recycling: Challenges and opportunities. *Philosophical Transactions of the Royal Society B: Biological Sciences, 364*(1526), 2115–2126.

Horton, A. A., Walton, A., Spurgeon, D. J., Lahive, E., & Svendsen, C. (2017). Microplastics in freshwater and terrestrial environments: Evaluating the current understanding to identify the knowledge gaps and future research priorities. *Science of the Total Environment, 586*, 127–141.

Horvatits, T., Tamminga, M., Liu, B., Sebode, M., Carambia, A., Fischer, L., Püschel, K., Huber, S., & Fischer, E. K. (2022). Microplastics detected in cirrhotic liver tissue. *EBioMedicine, 82*.

Hu, J., Qin, X., Zhang, J., Zhu, Y., Zeng, W., Lin, Y., & Liu, X. (2021). Polystyrene microplastics disturb maternal-fetal immune balance and cause reproductive toxicity in pregnant mice. *Reproductive Toxicology, 106*, 42–50.

Hussein, Z. A., Shakor, Z. M., Alzuhairi, M., & Al-Sheikh, F. (2023). Thermal and catalytic cracking of plastic waste: A review. *International Journal of Environmental Analytical Chemistry, 103*(17), 5920–5937.

Ibrahim, A. M. (2024). Plastics: photodegradations and mechanisms. In *Microplastics and pollutants: interactions, degradations and mechanisms* (pp. 25–49). Springer Nature Switzerland.

Iyare, P. U., Ouki, S. K., & Bond, T. (2020). Microplastics removal in wastewater treatment plants: A critical review. *Environmental Science: Water Research & Technology, 6*(10), 2664–2675.

Jewett, E., Arnott, G., Connolly, L., Vasudevan, N., & Kevei, E. (2022). Microplastics and their impact on reproduction—Can we learn from the *C. elegans* model? *Frontiers in Toxicology, 4*, 748912.

Jiang, X., Chen, H., Liao, Y., Ye, Z., Li, M., & Klobučar, G. (2019). Ecotoxicity and genotoxicity of polystyrene microplastics on higher plant Vicia faba. *Environmental Pollution, 250*, 831–838.

Kallenbach, E. M., Rødland, E. S., Buenaventura, N. T., & Hurley, R. (2022). Microplastics in terrestrial and freshwater environments. *Microplastic in the Environment: Pattern and Process*, 87–130.

Karami, A., Golieskardi, A., Keong Choo, C., Larat, V., Galloway, T. S., & Salamatinia, B. (2017). The presence of microplastics in commercial salts from different countries. *Scientific Reports, 7*(1), 46173.

Karlsson, T. M., Arneborg, L., Broström, G., Almroth, B. C., Gipperth, L., & Hassellöv, M. (2018). The unaccountability case of plastic pellet pollution. *Marine Pollution Bulletin, 129*(1), 52–60.

Katare, Y., Singh, P., Sankhla, M. S., Singhal, M., Jadhav, E. B., Parihar, K., Nikalje, B. T., Trpathi, A., & Bhardwaj, L. (2021). Microplastics in aquatic environments: sources, ecotoxicity, detection and remediation. *Biointerface Research in Applied Chemistry, 12*, 3407–3428.

Klemm, D., Heublein, B., Fink, H. P., & Bohn, A. (2005). Cellulose: Fascinating biopolymer and sustainable raw material. *Angewandte Chemie International Edition, 44*(22), 3358–3393.

Kole, P. J., Löhr, A. J., Van Belleghem, F. G., & Ragas, A. M. (2017). Wear and tear of tyres: A stealthy source of microplastics in the environment. *International Journal of Environmental Research and Public Health, 14*(10), 1265.

Kremer, A. M., Pal, T. M., Boleij, J. S., Schouten, J. P., & Rijcken, B. (1994). Airway hyperresponsiveness and the prevalence of work-related symptoms in workers exposed to irritants. *American Journal of Industrial Medicine, 26*(5), 655–669.

Latini, G., De Felice, C., Presta, G., Del Vecchio, A., Paris, I., Ruggieri, F., & Mazzeo, P. (2003). In utero exposure to di-(2-ethylhexyl) phthalate and duration of human pregnancy. *Environmental Health Perspectives, 111*(14), 1783–1785.

Law, R. J. (2009). Brominated flame retardants. In *Persistent organic pollutants* (pp. 5–24). Wiley.

Lebreton, L., Slat, B., Ferrari, F., Sainte-Rose, B., Aitken, J., Marthouse, R., Hajbane, S., Cunsolo, S., Schwarz, A., Levivier, A., Noble, K., Debeljak, P., Maral, H., Schoeneich-Argent, R., Brambini, R., & Reisser, J. (2018). Evidence that the Great Pacific Garbage Patch is rapidly accumulating plastic. *Scientific Reports, 8*(1), 4666.

Lehner, R., Weder, C., Petri-Fink, A., & Rothen-Rutishauser, B. (2019). Emergence of nanoplastic in the environment and possible impact on human health. *Environmental Science & Technology, 53*(4), 1748–1765.

Leifheit, E. F., Lehmann, A., & Rillig, M. C. (2021). Potential effects of microplastic on arbuscular mycorrhizal fungi. Frontiers in Plant Science, 12, 626709.

Leslie, H. A., Van Velzen, M. J., Brandsma, S. H., Vethaak, A. D., Garcia-Vallejo, J. J., & Lamoree, M. H. (2022). Discovery and quantification of plastic particle pollution in human blood. *Environment International, 163*, Article 107199.

Li, X., Chen, L., Mei, Q., Dong, B., Dai, X., Ding, G., & Zeng, E. Y. (2018). Microplastics in sewage sludge from the wastewater treatment plants in China. Water research, 142, 75–85

Li, L., Luo, Y., Li, R., Zhou, Q., Peijnenburg, W. J., Yin, N., Yang, J., Tu, C., & Zhang, Y. (2020). Effective uptake of submicrometre plastics by crop plants via a crack-entry mode. *Nature Sustainability, 3*(11), 929–937.

Li, Y., Tao, L., Wang, Q., Wang, F., Li, G., & Song, M. (2023). Potential health impact of microplastics: A review of environmental distribution, human exposure, and toxic effects. Environment & Health, 1(4), 249–257

Liu, E. K., He, W. Q., & Yan, C. R. (2014). 'White revolution' to 'white pollution'—agricultural plastic film mulch in China. *Environmental Research Letters, 9*(9), Article 091001.

Liu, K., Wang, X., Fang, T., Xu, P., Zhu, L., & Li, D. (2019). Source and potential risk assessment of suspended atmospheric microplastics in Shanghai. *Science of the Total Environment, 675*, 462–471.

Liu, M., Lu, S., Song, Y., Lei, L., Hu, J., Wu, W., Cao, C., & Shi, H. (2018). Microplastic and mesoplastic pollution in farmland soils in suburbs of Shanghai, China. *Environmental Pollution, 242*(Pt B), 855–862.

Lwanga, E. H., Gertsen, H., Gooren, H., Peters, P., Salánki, T., van der Ploeg, M., Besseling, E., Koelmans, A. A., & Geissen, V. (2017). Incorporation of microplastics from litter into burrows of Lumbricus terrestris. *Environmental Pollution, 220*, 523–531.

Makio, H., Terao, H., Iwashita, A., & Fujita, T. (2011). FI Catalysts for olefin polymerization—A comprehensive treatment. *Chemical Reviews, 111*(3), 2363–2449.

Mark, J. E. (Ed.) (2007). *Physical properties of polymers handbook* (Vol. 1076, p. 825). Springer.

Marvin, C. H., Tomy, G. T., Armitage, J. M., Arnot, J. A., McCarty, L., Covaci, A., & Palace, V. (2011). Hexabromocyclododecane: Current understanding of chemistry, environmental fate and toxicology and implications for global management. *Environmental Science & Technology, 45*(20), 8613–8623.

Mason, S. A., Garneau, D., Sutton, R., Chu, Y., Ehmann, K., Barnes, J., Fink, P, Papazissimos, D., & Rogers, D. L. (2016). Microplastic pollution is widely detected in US municipal wastewater treatment plant effluent. *Environmental Pollution, 218*, 1045–1054.

Masry, M., Rossignol, S., Roussel, B. T., Bourgogne, D., Bussière, P. O., R'mili, B., & Wong-Wah-Chung, P. (2021). Experimental evidence of plastic particles transfer at the water-air interface through bubble bursting. *Environmental Pollution, 280*, 116949.

Mesquita, M. D. S., & Abrantes, S. D. M. P. (2022). Potential antioxidant migration from polyethylene packaging to food: A systematic review. *Polímeros, 32*(4), e2022042.

Mohanan, N., Montazer, Z., Sharma, P. K., & Levin, D. B. (2020). Microbial and enzymatic degradation of synthetic plastics. *Frontiers in Microbiology, 11*, Article 580709.

Murphy, F., Ewins, C., Carbonnier, F., & Quinn, B. (2016). Wastewater treatment works (WwTW) as a source of microplastics in the aquatic environment. *Environmental Science & Technology, 50*(11), 5800–5808.

Najafi, V., Ahmadi, E., Ziaee, F., Omidian, H., & Sedaghat, H. (2019). Polyaniline-modified TiO^2, a highly effective photo-catalyst for solid-phase photocatalytic degradation of PVC. *Journal of Polymers and the Environment, 27*, 784–793.

Napper, I. E., Baroth, A., Barrett, A. C., Bhola, S., Chowdhury, G. W., Davies, B. F., Duncan, E. M., Kumar, S., Nelms, S. E., Niloy, M. N. H., Nishat, B., Maddalene, T., Thompson, R. C., & Koldewey, H. (2021). The abundance and characteristics of microplastics in surface water in the transboundary Ganges River. *Environmental Pollution, 274*, 116348.

Napper, I. E., & Thompson, R. C. (2016). Release of synthetic microplastic plastic fibres from domestic washing machines: Effects of fabric type and washing conditions. *Marine Pollution Bulletin, 112*(1–2), 39–45.

Nicholson, J. (2017). *The chemistry of polymers*. Royal Society of Chemistry.

Nizzetto, L., Futter, M., & Langaas, S. (2016). Are agricultural soils dumps for microplastics of urban origin?

Painter, P. C., & Coleman, M. M. (1997). *Fundamentals of polymer science: An introductory text*.

Pan, D., Su, F., Liu, C., & Guo, Z. (2020). Research progress for plastic waste management and manufacture of value-added products. *Advanced Composites and Hybrid Materials, 3*, 443–461.

Panko, J. M., Chu, J., Kreider, M. L., & Unice, K. M. (2013). Measurement of airborne concentrations of tire and road wear particles in urban and rural areas of France, Japan, and the United States. *Atmospheric Environment, 72*, 192–199.

Peterson, J. D., Vyazovkin, S., & Wight, C. A. (2001). Kinetics of the thermal and thermo-oxidative degradation of polystyrene, polyethylene and poly (propylene). *Macromolecular Chemistry and Physics, 202*(6), 775–784.

Pielichowski, K., Njuguna, J., & Majka, T. M. (2022). *Thermal degradation of polymeric materials*. Elsevier.

Pilapitiya, P. N. T., & Ratnayake, A. S. (2024). The world of plastic waste: A review. *Cleaner Materials, 11*, Article 100220.

Pitt, J. A., Aluru, N., Hahn, M. E., Shumway, S. E., & Ward, J. E. (2023). Microplastics in marine food webs. In *Plastics in the Sea: Occurrence and impacts*. Elsevier.

Prata, J. C. (2018). Airborne microplastics: Consequences to human health? *Environmental Pollution, 234*, 115–126.

Prata, J. C., Da Costa, J. P., Lopes, I., Duarte, A. C., & Rocha-Santos, T. (2020). Environmental exposure to microplastics: An overview on possible human health effects. *Science of the Total Environment, 702*, Article 134455.

Qi, Y., Yang, X., Pelaez, A. M., Lwanga, E. H., Beriot, N., Gertsen, H., Garbeva, P., & Geissen, V. (2018). Macro-and micro-plastics in soil-plant system: effects of plastic mulch film residues on wheat (Triticum aestivum) growth. *Science of the Total Environment, 645*, 1048–1056.

Rezaei, M., Abbasi, S., Pourmahmood, H., Oleszczuk, P., Ritsema, C., & Turner, A. (2022). Microplastics in agricultural soils from a semi-arid region and their transport by wind erosion. *Environmental Research, 212*, 113213.

Richardson, K., Wilcox, C., Vince, J., & Hardesty, B. D. (2019).

Rillig, M. C. (2012). Microplastic in terrestrial ecosystems and the soil?

Rillig, M. C., Lehmann, A., de Souza Machado, A. A., & Yang, G. (2019). Microplastic effects on plants. *New Phytologist, 223*(3), 1066–1070.

Russell, G. T. (2002). The kinetics of free-radical polymerization: Fundamental aspects. *Australian Journal of Chemistry, 55*(7), 399–414.

Schettler, T. E. D. (2006). Human exposure to phthalates via consumer products. *International Journal of Andrology, 29*(1), 134–139.

Scheurer, M., & Bigalke, M. (2018). Microplastics in Swiss floodplain soils. *Environmental Science & Technology, 52*(6), 3591–3598.

Schymanski, D., Goldbeck, C., Humpf, H. U., & Fürst, P. (2018). Analysis of microplastics in water by micro-Raman spectroscopy: Release of plastic particles from different packaging into mineral water. *Water Research, 129*, 154–162.

Sharma, V., & Kundu, P. P. (2008). Condensation polymers from natural oils. *Progress in Polymer Science, 33*(12), 1199–1215.

Soares, A., Guieysse, B., Jefferson, B., Cartmell, E., & Lester, J. N. (2008). Nonylphenol in the environment: A critical review on occurrence, fate, toxicity and treatment in wastewaters. *Environment International, 34*(7), 1033–1049.

Song, Y. K., Hong, S. H., Jang, M., Han, G. M., Rani, M., Lee, J., & Shim, W. J. (2015). A comparison of microscopic and spectroscopic identification methods for analysis of microplastics in environmental samples. *Marine Pollution Bulletin, 93*(1–2), 202–209.

Speight, J. G. (2016). *Handbook of petroleum refining*. CRC Press.

Sperling, L. H. (2015). *Introduction to physical polymer science*. Wiley.

Stephens, B., Azimi, P., El Orch, Z., & Ramos, T. (2013). Ultrafine particle emissions from desktop 3D printers. *Atmospheric Environment, 79*, 334–339.

Sun, J., Dai, X., Wang, Q., Van Loosdrecht, M. C., & Ni, B. J. (2019). Microplastics in wastewater treatment plants: Detection, occurrence and removal. *Water Research, 152*, 21–37.

Trainic, M., Flores, J. M., Pinkas, I., Pedrotti, M. L., Lombard, F., Bourdin, G., Gorsky, G., Boss, E., Rudich, Y., Vardi, A., & Koren, I. (2020). Airborne microplastic particles detected in the remote marine atmosphere. *Communications Earth & Environment, 1*(1), 64.

Turner, A. (2021). Paint particles in the marine environment: An overlooked component of microplastics. *Water rEsearch x, 12*, Article 100110.

Urbina, M. A., Correa, F., Aburto, F., & Ferrio, J. P. (2020). Adsorption of polyethylene microbeads and physiological effects on hydroponic maize. *Science of the Total Environment, 741*, Article 140216.

Vandenberg, L. N., Hauser, R., Marcus, M., Olea, N., & Welshons, W. V. (2007). Human exposure to bisphenol A (BPA). *Reproductive Toxicology, 24*(2), 139–177.

Vert, M., Doi, Y., Hellwich, K. H., Hess, M., Hodge, P., Kubisa, P., Rinaudo, M., & Schué, F. (2012). Terminology for biorelated polymers and applications (IUPAC Recommendations 2012). *Pure and Applied Chemistry, 84*(2), 377–410.

Vianello, A., Jensen, R. L., Liu, L., & Vollertsen, J. (2019). Simulating human exposure to indoor airborne microplastics using a Breathing Thermal Manikin. *Scientific Reports, 9*(1), 8670.

Vuong, H. (2023). Marine plastic pollution: A review of the scientific evidence, current policies, and potential solutions.

Wang, F., Wang, Q., Adams, C. A., Sun, Y., & Zhang, S. (2022a). Effects of microplastics on soil properties: Current knowledge and future perspectives. *Journal of Hazardous Materials, 424*, Article 127531.

Wang, J., Li, J., Wang, Q., & Sun, Y. (2020). Microplastics as a vector for HOC bioaccumulation in earthworm Eisenia fetida in soil: Importance of chemical diffusion and particle size. *Environmental Science & Technology, 54*(19), 12154–12163.

Wang, Q., Adams, C. A., Wang, F., Sun, Y., & Zhang, S. (2022b). Interactions between microplastics and soil fauna: A critical review. *Critical Reviews in Environmental Science and Technology, 52*(18), 3211–3243.

Wik, A., & Dave, G. (2009). Occurrence and effects of tire wear particles in the environment–A critical review and an initial risk assessment. *Environmental Pollution, 157*(1), 1–11.

Wright, S. L., & Kelly, F. J. (2017). Plastic and human health: A micro issue? *Environmental Science & Technology, 51*(12), 6634–6647.

Wright, S. L., Thompson, R. C., & Galloway, T. S. (2013). The physical impacts of microplastics on marine organisms: A review. *Environmental Pollution, 178*, 483–492.

Wright, S. L., Ulke, J., Font, A., Chan, K. L. A., & Kelly, F. J. (2020). Atmospheric microplastic deposition in an urban environment and an evaluation of transport. *Environment International, 136*, Article 105411.

Yu, H., Zhang, X., Hu, J., Peng, J., & Qu, J. (2020). Ecotoxicity of polystyrene microplastics to submerged carnivorous Utricularia vulgaris plants in freshwater ecosystems. *Environmental Pollution, 265*, Article 114830.

Zhang, G. S., & Liu, Y. F. (2018). The distribution of microplastics in soil aggregate fractions in southwestern China. *Science of the Total Environment, 642*, 12–20.

Zhang, Y., Kang, S., Allen, S., Allen, D., Gao, T., & Sillanpää, M. (2020). Atmospheric microplastics: A review on current status and perspectives. *Earth-Science Reviews, 203*, Article 103118.

Ziajahromi, S., Neale, P. A., Rintoul, L., & Leusch, F. D. (2017). Wastewater treatment plants as a pathway for microplastics: Development of a new approach to sample wastewater-based microplastics. *Water Research, 112*, 93–99.

Chapter 2
Microplastics in Air, Drinking Water and Food: Impacts on Human Health

Abdullah Al Mamun

Abstract The widespread use of plastic materials has brought significant benefits across various sectors of human life, marking a new historical era. However, their desirable properties—such as durability, low cost, and water resistance—have led to substantial misuse, resulting in an accumulation of plastic waste. This waste breaks down into microplastics (MPs), which infiltrate marine, terrestrial, and freshwater ecosystems, as well as the air, making their way into the human food chain. Major sources of microplastic exposure include foods from diverse origins, food additives, beverages, and plastic packaging. The presence of MPs raises serious public health concerns due to their tendency to accumulate in living tissues, potentially leading to gastrointestinal diseases, immune system disorders, respiratory issues, cancer, infertility, and chromosomal alterations. Additionally, MPs can act as carriers for toxic chemicals and harmful organisms. This chapter offers a comprehensive overview of MPs, their environmental transfer to the human food chain, and their potential toxic effects on health. It concludes with a discussion on public health policy implications related to microplastic exposure.

Keywords Food chain · Human health · Environmental safety · Food safety · Plastic waste · MPs in air · Water and food: impacts on health

Abbreviations

BPA	Bisphenol A
CE	Cellulose
DWTP	Drinking water treatment plant
HDPE	High density polyethylene
LDPE	Low density polyethylene
MP	Microplastic

A. Al Mamun (✉)
Doctoral Study Program in Public Health, Faculty of Public Health, Universitas Airlangga, Surabaya, Indonesia
e-mail: aamfst@gmail.com

V. Kumar et al. (eds.), *Microplastics Pollution*,
https://doi.org/10.1007/978-3-031-99774-7_2

MPs	Microplastics
NP	Nanoplastic
NPs	Nanoplastics
PA	Polyamide
PE	Polyethylene
PES	Polyester
PET	Polyethylene terephthalate
PP	Polypropylene
PTFE	Polytetrafluoroethylene
PS	Polystyrene
PMMA	Poly methyl methacrylate
PU	Polyurethane
PVC	Polyvinyl chloride
SBR	Styrene-butadiene rubber
WWTP	Wastewater treatment plant

2.1 Introduction

Plastics are multifunctional, inexpensive, and durable materials that play an essential role in the modern world. Their versatility make them widely used in various aspects of daily life. Since the invention of first plastic in nineteenth century, it has lived up to its undeniable reputations as "material of a thousand uses" and "plastic is the greatest finding in the millennium" (Gilbert, 2017; Nayanathara Thathsarani Pilapitiya & Ratnayake, 2024). People beings use plastics in every possible way in daily life, resulting in the massive expansion of the plastic industry which boosted approximately 400 million metric tons of global plastic production in 2022 (Statista, 2024). Excessive use of plastics creates a bulk amount of waste, leading to the pile-up of plastic debris in every corner of the world such as landfills, waterways, and oceans. This bulk volume of plastic waste easily transported and discarded to the environment, which will sustain for centuries. To date, we understand that plastic debris in water bodies and plastic waste on land gradually break down into smaller particles, eventually forming microplastics (MPs) and even nanoplastics (NPs). For years, researchers believed the ocean was the main global plastic sink. However, recent studies have shown that terrestrial ecosystems, particularly soil, also serve as significant sinks for MPs (Vitali et al., 2023).

The study of MPs in environment was the initiation of knowing a potential threat of MPs pollution. Recent studies reported the contaminations MPs in atmosphere, all kind of drinking water, and human food chain. Since then, knowledge on MPs in every corner of environment including air, water and foods is growing rapidly. Airborne MPs inhale by human from urban dust, textile, synthetic fabrics, and rubber tires. These MPs can also enter the human food chain and water resources, leading to ingestion. Additionally, dust accumulation in urban areas facilitates direct inhalation and ingestion of microplastics. Furthermore, washing clothes and using cosmetics that contain microplastics may have toxic effects on the human body.

Researcher also reporting the health hazards of MPs and it effects in different body parts. In addition, MPs can act as carrier for several toxic chemicals, additives, and pathogens. Given the rapidly growing number of documents on this research topic, this book chapter aims to provide knowledge and overview of MPs in air, drinking water, and foods to the academic and research community. This book chapter provides a comprehensive knowledge for novices to the field of MPs in air, drinking water, and human food chain, while guiding those with more experience to explore the latest research findings.

2.2 Production to Plastics Pollution

2.2.1 Global Production of Plastic

Plastics offer several advantages: (i) Environmental—polyethylene terephthalate (PET) packaging reduces energy use and emissions compared to glass and metal, while lighter composites in aircraft lower fuel consumption; (ii) Social—plastics enable clean drinking water supply and safe food packaging; (iii) Economic—plastics are cost-effective substitutes for traditional materials. As shortages of glass, wood, metal, and paper arise, plastics increasingly replace these materials. Plastic production involves four steps: raw material extraction, polymer synthesis, compounding, and molding. In 2022, global plastic production reached 400.3 million metric tons, up 1.6% from the previous year, reflecting a steady increase since the 1950s due to plastics' versatility (Fig. 2.1). The market value of plastics is also rising, with Asia leading production, particularly China, which produced 32% of the world's plastics in 2022 (Statista, 2024). North America follows, contributing 17% to global production. Global plastic production is expanding to meet demand and is expected to double in the next 20 years at the current growth rate. Plastics are now widely used in products, replacing materials like wood, metal, and glass. A key application is plastic packaging. Since the widespread production of plastic began in the mid-twentieth century, it has become the leading contributor to environmental pollution (Table 2.1).

2.2.2 Disposal of Plastics

The global increase in plastic production and consumption has led to a significant waste management challenge. Around one-third of plastic products are discarded within three years of production, contributing to the accumulation of persistent plastic waste in the environment (Karim et al., 2022). As of 2015, an estimated 6,300 million

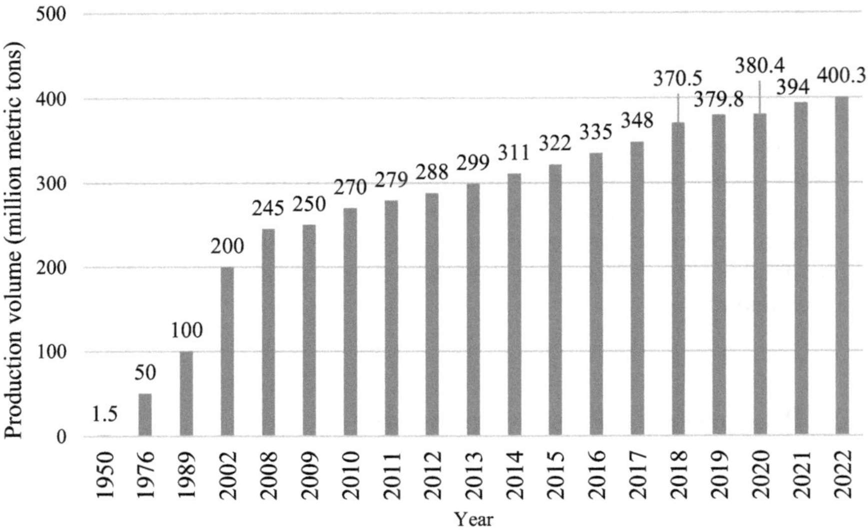

Fig. 2.1 Global plastic production (in million metric tons) in a year from 1950 to 2021 (Statista, 2024)

Table 2.1 Historical events of plastics

Year	Chronology
1860s	First artificial plastics invention • Invention of Parkesin • Invention of celluloid
1907	Invention of Bakelite (less flammable and more durable than celluloid): first synthetic thermosetting plastic
1912–1956	• Polyvinyl chloride (PVC) or vinyl production • Invention of various plastic types including cellulose (CE), low density polyethylene (LDPE), polyurethane (PU), nylon/polyamide (PA), polystyrene (PS), PET, polypropylene (PP), teflon
1960	Awareness of plastic pollution
1973	Biodegradable plastic
1974	Electrically conductive plastic
1983	Identify harmful effects of plasticizers
1988	International policies for plastic pollution
1997	Discovering the Great Pacific Garbage Patch

Adopted from: Nayanathara Thathsarani Pilapitiya and Ratnayake (2024) metric tons of plastic waste had been generated, with projections reaching 12,000 million metric tons by 2050 if current practices continue (Geyer et al., 2017). This massive accumulation is not only increasing the volume of global solid waste but also contributing to marine pollution, as a significant portion of plastic enters the oceans due to poor waste management. Although efforts such as recycling, incineration, and landfilling have been implemented, about 22% of plastic waste remains mismanaged (Evode et al., 2021; Haque & Fan, 2023). This includes improper disposal along coastlines, illegal dumping, and leakage during industrial operations and transportation. Such mismanaged plastics are the primary sources of environmental contamination and eventually degrade into microplastics, posing further ecological and health risks.

2.2.3 *Degradation and Fragmentation of Plastic*

Plastics in environment can be breakdown in two steps namely: degradation and fragmentation. Simply, fragmentation is the breakdown of larger polymer chain to small polymer fragments. Whereas, degradation is the chemical change of polymer properties caused by chemical, physical, or biological action (Mattsson et al., 2018).

- *Photodegradation* occurs due to exposure of sunlight which is more prominent on land surface and in air compared to in water. Ultraviolet (UV) radiation from sunlight is capable to cut carbon–carbon bonds in the polymer.
- *Physical* and/or *mechanical* degradation breaks the polymer molecule due to mechanical stress such as wave action, rock, sand, and other shearing forces.
- Hydrolysis can occur within plastic materials through contact with water, progression through condensation cycles, or moisture attack, or by exposure to steam.
- *Biodegradation* is the breakdown of organic substances due to living organisms such as bacteria. Biodegradable plastics can be broken down into water, carbon monoxide, and other biomaterials from microorganisms, including bacteria and fungi. *Salinity* significantly affects the distribution of plastic degradation bacteria.
- Biofilm formation is linked with plastic degradation, which enhances microbial and enzymatic interactions through direct surface contact with the polymer.
- *Thermo-oxidative* degradation occurs in moderate temperature which results slow oxidative breakdown of polymer.

2.2.3.1 Microplastics (MPs)

Generally, MPs constitutes different form of plastics in small size as a result of different factors such as heat, sunlight, water activity, physical stress, etc. MPs is defined as small plastic particles range from 100 nm $\leq x \leq$ 5 mm in size (Petersen, 2016). The major characteristics to consider for small particles classified as MPs are (Jadhav et al., 2021):

- Synthetic materials with a high polymer content
- Solid particles
- Smaller than 5 mm
- Insoluble in water
- Non-degradable.

Based on the origin, MPs are categorized as primary and secondary MPs. Primary MPs are defined as small-sized particles designed for commercial use, such as cosmetics and personal care products, as well as microfibers shed from textiles, including clothes and fishing nets. Whereas, secondary MPs are larger particles that result from the abrasion, degradation, and fragmentation of larger plastic items, such as bags, packaging, and water bottles.

2.2.3.2 Nanoplastics (NPs)

NPs are extremely small piece of plastics. Nanoplastics is defined as plastics particles smaller than microplastics with a size ranging from 1 nm $\leq x \leq$ 1 μm (Gigault et al., 2018). Because of their extremely small size, NPs might show increased toxicity and biological effects. Besides the NPs contamination in environment, NPs are widely used in certain products, including exfoliating beads in facial scrubs, beauty items, and industrial pellets (Alqahtani et al., 2023).

2.2.3.3 Types of Microplastics

Various types of MPs play significant contribution in MPs pollution and pose threat to ecosystems including human food chain. The MPs can be categorized based on size, shape, and characteristics (Table 2.2). The major types of MPs are:

- *Microbeads* are non-biodegradable polymers, measuring less than 1 mm in diameter, are found in personal care and hygiene items, facial cleansers, exfoliating soaps, and toothpaste. Their small size and indigestibility allow them to easily infiltrate the food chain, posing potential risks.
- *MP fragments*—larger plastic items, such as single-use products, lids, water bottles, plastic bags, and dining items, break down into smaller pieces through UV radiation from sunlight and wave action.
- *Foam*—Styrofoam is used in packaging, tea or coffee cups, and as cushioning material. Compounds from Styrofoam can leach into food and drinks, raising health concerns.
- *Fibers*—microfibers mainly come from laundry washing, woolen clothing, diapers, and cigarette butts. Synthetic fabrics, especially polyester (PES) found in athletic wear, release a higher volume of microfibers.

Table 2.2 Different types of microplastics and their major sources

MP type	Description	Polymer type	Major source
Microbead	Spherical, smooth shape, hard texture	PE, poly methyl methacrylate (PMMA), polytetrafluoroethylene (PTFE), PP, Nylon, and PET	Facial scrabs, body wash
MP fragment/flakes	Hard, jagged plastic particle	High-density polyethylene (HDPE) or LDPE	Beverage bottles, plastic lids, plastic cups, straws, stirrers, forks, knives, and spoons
Foam	Lightweight, sponge-like plastic	PS	Foam floats, styrofoam, and cushioning
Fiber	Thin, fibrous, straight plastic, soft, malleable	PS	Fishing items such as net and fabrics or textiles
Film	Thin plane of flimsy plastic	Polyolefins and PES	Plastic bags and food wrappers
Pellet	Hard, rounded plastic	PVC, PA or nylon, etc.	Virgin resin pellets, beauty items, facial cleaners

- *Film*—plastic films primarily originate from plasticizers often used in agriculture, which can contaminate the soil ecosystem, disrupt the balance of soil organisms, and harm the overall ecosystem.

2.2.4 Migration Pathways of Microplastics to Human Exposure

The human body is exposed to microplastics through three main routes: ingestion, inhalation, and skin contact (Prata et al., 2020). Ingestion occurs primarily through drinking water, beverages, different foods, and food additives. Common MPs found in these sources include PP, PET, and PS. Inhalation occurs when MPs are released into the air (both indoor and outdoor air) from various sources, such as textiles, tire wear, and surface resuspension. The types of MPs inhaled often include synthetic fibers and plastics like PES and acrylic. Skin contact, although less significant, can happen through dust, synthetic clothing, and microbeads in personal care items. MPs are pervasive in terrestrial, freshwater, and marine environments, often contaminating food products. Various factors, such as food processing and packaging, also contribute to this contamination.

Drinking water, primarily sourced from groundwater and surface water, is also at risk. Studies have shown that freshwater reservoirs are frequently contaminated with MPs. Additionally, environmental factors like weather, mechanical degradation, and microbial action can break down larger plastics into MPs. Airborne MPs, especially from textiles, pose further health risks, as inhaled fibers can harm lung health.

The soil–plant system plays a crucial role in the ecosystem and food supply. MPs can be taken up by plants from contaminated soils, entering the food chain. Various crops can absorb MPs, potentially affecting their growth and transferring these particles to human food sources (Sun et al., 2020). Moreover, ants' activity in the soil ecosystem affects the migration and distribution of microplastics, and also influences the exposure of other soil fauna (Liu et al. 2023b).

2.3 Microplastics in Air

MPs primarily end up in the hydrosphere due to human activities like tourism and wastewater treatment (Haque & Fan, 2023). They also contaminate the pedosphere through domestic and agricultural practices, where plastic waste accumulates on land. However, MPs can become airborne, where small, light particles can be carried long distances by wind currents, even reaching remote regions such as the Arctic (Allen et al., 2019). Atmospheric MPs can be transported to remote regions and deposited by rain and snow. Understanding how these precipitation events affect microplastic deposition and their relationship with climate conditions requires further research. MPs have been found in remote locations such as Arctic Sea Ice, the Fram Strait, the Italian Alps, and Iceland's Vatnajökull ice cap (Haque & Fan, 2023). In Arctic areas, snowfall is a key method of microplastic deposition. During ice formation, microplastics can become trapped as ice crystals form and accumulate, embedding the particles in deeper ice layers.

Strong winds and rainfall lead to significant atmospheric microplastic deposition in cities like Paris and Hamburg, with about 120 MP/m^2 and 275 MP/m^2 deposited daily, respectively (Dris et al., 2015; Klein & Fischer, 2019). Remote areas like the Pyrenees Mountains have lower deposition rates, averaging 365 MP/m^2 per day, due to the long-distance transport of microplastics (Allen et al., 2019). Cities with higher population density and industrial activity, like Shanghai, have greater deposition rates, with 4.18 MP/m^3 compared to Paris's 2.84 MP/m^3 (Ambrosini et al., 2019).

Overall, the destination and behavior of microplastics in both indoor and outdoor atmosphere are influenced by various factors, including (Madeła, 2023):

- Vertical pollutant concentration gradient
- Wind speed
- Wind direction
- Precipitation
- Temperature.

2.3.1 Sources and Pathways of MPs in Air

Several potential sources contribute to the presence of microplastics (MPs) in the air. While the exact sources of airborne microplastics are not yet fully quantified, they may include the interface between air and water, sea spray, wind-driven abrasion, resuspension from tire wear, emissions from nail salons, mechanical drying of synthetic materials, landfill sites, and human activity involving synthetic clothing (Fig 2.2). The potential sources of MPs in air are:

(a) ***Air water interface and sea spray***

The air–water interface (AWI) is a common reaction zone distinct from the bulk phase, where unique reactions and physical processes frequently occur. This interface encompasses interactions between air and cloud droplets, aerosol particles, ocean surfaces, and biological fluids like those lining human epithelia. Microplastics (average size 20 ± 13 μm) are aerosolized from the ocean's surface, likely through bubble bursts or sea spray ejection (Allen et al., 2020). Once airborne, microplastics can be transported by wind, aid the aggregation of particles, and influence cloud and droplet formation because of changes in hydrophobicity. Microplastics are suspended, transported, and ultimately deposited back into the ocean surface

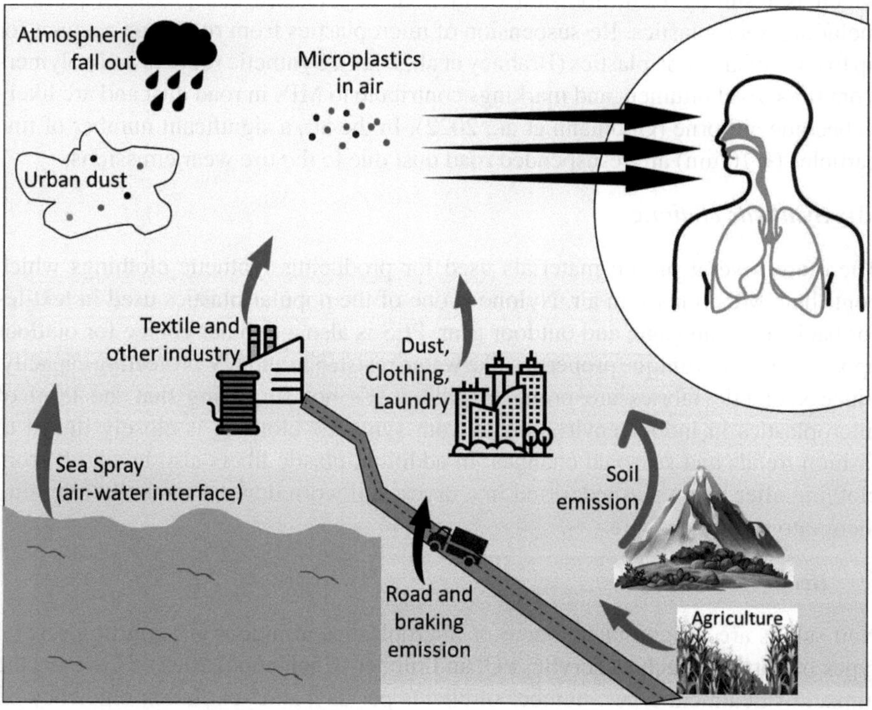

Fig. 2.2 Transportation of microplastics in atmospheric

due to a cyclical “grasshopper” effect caused by the combination of deposition, air transport, and aerosolization at the ocean surface (O’Brien et al., 2023).

NPs enter the atmosphere through bubble bursting at the air–water interface, with their aerosolization efficiency influenced by surface properties such as hydrophobicity. Despite the fact that particles tend to aggregate near the ocean’s surface, UV degradation, plastic age, and smaller bubble sizes increase this efficiency.

(b) ***Wind abrasion and wind erosion***

Wind abrasion involves wind-driven particles such as sand or dust, physically rub against and wear down microplastics. It can break down larger plastic fragments into smaller microplastics or cause existing microplastics to become more airborne by detaching them from surfaces like textiles or landfills. Air contamination can be occurred by wind erosion, which remove and lift microplastics from surfaces due to the action of the wind. It can cause microplastics on the ground, in landfills, or from other sources to become suspended in the air, contributing to atmospheric pollution.

(c) ***Tire wear***

The materials that make up rubber tires and wear on the road are varied, but approximately half of them are made of synthetic and natural polymers like carbon black, natural rubber, and styrene-butadiene rubber (SBR) (Jahandari, 2023). Tire wear, which is resulting from abrasion or thermal degradation can produces particles including microplastics. Re-suspension of microplastics from roads can account for up to 84% of airborne plastics (Brahney et al., 2021). Synthetic rubbers and polymers from tires, road bitumen, and markings contribute to MPs in road dust and are likely to become airborne (Goßmann et al., 2022). In the air, a significant number of fine particles (< 10 μm) are resuspended road dust due to the tire wear emissions.

(d) ***Synthetic clothing***

There are several plastic materials used for producing synthetic clothings which contribute MPs release in air. Nylone is one of the popular plastics used in textiles for backpacks, luggage, and outdoor gear. PES is also a popular choice for outdoor apparel due to it’s unique properties like water-resistance and UV protecting capacity. Since synthetic fabrics are prone to pilling, it’s not surprising that the level of microplastics in indoor environments from synthetic clothing is closely linked to fashion trends and seasonal changes. In addition, plastic fibers also break off from clothing after being washed, dried in a dryer, and worn down, eventually releasing them into the atmosphere.

(e) ***Beauty salon***

Nail salons are a significant source of microplastics in indoor air, emitting several types of particles such as acrylic, PU, and rubber (Chen et al., 2022). However, the impact of treatment types, number of clients per day, and ventilation efficiency on airborne microplastic concentrations remains unclear.

2.3.2 Behaviors of Atmospheric Microplastics

Monitoring MPs in air is vital for assessing microplastic pollution and its risks for human health, as inhalation is a major exposure route. MPs distribution in air is influenced by factors like plastic material prevalence, local weather conditions, ventilation, cleanliness, urban topography, and population density. Human activity increases microplastic sources, but population alone doesn't determine MPs levels. Airflow between indoor and outdoor spaces can resuspend particles, and ineffective ventilation systems can bring outdoor MPs indoors. Meteorological conditions, including wind, humidity, and precipitation, significantly affect MPs distribution in air, with seasonal variations in concentrations. High humidity and wet deposition can lead to larger particles settling out of the air. The residence time of MPs in the atmosphere depends on their size and environmental factors, with larger particles settling faster and lighter ones remaining airborne longer. Urban topography may affects air currents and pollutant dispersion. In addition, temperature inversions can trap atmospheric MPs near the ground, leading to localized pollution (Jahandari, 2023).

2.4 Microplastics in Drinking Water

The contamination of MPs in drinking water (DW) varies with different factors including geographical location, quality of source water and water processing treatments (Gambino et al., 2022). DW faces MP contamination from various processes of MPs transportation to environment including landfill without proper waste treatment, degradation of plastics, shedding synthetic fibers during textile washing, and fragmentation of larger plastics in the environment (Chen et al., 2023). In addition, DW can be contaminated by other contributors such as activated carbon, plastic pipes, chlorine treatment, and water storage (Sol et al., 2021).

The presence of MPs in every corner of the environment creates a situation where the drinking water is increasingly exposed to microplastic contamination. DW is at risk of contamination from microplastic particles and their related chemicals, which can come from multiple sources including groundwater, lakes, reservoirs, and rivers (Fig. 2.3). Microplastics enter drinking water through various pathways including wastewater discharge, storm runoff, atmospheric deposition, the breakdown of larger plastics, and microbeads from personal care products (Raza et al., 2022). Groundwater can also become contaminated through rainwater and plastic debris leaching (Panno et al., 2019). Lakes, rivers, and treated wastewater contribute to this contamination as microplastics are transported from different sources and can carry harmful pollutants, posing health risks.

It is important to consider wastewater treatment plant (WWTP) and drinking water treatment plant (DWTP) as hotspot for MPs and microfibres contamination in the DW (Sol et al., 2021). MPs from industrial and urban sources enter sewage systems and,

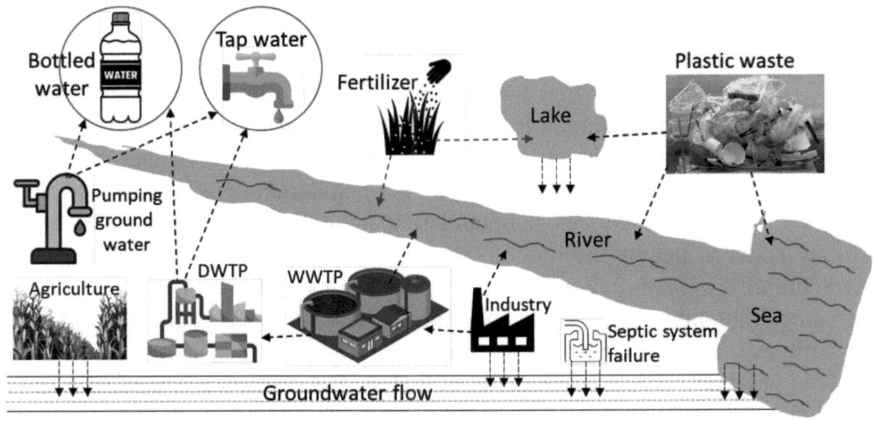

Fig. 2.3 Microplastics contamination in drinking water

despite WWTPs removing over 90% of MPs, millions still end up in the environment daily (Ali et al., 2021). Household microfibres come from clothing and furnishings, while industrial microfibres originate from sectors like automotive, construction, and textiles, with the clothing industry alone generating around 42 million tons annually (Kelly et al., 2019; Suaria et al., 2020). Laundry contributes 35% of global primary MP release, with each wash emitting up to 1.3×10^7 million fibres (Boucher & Friot, 2017; Sillanpää & Sainio, 2017). Wear and tear from garments also adds to atmospheric microfibres, which enter sewage systems via surface runoff (Bayo et al., 2020; De Falco et al., 2019).

WWTPs handle wastewater through pretreatment, primary, secondary, and sometimes tertiary stages. Pretreatment removes large debris to prevent operational issues, capturing 45% of MPs. Primary treatment, using clarifiers, can reduce MP concentration by 78–98%, depending on system specifics. Secondary treatment, involving biological processes, may release some MPs into the air and achieves variable removal rates, ranging from 28 to 67%. Tertiary treatment, if used, varies by local regulations and purposes, with processes like rapid sand filtration removing 45–97% of MPs, while membrane systems can remove over 99% (Sol et al., 2021). Photocatalytic degradation shows promise but needs further research for improved effectiveness.

Drinking water treatment plants (DWTPs) is plays important role for safe DW as it can be polluted and require treatment to meet safety standards and remove harmful pathogens and chemicals. DWTPs use various processes to ensure water quality, including coagulation-flocculation, sedimentation, filtration, and disinfection (Betancourt & Rose, 2004; Shen et al., 2020). Coagulation-flocculation involves adding chemicals to clump particles, which then settle out. The water is then filtered through materials like sand or activated carbon to remove microorganisms and reduce turbidity. Finally, disinfection methods such as chlorination, ozonation, or ultraviolet light are used to kill pathogens.

While there is less focus on microplastic (MP) contamination in DWTPs compared to WWTPs, it is a growing concern due to potential health risks (Eerkes-Medrano et al., 2019; Shen et al., 2020). Studies have found that people could ingest thousands of microplastics annually through tap water (Cox et al., 2019). The concentration of microplastics, including microfibers, in DWTPs varies significantly, ranging from none to over 6000 particles per liter in influent, and from none to about 930 particles per liter in effluent (Sol et al., 2021). The average concentration is similar to or higher in WWTPs. Sources of microplastics in drinking water include environmental degradation of plastics, industrial discharges, and airborne particles (Eerkes-Medrano et al., 2019).

Microfiber concentrations are generally lower in DWTP influents compared to WWTP influents, likely due to differences in water sources (Dalmau-Soler et al., 2021). The percentage of microfibers relative to total microplastics is similar in DWTP influents and effluents. Removal efficiency for microfibers in DWTPs ranges from 25 to 90%, with some plants achieving up to 90.4% removal (Sol et al., 2021). Compared to WWTPs, which can achieve removal efficiencies over 99%, DWTPs are less effective, often due to simpler treatment processes (Mintenig et al., 2019). Enhancing microplastic removal in DWTPs is crucial for reducing potential health risks associated with microplastic ingestion.

2.4.1 MPs in Ground Water

Groundwater is the primary source of freshwater used by more than 2 billion people for various purposes, such as drinking, household use, agriculture, and industry (Viaroli et al., 2022). Groundwater makes up 33% of the world's freshwater supply (Lee et al., 2024). Because it naturally filters through layers of rock and soil, it is usually of higher quality than surface water from sources like rivers, lakes, streams, and reservoirs, where contamination is less likely (Kayastha et al., 2022).

However, evidence is increasing that MPs are entering groundwater systems, raising concerns for both the environment and human health (Srihari et al., 2023; Viaroli et al., 2022). Traditionally, groundwater research has focused on chemical pollutants like heavy metals and pesticides. Now, MPs add a new layer of complexity to contamination issues. MPs enter groundwater through various pathways, including surface runoff, agricultural activities, atmospheric deposition, and wastewater (Cha et al., 2023; Lee et al., 2024; Viaroli et al., 2022). Some geological processes such as sea-water intrusion and groundwater—river/lake interaction can also be considered as the possible sources of MPs in the groundwater (Singh & Bhagwat, 2022). Microplastics enter soil from various sources, including wastewater treatment plant discharge, septic tank outflows, landfill leachate, and direct contamination (Re, 2019). Plastic mulching in agriculture also contributes as plastics break down under UV light. Additionally, microplastic-rich sludge from wastewater treatment is often spread on fields, further contaminating soil. When water seeps through contaminated soil, it carries microplastics deeper, potentially reaching groundwater.

Landfills, which contain much of the world's plastic waste, contribute significantly to microplastic pollution. Over time, landfill leachate breaks down plastics into smaller particles, which then contaminate groundwater (Singh & Bhagwat, 2022). Because of their resilience to degradation, these particles can travel great distances and gather in aquifers, where they can stay for long periods of time. By altering microbial communities, nitrogen cycling, and soil structure, MPs in groundwater can have an effect on ecosystems. For human health, MPs in groundwater pose risks such as physical harm, chemical exposure, and pathogen transmission.

The major transport pathways of MPs into the groundwater:

1. Groundwater—river interaction
2. Groundwater—lake interaction
3. Sea-water—groundwater interaction
4. Transport through soil
5. Transport from landfills.

2.4.2 MPs in Tap Water

The levels of microplastics in tap water are primarily influenced by the microplastic content of source water and the effectiveness of DWTP in removing them (Oladoja & Unuabonah, 2021). While DWTPs generally remove most microplastics, the distribution system can contribute additional microplastics, which end up in tap water (Barchiesi et al., 2021). Concerns exist regarding the release of microplastics from membrane filtration systems in DWTPs due to their plastic composition and the presence of similar microplastics in treated water (Fig. 2.4). Furthermore, there is uncertainty regarding microplastic leaks from distribution networks due to the widespread usage of plastics and their lengthy life cycle. Disinfection and cleaning processes, involving strong oxidants, high pressure, and ultraviolet light, are thought to age plastic pipes and contribute to microplastic release (Yang et al., 2024; Zhou et al., 2023).

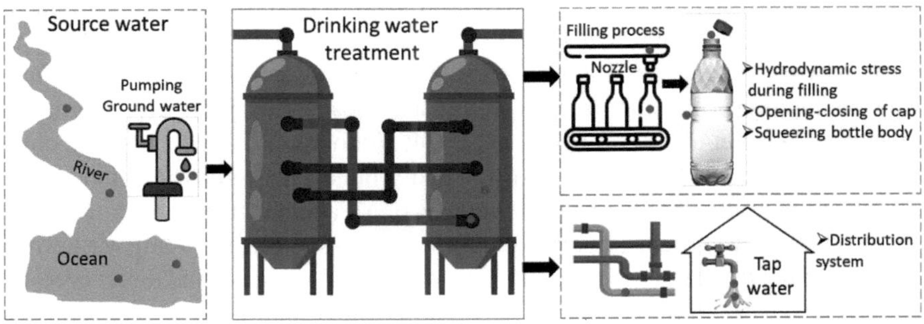

Fig. 2.4 Microplastics contamination in tap water and bottled water

2.4.3 MPs in Bottled Water

Bottled water industry is a juggernaut and one of the most dynamic sectors within the food and beverage industry, primarily due to it's status and "must have-daily item" value. However, drinking bottled water is not safe in context of MPs contamination including both single-use and reusable bottles. Microplastics in reusable bottled water are roughly double the amount found in single-use bottled mineral water (Oßmann et al., 2018). Bottled water can be contaminated by MPs due to several reason including quality of source water, production process, and the bottle itself (Li et al., 2020). Microplastics are more likely to be released from older returnable PET bottles compared to single-use plastic bottles, glass bottles, and new returnable PET bottles (Kankanige & Babel, 2020; Oßmann, 2021) (Fig. 2.4).

The process of bottling water starts with choosing a high-quality water source. While disinfection procedures like UV treatment or chlorination kill hazardous germs, filtering techniques like sand filters, activated carbon filters, nanofiltration, and reverse osmosis remove pollutants (Haleem et al., 2024). After being thoroughly cleaned, the water is packed into glass, aluminum, or plastic bottles or containers that have been sanitized and kept in controlled settings to preserve freshness until it is distributed.

Microplastics (MPs) often come from the degradation of PET bottles and PE caps during production, transport, and storage (Makhdoumi et al., 2021). The use of packaging materials can also release MPs into the water. Plastic bottles can release microplastics through various mechanisms such as opening and closing, squeezing, cracking and photodegradation (Julienne et al., 2019; Makhdoumi et al., 2021; Winkler et al., 2019; L. Yang et al., 2024). Studies have shown that repeated cap usage can significantly increase microplastic content in bottled water (Singh, 2021). Additionally, microplastics have been found in glass bottled water, suggesting potential contamination during the bottling process rather than solely from the plastic packaging (Oßmann et al., 2018).

2.4.4 Vector for Toxic Chemicals and Microorganisms

2.4.4.1 Leaching Toxic Chemicals

Microplastics (MPs) contain various chemicals such as plasticizers, stabilizers, UV filters, flame retardants, polycyclic aromatic hydrocarbons and polychlorinated biphenyls (Haleem et al., 2024). MPs can release these chemicals into water, which can be toxic to humans and an important factor in assessing health risks (Elkhatib & Oyanedel-Craver, 2020). The amount of chemical leaching from MPs depends on the type of plastic. For example, PVC tends to leach more dissolved organic carbon (DOC) than PE, PS, or PET (Lee et al., 2021). Some plastic products can release over

80% of their chemical content into water throughout their lifecycle (Sripada et al., 2022).

Leaching is influenced by factors such as temperature, pH, and UV exposure (Na et al., 2021). As plastics age, their additives degrade, increasing the release of dissolved organic matter (DOM) (Chen et al., 2021; Haleem et al., 2024). UV radiation, in particular, can lead to significant chemical changes in plastics, resulting in the formation of disinfection byproducts (DBPs) like trihalomethanes during chlorination (Lee et al., 2021). Different polymers degrade at different rates, with aliphatic hydrocarbons being more prone to degradation than aromatic compounds. UV light can significantly accelerate the breakdown of polymers, increasing DOC leaching by up to 100 times compared to dark conditions (Lee et al., 2020). Aging processes can affect MPs' surface properties, impacting how they interact with and release organic pollutants.

In DWTP, advanced oxidation processes such as photooxidation and ozonation can further break down MPs, releasing additional chemicals (Samarth & Mahanwar, 2021). Disinfectants like chlorine can cause MPs to fragment and leach additives more rapidly (Lin & Su, 2022). Overall, these processes can increase the breakdown and potential toxicity of MPs.

2.4.4.2 Factors Affecting Chemical Leaching from Microplastics

Several factors influence how chemicals leach from microplastics (MPs) into drinking water:

MP concentration and polymer type: Higher MP concentrations increase surface area, enhancing chemical leaching. Different polymers release chemicals at different rates, with polyethylene (PE) degrading more quickly than PP. Leaching rates rise with higher MP doses.

pH levels: Chemical solubility and stability on MPs are influenced by the pH of the water. Weakly bound compounds are often more soluble at lower pH values, but chemical breakdown can occur at higher pH values. The rate of chemical leaching is higher at alkaline pH compared to neutral or acidic conditions.

Contact time: Longer contact times allow more chemicals to dissolve from MPs into the water. Degraded MPs also promote microbial growth, which can further enhance chemical release.

2.4.4.3 Biofilm and Pathogen Carrier

Biofilms are clusters of microorganisms attached to surfaces—whether natural or synthetic (biotic or abiotic)—at the solid–liquid interface, encased in a self-produced, highly hydrated extracellular matrix of proteins, carbohydrates, and nucleic acids. These biofilms are found across various environments. Within this matrix, microbial cells often remain dormant and are shielded from higher concentrations of antibiotics and disinfectants (Karim et al., 2022).

MPs act as carriers for pathogens that can enter humans through inhalation, infection, or skin contact. Studies using metagenomics and bioinformatics have shown that MPs can facilitate bacterial transmission in aquatic environments, with potential pathogens lingering longer in water compared to natural substances like wood or feathers (Oberbeckmann et al., 2016). Their buoyancy allows MPs to float and provide a surface for biofilm formation. Plastics like PE and PET offer excellent surfaces for bacterial colonization and biofilm formation, allowing bacteria to survive longer compared to non-adhered forms.

2.5 Microplastic Contamination in Human Food Chain

2.5.1 MPs in Food Chain

Microplastics (MPs) have been found throughout the human food chain, from lower trophic organisms like zooplankton to higher levels, including invertebrates and vertebrates, and other food items. MPs enter these organisms through direct ingestion or trophic transfer. While MPs in fish digestive tracts pose less concern for human consumption, bioaccumulation in crustaceans is more problematic, as their digestive systems are eaten by humans. Mussels and mollusks, consumed whole, serve as important indicators of MP pollution, along with species like sardines and anchovies, which signal contamination in open sea ecosystems (Cverenkárová et al., 2021).

MPs have been detected in various aquatic biota, including coral and fish. Additionally, microplastics interacting with toxic algae can produce phycotoxins, posing indirect risks to human health. These algal toxins can accumulate in seafood, leading to harmful effects on humans. Given the widespread presence of microplastics in terrestrial, freshwater, and marine environments, the food chain is becoming highly contaminated and poses a potential threat to health and safety.

Microplastics (MPs), along with absorbed pollutants such as toxic chemicals and pathogens, are increasingly entering the human food chain due to extensive environmental contamination. Hazardous contaminants like persistent organic pollutants (POPs) are hydrophobic, which enables MPs to absorb these toxins. When aquatic organisms, such as fish, phytoplankton, or other seafood, ingest MPs, these harmful pollutants, including polychlorinated biphenyls (PCBs), polyaromatic hydrocarbons (PAHs), and organochlorine pesticides (like DDT), can be transferred into the marine food chain (Bouwmeester et al., 2015).

2.5.1.1 Fish and Fish Products

Fish is considered as the most MPs contaminated foods as they live in freshwater and aquatic environment which is highly polluted. Microplastics have been found in various cultured and captured marine species, including edible demersal, pelagic, and reef fish worldwide (Walkinshaw et al., 2020). Most major aquaculture species are fish, with research indicating that many contain ingested microplastics. For example, average microplastic ingestion per individual was reported as 2.5 ± 1.3 MPs for common carp, 1.9 ± 1.0 MPs for crucian carp, and 3.8 ± 2.0 MPs for silver carp (Biginagwa et al., 2016; Rochman et al., 2015). Fibers were identified as the most common type of microplastic, constituting 57.6–86.5% of the observed particles (Walkinshaw et al., 2020). Species such as skipjack tuna, Atlantic herring, and yellowfin tuna were frequently found with microplastics in their gastrointestinal tracts.

Fish products like canned fish also contaminated by MPs. A study on various brands of canned sardines and sprats from multiple countries found microplastics in several brands, with PP and PET being the most prevalent types (Karami et al., 2018). The contamination likely results from microplastic particles transferring into edible tissues, inadequate gutting, or contamination during processing. Additionally, dried fish products also contain microplastics, raising concerns about human exposure, as these are often consumed in whole form rather than raw (Kwon et al., 2020).

2.5.1.2 Shellfish, Crustaceans, and Other Seafoods

Seafood serves as a key dietary component, providing significant amount of animal protein for around three billion people. Many marine organisms can easily ingest plastic particles, contributing to trophic transfer within the food web, as microplastics have been detected in various marine life forms, from plankton to larger fish. Shellfish and crustaceans are diverse and significant food sources worldwide, playing a crucial role in food security. Among the most aquacultured shellfish species are cupped oysters and Japanese carpet shells, both of which have been found to contain microplastics, with reported levels ranging from 0.18 to 3.84 particles/g in cupped oysters and 0.9 to 2.5 particles/g in Japanese carpet shells (Walkinshaw et al., 2020). Mussels from the Mytilidae family have also shown varying levels of microplastics, averaging from 0.2 to 5.36 particles per gram (Mamun et al., 2023). Crustaceans, including crabs, lobsters, shrimp, and prawns, are vital for human consumption, but many species have been found to ingest microplastics. For example, brown shrimp and tiger prawns have been reported to contain microplastics at average levels of 0.68 and 1.5 particles/g, respectively (Mamun et al., 2023).

Coral reefs are vital to marine ecosystems, offering coastal protection and habitats for diverse marine species. However, microplastic contamination adversely affects coral health by accumulating in their digestive systems, leading to tissue damage. Because of its therapeutic qualities, both soft and hard corals are employed in traditional medicine; therefore, this contamination puts human health at risk in addition to harming coral (Cooper et al., 2014).

2.5.1.3 Salt and Sugar

Sugar is a widely consumed ingredient, with global consumption expected to rise due to urbanization and population growth. While studies on microplastic contamination in sugar are limited, some research has found significant levels of microplastics in various sugar samples. Investigations of MPs in both branded and open market sugar samples from Bangladesh and Iran showed that all samples were contaminated (Afrin et al., 2022; Makhdoumi et al., 2023). Contamination may occur during processing, packaging, and through machinery, raising concerns about human exposure.

Salt is another significant source of MPs contamination as commercial salts from different origins, such as sea salt, rock salt, lake salt, well salt, are widely consumed by human. Research has consistently shown microplastics in different types of salt, with varying particle counts across sea, lake, and rock salts (Deswati et al., 2023; Makhdoumi et al., 2023; Özçifçi et al., 2023). The contamination likely arises from polluted sea water sources, harvesting techniques, and processing methods (Thiele et al., 2023).

2.5.1.4 Honey

Honey is an important natural product due to its nutritional, medicinal, and industrial values. It can be contaminated by MPs as reported in several studies, where fibers and fragments in honey are found in significant quantities (Basaran et al., 2024; Mamun et al., 2023). Possible sources of contamination include environmental factors, honey types such as monofloral honey, bee contamination by ingestion and by adhesion to body processing, and packaging practices.

2.5.1.5 Milk

Milk and dairy products are essential for nutrition but have been found to contain MPs (Basaran et al., 2023; Kutralam-Muniasamy et al., 2020). MPs contamination in various milk brands, highlighting hygiene concerns, and processing steps particularly since milk is a common infant food.

2.5.1.6 Beverage and Drinks

Soft drink—MPs are detected in the beverages and soft drinks. Contamination can originate from various sources, including the water used in beverage production, washing procedures at the plants, contamination of the working area, and airborne particles (Crosta et al., 2023). A study showed microplastics were detected in PET-packaged bottles than in Tetra Pak containers, and the predominant polymer types were PA and PET (Altunışık, 2023).

Beer—MPs contamination has also been detected in alcoholic drinks like beer (Kosuth et al., 2018). Although no direct link to municipal water supplies was established, beer processing methods may contribute to contamination levels.

2.5.1.7 Fruits, Vegetables, and Seaweed

Fruits and vegetables—Plants act as pathways for plastics in the food chain, absorbing MPs or NPs through their roots and surfaces (Lazăr et al., 2024). These particles can already be found in fruits and vegetables or may enter through seedlings, leading to various impacts. Recent investigations into common fruits and vegetables, such as pears, apples, tomatoes, onions, potatoes, and cucumbers, revealed presence of MPs particles in all the samples (Aydın et al., 2023; Oliveri Conti et al., 2020). The presence of larger microplastics, which cannot be transported through the xylem, indicates that contamination often occurs during production, agricultural practices, and marketing processes, including transport and sale (Aydın et al., 2023).

Seaweed—Seaweed is consider as sea vegetables and consumed all over the world due to its nutritional content including vitamins and minerals, iodine, calcium, and iron. However, MPs contamination in seaweed is a growing concern, particularly in commercial seaweed, with roasted varieties containing higher levels than dried and seasoned ones (Kutralam-Muniasamy et al., 2024). Seaweeds are contaminated with microplastics from the source and farming site (Vitali et al., 2023).

2.5.1.8 Other Food Products

Rice—Rice, a staple for over half the global population, has been shown to absorb microplastics from soil and packaging (Amirhosseini et al., 2023; Gong et al., 2023). Potential microplastic ingestion in servings of rice, depending on washing practices also.

Meat—Microplastics have been detected in various meats, including chicken, due to soil contamination and packaging materials (Habib et al., 2022; Kedzierski et al., 2020).

MPs have been found in a range of processed foods, including dried food items, traditional medicines, and tea bags, indicating a broad scope of contamination across the food supply chain (Mamun et al., 2023).

2.5.2 *Microplastic Contamination from Food Packaging*

People commonly consume water, food, and beverages stored in plastic containers, cans, bottles, wraps, and cups for easy transport, storage, and longer shelf life. Aluminum cans and glass bottles with plastic linings are also widely used. The extensive use of plastic packaging, including baby feeders and disposable cups, increases the risk of microplastic contamination in food (Muhib et al., 2023) (Fig. 2.5). MPs in food packaging, particularly in take-out containers, a trend that has increased with the rise in plastic use during the COVID-19 pandemic. Common polymers in take-out packages include PP, PS, PE, and PET. PS packages exhibited the highest levels of microplastic contamination, compared to lower levels in other types of packaging. Microplastics may also interact with antimicrobial agents like triclosan, posing additional health risks. Millions of microplastic particles can be released from plastic packets, disposable cups, or food boxes when exposed to hot foods or liquids. High temperatures can release substances like bisphenol A (BPA), fluorescent additives, and other endocrine disruptors alongside microplastics (Muhib et al., 2023). Repeated use of food containers, especially those made from PET, PP, PE, and PS, can lead to significant microplastic leaching. The design and manufacturing process of these packages affect their susceptibility to leaching. Factors such as oil in fried foods and improper use of packaging (e.g., microwave heating, cryopreservation) can further contribute to microplastic contamination (Du et al., 2020). In addition, a coffee bag could release nearly 12,000 microplastic particles within five minutes at 95 °C (Wang et al., 2023).

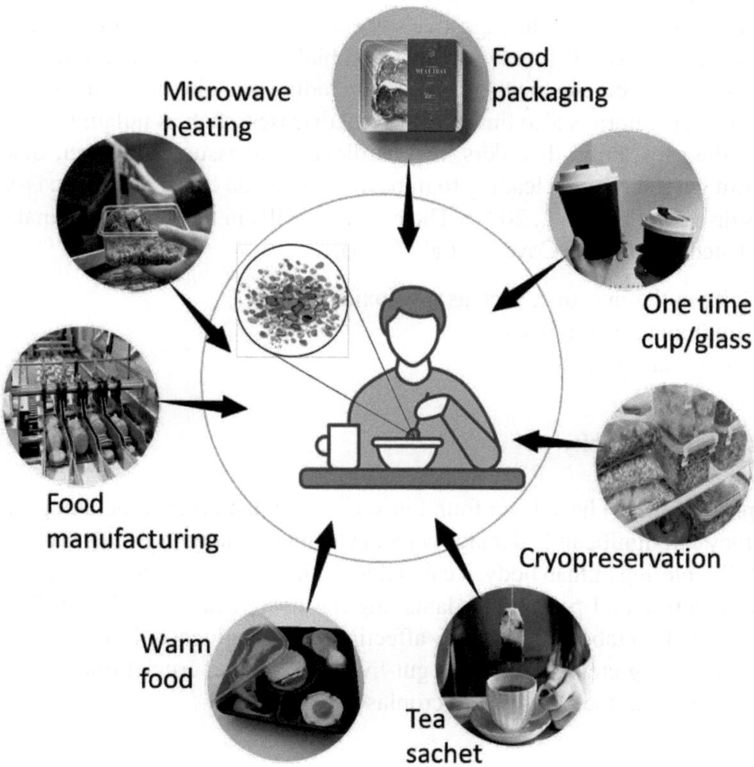

Fig. 2.5 Microplastics contamination in food from food packaging

2.6 Impact of Micropastics on Human Health

2.6.1 *Potential Health Problems*

MPs pose potential health risks to humans, influenced by exposure levels and individual vulnerability. Their varied sizes and shapes can lead to several health issues such as oxidative stress, liver toxicity, immune disorders, increased cancer risk, etc. Additionally, MPs may release harmful additives, toxic chemicals, and transport dangerous microorganisms, poses addition health risks.

2.6.1.1 Gastrointestinal Exposure

Ingestion is the primary route through which humans consume microplastics, primarily via contaminated food and beverages. Continuous environmental microplastic contamination means ongoing human exposure. Once ingested, MPs can penetrate the gut lining, affecting digestion and absorption by interacting with various molecules in the GI tract including nutrients and water (Yee et al., 2021). The intestine is a crucial component of the digestive system, is responsible for digestion, absorption, and metabolism. It functions as a microbial ecosystem, where intestinal flora regulates various biochemical processes, including protein synthesis and mineral absorption. This flora is also linked to several diseases, such as inflammatory bowel disease, diabetes, brain disorders, and cardiovascular issues. However, disruption may harm gut microbiota, leading to imbalances that can cause metabolic issues and various diseases (Yin et al., 2021). The effects of MPs in the gastrointestinal system can be listed as follows (Covello et al., 2024):

- Toxicity in the intestine, such as cytotoxicity
- Gut immunological impact
- Effect on gut microbiota.

2.6.1.2 Liver Toxicity

Microplastics (MPs) have been found in various human organs, including the liver, where they can significantly damage hepatocyte microstructures (Wang et al., 2024b). Once MPs entering human body, it can induce oxidative stress, impairing antioxidant enzyme function and potentially damaging the liver (Liu et al., 2021). They may disrupt crucial metabolic processes, affecting energy, glucose, lipid, and bile acid metabolism (Feng et al., 2021). The gut-liver axis plays a crucial role in mediating the liver injuries associated with microplastic exposure.

2.6.1.3 Respiratory Tract Exposure

Inhalation of airborne microplastics is a notable exposure pathway, with smaller particles penetrating deep into the lungs. This exposure can lead to respiratory issues, including irritation, reduced lung capacity, and symptoms resembling allergic reactions (Gasperi et al., 2018). High concentrations of airborne MPs may pose significant risks to respiratory health.

The key factors and mechanisms that influencing on particle deposition into the respiratory tract are (Saha & Saha, 2024):

- *MPs size*—Smaller particles are more likely to penetrate deeper into the lungs, while larger particles trend to accumulate in the upper respiratory tract.
- *Inhalation pattern*—Inhaling at a slower flow rate and taking deeper breaths can enhance deposition in the lower airways.
- *Airway geometry (diameter and branching angles)*—Narrower and more convoluted airways trend to deposit particles more effectively than wider, straighter ones.
- *Gravity*—Particles are more likely to settle in the lower regions of the lungs due to gravity.
- *MPs shape*—Irregularly shaped particles more likely to accumulate in the airways than spherical ones.
- *Lung morphology*—Lung surfactant and mucus movement are crucial for airflow, particle retention, and the compliance and recoil of lung alveoli.
- *Boundary conditions*—The wall boundary (inlet boundary: mouth, nose, or trachea; outlet boundary: bronchioles' end) condition used is the "Trap" condition, meaning particles that contact the walls are deposited, reflecting the sticky mucus layer lining the airways.

2.6.1.4 Adverse Effect on Reproductive System

MPs disrupt the neuroendocrine system by affecting sex hormone production through the hypothalamic-pituitary–gonadal (HPG) axis. In males, they compromise spermatogenesis by interfering with the blood-testis barrier, while in females, they lead to placental dysfunction, ovarian atrophy, endometrial hyperplasia, and fibrosis (Wang et al., 2024a). Additionally, microplastics may influence the lipid metabolism and reproductive functions of offspring. Studies show that microplastics can reduce sperm quality and testosterone levels, and in animal models, exposure has caused inflammation and decreased egg quality in females, highlighting the need for further exploration of their reproductive toxicity in humans (Mamun et al., 2023).

2.6.1.5 Effect on Nervous System, Brain Function, and Neurological Disorders

Nervous system—MPs pose serious health risks by being absorbed by cells and affecting cellular and organ functions. Ingestion of MPs can allow them to enter the bloodstream, potentially harming the nervous system. Research shows that exposure can alter neurotransmitter function and increase oxidative stress, which may affect brain health (Mamun et al., 2023).

Brain function—MPs also infiltrate the brain, impacting brain function. These substances can cross the blood–brain barrier through olfactory pathways and blood vessels, disrupting the secretion of neuroinflammatory (cytokines and chemokines) transporters, and receptor markers (Kaushik et al., 2024). This neurotoxicity and inflammation raise concerns about potential brain injuries and disorders.

Neurological disorders—Small-sized MPs can cross the blood–brain barrier and accumulate in the human brain, leading to inflammation, reduced synaptic proteins, and cognitive deficits, particularly in the context of Alzheimer's disease (Zhi et al., 2024).

2.6.1.6 Disorder in Body Immunity

MPs infiltrate to human body may impair immune function, leading to autoimmune diseases or immunosuppression. Several immune disruption mechanisms including inflammation induction, oxidative stress, and altered gene expression can increase the risk of neurological disorders (Prata et al., 2020; Zhi et al., 2024). The major impacts of MPs on immune system are (Yang et al., 2022; Zhi et al., 2024):

- *Intestinal immunity disruption*—A false sense of fullness, causing the organism to reduce food intake, which can ultimately result in malnutrition and even starvation.
- *Metabolic response*—MPs interfere with essential genes in the intestines, resulting in dysfunction and metabolic irregularities.
- *Immunotoxicity*—Through production of reactive oxygen species (ROS) and danger associated molecular patterns (DAMPs).
- *Immune response*—Inhibiting Ca^{2+} signaling and apoptosis-related molecular pathways.

2.6.1.7 Cellular Uptake and Intracellular Fate

MPs can contact the skin, penetrating through sweat glands or hair follicles and potentially disrupting skin function. Once absorbed, MPs may interact with various cell types, and their effects depend on their size, composition, and surface chemistry (Yee et al., 2021). When exposed to biological fluids, plastic particles form a 'protein corona' from surrounding proteins, which potentially increasing toxicity. Cellular uptake occurs through endocytosis pathways involving interactions with transport proteins. This includes phagocytosis, micropinocytosis, and clathrin- or caveolae-mediated endocytosis (Ali et al., 2024). The mode of uptake varies by cell type and particle properties.

2.6.1.8 Cancer

MPs themselves, along with the carcinogenic substances they carry, increase cancer risks for humans. Studies reveal that hazardous chemicals adhere to MPs, raising the lifetime cancer risk significantly beyond recommended safety levels. Particles smaller than 1 μm can enhance cell migration, potentially promoting metastasis, influence cellular behavior and disease progression (Brynzak-Schreiber et al., 2024; Park et al., 2023). MPs persist and bioaccumulate in colorectal cancer cell lines, Microplastic particles accumulate in non-proliferating areas of tumor spheroids and transfer between cells during division. Study showed that irregularly shaped plastic particles significantly affected human breast cancer cells, increasing cell cycle-related gene expression (Park et al., 2023).

2.6.1.9 Breastmilk

MPs have been detected in human breast milk, with their presence linked to maternal hygiene practices, particularly handwashing, as well as behaviors that increase exposure to contaminated products, such as food packaging and frequent seafood consumption (Saraluck et al., 2024). Additionally, the presence of microbiota in breast milk may play a role in MP detection. In a study, microplastics were found in all breast milk storage bags tested, with particles in the sub-micron range displaying spherical or oval shapes. Infants could be exposed to approximately 0.61–0.89 mg of microplastics per day from these bags (Liu et al., 2023a).

2.6.2 Potential Toxicity of Contaminants Carried by MPs

Microplastics can serve as carriers for harmful chemical and contaminants, which can pose risks such as toxicity and carcinogenicity. For example, MPs have been shown to enhance the toxicity of phthalates by aiding their transport and release in the intestines of mice. Many polymers consist of monomers with significant bioeffects, including carcinogenic PVC (Sun & Wang, 2023). MPs can adsorb persistent organic pollutants (POPs) and facilitate their ingestion by organisms. Additionally, toxic metals and antibiotics can attach to MPs, leading to unpredictable bioeffects. Moreover, microplastics can provide a habitat for pathogens, which may alter gut microbiomes and evade immune detection (Mamun et al., 2023).

2.7 Policy Considerations for Public Health

MPs have brought several issues that affect the health of human, animal and environment. For example, MPs have the potential to negatively affect animals and other creatures, disrupt the entire food chain, and endanger human health by increasing the frequency of algal blooms and the number of bacteria resistant to antibiotics. However, policies on MPs have largely neglected in several aspects including agricultural land pollution via sewage, plastic-coated fertilizers, and ground water contamination, highlighting the need for governance measures to prevent contamination and ensure food safety (Usman et al., 2022). Another serious risk to food security and availability is microplastic contamination, which can result in nutritional deficits. As a result, microplastic pollution poses a serious threat to public health in addition to the environment. Governments and local authorities are responsible for implementing global and domestic policies that protect the environment and public health. Environmental organizations, researchers, and individuals all have important roles to play. Industries have a crucial role in lowering the prevalence of MPs as they are involved in manufacture, consumption, reuse, and waste management of plastic. NGOs and civil societies can also raise awareness and promote mitigation efforts.

2.8 Summary and Future Research Direction

MPs are found everywhere, all corner of environment, including air, all types of water sources, soil, and most importantly in human food chain. Through different pathways, MPs are easily entering and posing threat to human health. Since there is clear evidence of MPs accumulation in human body, it is necessary to study the unknown facts of MPs interactions in the different environmental settings, existing critical points in human food chain, and potential side effects on human health. Further studies need to focus on several topics to discover the new knowledge:

- More research is needed to understand how MPs are released during the production and packaging of bottled water.
- Empirical data on MPs release at each food and water production step is lacking, and variations in microplastic levels across different packaging and production dates could offer insights.
- Simulation experiments under various conditions (e.g., exposure to sunlight, shaking) are also crucial to determine MPs release sources.
- Other contaminants like pigments, bisphenol A, and phthalates in water should be investigated.
- A thorough understanding of MPs' characteristics and fate throughout the water supply chain is essential.
- There is an urgent need to develop reliable methods for rapid sampling and analyzing nanoplastics in drinking water.
- MPs in groundwater needs standardized sampling methods.

- The contamination of MPs in milk during the production process from farms to the dairy sector is still unclear.

References

Afrin, S., Rahman, Md. M., Hossain, Md. N., Uddin, Md. K., & Malafaia, G. (2022). Are there plastic particles in my sugar? A pioneering study on the characterization of microplastics in commercial sugars and risk assessment. *Science of the Total Environment, 837*, Article 155849. https://doi.org/10.1016/j.scitotenv.2022.155849

Ali, I., Ding, T., Peng, C., Naz, I., Sun, H., Li, J., & Liu, J. (2021). Micro- and nanoplastics in wastewater treatment plants: Occurrence, removal, fate, impacts and remediation technologies—A critical review. *Chemical Engineering Journal, 423*, Article 130205. https://doi.org/10.1016/j.cej.2021.130205

Ali, N., Katsouli, J., Marczylo, E. L., Gant, T. W., Wright, S., & Bernardino de la Serna, J. (2024). The potential impacts of micro-and-nano plastics on various organ systems in humans. *eBioMedicine, 99*, Article 104901. https://doi.org/10.1016/j.ebiom.2023.104901

Allen, S., Allen, D., Moss, K., Le Roux, G., Phoenix, V. R., & Sonke, J. E. (2020). Examination of the ocean as a source for atmospheric microplastics. *PLoS ONE, 15*(5), Article e0232746. https://doi.org/10.1371/journal.pone.0232746

Allen, S., Allen, D., Phoenix, V. R., Le Roux, G., Durántez Jiménez, P., Simonneau, A., Binet, S., & Galop, D. (2019). Atmospheric transport and deposition of microplastics in a remote mountain catchment. *Nature Geoscience, 12*(5), 339–344. https://doi.org/10.1038/s41561-019-0335-5

Alqahtani, S., Alqahtani, S., Saquib, Q., & Mohiddin, F. (2023). Toxicological impact of microplastics and nanoplastics on humans: understanding the mechanistic aspect of the interaction. *Frontiers in Toxicology, 5*.https://doi.org/10.3389/ftox.2023.1193386

Altunışık, A. (2023). Prevalence of microplastics in commercially sold soft drinks and human risk assessment. *Journal of Environmental Management, 336*, Article 117720. https://doi.org/10.1016/j.jenvman.2023.117720

Ambrosini, R., Azzoni, R. S., Pittino, F., Diolaiuti, G., Franzetti, A., & Parolini, M. (2019). First evidence of microplastic contamination in the supraglacial debris of an alpine glacier. *Environmental Pollution, 253*, 297–301. https://doi.org/10.1016/j.envpol.2019.07.005

Amirhosseini, K., Haghani, Z., & Alikhani, H. A. (2023). Microplastics pollution in rice fields: A case study of Pir Bazar rural district of Gilan, Iran. *Environmental Monitoring and Assessment, 195*(12), 1473. https://doi.org/10.1007/s10661-023-12111-x

Aydın, R. B., Yozukmaz, A., Şener, İ, Temiz, F., & Giannetto, D. (2023). Occurrence of microplastics in most consumed fruits and vegetables from Turkey and public risk assessment for consumers. *Life, 13*(8), 1686. https://doi.org/10.3390/life13081686

Barchiesi, M., Chiavola, A., Di Marcantonio, C., & Boni, M. R. (2021). Presence and fate of microplastics in the water sources: Focus on the role of wastewater and drinking water treatment plants. *Journal of Water Process Engineering, 40*, Article 101787. https://doi.org/10.1016/j.jwpe.2020.101787

Basaran, B., Özçifçi, Z., Akcay, H. T., & Aytan, Ü. (2023). Microplastics in branded milk: Dietary exposure and risk assessment. *Journal of Food Composition and Analysis, 123*, Article 105611. https://doi.org/10.1016/j.jfca.2023.105611

Basaran, B., Özçifçi, Z., Kanbur, E. D., Akçay, H. T., Gül, S., Bektaş, Y., & Aytan, Ü. (2024). Microplastics in honey from Türkiye: Occurrence, characteristic, human exposure, and risk assessment. *Journal of Food Composition and Analysis, 135*, Article 106646. https://doi.org/10.1016/j.jfca.2024.106646

Bayo, J., Olmos, S., & López-Castellanos, J. (2020). Microplastics in an urban wastewater treatment plant: The influence of physicochemical parameters and environmental factors. *Chemosphere, 238*, Article 124593. https://doi.org/10.1016/j.chemosphere.2019.124593

Betancourt, W. Q., & Rose, J. B. (2004). Drinking water treatment processes for removal of Cryptosporidium and Giardia. *Veterinary Parasitology, 126*(1–2), 219–234. https://doi.org/10.1016/j.vetpar.2004.09.002

Biginagwa, F. J., Mayoma, B. S., Shashoua, Y., Syberg, K., & Khan, F. R. (2016). First evidence of microplastics in the African Great Lakes: Recovery from Lake Victoria Nile perch and Nile tilapia. *Journal of Great Lakes Research, 42*(1), 146–149. https://doi.org/10.1016/j.jglr.2015.10.012

Boucher, J., & Friot, D. (2017). Primary microplastics in the oceans: A global evaluation of sources. *IUCN International Union for Conservation of Nature*. https://doi.org/10.2305/IUCN.CH.2017.01.en

Bouwmeester, H., Hollman, P. C. H., & Peters, R. J. B. (2015). Potential health impact of environmentally released micro- and nanoplastics in the human food production chain: Experiences from nanotoxicology. *Environmental Science & Technology, 49*(15), 8932–8947. https://doi.org/10.1021/acs.est.5b01090

Brahney, J., Mahowald, N., Prank, M., Cornwell, G., Klimont, Z., Matsui, H., & Prather, K. A. (2021). Constraining the atmospheric limb of the plastic cycle. *Proceedings of the National Academy of Sciences, 118*(16). https://doi.org/10.1073/pnas.2020719118

Brynzak-Schreiber, E., Schögl, E., Bapp, C., Cseh, K., Kopatz, V., Jakupec, M. A., Weber, A., Lange, T., Toca-Herrera, J. L., del Favero, G., Wadsak, W., Kenner, L., & Pichler, V. (2024). Microplastics role in cell migration and distribution during cancer cell division. *Chemosphere, 353*, Article 141463. https://doi.org/10.1016/j.chemosphere.2024.141463

Cha, J., Lee, J.-Y., & Chia, R. W. (2023). Microplastics contamination and characteristics of agricultural groundwater in Haean Basin of Korea. *Science of the Total Environment, 864*, Article 161027. https://doi.org/10.1016/j.scitotenv.2022.161027

Chen, E.-Y., Lin, K.-T., Jung, C.-C., Chang, C.-L., & Chen, C.-Y. (2022). Characteristics and influencing factors of airborne microplastics in nail salons. *Science of the Total Environment, 806*, Article 151472. https://doi.org/10.1016/j.scitotenv.2021.151472

Chen, X., Lian, X., Wang, Y., Chen, S., Sun, Y., Tao, G., Tan, Q., & Feng, J. (2023). Impacts of hydraulic conditions on microplastics biofilm development, shear stresses distribution, and microbial community structures in drinking water distribution pipes. *Journal of Environmental Management, 325*, Article 116510. https://doi.org/10.1016/j.jenvman.2022.116510

Chen, Y., Awasthi, A. K., Wei, F., Tan, Q., & Li, J. (2021). Single-use plastics: Production, usage, disposal, and adverse impacts. *Science of the Total Environment, 752*, Article 141772. https://doi.org/10.1016/j.scitotenv.2020.141772

Cooper, E. L., Hirabayashi, K., Strychar, K. B., & Sammarco, P. W. (2014). Corals and their potential applications to integrative medicine. *Evidence-Based Complementary and Alternative Medicine, 2014*(1). https://doi.org/10.1155/2014/184959

Covello, C., Di Vincenzo, F., Cammarota, G., & Pizzoferrato, M. (2024). Micro(nano)plastics and their potential impact on human gut health: A narrative review. *Current Issues in Molecular Biology, 46*(3), 2658–2677. https://doi.org/10.3390/cimb46030168

Cox, K. D., Covernton, G. A., Davies, H. L., Dower, J. F., Juanes, F., & Dudas, S. E. (2019). Human consumption of microplastics. *Environmental Science & Technology, 53*(12), 7068–7074. https://doi.org/10.1021/acs.est.9b01517

Crosta, A., Parolini, M., & De Felice, B. (2023). Microplastics contamination in nonalcoholic beverages from the Italian market. *International Journal of Environmental Research and Public Health, 20*(5), 4122. https://doi.org/10.3390/ijerph20054122

Cverenkárová, K., Valachovičová, M., Mackuľak, T., Žemlička, L., & Bírošová, L. (2021). Microplastics in the food chain. *Life, 11*(12), 1349. https://doi.org/10.3390/life11121349

Dalmau-Soler, J., Ballesteros-Cano, R., Boleda, M. R., Paraira, M., Ferrer, N., & Lacorte, S. (2021). Microplastics from headwaters to tap water: Occurrence and removal in a drinking water treatment plant in Barcelona Metropolitan area (Catalonia, NE Spain). *Environmental Science and Pollution Research, 28*(42), 59462–59472. https://doi.org/10.1007/s11356-021-13220-1

De Falco, F., Di Pace, E., Cocca, M., & Avella, M. (2019). The contribution of washing processes of synthetic clothes to microplastic pollution. *Scientific Reports, 9*(1), 6633. https://doi.org/10.1038/s41598-019-43023-x

Deswati, D., Kurnia Hamzani, B., Yusuf, Y., Elsa Fitri, W., & Putra, A. (2023). Detection of microplastic contamination in table salts in Padang City, Indonesia, and control strategies for choosing healthy salt. *International Journal of Environmental Analytical Chemistry*, 1–16.https://doi.org/10.1080/03067319.2023.2268523

Dris, R., Gasperi, J., Rocher, V., Saad, M., Renault, N., & Tassin, B. (2015). Microplastic contamination in an urban area: A case study in Greater Paris. *Environmental Chemistry, 12*(5), 592. https://doi.org/10.1071/EN14167

Du, F., Cai, H., Zhang, Q., Chen, Q., & Shi, H. (2020). Microplastics in take-out food containers. *Journal of Hazardous Materials, 399*, Article 122969. https://doi.org/10.1016/j.jhazmat.2020.122969

Eerkes-Medrano, D., Leslie, H. A., & Quinn, B. (2019). Microplastics in drinking water: A review and assessment. *Current Opinion in Environmental Science & Health, 7*, 69–75. https://doi.org/10.1016/j.coesh.2018.12.001

Elkhatib, D., & Oyanedel-Craver, V. (2020). A critical review of extraction and identification methods of microplastics in wastewater and drinking water. *Environmental Science & Technology, 54*(12), 7037–7049. https://doi.org/10.1021/acs.est.9b06672

Evode N, Qamar SA, Bilal M, Barceló D, Iqbal HMN (2021) Plastic waste and its management strategies for environmental sustainability. Case Stud Chem Environ Eng 4:100142. https://doi.org/10.1016/j.cscee.2021.100142

Feng, S., Zeng, Y., Cai, Z., Wu, J., Chan, L. L., Zhu, J., & Zhou, J. (2021). Polystyrene microplastics alter the intestinal microbiota function and the hepatic metabolism status in marine medaka (Oryzias melastigma). *Science of the Total Environment, 759*, Article 143558. https://doi.org/10.1016/j.scitotenv.2020.143558

Gambino, I., Bagordo, F., Grassi, T., Panico, A., & De Donno, A. (2022). Occurrence of microplastics in tap and bottled water: current knowledge. *International Journal of Environmental Research and Public Health, 19*(9), 5283. https://doi.org/10.3390/ijerph19095283

Gasperi, J., Wright, S. L., Dris, R., Collard, F., Mandin, C., Guerrouache, M., Langlois, V., Kelly, F. J., & Tassin, B. (2018). Microplastics in air: Are we breathing it in? *Current Opinion in Environmental Science & Health, 1*, 1–5. https://doi.org/10.1016/j.coesh.2017.10.002

Geyer R, Jambeck JR, Law KL (2017) Production use and fate of all plastics ever made. Sci Adv 3(7). https://doi.org/10.1126/sciadv.1700782

Gigault, J., Halle, A. Ter, Baudrimont, M., Pascal, P.-Y., Gauffre, F., Phi, T.-L., El Hadri, H., Grassl, B., & Reynaud, S. (2018). Current opinion: What is a nanoplastic? *Environmental Pollution (Barking, Essex : 1987), 235*, 1030–1034. https://doi.org/10.1016/j.envpol.2018.01.024

Gilbert, M. (2017). Plastics materials: Introduction and historical development. In *Brydson's Plastics Materials* (pp. 1–18). Elsevier. https://doi.org/10.1016/B978-0-323-35824-8.00001-3

Gong, R., Wang, J., Wang, X., Liu, Y., & Shan, J. (2023). Analysis of microplastics release from rice package in combination with machine learning and hyperspectral imaging technique. *Food Packaging and Shelf Life, 39*, Article 101152. https://doi.org/10.1016/j.fpsl.2023.101152

Goßmann, I., Süßmuth, R., & Scholz-Böttcher, B. M. (2022). Plastic in the air?!—Spider webs as spatial and temporal mirror for microplastics including tire wear particles in urban air. *Science of the Total Environment, 832*, Article 155008. https://doi.org/10.1016/j.scitotenv.2022.155008

Habib, R. Z., Kindi, R. A., Salem, F. A., Kittaneh, W. F., Poulose, V., Iftikhar, S. H., Mourad, A.-H.I., & Thiemann, T. (2022). Microplastic contamination of chicken meat and fish through plastic cutting boards. *International Journal of Environmental Research and Public Health, 19*(20), 13442. https://doi.org/10.3390/ijerph192013442

Haleem, N., Kumar, P., Zhang, C., Jamal, Y., Hua, G., Yao, B., & Yang, X. (2024). Microplastics and associated chemicals in drinking water: A review of their occurrence and human health implications. *Science of the Total Environment, 912*, Article 169594. https://doi.org/10.1016/j.scitotenv.2023.169594

Haque, F., & Fan, C. (2023). Fate and impacts of microplastics in the environment: Hydrosphere, pedosphere, and atmosphere. *Environments, 10*(5), 70. https://doi.org/10.3390/environments10050070

Jadhav, E. B., Sankhla, M. S., Bhat, R. A., & Bhagat, D. S. (2021). Microplastics from food packaging: An overview of human consumption, health threats, and alternative solutions. In *Environmental Nanotechnology, Monitoring and Management* (Vol. 16). Elsevier B.V. https://doi.org/10.1016/j.enmm.2021.100608

Jahandari, A. (2023). Microplastics in the urban atmosphere: Sources, occurrences, distribution, and potential health implications. *Journal of Hazardous Materials Advances, 12*, Article 100346. https://doi.org/10.1016/j.hazadv.2023.100346

Julienne, F., Delorme, N., & Lagarde, F. (2019). From macroplastics to microplastics: Role of water in the fragmentation of polyethylene. *Chemosphere, 236*, Article 124409. https://doi.org/10.1016/j.chemosphere.2019.124409

Kankanige, D., & Babel, S. (2020). Smaller-sized micro-plastics (MPs) contamination in single-use PET-bottled water in Thailand. *Science of the Total Environment, 717*, Article 137232. https://doi.org/10.1016/j.scitotenv.2020.137232

Karami, A., Golieskardi, A., Choo, C. K., Larat, V., Karbalaei, S., & Salamatinia, B. (2018). Microplastic and mesoplastic contamination in canned sardines and sprats. *Science of the Total Environment, 612*, 1380–1386. https://doi.org/10.1016/j.scitotenv.2017.09.005

Karim, M. M., Shivappa, P., Dileep, N., Jhuma, T. A., & Hossain, A. (2022). Microplastic toxicity and the gut microbiome. In D. Bagchi & B. W. Downs (Eds.), *Microbiome, immunity, digestive health and nutrition: Epidemiology, pathophysiology, prevention and treatment* (pp. 345–358). Nikki P. Levy.

Kaushik, A., Singh, A., Kumar Gupta, V., & Mishra, Y. K. (2024). Nano/micro-plastic, an invisible threat getting into the brain. *Chemosphere, 361*, Article 142380. https://doi.org/10.1016/j.chemosphere.2024.142380

Kayastha, V., Patel, J., Kathrani, N., Varjani, S., Bilal, M., Show, P. L., Kim, S.-H., Bontempi, E., Bhatia, S. K., & Bui, X.-T. (2022). New Insights in factors affecting ground water quality with focus on health risk assessment and remediation techniques. *Environmental Research, 212*, Article 113171. https://doi.org/10.1016/j.envres.2022.113171

Kedzierski, M., Lechat, B., Sire, O., Le Maguer, G., Le Tilly, V., & Bruzaud, S. (2020). Microplastic contamination of packaged meat: Occurrence and associated risks. *Food Packaging and Shelf Life, 24*, Article 100489. https://doi.org/10.1016/j.fpsl.2020.100489

Kelly, M. R., Lant, N. J., Kurr, M., & Burgess, J. G. (2019). Importance of water-volume on the release of microplastic fibers from laundry. *Environmental Science & Technology, 53*(20), 11735–11744. https://doi.org/10.1021/acs.est.9b03022

Klein, M., & Fischer, E. K. (2019). Microplastic abundance in atmospheric deposition within the Metropolitan area of Hamburg, Germany. *Science of the Total Environment, 685*, 96–103. https://doi.org/10.1016/j.scitotenv.2019.05.405

Kosuth, M., Mason, S. A., & Wattenberg, E. V. (2018). Anthropogenic contamination of tap water, beer, and sea salt. *PLoS ONE, 13*(4), Article e0194970. https://doi.org/10.1371/journal.pone.0194970

Kutralam-Muniasamy, G., Pérez-Guevara, F., Elizalde-Martínez, I., & Shruti, V. C. (2020). Branded milks—Are they immune from microplastics contamination? *Science of the Total Environment, 714*, Article 136823. https://doi.org/10.1016/j.scitotenv.2020.136823

Kutralam-Muniasamy, G., Shruti, V. C., & Pérez-Guevara, F. (2024). Microplastic contamination in commercially packaged edible seaweeds and exposure of the ethnic minority and local population in Mexico. *Food Research International, 176*, Article 113840. https://doi.org/10.1016/j.foodres.2023.113840

Kwon, J.-H., Kim, J.-W., Pham, T. D., Tarafdar, A., Hong, S., Chun, S.-H., Lee, S.-H., Kang, D.-Y., Kim, J.-Y., Kim, S.-B., & Jung, J. (2020). Microplastics in food: A review on analytical methods and challenges. *International Journal of Environmental Research and Public Health, 17*(18), 6710. https://doi.org/10.3390/ijerph17186710

Lazăr, N.-N., Călmuc, M., Milea, Ş-A., Georgescu, P.-L., & Iticescu, C. (2024). Micro and nano plastics in fruits and vegetables: A review. *Heliyon, 10*(6), Article e28291. https://doi.org/10.1016/j.heliyon.2024.e28291

Lee, J.-Y., Cha, J., Ha, K., & Viaroli, S. (2024). Microplastic pollution in groundwater: A systematic review. *Environmental Pollutants and Bioavailability, 36*(1). https://doi.org/10.1080/26395940.2023.2299545

Lee, Y. K., Hong, S., & Hur, J. (2021). Copper-binding properties of microplastic-derived dissolved organic matter revealed by fluorescence spectroscopy and two-dimensional correlation spectroscopy. *Water Research, 190*, Article 116775. https://doi.org/10.1016/j.watres.2020.116775

Lee, Y. K., Romera-Castillo, C., Hong, S., & Hur, J. (2020). Characteristics of microplastic polymer-derived dissolved organic matter and its potential as a disinfection byproduct precursor. *Water Research, 175*, Article 115678. https://doi.org/10.1016/j.watres.2020.115678

Li, Y., Li, W., Jarvis, P., Zhou, W., Zhang, J., Chen, J., Tan, Q., & Tian, Y. (2020). Occurrence, removal and potential threats associated with microplastics in drinking water sources. *Journal of Environmental Chemical Engineering, 8*(6), Article 104527. https://doi.org/10.1016/j.jece.2020.104527

Lin, T., & Su, J. (2022). The fate of microplastics and organic matter leaching behavior during chlorination. *Chemosphere, 302*, Article 134892. https://doi.org/10.1016/j.chemosphere.2022.134892

Liu, J., Dong, C., Zhai, Z., Tang, L., & Wang, L. (2021). Glyphosate-induced lipid metabolism disorder contributes to hepatotoxicity in juvenile common carp. *Environmental Pollution, 269*, Article 116186. https://doi.org/10.1016/j.envpol.2020.116186

Liu, L., Zhang, X., Jia, P., He, S., Dai, H., Deng, S., & Han, J. (2023a). Release of microplastics from breastmilk storage bags and assessment of intake by infants: A preliminary study. *Environmental Pollution, 323*, Article 121197. https://doi.org/10.1016/j.envpol.2023.121197

Liu, X., Wang, J., Zhang, L., & Zhu, Y. (2023b). The transport of microplastics by ants cannot be neglected in the soil ecosystem. *Environmental Pollution, 317*, Article 120796. https://doi.org/10.1016/j.envpol.2022.120796

Madeła, M. (2023). Microplastic toxicity to humans. In *Microplastics in the ecosphere* (pp. 381–390). Wiley. https://doi.org/10.1002/9781119879534.ch23

Makhdoumi, P., Amin, A. A., Karimi, H., Pirsaheb, M., Kim, H., & Hossini, H. (2021). Occurrence of microplastic particles in the most popular Iranian bottled mineral water brands and an assessment of human exposure. *Journal of Water Process Engineering, 39*, Article 101708. https://doi.org/10.1016/j.jwpe.2020.101708

Makhdoumi, P., Pirsaheb, M., Amin, A. A., Kianpour, S., & Hossini, H. (2023). Microplastic pollution in table salt and sugar: Occurrence, qualification and quantification and risk assessment. *Journal of Food Composition and Analysis, 119*, Article 105261. https://doi.org/10.1016/j.jfca.2023.105261

Mamun, A. A., Prasetya, T. A. E., Dewi, I. R., & Ahmad, M. (2023). Microplastics in human food chains: Food becoming a threat to health safety. *Science of the Total Environment, 858*, Article 159834. https://doi.org/10.1016/J.SCITOTENV.2022.159834

Mattsson, K., Jocic, S., Doverbratt, I., & Hansson, L.-A. (2018). Nanoplastics in the aquatic environment. In *Microplastic contamination in aquatic environments* (pp. 379–399). Elsevier. https://doi.org/10.1016/B978-0-12-813747-5.00013-8

Mintenig, S. M., Löder, M. G. J., Primpke, S., & Gerdts, G. (2019). Low numbers of microplastics detected in drinking water from ground water sources. *Science of the Total Environment, 648*, 631–635. https://doi.org/10.1016/j.scitotenv.2018.08.178

Muhib, M. I., Uddin, M. K., Rahman, M. M., & Malafaia, G. (2023). Occurrence of microplastics in tap and bottled water, and food packaging: A narrative review on current knowledge. In *Science of the Total Environment* (Vol. 865). Elsevier B.V. https://doi.org/10.1016/j.scitotenv.2022.161274

Na, S.-H., Kim, M.-J., Kim, J.-T., Jeong, S., Lee, S., Chung, J., & Kim, E.-J. (2021). Microplastic removal in conventional drinking water treatment processes: Performance, mechanism, and potential risk. *Water Research, 202*, Article 117417. https://doi.org/10.1016/j.watres.2021.117417

Nayanathara Thathsarani Pilapitiya, P. G. C., & Ratnayake, A. S. (2024). The world of plastic waste: A review. *Cleaner Materials, 11*, 100220. https://doi.org/10.1016/j.clema.2024.100220

Oberbeckmann, S., Osborn, A. M., & Duhaime, M. B. (2016). Microbes on a bottle: Substrate, season and geography influence community composition of microbes colonizing marine plastic debris. *PLoS ONE, 11*(8), Article e0159289. https://doi.org/10.1371/journal.pone.0159289

O'Brien, S., Rauert, C., Ribeiro, F., Okoffo, E. D., Burrows, S. D., O'Brien, J. W., Wang, X., Wright, S. L., & Thomas, K. V. (2023). There's something in the air: A review of sources, prevalence and behaviour of microplastics in the atmosphere. *Science of the Total Environment, 874*, Article 162193. https://doi.org/10.1016/j.scitotenv.2023.162193

Oladoja, N. A., & Unuabonah, I. E. (2021). The pathways of microplastics contamination in raw and drinking water. *Journal of Water Process Engineering, 41*, Article 102073. https://doi.org/10.1016/j.jwpe.2021.102073

Oliveri Conti, G., Ferrante, M., Banni, M., Favara, C., Nicolosi, I., Cristaldi, A., Fiore, M., & Zuccarello, P. (2020). Micro- and nano-plastics in edible fruit and vegetables. The first diet risks assessment for the general population. *Environmental Research, 187*, 109677. https://doi.org/10.1016/j.envres.2020.109677

Oßmann, B. E. (2021). Microplastics in drinking water? Present state of knowledge and open questions. *Current Opinion in Food Science, 41*, 44–51. https://doi.org/10.1016/j.cofs.2021.02.011

Oßmann, B. E., Sarau, G., Holtmannspötter, H., Pischetsrieder, M., Christiansen, S. H., & Dicke, W. (2018). Small-sized microplastics and pigmented particles in bottled mineral water. *Water Research, 141*, 307–316. https://doi.org/10.1016/j.watres.2018.05.027

Özçifçi, Z., Basaran, B., & Akçay, H. T. (2023). Microplastic contamination and risk assessment in table salts: Turkey. *Food and Chemical Toxicology, 175*, Article 113698. https://doi.org/10.1016/j.fct.2023.113698

Panno, S. V., Kelly, W. R., Scott, J., Zheng, W., McNeish, R. E., Holm, N., Hoellein, T. J., & Baranski, E. L. (2019). Microplastic contamination in Karst groundwater systems. *Groundwater, 57*(2), 189–196. https://doi.org/10.1111/gwat.12862

Park, J. H., Hong, S., Kim, O.-H., Kim, C.-H., Kim, J., Kim, J.-W., Hong, S., & Lee, H. J. (2023). Polypropylene microplastics promote metastatic features in human breast cancer. *Scientific Reports, 13*(1), 6252. https://doi.org/10.1038/s41598-023-33393-8

Petersen, A. (2016). Presence of microplastics and nanoplastics in food, with particular focus on seafood. *EFSA Journal, 14*(6). https://doi.org/10.2903/j.efsa.2016.4501

Prata, J. C., da Costa, J. P., Lopes, I., Duarte, A. C., & Rocha-Santos, T. (2020). Environmental exposure to microplastics: An overview on possible human health effects. *Science of the Total Environment, 702*, Article 134455. https://doi.org/10.1016/j.scitotenv.2019.134455

Raza, M., Lee, J.-Y., & Cha, J. (2022). Microplastics in soil and freshwater: Understanding sources, distribution, potential impacts, and regulations for management. *Science Progress, 105*(3), 003685042211266. https://doi.org/10.1177/00368504221126676

Re, V. (2019). Shedding light on the invisible: Addressing the potential for groundwater contamination by plastic microfibers. *Hydrogeology Journal, 27*(7), 2719–2727. https://doi.org/10.1007/s10040-019-01998-x

Rochman, C. M., Tahir, A., Williams, S. L., Baxa, D. V., Lam, R., Miller, J. T., Teh, F.-C., Werorilangi, S., & Teh, S. J. (2015). Anthropogenic debris in seafood: Plastic debris and fibers from textiles in fish and bivalves sold for human consumption. *Scientific Reports, 5*(1), 14340. https://doi.org/10.1038/srep14340

Saha, S. C., & Saha, G. (2024). Effect of microplastics deposition on human lung airways: A review with computational benefits and challenges. *Heliyon, 10*(2), Article e24355. https://doi.org/10.1016/j.heliyon.2024.e24355

Samarth, N. B., & Mahanwar, P. A. (2021). Degradation of polymer and elastomer exposed to chlorinated water—A Review. *Open Journal of Organic Polymer Materials, 11*(01), 1–50. https://doi.org/10.4236/ojopm.2021.111001

Saraluck, A., Techarang, T., Bunyapipat, P., Boonchuwong, K., Pullaput, Y., & Mordmuang, A. (2024). Detection of microplastics in human breast milk and its association with changes in human milk bacterial microbiota. *Journal of Clinical Medicine, 13*(14), 4029. https://doi.org/10.3390/jcm13144029

Shen, M., Song, B., Zhu, Y., Zeng, G., Zhang, Y., Yang, Y., Wen, X., Chen, M., & Yi, H. (2020). Removal of microplastics via drinking water treatment: Current knowledge and future directions. *Chemosphere, 251*, Article 126612. https://doi.org/10.1016/j.chemosphere.2020.126612

Sillanpää, M., & Sainio, P. (2017). Release of polyester and cotton fibers from textiles in machine washings. *Environmental Science and Pollution Research, 24*(23), 19313–19321. https://doi.org/10.1007/s11356-017-9621-1

Singh, S., & Bhagwat, A. (2022). Microplastics: A potential threat to groundwater resources. *Groundwater for Sustainable Development, 19*, Article 100852. https://doi.org/10.1016/j.gsd.2022.100852

Singh, T. (2021). Generation of microplastics from the opening and closing of disposable plastic water bottles. *Journal of Water and Health, 19*(3), 488–498. https://doi.org/10.2166/wh.2021.025

Sol, D., Laca, A., Laca, A., & Díaz, M. (2021). Microplastics in wastewater and drinking water treatment plants: Occurrence and removal of microfibres. *Applied Sciences, 11*(21), 10109. https://doi.org/10.3390/app112110109

Srihari, S., Subramani, T., Prapanchan, V. N., & Peiyue, L. (2023). Human health risk perspective study on characterization, quantification and spatial distribution of microplastics in surface water, groundwater and coastal sediments of thickly populated Chennai coast of South India. *Human and Ecological Risk Assessment: An International Journal, 29*(1), 222–244. https://doi.org/10.1080/10807039.2022.2154635

Sripada, K., Wierzbicka, A., Abass, K., Grimalt, J. O., Erbe, A., Röllin, H. B., Weihe, P., Díaz, G. J., Singh, R. R., Visnes, T., Rautio, A., Odland, J. Ø., & Wagner, M. (2022). A children's health perspective on nano- and microplastics. *Environmental Health Perspectives, 130*(1). https://doi.org/10.1289/EHP9086

Statista. (2024). *Annual production of plastics worldwide from 1950 to 2022*. Statista Research Department. https://www.statista.com/statistics/282732/global-production-of-plastics-since-1950/

Suaria, G., Achtypi, A., Perold, V., Lee, J. R., Pierucci, A., Bornman, T. G., Aliani, S., & Ryan, P. G. (2020). Microfibers in oceanic surface waters: A global characterization. *Science Advances, 6*(23). https://doi.org/10.1126/sciadv.aay8493

Sun, A., & Wang, W.-X. (2023). Human exposure to microplastics and its associated health risks. *Environment & Health, 1*(3), 139–149. https://doi.org/10.1021/envhealth.3c00053

Sun, X.-D., Yuan, X.-Z., Jia, Y., Feng, L.-J., Zhu, F.-P., Dong, S.-S., Liu, J., Kong, X., Tian, H., Duan, J.-L., Ding, Z., Wang, S.-G., & Xing, B. (2020). Differentially charged nanoplastics demonstrate distinct accumulation in Arabidopsis thaliana. *Nature Nanotechnology, 15*(9), 755–760. https://doi.org/10.1038/s41565-020-0707-4

Thiele, C. J., Grange, L. J., Haggett, E., Hudson, M. D., Hudson, P., Russell, A. E., & Zapata-Restrepo, L. M. (2023). Microplastics in European sea salts—An example of exposure through consumer choice and of interstudy methodological discrepancies. *Ecotoxicology and Environmental Safety, 255*, Article 114782. https://doi.org/10.1016/j.ecoenv.2023.114782

Usman, S., Abdull Razis, A. F., Shaari, K., Azmai, M. N. A., Saad, M. Z., Mat Isa, N., & Nazarudin, M. F. (2022). The Burden of microplastics pollution and contending policies and regulations. *International Journal of Environmental Research and Public Health, 19*(11), 6773. https://doi.org/10.3390/ijerph19116773

Viaroli, S., Lancia, M., & Re, V. (2022). Microplastics contamination of groundwater: Current evidence and future perspectives. A review. *Science of The Total Environment, 824*, 153851. https://doi.org/10.1016/j.scitotenv.2022.153851

Vitali, C., Peters, R. J. B., Janssen, H.-G., & Nielen, M. W. F. (2023). Microplastics and nanoplastics in food, water, and beverages; part I. occurrence. *TrAC Trends in Analytical Chemistry, 159*, 116670. https://doi.org/10.1016/j.trac.2022.116670

Walkinshaw, C., Lindeque, P. K., Thompson, R., Tolhurst, T., & Cole, M. (2020). Microplastics and seafood: Lower trophic organisms at highest risk of contamination. *Ecotoxicology and Environmental Safety, 190*, Article 110066. https://doi.org/10.1016/j.ecoenv.2019.110066

Wang, H.-P., Huang, X.-H., Chen, J.-N., Dong, M., Zhang, Y.-Y., & Qin, L. (2023). Pouring hot water through drip bags releases thousands of microplastics into coffee. *Food Chemistry, 415*, Article 135717. https://doi.org/10.1016/j.foodchem.2023.135717

Wang, M., Wu, Y., Li, G., Xiong, Y., Zhang, Y., & Zhang, M. (2024a). The hidden threat: Unraveling the impact of microplastics on reproductive health. *Science of the Total Environment, 935*, Article 173177. https://doi.org/10.1016/j.scitotenv.2024.173177

Wang, X., Deng, K., Zhang, P., Chen, Q., Magnuson, J. T., Qiu, W., & Zhou, Y. (2024b). Microplastic-mediated new mechanism of liver damage: From the perspective of the gut-liver axis. *Science of the Total Environment, 919*, Article 170962. https://doi.org/10.1016/j.scitotenv.2024.170962

Winkler, A., Santo, N., Ortenzi, M. A., Bolzoni, E., Bacchetta, R., & Tremolada, P. (2019). Does mechanical stress cause microplastic release from plastic water bottles? *Water Research, 166*, Article 115082. https://doi.org/10.1016/j.watres.2019.115082

Yang, L., Kang, S., Luo, X., & Wang, Z. (2024). Microplastics in drinking water: A review on methods, occurrence, sources, and potential risks assessment. *Environmental Pollution, 348*, Article 123857. https://doi.org/10.1016/j.envpol.2024.123857

Yang, W., Jannatun, N., Zeng, Y., Liu, T., Zhang, G., Chen, C., & Li, Y. (2022). Impacts of microplastics on immunity. *Frontiers in Toxicology, 4*.https://doi.org/10.3389/ftox.2022.956885

Yee, M.S.-L., Hii, L.-W., Looi, C. K., Lim, W.-M., Wong, S.-F., Kok, Y.-Y., Tan, B.-K., Wong, C.-Y., & Leong, C.-O. (2021). Impact of microplastics and nanoplastics on human health. *Nanomaterials, 11*(2), 496. https://doi.org/10.3390/nano11020496

Yin, K., Wang, Y., Zhao, H., Wang, D., Guo, M., Mu, M., Liu, Y., Nie, X., Li, B., Li, J., & Xing, M. (2021). A comparative review of microplastics and nanoplastics: Toxicity hazards on digestive, reproductive and nervous system. *Science of the Total Environment, 774*, Article 145758. https://doi.org/10.1016/j.scitotenv.2021.145758

Zhi, L., Li, Z., Su, Z., & Wang, J. (2024). Immunotoxicity of microplastics: Carrying pathogens and destroying the immune system. *TrAC Trends in Analytical Chemistry, 177*, Article 117817. https://doi.org/10.1016/j.trac.2024.117817

Zhou, G., Wu, Q., Wei, X.-F., Chen, C., Ma, J., Crittenden, J. C., & Liu, B. (2023). Tracing microplastics in rural drinking water in Chongqing, China: Their presence and pathways from source to tap. *Journal of Hazardous Materials, 459*, Article 132206. https://doi.org/10.1016/j.jhazmat.2023.132206

Chapter 3
Route of Microplastic Contamination to Agricultural Food

Kalani Perera, Keshani Jayathilaka, Sachini Hansika Senevirathne, Manura Weerasinghe, Meththika Vithanage, and Anushka Upamali Rajapaksha

Abstract Microplastic (MP) contamination of agricultural food is a concern for food safety, human health, and agroecosystems. Agricultural ecosystems can be affected by MP contamination via plastic mulch, sewage sludge, fertilizer, compost, irrigation water, municipal solid waste, and atmospheric fallout (tire wear, and dump sites). When these particles enter the soil, they can be absorbed by plants, and accumulate in soil organisms leading to the disruption of soil health. These contaminants can accumulate in edible crop parts and reach human food chains, causing a huge threat to human health and food security. The way to develop mechanisms for mitigating the effects on health and food security is through being aware of the MP contamination route in agricultural systems. However, a comprehensive understanding of the impacts of MPs on soil and agricultural crops is unclear. Hence, this chapter highlights the sources, abundance, and pathways of MP contamination in agricultural ecosystems, plant uptake, the impact of MPs on agricultural food productivity, MPs in the food chain, and possible risks for food safety and human health.

Keywords Microplastic · Agricultural ecosystem · Food chain · Food productivity

K. Perera · K. Jayathilaka · M. Weerasinghe · M. Vithanage · A. U. Rajapaksha (✉)
Ecosphere Resilience Research Centre, Faculty of Applied Sciences, University of Sri Jayewardenepura, Nugegoda, Sri Lanka
e-mail: anurajapaksha@sjp.ac.lk

S. H. Senevirathne
Food Science and Technology, Faculty of Applied Sciences, Sabaragamuwa University of Sri Lanka, Belihuloya, Sri Lanka

A. U. Rajapaksha
Instrument Center, Faculty of Applied Sciences, University of Sri Jayewardenepura, Nugegoda, Sri Lanka

V. Kumar et al. (eds.), *Microplastics Pollution*,
https://doi.org/10.1007/978-3-031-99774-7_3

3.1 Introduction

Microplastics (MPs), considered to be an emerging environmental issue are also a concern in agricultural ecosystems since they not only impact the quality of the soil and the productivity of crops but also have the potential to make their way into the food chain, which is a concern to human health and food safety (Aparna et al., 2019). MPs can enter the food chain directly through ingestion of contaminated plants or indirectly through bioaccumulation in animals consuming contaminated feed (Aparna et al., 2019). MPs have been found in several food items, including vegetables, poultry, seafood, and honey, as per studies, which show the pervasiveness of MP contamination in foods (Okeke et al., 2022; Toussaint et al., 2019). MPs may accumulate in animals and transfer through various trophic levels before finally accumulating in humans (Aparna et al., 2019). While it is not yet established how MPs affect the health of human beings, it is quite unsettling with endocrine interference and immune system disturbances (Li et al., 2023; Silva et al., 2018).

MPs lead to degradation of the soil by changing the chemical and physical characteristics of the soil, which affects plant growth and yield. Such alterations may lower the porosity, water content, and nutrient levels of the soil, all of which are essential in maintaining the healthy crop development (Rüthi et al., 2020; Wang et al., 2021). MPs also block microbial activity in the soil responsible for breaking down organic matter and nutrient cycling, thereby causing soil fertility decline and crop yield loss. MPs are also found to adsorb and carry poisonous chemicals such as pesticides and heavy metals, which are absorbed mainly through plant roots, thereby reducing the health of the plants and the safety of food (Silva et al., 2018).

The long-term contact with MPs impacts not just the soil system but also the agricultural ecosystem as a whole. Soil erosion, disruption of the fauna of the soil, and compromised pollination greatly decrease food yields. MPs can also congest water filters in treatment plants, therefore undermining water security, and consequently food security indirectly (Okeke et al., 2022). The bioaccumulation of MPs in the food chain, particularly among animals, presents additional human health risks by way of chemical leakage as well as long-term complications such as diabetes, cancer, and obesity (Aparna et al., 2019; Hlatshwayo et al., 2023). Animals that are exposed to MPs get their gut microbiota affected, making them suffer from health issues and lowered productivity. As MPs tend to adsorb toxic chemicals and heavy metals, they also enhance the likelihood of food chain contamination (Corrêa et al., 2023).

Since MPs are detrimental to agroecosystems, food safety, and human health, this pollution needs to be controlled in the interests of food security and environmental sustainability. The dissemination of MPs in agroecosystems not only poses a threat to food productivity but also has significant long-term impacts on human health, and hence it is necessary to acquire knowledge regarding MP contamination pathways and develop comprehensive risk assessment and control measures.

3.2 Sources of Microplastics in Agricultural Ecosystems

Microplastics (MPs) can be introduced into farmland soil directly (primary MPs) or indirectly (secondary MPs). MPs that are mainly derived from directly manufacturing small plastic particles; plastic nurdles, pellets, microbeads, and microfibers are known as primary MPs (Atugoda et al., 2021). When large plastics deteriorate owing to abiotic and biotic processes such as oxidation, hydrolysis, biodegradation, and photodegradation, secondary MPs are generated (Sharma & Chatterjee, 2017). MP accumulation in agricultural environments can occur through multiple sources; plastic mulches (Qiang et al., 2023), sewage sludge (Galafassi et al., 2019), composts, and fertilizers (Lwanga et al., 2022; Vithanage et al., 2021), irrigation water (Bläsing & Amelung, 2018), indiscriminate disposal of municipal solid waste; household items, personal care products, and electronics in landfills (21–42% of global plastic waste) (Galafassi et al., 2019; Nizzetto et al., 2016), atmospheric fallout (over 0.3 million tons per year) globally (Evangeliou et al., 2020; Luo et al., 2021) (Fig. 3.1).

3.2.1 Plastic Mulches

Plastic mulches are mainly used to promote seed germination, provide thermal insulation to protect seedlings, and stop evaporation in dry and cold conditions (Huang et al., 2021). They suppress weed growth and soil erosion, promote agricultural output, and lower pest pressure. (Kaur et al., 2022). These benefits have made plastic mulches (with thicknesses of 6–20 μm) a widespread agricultural technology used in

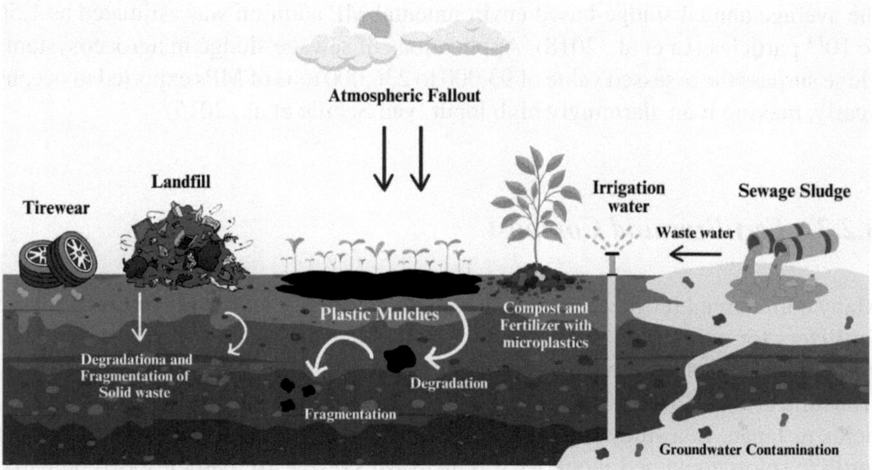

Fig. 3.1 Sources of microplastics in agroecosystems

systems of intensive cultivation (Ng et al., 2018). However, plastic mulches only stay in agricultural soil for a few months. It can be lowered even further due to physical fragmentation, chemical degradation and aging when mulch is subjected to harsh weather events such as storms, rain, and hail (Li et al., 2018). Finally, it breaks down into tiny pieces, and plastic mulch becomes a major source of MPs pollution (Kaur et al., 2022). Therefore, the recovery of mulch MPs from agricultural lands is not practically feasible, and up to 25–33% of the residue from plastic mulch films may still be present in agricultural soils each year (Yang et al., 2022).

3.2.2 Sewage Sludge

Large quantities (more than 90% of MPs) are accumulated in sewage sludge due to wastewater treatment's ineffective MP removal process (Corradini et al., 2019). Sewage sludge is enriched with nutrients and organic matter. Therefore, treated sewage sludge is utilized for farmland MPs that enter the soil and are transported throughout the terrestrial environment. With the help of rain, waning wind, barriers, wind erosion, and transport, MPs spread from their source to far-flung locations. However, these sewage sludge applications introduce MPs into the agricultural soil. A study conducted in wastewater treatment plants (WWTPs) in Norway found that 500 billion MPs were released into the environment, whereas in Canada, 32.4% of suspected MP materials from the WWTPs enter the soil ecosystem (Sa'adu & Farsang, 2023). Approximately, a total input of 63,000 to 430,000 tons of MPs are released annually to European farmlands, whereas 44,000 to 300,000 tons are released to North American farmlands yearly through the discharge of sewage sludge (Nizzetto et al., 2016). It is also estimated that 2.8×10^3 to 1.9×10^4 tons of MPs will be introduced to the farmland soil in Australia annually (Arab et al., 2024). In China, the average annual sludge-based environmental MP addition was estimated as 1.56×10^{14} particles (Li et al., 2018). Applications of sewage sludge in agroecosystems alone surpass the assessed value of 93,000 to 236,000 tons of MPs exported to oceans yearly, making it an alarmingly high input (van Sebille et al., 2015).

3.2.3 Fertilizer and Compost

Many commercial fertilizer manufacturing industries use plastic polymers to encase fertilizers for delayed discharge into the soil. After the successful release of these compounds, the plastic coating of the fertilizer pellet persists in the soil. A survey by (Katsumi et al., 2020) reported 6 to 369 mg/kg of microcapsules presence in all paddy fields in Japan. In a study conducted in China, spanning 32 years, it was found that fertilized plots contained about twice as many MPs (3.7×10^6 particles/kg) compared to non-fertilized plots (2.2×10^6 particles/kg), indicating that fertilizer usage in agriculture can exacerbate MP pollution in both surface and subsurface soils (Li et al.,

2022). Due to the inadequate waste segregation, most municipal solid waste is tainted with plastics. Studies indicate that urban runoff can carry these plastics into nearby farmlands, where they become integrated into the soil matrix (Sa'adu & Farsang, 2023). High temperatures, microbiological and enzymatic activity, and mechanical crushing speed up the breakdown of plastic during municipal solid waste composting, resulting in the creation of MPs that are then sent to agricultural areas (Weithmann et al., 2018). According to reports, Germany could reach 35 billion to 2.2 trillion MPs annually entering agricultural ecosystems through the use of organic fertilizer (Weithmann et al., 2018).

3.2.4 Atmospheric Fallout

Atmospheric fallout is an emerging source of MPs pollution that has garnered increasing attention in recent years. MPs are particulate matter that can be carried by wind over long distances from surfaces like roadways and landfills and deposited on land surfaces by air settling or precipitation (Bläsing & Amelung, 2018). Research indicates that urban areas with high traffic density experience significant deposition rates of airborne MPs (Sa'adu & Farsang, 2023). Studies have detected various types of MPs in atmospheric samples collected near urban centers, which subsequently settle on agricultural fields. These airborne particles can originate from multiple sources, including tire wear, synthetic textiles, and industrial emissions (Mehmood et al., 2023). The accumulation of atmospheric fallout on soils poses risks not only for crop health but also for broader ecological systems due to potential bioaccumulation within food webs.

3.3 Abundance of Microplastics in Agricultural Soil

Plastic mulch covers approximately 22 million hectares of agricultural area worldwide and contaminates the soil with its residues in the field (Abbate et al., 2023). China alone produced over 2.46 million tons of agricultural plastic mulching film in 2019 covering an area of over 25 million hectares, accounting for 61% of global usage (Hu et al., 2020). This has made plastic mulching the primary source of MPs in China's agroecosystems (Gao et al., 2019; Huang et al., 2021). In confirmation, research conducted in Guizhou, China, revealed that polyethylene (PE) is the predominant type of MP found in mulched agricultural soils, with concentrations ranging from 143.28 to 3283.46 items/kg, averaging 1,150.60 ± 647.86 items/kg (Table 3.1). In Korea, 160 items/kg of MPs were found in rice cultivation fields (Park et al., 2023). In contrast, in vegetation fields in Murcia, Spain, 2116 items/kg of MPs were detected, both originating from plastic mulch films (Beriot et al., 2021) (Table 3.1). A study carried out in Argentina found 69 ± 30 particles/kg of MPs while in Hainan Island, China it was 15,461.52 particles/kg (Table 3.1). Sewage sludge transfers into Chilean agroecosystems revealed high MP loads of 18,000–41,000 particles/kg (Corradini et al., 2019) (Table 3.1). Studies indicated that countries like Italy and Germany report concentrations between 120 and 465 particles/kg (Liu et al., 2019), while Australia shows variations from 37.8×10^3 to 46.1×10^3 particles/kg (Ziajahromi et al., 2021) of MPs in agricultural soils (Table 3.1). MP content in the sludge in Spain varied from 0.59 to 1.58 particles/kg (Menéndez-Manjón et al., 2022) (Table 3.1). According to the literature, the number of MPs in sludge from WWTPs worldwide spanned from > 1000 particles/kg to 301,400 particles/kg (Hassan et al., 2023). The type of raw materials, the length of the procedure, and the pretreatment used to extract plastics from raw materials are the key factors influencing the number of MPs in compost. Due to improperly sorted municipal trash, compost made from municipal organic waste (MOW) was found to contain 2800 ± 616 particles/kg of MPs (van Schothorst et al., 2021) (Table 3.1). The mean abundance of MPs in compost derived from rural solid waste (RSW) was 2400 ± 358 items/kg (Gui et al., 2021) (Table 3.1). Bläsing and Amelung (2018) discovered up to 2.38–1200 mg/kg plastic fragments in organic fertilizers. A total of 310, 400, and 410 MPs/kg were found in compost made from MSW gathered from Muthurajawela, Karadiyana, and Kaluthara, Sri Lanka, respectively (Premarathna et al., 2023) (Table 3.1).

3.4 Pathways of Microplastic Transfer in the Agricultural Environment

To evaluate MPs' impacts on agriculture, the environment, and human health, it is essential to comprehend their pathways. The most frequent sources of MPs in farmlands were found to be plastic mulch, sewage sludge, fertilizer and compost. In China, MPs were found to be abundant in agricultural lands in the range of 23–41,741 particles/kg, while studies in Pakistan indicate a concentration of about 876 particles/kg. The plastic contaminants released to the environment breaks down to form MPs. UV light exposure causes fragmentation to occur. The oxidation and shortening of polymer chains caused by UV causes the plastic to become brittle and break apart to create MPs (Sorasan et al., 2022). In addition, MPs are produced when water hydrolyses plastics (Gerritse et al., 2020). Sand-induced friction and other mechanical stressors play a part in the fragmentation process (Song et al., 2023), while microorganisms too help break down plastics into MPs (Dhiman et al., 2024).

MPs that get into the soil are dispersed throughout the terrestrial environment. With the help of rain, wind erosion, and barriers, MPs spread from their source to far-flung locations (Dris et al., 2017). Rain is involved in MP movement through surface runoff leading to their redistribution through agricultural lands (Zhao et al., 2022). Wind which plays a vital role in the transfer of MPs, helps in lifting and carrying MPs over significant distances contributing to their widespread distribution (Li et al., 2024). When erosion takes place, MPs are pushed and rolled across the surface as part of the surface creep method (Piehl et al., 2018). Prior research has

Table 3.1 Sources, abundance, and type of microplastics in sources and agricultural soil

Source	Abundance	Microplastic type	Location	References
Plastic Mulch	143.28–3283.46 items/kg	PE	Guizhou Province, China	Miao et al. (2024)
	160 items/kg	PE, PP, PET	Korea	Park et al. (2023)
	2116 particles/kg	PE	Murcia, Spain	Beriot et al. (2021)
	69 ± 30 particles/kg	PS	Cuartel V, Moreno District, Buenos Aires, Argentina	Berenstein et al. (2024)
	15,461.52 particles/kg	PE, PP, PS, PVC	Hainan Island, China	Khan et al. (2023)
Sewage sludge	18,000–41,000 particles/kg (DW)	PE, PP	Chile	Corradini et al. (2019)
	37.8–46.1 × 10^3 particles/kg (DW)	PE	Australia	Ziajahromi et al. (2021)
	0.59–1.58 particles/kg (DW)	PET, PA, PS	Spain	Menéndez-Manjón et al. (2022)
	20–57,180 particles/kg (DW)	PE, PP	Cádiz, Spain	Franco et al. (2023)
	44.4–750.0 particles/kg (DW)	PE, PP, PS	Chengdu, China	Wei et al. (2022)
Compost	2400 ± 358 items/kg (RDW)	PP, PE	Yunnan, China	Gui et al. (2021)
	2800 ± 616 particles/kg (MOW)	–	Dutch and Spain	van Schothorst et al. (2021)
	310 items/kg (MSW)	PE, PP, PS	Muthurajawela, Sri Lanka	Premarathna et al. (2023)
	400 items/kg (MSW)		Karadiyana, Sri Lanka	
	410 items/kg (MSW)		Kaluthara, Sri Lanka	

Polyethylene (PE), Polypropylene (PP), Polyethylene terephthalate (PET), polyamide (PA), polystyrene (PS), Polyvinyl Chloride (PVC) Rural domestic waste (RDW), Municipal organic waste (MOW), Municipal solid waste (MSW), Dry Weight (DW)

also shown that earthworms and preferential flow pathways can carry MPs along the soil profile (Huerta et al., 2017). Earthworms were found to transport microbeads on the soil surface to deeper layers and also, they were found to ingest MPs on the soil surface and transport them into the soil (Rillig et al., 2017). The burrowing activities of earthworms also help transport and redistribute MPs in soil as these channels help the MPs bypass the soil matrix (Yu et al., 2020). MPs have the capability to penetrate the soil structure and spread throughout various microenvironments within soil conditions since the soil is typically heterogeneous (Zhang & Liu, 2018).

In contrast to uncolonized MPs, the development of the plastisphere changes the physical properties of MPs, which in turn changes how they move. The microbial community in the plastisphere may have an impact on the fate and movement of these contaminants in soil (Cverenkárová et al., 2021). Natural barriers, such as vegetation and topography, can also influence the accumulation and transport pathways of MPs, affecting their final deposition sites (Li et al., 2024).

3.4.1 Soil Contamination

When MPs in soil interact with other soil constituents, they cause contamination, which affects the functionality and health of the soil. MPs debris can cause mechanical and physiological harm due to its size of less than 5 mm, which makes it easy for soil organisms to consume it and for it to move up the food chain. These MPs pose serious risks to soil health, microbial diversity, and the integrity of the ecosystem as a whole given that they continuously build up and fragment in the soil (Hoang et al., 2024). Depending on the type, concentration, and length of exposure, MPs can have a complex effect on properties of soil. MPs can modify soil aggregates, water retention capacity, and nutrient availability, among other physicochemical changes. Depending on their properties, they may promote or prevent soil aggregation (Chen et al., 2024). MPs, on the other hand, give microorganisms new habitats, which may upset already-existing microbial communities. Different kinds of MPs have been shown to support unique microbial populations, which may have an impact on the deterioration of organic matter and the cycling of nutrients. Research has demonstrated that, in comparison to polypropylene (PP) granules, the addition of expanded polystyrene (ePS) granules to soil substantially changed microbial diversity (Kublik et al., 2022).

MPs' transport and interactivity with other contaminants: organic matter, heavy metals, and microorganisms, are greatly impacted by their numerous functional groups: carbonyl, carboxyl, and hydroxyl. Changes in soil ecosystems are also brought about by coexisting contaminants and the intermediates and additives that are released when MPs depolymerize. The mobility of coexisting contaminants is facilitated by MPs' high transport and adsorption capacity, which may lead to widespread soil contamination (Luo et al., 2023). Numerous ecological hazards to soil ecosystems are made possible by MPs contamination in the soil. MPs may carry pathogens and dangerous chemicals such as polycyclic aromatic hydrocarbons (PAHs), polychlorinated biphenyls (PCBs), and dichlorodiphenyltrichloroethane (DDT) that seep into the groundwater and soil (Yuan et al., 2022). Additionally, MPs are connected to detrimental impacts on beneficial fauna and soil-dwelling organisms like earthworms. Reduced soil health could result from the decline of these organisms, which are essential for preserving soil fertility and structure (Sajjad et al., 2022). Furthermore, as plastics are durable, MPs contamination is likely to continue over time, making cleanup more difficult and possibly having long-term ecological effects (Chen et al., 2024). Overall, by changing physicochemical characteristics, raising microbial biomass, lowering microbial diversity, and preventing plant and animal development, MPs and their depolymerization products have the potential to upset soil ecosystems (Zhai et al., 2024).

MPs have become widespread contaminants in many ecosystems including agricultural environments. Their origin is from multiple sources such as sewage sludge application, plastic mulch application, irrigation with contaminated water, and atmospheric deposition (Briassoulis, 2023; Zhu et al., 2023). After contaminating the soil, movements of MPs in the soil layers are affected by several factors, including soil structure, interaction with soil organisms, and agricultural activities (Cverenkárová et al., 2021). Once in the soil movement and accumulation of MPs are influenced by several agricultural activities, according to a study done by (Yang et al., 2021) tilling and plowing could mix the soil layers which leads the MPs to penetrate deeper soil layers. Crop harvesting practices and root growth of certain crops such as carrots and potatoes could drive MPs to deeper layers (Cverenkárová et al., 2021; Li et al., 2020a). Moreover, dry weather conditions that cause soil cracking are another way of MPs movement into deeper soils. Furthermore, a microbial biofilm could form around these MPs creating a plastisphere that contains potential human pathogens, enhancing their presence and movement in the agriculture soils (Quilliam et al., 2023). When microbes attach to MPs, these plastic particles stay in the soil for longer periods and can help transport harmful pathogens to plants through water irrigation or rain splashing contaminated soil. This way of spreading pathogens onto plants and into our food supply creates potential health concerns for people who eat the contaminated products.

3.4.2 Plant Uptake and Translocation of Microplastics

Studies have showcased that MP can be absorbed by plant roots and transported throughout the plant. Smaller MPs, because of their size can enter plants through root cracks or via stomatal openings on leaves while larger MPs tend to accumulate on the plant surface (Azeem et al., 2021). Once in contact with the plant roots, MPs can be absorbed through the root epidermis and transported to the root xylem. From the xylem, the particles can translocate to other parts of the plant, including the stem, leaves, and potentially edible portions like fruits (Azeem et al., 2021). In laboratory studies, MPs were identified to be accumulated in the roots of radish (*Raphanus sativus*) showcasing their potential to enter crop plant systems (Cverenkárová et al., 2021). Plant leaf structure and plant morphology further influence MP adherence, for example, crops with trichomes or specialized leaf structures can retain more MPs, as they can trap and secure the particles on crop surface (Bi et al., 2020).

MP translocation in plants depends on several factors. This includes the physical properties of MP particles including their size, polymer type, surface charge, and interaction strength with crop tissues. For instance, negatively charged MP particles can bind to positively charged plant tissue surfaces making them difficult to remove from the plant surface even by washing (He et al., 2023). Furthermore, polymers with high adhesion characteristics, such as polyethylene and polypropylenes also resist removal from the plant surfaces persisting these MP on edible plant surfaces during harvesting and retailing processes that could ultimately lead to contaminate of the human diet (Quilliam et al., 2023). Moreover, MP uptake depends on plant physiology. Plants with dense root systems and larger surface areas for nutrient absorption might absorb more MPs. Trichomes, mucilage, and exudates on the plant surface can also trap particles, allowing them to adhere to the plant and remain attached even after washing (Mateos-Cárdenas et al., 2021). Some microplastics, once inside the plant tissue, can inhibit plant growth by obstructing pore spaces, which disrupts water uptake and nutrient flow, impacting overall plant health and productivity (Li et al., 2020b).

3.5 Microplastics in Agri Food Chain

Therefore, in agricultural environment MP raise concerns as they not only affect the soil health and productivity of crops but also could enter into food chains affecting human health and food safety. They can enter the food chain either directly through the consumption of contaminated plants or indirectly through bioaccumulation in animals that consume contaminated feed (Aparna et al., 2019). Research has documented MPs in various food items, including vegetables, poultry, seafood, honey, and bottled water, highlighting the prevalence of MP contamination across multiple food sources (Aparna et al., 2019; Okeke et al., 2022; Toussaint et al., 2019) as that could be seen in Table 3.2 they are being found in different agricultural commodities and crops in different countries. Studies have shown that MPs can bioaccumulate in animal tissues, leading to their transfer through different trophic levels until they reach humans as showcased in Fig. 3.2. Once ingested, the impact of MPs on human health remains uncertain. These health risks underscore the need to better understand MP behavior within agroecosystems and the potential for human exposure.

3.6 Impact of Microplastics on Agricultural Food Productivity

Food productivity is the capacity of farmers to optimize resource allocation, maximizing economic outputs while maintaining sustainable practices (Hlatshwayo et al., 2023). It depends on factors like resource allocation, technology adaptation, soil health, climate conditions, policy support, etc. MPs begin breaking down in the soil, impacting soil fertility by altering its physical and chemical properties. These changes degrade soil structure, making it less porous and reducing its ability to retain water and nutrients, both of which are essential for healthy plant growth. MPs interfere with microbial activity in the soil as well. Microbial activity is major in breaking

Table 3.2 Occurrence of microplastics in different vegetables, fruits and agricultural commodities

Species/Commodity	Abundance	Country	Reference
Pear	38 particles g^{-1}	Turkey	Aydın et al. (2023)
Potatoes	17 particles g^{-1}		
Apple	37 particles g^{-1}		
Tomato	44 particles g^{-1}		
Cucumber	43 particles g^{-1}		
Onion	31 particles g^{-1}		
Uncooked rice	67 ± 26 μg g^{-1} (DW)	Australia	Dessì et al. (2021)
Instant rice	283 ± 50 μg g^{-1} (DW)		
Honey	314 ± 353 (0–1280) particles kg^{-1}	Turkey	Basaran et al. (2024)
Vinegar	51.35 ± 20.73 particles L^{-1}	Iran	Makhdoumi et al. (2021)
Milk	6.5 ± 2.3 particles L^{-1}	Mexico	Kutralam-Muniasamy et al. (2020)
Chicken and turkey	4.0 -18.7 particles kg^{-1}	France	Kedzierski et al. (2020)
White wine	2563–5857 suspected particles L^{-1}	Italy	Prata et al. (2020)
Cold tea	11 ± 5.26 particles/drink	Mexico	Shruti et al. (2020)
Beer	152 ± 50.97 particles/drink		

Dry Weight (DW)

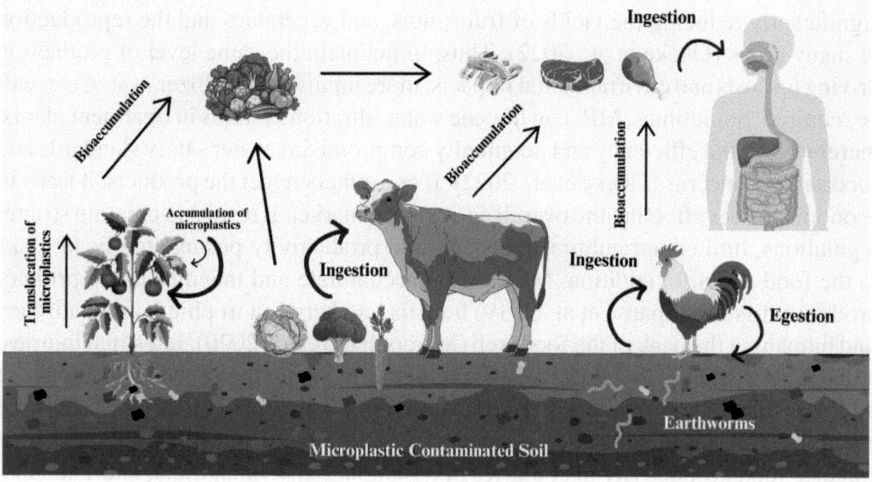

Fig. 3.2 Microplastic contamination in the food chain

down organic matter and recycling nutrients essential for crop growth. The reduction of microbial activity severely affects lowering soil fertility and diminishing crop yields. It is further increased by the ability of MPs to absorb and transport harmful chemicals, such as pesticides and heavy metals, which can be taken up by plant roots (Hore et al., 2024). Ultimately these contaminants can inhibit plant growth or lead to the production of crops that are unsafe for consumption (Silva et al., 2018).

This soil structure and composition directly affect the water quality. MPs decrease the soil's water-holding capacity, leaving crops more vulnerable to drought and water stress (Jayasinghe et al., 2023). When soil plastisphere forms they can alter the soil's physical characteristics such as structure, porosity, water holding capacity, and chemical characteristics like pH, nutrient content and organic matter (Silva et al., 2018). Soil porosity regulates the distribution of water and the aeration within the soil. Therefore, due to the formation of soil plastisphere high porosity can occur and that could lead to reduced water infiltration and retention, having an impact on plant growth (Rüthi et al., 2020). This vulnerability leads to reduced yields, especially in regions where irrigation is essential for maintaining productivity. Moreover, MPs change pH in the soil, which alters the microbial communities and has an effect on availability of soil nutrient leading to growth of certain plants and crop productivity issues by affecting seed germination as well as growth of roots (Wang et al., 2021).

Agricultural crops are highly impacted by long-term exposure to MPs. Soil degradation and disruption of soil soil biota or edaphon like earthworms is crucial in enhancing soil aeration as well as nutrient recycling (Silva et al., 2018) leading to chronic soil health issues, and reducing the crop yield, making it less productive for farming (Aparna et al., 2019). MPs effect could also lead to inadequate pollination, significantly reducing the yields of fruits, nuts, and vegetables and the reproduction of many crops (Okeke et al., 2022). Thus, to maintain the same level of production driving up costs and environmental impacts, more inputs like fertilizer, water, etc. will be required. Sometimes, MPs can impede water filtration systems in treatment plants, thereby reducing efficiency and potentially compromising water safety standards and food safety concerns (Okeke et al., 2022). If consumers reject the products, it leads to economic losses affecting the overall agricultural market. It results in creating stricter regulations, limited agricultural practices, and productivity posing significant risks to the food chain. In addition, MPs can bioaccumulate and magnify in the primary producers tissues (Aparna et al., 2019) transferring between trophic levels until they find humans at the peak of the food web (Mercogliano et al., 2020). Intestinal injuries, reduced food intake, and bowel obstructions are the main consequences of leakage of chemicals like DDT and polybrominated diphenyl ethers (PBDE) (Hlatshwayo et al., 2023) released in farming systems especially rice-fish integrated systems that enhance food productivity in countries like Bangladesh, China, India, and Thailand. These chemical and physical disruptions severely affect their physiological functions and may ultimately infiltrate the human food chain.

The fate of MPs in the human gut lumen following ingestion remains unclear (Silva et al., 2018). Nevertheless, elements of synthetic MPs like phthalates, and bisphenols are linked to endocrine disruption (Okeke et al., 2022). These chemicals cause health issues, including diabetes, cancer, and obesity (Aparna et al., 2019). Therefore, in addition to low yields across the food chain, MPs also present a significant health risk to humans, particularly to the immune system. According to bird studies (Li et al., 2023), the disturbance in intestinal microbial homeostasis led to reduced microbial diversity and changes in taxonomy, affecting D-amino acid metabolism, ATP-binding cassette (ABC) transporters, vitamin and mineral absorption, and histidine metabolism. Studies also reveal that plastic utensils like water hoses, drinking fonts, storage containers, and feed supply lines can undergo chemical and automatically supply wear, leading in partitioning into minor plastic pieces and production of MPs (Corrêa et al., 2023). These MPs accumulate in the intestines of cattle, pigs, poultry, and other livestock also, disrupting the intestinal microbiota which consists of trillions of living organisms causing negative health effects. The intestinal microbiota is crucial for maintaining the physiological balance of the host, producing antimicrobial peptides, short-chain fatty acids, vitamins, and enzymes during metabolism. Especially in ruminants, specific microbiota helps to break down plant fibres into volatile fatty acids and other essential compounds, contributing up to 70% of cumulative dietary fizz (Corrêa et al., 2023). Due to the large surface-to-volume ratio, MPs are effective at absorbing toxic chemicals and heavy metals. They can adsorb both hydrophilic contaminants, which are attracted to the negatively charged areas of plastics (Corrêa et al., 2023), and hydrophobic compounds, which are drawn to the neutral areas. As a result, plastics not only release toxins from their components but also act as carriers for absorbed toxic chemicals (Corrêa et al., 2023). In addition, some studies (Ross et al., 2012) revealed that livestock feed and water can lead to rumen blockage and reduced protozoal activity (Priyanka & Dey, 2018), causing indigestion and, in severe cases, death. Among the limited studies on MPs contamination in food, (Toussaint et al., 2019) data have been compiled on 201 consumable animal types, including 164 saltwater fishes, 23 molluscs, 7 crustaceans, 2 freshwater fishes, 2 turtles, 2 birds, 2 chickens, along with various food production like sprats, canned sardines, sea salt, honey, sugar, as well as water and beer (Hlatshwayo et al., 2023).

Food productivity is essential for global food security, yet the increasing presence of MPs in agricultural ecosystems presents a serious threat (Hore et al., 2024). MPs degrade soil quality, impede plant growth, disrupt soil and water systems, and harm livestock, collectively undermining the efficiency of food production. Tackling MPs pollution in agriculture is crucial for safeguarding and enhancing food productivity in the face of global environmental challenges.

3.7 Future Perspectives

Even though substantial progress has been made in understanding the routes of MP contamination in agricultural food systems, critical gaps remain which sets back comprehensive risk assessment and mitigation. Sufficient insights must be spread among the general public on how MPs migrate across various environmental matrices which supports to development of precise predictions on contamination levels in food products. Also, comprehensive toxicological assessments are needed to evaluate the health risks associated with consuming crops contaminated with MPs (Sharma & Vidyarthi, 2024). This can be done by studying the potential for MPs to act as vectors for harmful chemicals or pathogens (Sharma & Vidyarthi, 2024). Ineffective regulations are another important consideration that poorly addresses the long-term consequences of MPs on food security (Jayasinghe et al., 2023). There is a pressing need for research to inform policy that diminishes MP pollution while concerning agricultural productivity. But countries like Germany and the EU through ordinances, regulate the discharge of plastic pollutants into the environment through fertilizer applications (Wiesner et al., 2023). Reliable and rapid analytical methods form the foundation of legal regulations. Unfortunately, only macro plastics and larger MP contents (> 1 mm) are monitored in current methods (Wiesner et al., 2023). MPs (1–1000 μm), are not yet monitored. To improve the sustainability of the Agri-food systems, usage of biodegradable materials as alternatives to plastic mulches, development and application of advanced methods and systems to remove MPs from water treatment plants, proper maintenance of existing systems with regular inspection, efficient removal of MPs during compost preparation from municipal waste can be used. Thus, collaborative efforts among researchers, industry stakeholders and policy makers are needed to navigate these advancements and develop robust framework to handle the routes of MPs contamination in agricultural foods.

References

Abbate, C., Scavo, A., Pesce, G. R., Fontanazza, S., Restuccia, A., & Mauromicale, G. (2023). Soil bioplastic mulches for agroecosystem sustainability: a comprehensive review. *Agriculture, 13*(1).

Aparna, K., Valachovičová, M., Mackuľak, T., Žemlička, L., & Bírošová, L. (2019). Microplastics in food chain. *Health Action, 2019*, 27–28. https://doi.org/10.3390/life11121349

Arab, M., Yu, J., & Nayebi, B. (2024). Microplastics in sludges and soils: A comprehensive review on distribution, characteristics, and effects. *ChemEngineering*, *8*(5).

Atugoda, T., Vithanage, M., Wijesekara, H., Bolan, N., Sarmah, A. K., Bank, M. S., You, S., & Ok, Y. S. (2021). Interactions between microplastics, pharmaceuticals and personal care products: Implications for vector transport. *Environment International, 149*, Article 106367. https://doi.org/10.1016/j.envint.2020.106367

Aydın, R., Yozukmaz, A., şener, İ., Temiz, F., & Giannetto, D. (2023). Occurrence of microplastics in most consumed fruits and vegetables from Turkey and public risk assessment for consumers. *Life, 13*, 1686. https://doi.org/10.3390/life13081686

Azeem, I., Adeel, M., Ahmad, M. A., Shakoor, N., Jiangcuo, G. D., Azeem, K., Ishfaq, M., Shakoor, A., Ayaz, M., Xu, M., & Rui, Y. (2021). Uptake and accumulation of nano/microplastics in plants: A critical review. *Nanomaterials, 11*(11), 2935. https://www.mdpi.com/2079-4991/11/11/2935

Basaran, B., Özçifçi, Z., Kanbur, E. D., Akçay, H. T., Gül, S., Bektaş, Y., & Aytan, Ü. (2024). Microplastics in honey from Türkiye: Occurrence, characteristic, human exposure, and risk assessment. *Journal of Food Composition and Analysis, 135*, Article 106646. https://doi.org/10.1016/j.jfca.2024.106646

Berenstein, G., Córdoba, P., Díaz, Y. B., González, N., Ponce, M. B., & Montserrat, J. M. (2024). Macro, meso, micro and nanoplastics in horticultural soils in Argentina: Abundance, size distribution and fragmentation mechanism. *Science of the Total Environment, 906*, Article 167672. https://doi.org/10.1016/j.scitotenv.2023.167672

Beriot, N., Peek, J., Zornoza, R., Geissen, V., & Huerta Lwanga, E. (2021). Low density-microplastics detected in sheep faeces and soil: A case study from the intensive vegetable farming in Southeast Spain. *Science of the Total Environment, 755*, Article 142653. https://doi.org/10.1016/j.scitotenv.2020.142653

Bi, M., He, Q., & Chen, Y. (2020). What roles are terrestrial plants playing in global microplastic cycling? *Environmental Science & Technology, 54*(9), 5325–5327. https://doi.org/10.1021/acs.est.0c01009

Bläsing, M., & Amelung, W. (2018). Plastics in soil: Analytical methods and possible sources. *Science of the Total Environment, 612*, 422–435. https://doi.org/10.1016/j.scitotenv.2017.08.086

Briassoulis, D. (2023). Agricultural plastics as a potential threat to food security, health, and environment through soil pollution by microplastics: Problem definition. *Science of the Total Environment, 892*, Article 164533. https://doi.org/10.1016/j.scitotenv.2023.164533

Chen, Y., Li, Y., Liang, X., Lu, S., Ren, J., Zhang, Y., Han, Z., Gao, B., & Sun, K. (2024). Effects of microplastics on soil carbon pool and terrestrial plant performance. *Carbon Research, 3*(1), 37. https://doi.org/10.1007/s44246-024-00124-1

Corradini, F., Meza, P., Eguiluz, R., Casado, F., Huerta-Lwanga, E., & Geissen, V. (2019). Evidence of microplastic accumulation in agricultural soils from sewage sludge disposal. *Science of the Total Environment, 671*, 411–420. https://doi.org/10.1016/j.scitotenv.2019.03.368

Corrêa, T. R., Gaylarde, C. C., Baptista Neto, J. A., Delgado, J. d. F., Lima, L. d. S., Cunha, D. L., & da Fonseca, E. M. (2023). The impact of microplastics on global food production: A brief overview of this complex sector. *Microplastics, 2*(4), 371–388. https://doi.org/10.3390/microplastics2040028

Cverenkárová, K., Valachovičová, M., Mackuľak, T., Žemlička, L., & Bírošová, L. (2021). Microplastics in the Food Chain. *Life (Basel), 11*(12). https://doi.org/10.3390/life11121349

Dessì, C., Okoffo, E. D., O'Brien, J. W., Gallen, M., Samanipour, S., Kaserzon, S., Rauert, C., Wang, X., & Thomas, K. V. (2021). Plastics contamination of store-bought rice. *Journal of Hazardous Materials, 416*, Article 125778.

Dhiman, S., Gupta, R., Rajput, S., Raj, R., & Babu, R. (2024). Microbes in plastic degradation. *International Journal of Current Science Research and Review, 07*. https://doi.org/10.47191/ijcsrr/V7-i7-49

Dris, R., Gasperi, J., Mirande, C., Mandin, C., Guerrouache, M., Langlois, V., & Tassin, B. (2017). A first overview of textile fibers, including microplastics, in indoor and outdoor environments. *Environmental Pollution, 221*, 453–458. https://doi.org/10.1016/j.envpol.2016.12.013

Evangeliou, N., Grythe, H., Klimont, Z., Heyes, C., Eckhardt, S., Lopez-Aparicio, S., & Stohl, A. (2020). Atmospheric transport is a major pathway of microplastics to remote regions. *Nature Communications, 11*(1), 3381. https://doi.org/10.1038/s41467-020-17201-9

Franco, A. A., Martín-García, A. P., Egea-Corbacho, A., Arellano, J. M., Albendín, G., Rodríguez-Barroso, R., Quiroga, J. M., & Coello, M. D. (2023). Assessment and accumulation of microplastics in sewage sludge at wastewater treatment plants located in Cádiz, Spain. *Environmental Pollution, 317*, Article 120689. https://doi.org/10.1016/j.envpol.2022.120689

Galafassi, S., Nizzetto, L., & Volta, P. (2019). Plastic sources: A survey across scientific and grey literature for their inventory and relative contribution to microplastics pollution in natural environments, with an emphasis on surface water. *Science of the Total Environment, 693*, 133–499. https://doi.org/10.1016/j.scitotenv.2019.07.305

Gao, M., Liu, Y., & Song, Z. (2019). Effects of polyethylene microplastic on the phytotoxicity of di-n-butyl phthalate in lettuce (Lactuca sativa L. var. ramosa Hort). *Chemosphere, 237*, 124–482. https://doi.org/10.1016/j.chemosphere.2019.124482

Gerritse, J., Leslie, H. A., de Tender, C. A., Devriese, L. I., & Vethaak, A. D. (2020). Fragmentation of plastic objects in a laboratory seawater microcosm. *Scientific Reports, 10*(1), 10945. https://doi.org/10.1038/s41598-020-67927-1

Gui, J., Sun, Y., Wang, J., Chen, X., Zhang, S., & Wu, D. (2021). Microplastics in composting of rural domestic waste: Abundance, characteristics, and release from the surface of macroplastics. *Environmental Pollution, 274*, 116–553. https://doi.org/10.1016/j.envpol.2021.116553

Hassan, F., Prasetya, K. D., Hanun, J. N., Bui, H. M., Rajendran, S., Kataria, N., Khoo, K. S., Wang, Y.-F., You, S.-J., & Jiang, J.-J. (2023). Microplastic contamination in sewage sludge: Abundance, characteristics, and impacts on the environment and human health. *Environmental Technology & Innovation, 31*, Article 103176. https://doi.org/10.1016/j.eti.2023.103176

He, D., Guo, T., Li, J., & Wang, F. (2023). Optimize lettuce washing methods to reduce the risk of microplastics ingestion: The evidence from microplastics residues on the surface of lettuce leaves and in the lettuce washing wastewater. *Science of the Total Environment, 868*, Article 161726. https://doi.org/10.1016/j.scitotenv.2023.161726

Hlatshwayo, S. I., Ngidi, M. S. C., Ojo, T. O., Modi, A. T., Mabhaudhi, T., & Slotow, R. (2023). The determinants of crop productivity and its effect on food and nutrition security in rural communities of South Africa. *Frontiers in Sustainable Food Systems, 7*, 1091333.

Hoang, V.-H., Nguyen, M.-K., Hoang, T.-D., Ha, M. C., Huyen, N. T. T., Bui, V. K. H., Pham, M.-T., Nguyen, C.-M., Chang, S. W., & Nguyen, D. D. (2024). Sources, environmental fate, and impacts of microplastic contamination in agricultural soils: A comprehensive review. *Science of the Total Environment, 950*, Article 175276. https://doi.org/10.1016/j.scitotenv.2024.175276

Hore, M., Bhattacharyya, S., Roy, S., Sarkar, D., & Biswas, J. K. (2024). Human exposure to dietary microplastics and health risk: A comprehensive review. *Reviews of Environmental Contamination and Toxicology, 262*(1), 14. https://doi.org/10.1007/s44169-024-00066-0

Hu, C., Wang, X., Wang, S., Lu, B., Guo, W., Liu, C., & Tang, X. (2020). Impact of agricultural residual plastic film on the growth and yield of drip-irrigated cotton in arid region of Xinjiang, China. *International Journal of Agricultural and Biological Engineering, 13*, 160–169. https://doi.org/10.25165/j.ijabe.20201301.5084

Huang, D., Tao, J., Cheng, M., Deng, R., Chen, S., Yin, L., & Li, R. (2021). Microplastics and nanoplastics in the environment: Macroscopic transport and effects on creatures. *Journal of Hazardous Materials, 407*, Article 124399. https://doi.org/10.1016/j.jhazmat.2020.124399

Huerta, L., Esperanza, Gertsen, H., Gooren, H., Peters, P., Salánki, T., van der Ploeg, M., Besseling, E., Koelmans, A. A., & Geissen, V. (2017). Incorporation of microplastics from litter into burrows of Lumbricus terrestris. *Environmental Pollution, 220*, 523–531. https://doi.org/10.1016/j.envpol.2016.09.096

Jayasinghe, C. V. L., Jayatilake, S., Rajapakse, H. U. K. D. Z., Kithmini, N. K. S., & Kulathunga, K. M. P. M. (2023). Microplastics in the food chain. In *Microplastics in the ecosphere* (pp. 339–355). https://doi.org/10.1002/9781119879534.ch20

Katsumi, N., Kusube, T., Nagao, S., & Okochi, H. (2020). The role of coated fertilizer used in paddy fields as a source of microplastics in the marine environment. *Marine Pollution Bulletin, 161*, 111–727. https://doi.org/10.1016/j.marpolbul.2020.111727

Kaur, K., Reddy, S., Barathe, P., Oak, U., Shriram, V., Kharat, S. S., Govarthanan, M., & Kumar, V. J. C. (2022). Microplastic-associated pathogens and antimicrobial resistance in environment. *Chemosphere, 291*, Article 133005. https://doi.org/10.1016/j.chemosphere.2021.133005

Kedzierski, M., Lechat, B., Sire, O., Le Maguer, G., Le Tilly, V., & Bruzaud, S. (2020). Microplastic contamination of packaged meat: Occurrence and associated risks. *Food Packaging and Shelf Life, 24*, Article 100489.

Khan, M. B., Urmy, S. Y., Setu, S., Kanta, A. H., Gautam, S., Eti, S. A., Rahman, M. M., Sultana, N., Mahmud, S., & Baten, M. A. (2023). Abundance, distribution and composition of microplastics in sediment and fish species from an Urban River of Bangladesh. *Science of the Total Environment, 885*, Article 163876. https://doi.org/10.1016/j.scitotenv.2023.163876

Kublik, S., Gschwendtner, S., Magritsch, T., Radl, V., Rillig, M. C., & Schloter, M. (2022). Microplastics in soil induce a new microbial habitat, with consequences for bulk soil microbiomes [Original Research]. *Frontiers in Environmental Science, 10*. https://doi.org/10.3389/fenvs.2022.989267

Kutralam-Muniasamy, G., Pérez-Guevara, F., Elizalde-Martínez, I., & Shruti, V. (2020). Branded milks–are they immune from microplastics contamination? *Science of the Total Environment, 714*, Article 136823.

Li, A., Wang, Y., Kulyar, M. F.-e.-A., Iqbal, M., Lai, R., Zhu, H., & Li, K. (2023). Environmental microplastics exposure decreases antioxidant ability, perturbs gut microbial homeostasis and metabolism in chicken. *Science of the Total Environment, 856*, 159089.

Li, J., Liu, H., & Paul Chen, J. (2018). Microplastics in freshwater systems: A review on occurrence, environmental effects, and methods for microplastics detection. *Water Research, 137*, 362–374. https://doi.org/10.1016/j.watres.2017.12.056

Li, J., Song, Y., & Cai, Y. (2020a). Focus topics on microplastics in soil: Analytical methods, occurrence, transport, and ecological risks. *Environmental Pollution, 257*, Article 113570. https://doi.org/10.1016/j.envpol.2019.113570

Li, M., Wang, Z., Zhu, L., Zhu, Y., Yi, J., & Fu, X. (2024). Research advances on microplastics contamination in terrestrial geoenvironment: A review. *Science of the Total Environment, 935*, Article 173259. https://doi.org/10.1016/j.scitotenv.2024.173259

Li, P., Wang, X., Su, M., Zou, X., Duan, L., & Zhang, H. (2020b). Characteristics of plastic pollution in the environment: A review. *Bulletin of Environmental Contamination and Toxicology*, 1–8.

Li, S., Ding, F., Flury, M., Wang, Z., Xu, L., Li, S., Jones, D. L., & Wang, J. (2022). Macro- and microplastic accumulation in soil after 32 years of plastic film mulching. *Environmental Pollution, 300*, Article 118945. https://doi.org/10.1016/j.envpol.2022.118945

Liu, X., Yuan, W., Di, M., Li, Z., & Wang, J. (2019). Transfer and fate of microplastics during the conventional activated sludge process in one wastewater treatment plant of China. *Chemical Engineering Journal, 362*, 176–182. https://doi.org/10.1016/j.cej.2019.01.033

Luo, H., Tu, C., He, D., Zhang, A., Sun, J., Li, J., Xu, J., & Pan, X. (2023). Interactions between microplastics and contaminants: A review focusing on the effect of aging process. *Science of the Total Environment, 899*, Article 165615. https://doi.org/10.1016/j.scitotenv.2023.165615

Luo, Z., Zhou, X., Su, Y., Wang, H., Yu, R., Zhou, S., Xu, E. G., & Xing, B. (2021). Environmental occurrence, fate, impact, and potential solution of tire microplastics: Similarities and differences with tire wear particles. *Science of the Total Environment, 795*, Article 148902. https://doi.org/10.1016/j.scitotenv.2021.148902

Lwanga, E. H., Beriot, N., Corradini, F., Silva, V., Yang, X., Baartman, J., Rezaei, M., van Schaik, L., Riksen, M., & Geissen, V. (2022). Review of microplastic sources, transport pathways and correlations with other soil stressors: A journey from agricultural sites into the environment. *Chemical and Biological Technologies in Agriculture, 9*(1), 20. https://doi.org/10.1186/s40538-021-00278-9

Makhdoumi, P., Naghshbandi, M., Ghaderzadeh, K., Mirzabeigi, M., Yazdanbakhsh, A., & Hossini, H. (2021). Micro-plastic occurrence in bottled vinegar: Qualification, quantification and human risk exposure. *Process Safety and Environmental Protection, 152*, 404–413.

Mateos-Cárdenas, A., van Pelt, F. N. A. M., O'Halloran, J., & Jansen, M. A. K. (2021). Adsorption, uptake and toxicity of micro- and nanoplastics: Effects on terrestrial plants and aquatic macrophytes. *Environmental Pollution, 284*, Article 117183. https://doi.org/10.1016/j.envpol.2021.117183

Mehmood, T., Mustafa, B., Mackenzie, K., Ali, W., Sabir, R. I., Anum, W., Gaurav, G. K., Riaz, U., Liu, X., & Peng, L. (2023). Recent developments in microplastic contaminated water treatment: Progress and prospects of carbon-based two-dimensional materials for membranes separation. *Chemosphere, 316*, Article 137704. https://doi.org/10.1016/j.chemosphere.2022.137704

Menéndez-Manjón, A., Martínez-Díez, R., Sol, D., Laca, A., Laca, A., Rancaño, A., & Díaz, M. (2022). Long-term occurrence and fate of microplastics in WWTPs: A case study in Southwest Europe. *Applied Sciences, 12*(4).

Mercogliano, R., Avio, C. G., Regoli, F., Anastasio, A., Colavita, G., & Santonicola, S. (2020). Occurrence of microplastics in commercial seafood under the perspective of the human food chain. A review. *Journal of agricultural and food chemistry, 68*(19), 5296–5301.

Miao, H., Zhang, S., Gao, W., Zhou, J., Cai, H., Wu, L., Liu, J., Wang, Z., & Liu, T. (2024). Microplastics occurrence and distribution characteristics in mulched agricultural soils of Guizhou province. *Scientific Reports, 14*(1), 21505. https://doi.org/10.1038/s41598-024-72829-7

Ng, E. L., Huerta Lwanga, E., Eldridge, S. M., Johnston, P., Hu, H. W., Geissen, V., & Chen, D. (2018). An overview of microplastic and nanoplastic pollution in agroecosystems. *Science of the Total Environment, 627*, 1377–1388. https://doi.org/10.1016/j.scitotenv.2018.01.341

Nizzetto, L., Futter, M., & Langaas, S. (2016). Are agricultural soils dumps for microplastics of urban origin? *Environmental Science & Technology, 50*(20), 10777–10779. https://doi.org/10.1021/acs.est.6b04140

Okeke, E. S., Okoye, C. O., Atakpa, E. O., Ita, R. E., Nyaruaba, R., Mgbechidinma, C. L., & Akan, O. D. (2022). Microplastics in agroecosystems-impacts on ecosystem functions and food chain. *Resources, Conservation and Recycling, 177*, Article 105961. https://doi.org/10.1016/j.resconrec.2021.105961

Park, B., Kim, S.-K., Joo, S., Kim, J.-S., Jo, K., Song, N.-S., Im, J., Lee, H.-J., Kim, S. W., Lee, S. B., Kim, S., Lee, Y., Kim, B.-Y., & Kim, T. W. (2023). Microplastics in large marine animals stranded in the Republic of Korea. *Marine Pollution Bulletin, 189*, Article 114734. https://doi.org/10.1016/j.marpolbul.2023.114734

Piehl, S., Leibner, A., Löder, M. G. J., Dris, R., Bogner, C., & Laforsch, C. (2018). Identification and quantification of macro- and microplastics on an agricultural farmland. *Scientific Reports, 8*(1), 17950. https://doi.org/10.1038/s41598-018-36172-y

Prata, J. C., Paço, A., Reis, V., da Costa, J. P., Fernandes, A. J. S., da Costa, F. M., Duarte, A. C., & Rocha-Santos, T. (2020). Identification of microplastics in white wines capped with polyethylene stoppers using micro-Raman spectroscopy. *Food Chemistry, 331*, Article 127323. https://doi.org/10.1016/j.foodchem.2020.127323

Premarathna, K. S. D., Ramanayaka, S., Navaratne, A., Wijesekara, H., Jayasanka, J., & Vithanage, M. (2023). Compost-hosted microplastics—Municipal solid waste compost. In *Microplastics in the ecosphere* (pp. 433–448). https://doi.org/10.1002/9781119879534.ch27

Priyanka, M., & Dey, S. (2018). Ruminal impaction due to plastic materials-An increasing threat to ruminants and its impact on human health in developing countries. *Veterinary World, 11*(9), 1307.

Qiang, L., Hu, H., Li, G., Xu, J., Cheng, J., Wang, J., & Zhang, R. (2023). Plastic mulching, and occurrence, incorporation, degradation, and impacts of polyethylene microplastics in agroecosystems. *Ecotoxicology and Environmental Safety, 263*, Article 115274. https://doi.org/10.1016/j.ecoenv.2023.115274

Quilliam, R. S., Pow, C. J., Shilla, D. J., Mwesiga, J. J., Shilla, D. A., & Woodford, L. (2023). Microplastics in agriculture—A potential novel mechanism for the delivery of human pathogens onto crops [Perspective]. *Frontiers in Plant Science, 14*. https://doi.org/10.3389/fpls.2023.1152419

Rillig, M., Ziersch, L., & Hempel, S. (2017). Microplastic transport in soil by earthworms. *Scientific Reports, 7*. https://doi.org/10.1038/s41598-017-01594-7

Ross, E. M., Moate, P. J., Bath, C. R., Davidson, S. E., Sawbridge, T. I., Guthridge, K. M., Cocks, B. G., & Hayes, B. J. (2012). High throughput whole rumen metagenome profiling using untargeted massively parallel sequencing. *BMC Genetics, 13*, 1–14.

Rüthi, J., Bölsterli, D., Pardi-Comensoli, L., Brunner, I., & Frey, B. (2020). The "plastisphere" of biodegradable plastics is characterized by specific microbial taxa of alpine and arctic soils. *Frontiers in Environmental Science, 8*, Article 562263. https://doi.org/10.3389/fenvs.2020.562263

Sa'adu, I., & Farsang, A. (2023). Plastic contamination in agricultural soils: A review. *Environmental Sciences Europe, 35*(1), 13. https://doi.org/10.1186/s12302-023-00720-9

Sajjad, M., Huang, Q., Khan, S., Khan, M. A., Liu, Y., Wang, J., Lian, F., Wang, Q., & Guo, G. (2022). Microplastics in the soil environment: A critical review. *Environmental Technology & Innovation, 27*, Article 102408. https://doi.org/10.1016/j.eti.2022.102408

Sharma, P., & Vidyarthi, V. K. (2024). Impact of microplastic intake via poultry products: Environmental toxicity and human health. *Journal of Hazardous Materials Advances, 14*, Article 100426. https://doi.org/10.1016/j.hazadv.2024.100426

Sharma, S., & Chatterjee, S. (2017). Microplastic pollution, a threat to marine ecosystem and human health: A short review. *Environmental Science and Pollution Research International, 24*(27), 21530–21547. https://doi.org/10.1007/s11356-017-9910-8

Shruti, V., Pérez-Guevara, F., Elizalde-Martínez, I., & Kutralam-Muniasamy, G. (2020). First study of its kind on the microplastic contamination of soft drinks, cold tea and energy drinks-Future research and environmental considerations. *Science of the Total Environment, 726*, Article 138580.

Silva, A. B., Bastos, A. S., Justino, C. I. L., da Costa, J. P., Duarte, A. C., & Rocha-Santos, T. A. P. (2018). Microplastics in the environment: Challenges in analytical chemistry—A review. *Analytica Chimica Acta, 1017*, 1–19. https://doi.org/10.1016/j.aca.2018.02.043

Song, Y. K., Hong, S. H., Eo, S., & Shim, W. J. (2023). Fragmentation of nano- and microplastics from virgin- and additive-containing polypropylene by accelerated photooxidation. *Environmental Pollution, 327*, Article 121590. https://doi.org/10.1016/j.envpol.2023.121590

Sorasan, C., Edo, C., González-Pleiter, M., Fernández-Piñas, F., Leganés, F., Rodríguez, A., & Rosal, R. (2022). Ageing and fragmentation of marine microplastics. *Science of the Total Environment, 827*, Article 154438. https://doi.org/10.1016/j.scitotenv.2022.154438

Toussaint, B., Raffael, B., Angers-Loustau, A., Gilliland, D., Kestens, V., Petrillo, M., Rio-Echevarria, I. M., & Van den Eede, G. (2019). Review of micro- and nanoplastic contamination in the food chain. *Food Additives & Contaminants: Part A, 36*(5), 639–673. https://doi.org/10.1080/19440049.2019.1583381

van Schothorst, B., Beriot, N., Huerta Lwanga, E., & Geissen, V. (2021). Sources of light density microplastic related to two agricultural practices: The use of compost and plastic mulch. *Environments, 8*(4).

van Sebille, E., Wilcox, C., Lebreton, L., Maximenko, N., Hardesty, B. D., van Franeker, J. A., Eriksen, M., Siegel, D., Galgani, F., & Law, K. L. (2015). A global inventory of small floating plastic debris. *Environmental Research Letters, 10*(12), Article 124006. https://doi.org/10.1088/1748-9326/10/12/124006

Vithanage, M., Ramanayaka, S., Hasinthara, S., & Navaratne, A. (2021). Compost as a carrier for microplastics and plastic-bound toxic metals into agroecosystems. *Current Opinion in Environmental Science & Health, 24*, Article 100297. https://doi.org/10.1016/j.coesh.2021.100297

Wang, C., Zhao, J., & Xing, B. (2021). Environmental source, fate, and toxicity of microplastics. *Journal of Hazardous Materials, 407*, Article 124357. https://doi.org/10.1016/j.jhazmat.2020.124357

Wei, F., Xu, C., Chen, C., Wang, Y., Lan, Y., Long, L., Xu, M., Wu, J., Shen, F., Zhang, Y., Xiao, Y., & Yang, G. (2022). Distribution of microplastics in the sludge of wastewater treatment plants in Chengdu, China. *Chemosphere, 287*, Article 132357. https://doi.org/10.1016/j.chemosphere.2021.132357

Weithmann, N., Möller, J. N., Löder, M. G. J., Piehl, S., Laforsch, C., & Freitag, R. (2018). Organic fertilizer as a vehicle for the entry of microplastic into the environment. *Science Advances, 4*(4), eaap8060. https://doi.org/10.1126/sciadv.aap8060

Wiesner, Y., Bednarz, M., Braun, U., Bannick, C. G., Ricking, M., & Altmann, K. (2023). A promising approach to monitor microplastic masses in composts. *Frontiers in Environmental Chemistry, 4*, 1281558.

Yang, L., Zhang, Y., Kang, S., Wang, Z., & Wu, C. (2021). Microplastics in soil: A review on methods, occurrence, sources, and potential risk. *Science of the Total Environment, 780*, Article 146546. https://doi.org/10.1016/j.scitotenv.2021.146546

Yang, Y., Li, Z., Yan, C., Chadwick, D., Jones, D. L., Liu, E., Liu, Q., Bai, R., & He, W. (2022). Kinetics of microplastic generation from different types of mulch films in agricultural soil. *Science of the Total Environment, 814*, Article 152572. https://doi.org/10.1016/j.scitotenv.2021.152572

Yu, M., van der Ploeg, M., Ma, X., Ritsema, C. J., & Geissen, V. (2020). Effects of microplastics and earthworm burrows on soil macropore water flow within a laboratory soil column setup. *Vadose Zone Journal, 19*(1), Article e20059. https://doi.org/10.1002/vzj2.20059

Yuan, Z., Nag, R., & Cummins, E. (2022). Human health concerns regarding microplastics in the aquatic environment - From marine to food systems. *Science of the Total Environment, 823*, Article 153730. https://doi.org/10.1016/j.scitotenv.2022.153730

Zhai, Y., Bai, J., Chang, P., Liu, Z., Wang, Y., Liu, G., Cui, B., Peijnenburg, W., & Vijver, M. G. (2024). Microplastics in terrestrial ecosystem: Exploring the menace to the soil-plant-microbe interactions. *TrAC Trends in Analytical Chemistry, 174*, Article 117667. https://doi.org/10.1016/j.trac.2024.117667

Zhang, G. S., & Liu, Y. F. (2018). The distribution of microplastics in soil aggregate fractions in southwestern China. *Science of the Total Environment, 642*, 12–20. https://doi.org/10.1016/j.scitotenv.2018.06.004

Zhao, J., Zhang, J., Hu, Y., Li, Y., Tang, P., Gusarov, A. V., & Yu, Y. (2022). Effects of land uses and rainfall regimes on surface runoff and sediment yield in a nested watershed of the Loess Plateau, China. *Journal of Hydrology: Regional Studies, 44*, Article 101277. https://doi.org/10.1016/j.ejrh.2022.101277

Zhu, J., Dong, G., Feng, F., Ye, J., Liao, C.-H., Wu, C.-H., & Chen, S.-C. (2023). Microplastics in the soil environment: Focusing on the sources, its transformation and change in morphology. *Science of the Total Environment, 896*, Article 165291. https://doi.org/10.1016/j.scitotenv.2023.165291

Ziajahromi, S., Neale, P. A., Telles Silveira, I., Chua, A., & Leusch, F. D. L. (2021). An audit of microplastic abundance throughout three Australian wastewater treatment plants. *Chemosphere, 263*, Article 128294. https://doi.org/10.1016/j.chemosphere.2020.128294

Chapter 4
Microplastics Pollution: A Threat to Marine Ecosystems

Kalani Perera, Manura Weerasinghe, Ruwandi Nilakshi, Nalin Madusanka Gunathilaka, Sahan Nishada Perera, Meththika Vithanage, and Anushka Rajapaksha

Abstract Microplastic (MP) contamination in marine ecosystems is a rising threat that seriously affects ecosystem health and biodiversity. From sources like industrial wastes, personal care products, and the degradations of larger plastics, MP enters to the oceans via wastewater, runoff, mismanaged waste dumping, and atmospheric deposition. Once such tiny particles enter the marine ecosystem, they could get ingested by marine organisms, which might cause physical blockages chemical toxicity, bioaccumulation, and biomagnification in food chains. MP could alter the habitats chemically and physically within the water column and sediments and alter how an ecosystem functions and the species composition in habitats. This is crucial in coral reef and seagrass bed ecosystems, estuaries and salt marshes, which are most important and sensitive concerning MP pollution. Therefore, habitat degradation, disruption of marine food webs, negative effects on biodiversity, and potential human health risks via bioaccumulation and biomagnification were identified as the most prominent issues that could be led by MP contamination. This requires urgent attention and actions to mitigate MP pollution in marine ecosystems. The chapter highlights the sources and pathways of MP in marine environments, their impact on marine organisms and the ecosystem-level impact, and highlights global hotspots of MP pollution in marine ecosystems.

Keywords Microplastic · Marine pollution · Ecosystem impacts · Food web · Global hotspots

---

K. Perera · M. Weerasinghe · M. Vithanage · A. Rajapaksha (✉)
Ecosphere Resilience Research Centre, Faculty of Applied Sciences, University of Sri Jayewardenepura, Nugegoda, Sri Lanka
e-mail: anurajapaksha@sjp.ac.lk

R. Nilakshi · N. M. Gunathilaka · S. N. Perera
Faculty of Graduate Studies, University of Kelaniya, Dalugama, Kelaniya, Sri Lanka

A. Rajapaksha
Instrument Center, Faculty of Applied Sciences, University of Sri Jayewardenepura, Nugegoda, Sri Lanka

V. Kumar et al. (eds.), *Microplastics Pollution*,
https://doi.org/10.1007/978-3-031-99774-7_4

4.1 Introduction

Marine microplastics (MPs) have recently become a growing environmental concern, originating from both primary sources like cosmetics and industrial products (pellets) as well as secondary sources resulting from the degradation of larger plastics (Bashir et al., 2021; Lei et al., 2017; Singh & Mishra, 2023). These particles reach marine ecosystems via a variety of routes, including rivers, drainage systems, wastewater treatment facilities, and activities such as shipping and fishing, while airborne MPs spread them farther to isolated places (Lebreton et al., 2017; Sharma et al., 2021; Yang et al., 2021). In degradation of MPs in marine ecosystem, size and density decides whether the MP keep floating, suspended or sink in sediments. Fragmentation of MP into nanoparticles would enhance their chances of ingestion by marine organisms (Chubarenko et al., 2018; Kaiser et al., 2017). Thus, the cause of MP sources, spread, and marine environmental impacts must be understood to avoid damaging marine environmental pollution.

MPs are becoming increasingly prevalent in marine environments, from coral reefs to polar regions, because of industrial pollution and human activities such as shipping, and fishing. MPs are widely found in a variety of ecosystems, such as coral reefs (Hansani et al., 2023; Soares et al., 2023), seagrass beds (Dahl et al., 2021; Ramili & Umasangaji, 2022), estuaries (Mendes et al., 2024; Tan & Mohd Zanuri, 2023), salt marshes (Lloret et al., 2021; Trusler et al., 2025) and even in remote polar marine environments (Emberson-Marl et al., 2023; Huang et al., 2023). These ecosystems which are very vital and have important ecological roles are under threat because of accumulation of MPs that alters natural processes, affects marine life and has potential of causing future environmental harm. The increasing database indicates that there is the need for intensified and aggressive management to combat the increasing rates of plastic pollution in these sensitive areas.

MPs are increasingly recognized for their negative effects on marine life, including phytoplankton, bigger fish, and invertebrates (Acarer Arat, 2024; Nanthini devi et al., 2022). The presence of MPs in marine environments poses both physical and chemical hazards, affecting biological functioning, eating habits, and reproductive health (Jovanović, 2017; Li et al., 2021; Yao et al., 2024). MPs can build up in the tissues of marine organisms, resulting in toxicity, oxidative stress, immune system problems, and alterations in metabolic processes (Guzzetti et al., 2018; Miller et al., 2020). The physical harm caused by these particles includes intestinal blockages and histopathological changes (Jovanović, 2017). As MPs invade the food chain, their negative impacts can be magnified, posing major threats to biodiversity and marine ecosystem health (Carbery et al., 2018). Understanding these many effects is critical for combating the rising pollution of our seas and conserving marine life.

Biodiversity, ecosystem processes and food webs are affected by MP pollution in marine habitats (Cole et al., 2011; Corinaldesi et al., 2021; Hale et al., 2020). Including both primary producers such as phytoplankton and predators such as baleen whales, many aspects of marine life is affected by these tiny plastics that are wide spread in the ocean (Sjollema et al., 2016; Troost et al., 2018; Werth et al., 2024). MP disrupt the nutrient cycle, diminish carbon storage, and eliminate key basal species that are essential to the ecosystem (Cole et al., 2011; Hale et al., 2020; Shen et al., 2020). When marine species ingest or become entangled in MPs, it can result in physical injuries, reproductive problems, and altered behaviors, which may create a cascading impact throughout the food web (Bhardwaj et al., 2024; Roy et al., 2022; Yin et al., 2019). The vital services that the marine ecosystem provides to humans, including food, coastal protection, and climate control, are therefore at risk due to changes in species composition, habitat loss, and waning resistance to climate change (Beaumont et al., 2019).

While some regions are regarded major hotspots for plastic contamination, including the Great Pacific Garbage Patch, coastal waters of Southeast Asia, and the Mediterranean Ocean (Alomar et al., 2016; Amelia et al., 2021; Munhoz et al., 2022; Peng et al., 2020), this pollution threatens biodiversity, ecological services, and people who rely on marine resources (Beaumont et al., 2019). MP accumulation and dispersal is affected by ocean currents, urbanization, waste management practices and population density (Nakakuni et al., 2024). In addition to widely known hotspots, MP pollution have an impact on secluded locations such as the Arctic and Antarctic, signifying the global impact of plastic waste (Kelly et al., 2020). This issue will be solved with the help of a comprehensive approach that will include the enhanced surveillance, the creation of proper management plans and the cooperation on the international level in order to minimize the impacts of MPs on marine organisms and environments.

4.2 Sources and Pathways of Microplastics in Marine Environments

There are primary and secondary sources of MPs in the marine environment. Primary MPs are mainly plastic beads along with industrial production pellets or powder that are used to manufacture cosmetics and personal hygiene products, like sunscreen, varnish, nail polish, shower gel, and other care products that may contain scrubs and abrasives like toothpaste and facial cleansers (Bashir et al., 2021; Lei et al., 2017; Singh & Mishra, 2023). In addition to that, fibers generated during the production of synthetic textiles and garments are also considered primary MPs (Acharya et al., 2021) These MPs can move down bathroom drains into drainage systems and further escape into wastewater treatment facilities, ultimately ending up in the aquatic ecosystems (Chen et al., 2024; Lasee et al., 2017). Furthermore, during shipment and handling plastic resin pellets may accidentally leak or spill. In addition, operational and accidental spills by cargo and big transport boats might be considered major sources of resin pellets (Sewwandi et al., 2023; Wang et al., 2018).

Secondary MPs, primarily originate from large plastic particles' fragmentation and degradation due to high solar UV-radiation and mechanical abrasion via physical, chemical, and biological processes. These MPs are directly transported into marine

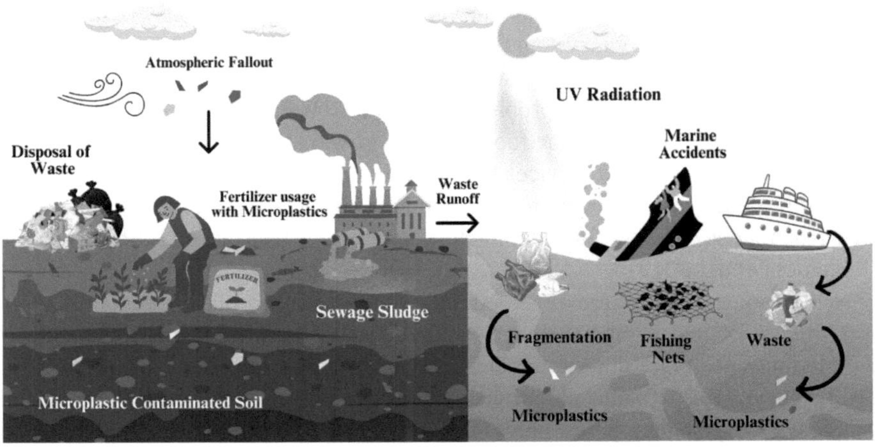

Fig. 4.1 Sources of microplastics in the marine environment

environments through shorelines, rivers, and sewage systems. Fishing, shipping, and offshore industry sectors are sea-based sources (Sharma et al., 2021). Introduction of secondary MP particles can sometimes occur from lost, abandoned, worn-out fishing and aquaculture gears and other equipment during use (Gilman et al., 2023; Wang et al., 2018). Figure 4.1, showcases some of the sources of MPs in marine ecosystems as discussed.

Marine MP can originate from diverse sources: inland-based, sea-based, and air-based sources (Auta et al., 2017; Yang et al., 2021). Among them, land-based sources are identified as the dominant ways of releasing MP into the ocean (Auta et al., 2017; Wang et al., 2024). Rivers are generally thought to act as the major pathway of transport from inland into the ocean (Lebreton et al., 2017). Similarly, 80% of marine plastic particles could originate from the land system in a given marine environment (Wang et al., 2024). While rivers transport plastic debris from municipal drainage systems, sewage effluents, or poorly managed inland areas to the sea, in the case of plastic wastes from beach tourism, they are often directly thrown into the environment (Dey et al., 2024).

Offshore industries, shipping, and fishing are sea-based sources (Bell et al., 2017). Emissions and leaks of shipping vessels are significant sources of MP (Folbert et al., 2022). Another method that plastic particles could be easily introduced into the marine environment is loss and damage to fishing and aquaculture equipment (Richardson et al., 2019). In addition to that, the primary offshore source of MPs is the global fishing fleet and illegally discarded waste from ships and offshore platforms, and a large proportion of items come from lost containers. Moreover, airborne MPs are also a significant source (Yang et al., 2021).

MP transport in marine environments is affected by integrating physical, chemical, and biological processes, while their physical properties, size, shape, and density are important factors. The range of MP densities is wide: from very light polystyrene foam (less than 0.05 g cm^{-3}) to much heavier polytetrafluoroethylene (2.1–2.3 g cm^{-3}) (Chubarenko et al., 2016). These density differences determine where the MPs end up in the ocean. Lighter MPs float on top or stay suspended in the water, while heavier ones sink to the ocean floor and settle in benthic environments (Chubarenko et al., 2018; Kaiser et al., 2017). Over time, these MPs may degrade into tiny nanoparticles. Both MPs and nano plastics are more easily swallowed by marine life and are likely to have long-term harmful effects on marine ecosystem, causing widespread public concern.

4.3 Abundance of Microplastics in Different Marine Ecosystems

4.3.1 Coral Ecosystem

In coral reef ecosystems, the presence of MPs can vary based on the species, their structural characteristics, and the surrounding environment. A study on reef-building corals in the Maldives revealed that 58% of the colonies examined contained MPs, with particle sizes ranging from 25 to 150 μm, and polyethylene (PE) and polypropylene (PP) being the most frequently identified types of plastic. The maximum concentration (8.9 particles/g of coral) was encountered in a *Pocillopora verrucosa* colony sampled from a shallow inner reef (Raguso et al., 2022). Another study on Hainan Island, Sanya Bay, China detected 0.01 to 3.60 items $polyp^{-1}$ in corals (Lei et al., 2021). Surface waters gathered from several coral reef regions in Peninsular Malaysia demonstrated an average abundance of 0.344 ± 0.457 MP/m^3 (Lim et al., 2024). A considerable portion of these MPs were small, less than 1 mm, with the 0.05–0.5 mm size class being the most common at 29.2%. Among them, PE and rayon MPs were the most frequently found. A study on the Southern and Western coasts of Sri Lanka, found blue-colored fibre type MPs to dominate the coral ecosystems. The average MPs in the corals were determined as 806.64 ± 93.72 particles kg^{-1} (wet weight) while 23.42 ± 5.01 particles m^{-3} and 54.11 ± 8.04 particles kg^{-1} (dry weight) of MPs were in surface water and sediment respectively (Abeysinghe et al., 2024). A study conducted in Sri Lanka revealed an average abundance of 546.7 ± 170.3 items/kg of MPs in the corals, with Acropora corals showing significantly higher concentrations compared to other species. (Hansani et al., 2023). Research on *Acropora millepora* showed an accumulation of only 0.27 ± 0.26 items/kg, while studies on *Galaxea fascicularis* indicated a much higher rate of 2.32 ± 0.86 MP items/kg, highlighting a notable difference in accumulation rates among coral species (Hansani et al., 2023). In three Mascarene islands spanning more than 1200 km along the Western Indian Ocean, an evaluation of hard and soft corals revealed that 78% of the MPs found were PP and 18% were PE (Hamman et al., 2024).

4.3.2 Seagrass Beds

In studies conducted in several global locations, the huge number of MPs in seagrass debris is demonstrated, such as the Baltic Sea, Spain, Almeria, Indonesia and Maregam and Marekofo villages in North Maluku (Dahl et al., 2021; Ramili & Umasangaji, 2022). The concentration of MP in the Baltic Sea varied from 0 to 1,817 particles kg^{-1}, with an average of 208 particles kg^{-1}. This is notably higher than the level observed in nearby unvegetated areas (Martinez et al., 2024). The same trends were noted in North Maluku, where the lower portions of the seagrass beds had up to 39,400 particles kg^{-1}, while Maregam Village and Marekofo Village documented a mean of 20,754 and 17,705 particles kg^{-1}, respectively (Ramili & Umasangaji, 2022). MP pollution, which is mostly composed of polyvinyl chloride (PVC) and polystyrene (PS), has increased in Almería, also referred to as "the sea of plastic," as a result of decades of intensive agricultural practices that have contaminated seagrass beds (Dahl et al., 2021). Apart from fibers, fragments comprise roughly 29.46% to 31.37% of the overall MP load, whereas pellets are less prevalent, accounting for only 1% to 8% of the total MPs (Li et al., 2023). The majority of commonly found polymer forms are PP, PVC, and PE with PE becoming the material of choice in seagrass environments (Ramili & Umasangaji, 2022).

4.3.3 Estuaries

MPs accumulate in estuarine environments due to a combination of natural processes and human activities. Recent studies have highlighted alarming levels of MP pollution in these areas. For instance, research carried out in the mangrove regions of the Amazon found the highest concentration of MPs ever recorded at 1.03 items m^{-3} in the inner estuary, while the outer estuary had 0.82 items m^{-3}. These findings are linked to the population density and fishing activities in these areas. MPs are mostly blue (55.2%), fiber-shaped (89.8%), and vary in size from 1000 to 2000 μm (31.7%) (Mendes et al., 2024). In tropical estuarine mangrove regions in Penan, Malaysia (Penaga, Kuala Muda, Seberang Perai, and Balik Pulau), MP concentrations in coastal surface water ranged from 201 ± 21.214 to 1407 ± 124.265 pieces L^{-1}, with sediment samples from the estuaries showing even higher levels, reaching up to 4000 pieces kg^{-1} (Tan & Mohd Zanuri, 2023). Seberang Perai had the highest abundance of MP at 1407 ± 124.265 pieces L^{-1}, with MP fragments making up 677 pieces L^{-1} (48%) (Tan & Mohd Zanuri, 2023). PE was the most widely recognized type of polymer, while dangerous substances such as dimethylmercury were discovered, posing risks to both species and the environment. (Tan & Mohd Zanuri, 2023). The estuary water in Pemalang, Central Java, Indonesia contained 19.10×10^2 particles m^{-3} of MPs. The most prevalent type of MP found in the waters of Pemalang was films. (Rifandi & Ratnasari, 2023).

4.3.4 Salt Marshes

Salt marshes rise and fall with the tides, and MPs that drift on the water's surface become ensnared in branches and roots, eventually settling into the marsh soil. According to the Marine Biological Laboratory (MBL) at the University of Chicago, MP concentrations in salt marsh sediments can vary significantly depending on land use and urbanization. Sediment core samples from the Waquoit Bay system on Cape Cod revealed that MP abundance increased fast when urbanization exceeded 50%. MP reached up to 267.1 ± 10.2 particles per 100 g of dry weight in high-marsh vegetation zones (Lloret et al., 2021). Sediment cores from salt marshes; Swanscombe, Kent, and Rainham, Essex dominated by low-density MPs, PP, PE, and PS with the greatest abundance (mean 267.1 ± 10.2 particles 100 g per dry weight at 0–10 cm depth) observed at Swanscombe in the high-marsh vegetation zone (Trusler et al., 2025). A case study from Southern Brazil found that MP contamination in the salt marsh, reached average levels of 279.63 ± 410.12 items kg^{-1}, 366.92 ± 975.18 items kg^{-1} and 8.89 ± 8.75 items L^{-1} in surface sediment, sediment cores, and water, respectively MPs were found in layers reaching depths of 66 cm, with higher concentrations in the upper sediments (Pinheiro et al., 2022). (Pinheiro et al., 2022).

4.3.5 Polar Marine Ecosystems

MP in polar regions is typically affected by a range of factors, such as shifting ocean current patterns, river runoff, shipping and fishing operations, emissions from nearby communities, and atmospheric deposition (Huang et al., 2023). Recent research discovered that the Barents Sea has an average MP abundance of 0.011 MPs m^{-3}), with concentrations varying between 0.007 to 0.015 m^3. Notably, higher levels were found near land masses and ice edges, reaching up to 0.015 m^3 during transect studies (Emberson-Marl et al., 2023). The majority of the abundant MPs were fibrous in shape (92.1%) and were usually blue (79%) or red (17%) in color (Emberson-Marl et al., 2023). In the Antarctic marine system, data are still scarce. Therefore, the Antarctic marine system is considered pristine compared to other regions. However, the Adventure Science organization found an average value of 22 particles L^{-1} in seawater samples from the western Antarctic Peninsula, with a high concentration of 117 particles L^{-1} (Waller et al., 2017).

4.4 Impact of Microplastics on Marine Organisms

MPs influence marine habitats through their physical presence and as carriers of chemical pollutants. Nevertheless, the effects of MP accumulation in marine organisms are still limited (Sui et al., 2024). MPs can negatively affect marine organisms such as phytoplankton, zooplankton, fish, mussels and corals (Acarer Arat, 2024; Nanthini devi et al., 2022).

The responses of phytoplankton, *Chlorella pyrenoidosa* was exposed to PS MPs and studied throughout its entire growth period, using MPs in two sizes of 0.1 mm and 1.0 mm and at three different concentration levels, which covered (10 and 50 mg/L) and exceeded (100 mg/L) its environmental concentrations, respectively (Mao et al., 2018). According to that, PS MPs caused harmful dose-dependent effects on *Chlorella pyrenoidosa*, particularly during its early growth phases, form the lag phase to the initial logarithmic phases, but showed a slight difference in the maximum inhibition ratio with respect to the two different MP sizes. The study examined the egestion and ingestion patterns, as well as the toxicity of various sizes and shapes of MPs, on two marine zooplankton species, *Brachionus koreanus* and *Diaphanosoma celebensis*, under different prey conditions (Park et al., 2024a). When exposed to 0.01 μg/mL of MPs, *B. koreanus* showed no noticeable fluorescence in its body in the absence of prey. However, significant fluorescence was observed after exposure to 0.1, 1, and 10 μg/mL of MPs. This confirms the ingestion of MPs to marine zooplankton and reduction of the ingestion of MPs occurs in zooplankton by the prey conditions. During the examination of digestive tracts of 100 individuals belonging to 10 species for MPs in commercial fish species from the Bay of Bengal, off the coast of Bangladesh showed, from the 215 MPs which were extracted has the average abundance range of 2.2 and 0.89 n individual^{-1}. (Ghosh et al., 2021). The behavior, metabolism, and energy reserve of marine demersal fish, *Sebastes schlegelii*, are altered by exposure to PS MPs with sizes ranging from 0.5 to 15 μm. Fish exposed to 15-μm PS showed considerably more behavioral abnormalities (such as clustering, decreased swimming speed, and decreased range of motion), which could have an impact on their ability to seek and explore (Yin et al., 2019). In the study, (Yu et al., 2024) the gastrointestinal (GI) tract was the organ that MPs were most frequently found. MP quantity varied by fish tissue in all fish samples ($p < 0.01$). Compared to fish's gills (mean 15.32 ± 16.76 n individual^{-1}) and muscle (mean 4.1 ± 4.4 n individual^{-1}), the number of MPs found in GI tracts (mean 24.15 ± 16.27 n individual-1) was considerably higher. Comparative research was done on the risk that mussels face from MPs and the herbicide chlorpyrifos (CPF) both by itself and in association with MPs and phytoplankton components of microalgae (MA) (Fernández et al., 2024). The synergistic toxicity of combined MPs and CPF is higher compared to either of the pollutants alone, MPs or CPF, or MA and CPF together. The microalgae species were *Isochrisis galbana* 4 to 8 μm; MPs consisted of virgin HDPE microparticles $\leq$ 10 μm in diameter. Animals were exposed for 21 days to MPs. Overall, oxidative damage, oxidative stress, and neurotoxicity did not show significant changes related to MPs alone. On the contrary, in mussel groups treated with CPF, MA-CPF, and MPs-CPF, on day 21, the development of immunological imbalance and adverse physiological state was recorded, while on day 7, neurotoxicity and oxidant imbalance were identified. During the later period of time, more changes were observed in mussels with MPs-CPF compared to the CPF or MA-CPF. The Mediterranean mussel *Mytilus galloprovincialis*, a marine benthic creature, has been used to study the metabolic effects of exposure to various PET

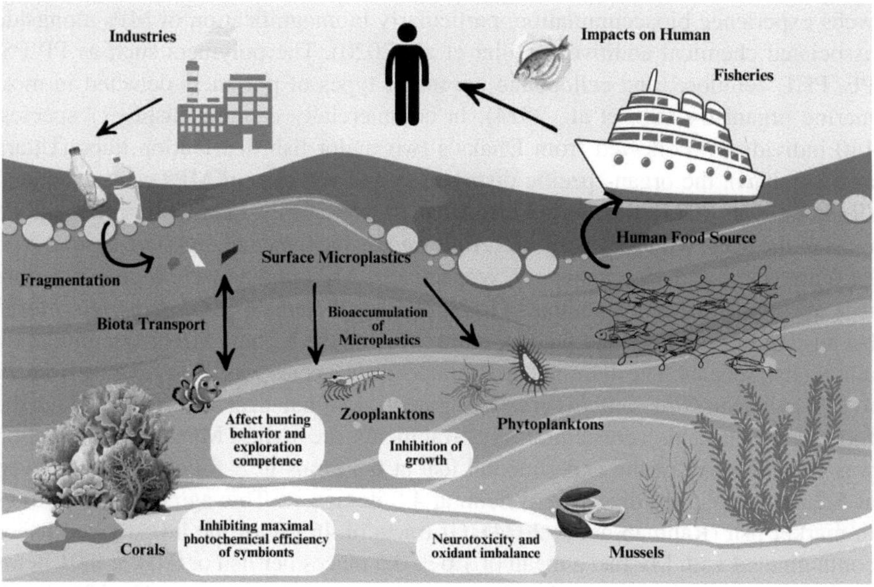

Fig. 4.2 Effects of microplastics on marine organisms

MP sizes and shapes (Park et al., 2024b). After 14 days of exposure to high concentrations (100 mg L^{-1}) of MPs, mussels exhibited a significant accumulation of round shape and fiber-type MPs.150 metabolites underwent substantial alterations as a result of exposure to round and fiber-type MPs. These findings offer fresh insights on the possible mechanisms via which MPs of particular sizes and shapes influence mussels' metabolic processes. The effects of MPs on corals in situ were assessed, and the occurrence characteristics of MPs in the environment media and reef-dwelling species were examined in coral reef sites from southern Hainan Island (Bian et al., 2024). The mean MP abundance ranged 0.36 items g^{-1} within coral, 1.50 items g^{-1} within shellfish, 0.48 items g^{-1} within the fish gills, and 1.71 items g^{-1} within the fish gastrointestinal tract. There are significant correlations between MP abundance and symbiont photosynthetic activity with important biological markers of corals, which would indicate that MPs can adversely affect corals by inhibiting symbionts to reach their optimum photochemical efficiency, eliciting coral immune responses, and the antioxidant system (Bian et al., 2024). Furthermore, Fig. 4.2 showcases some of the effects of MPs on marine organisms.

4.4.1 Bioaccumulation and Biomagnification

Several organisms in marine ecosystems and a variety of habitats have been shown to be contaminated by MP. Consequently, it is frequently assumed that marine food webs experience bioaccumulation, particularly biomagnification of MPs alongside associated chemical additives (Miller et al., 2020). The polymers such as PP PS, PE, PET, cellulose, and cellophane are major types of polymers detected in most marine organisms (Sui et al., 2024). In commercially captured fishes (7 species, 140 individuals) collected from Dhaka's two major fish distribution hubs (Uttara and Jatrabari), the organ-specific distribution and variance of MPs were examined (Rahman et al., 2024). Results indicated that, on average, the kidneys bioaccumulated most of the MPs, with 59.1 items g^{-1}, followed by the liver and digestive system with 24.6 items g^{-1} and 18.6 items g^{-1}, respectively. MP burdens were higher in Uttara fish, averaging 36 MPs/g, compared to Jatrabari at 25 items g^{-1}. *Glossogobius giuris*, because of morphology and feeding habit, showed the highest MP bioaccumulation among fish species. The dominant MPs shape in all species was fiber, ranging from 79 to 93%, while in *Glossogobius giuris*, fragments dominated, at 51%. Even though the food source had a high concentration of fragments, the highest MPs bioaccumulation was in the kidney of the experimental fish at 6.5 items g^{-1}, followed by the liver, at 5.2 items g^{-1}, and digestive system, at 4.8 items g-1. This agreed with findings in market fish (Rahman et al., 2024). GI tracts of 46.9% of the investigated fishes contaminated with MP had a mean of 1.6 ± 0.5 pieces per fish or 0.04 ± 0.01 items g^{-1} of fish tissue in the marine feeding fishes from the upper Gulf of Thailand. An approximation indicated that human consumers of marine fishes are exposed to 0.03–0.1 pieces of MPs each day via such fishes (Srisiri et al., 2024). Further, Table 4.1 shows the bioaccumulation of MPs in various marine organisms and Fig. 4.3 illustrates biomagnification in a marine ecosystem.

4.4.2 Toxicological Effects

Birds readily absorb the MPs, which causes bodily disruptions such as inflammation of the skin, decreased body weight, fledgling success, and breeding success (Singh & Upadhyay, 2025). In nautical biotas, MPs may contribute to cancer, oxidative and pathological pressure, and compromised immune responses (Guzzetti et al., 2018). Other consequences include a decline in nourishing action, fertility, survival, improvement, and energetic reserves (Guzzetti et al., 2018). The total quantity of eggs ovulated by *Thecrassostrea gigas* decreases substantially under the influence of PS MPs, and the level of movement of sperm is also significantly decreased. This indicates that MPs would significantly hinder the species' ability to reproduce (Li et al., 2021). MPs with a particle size of 40 to 150 μm may bring about oxidative impairment of the white blood cells of *Sparus aurata* and *Dicentrarchus labrax*, leading to immunotoxicity. Once the MPs have entered the living tissues and organs of the the oceans fishery, they will cause a number of immune-mediated reactions, including PVC and PE (Li et al., 2021).

Table 4.1 Bioaccumulation of microplastics in different marine organisms

Sampling site	Species	Bioaccumulation (items g^{-1})	References
Kuakata, Bay of Bengal	Bombay duck (loitta), *Harpadon nehereus* (N = 20)	41.33	Hasan et al. (2022)
	Ribbon fish (churi), *Trichiurus lepturus* (N = 20)	46.00	
Cox's bazar, Bay of Bengal	Bombay duck (loitta), *Harpadon nehereus*	28.54	
	Ribbon fish (churii), *Trichiurus lepturus*	34.17	
Labrador Sea	seabird *Fulmarus glacialis*	12.79	Provencher et al. (2018)
Mexican Caribbean	Sea turtle *Chelonia mydas*	10 ± 2 to 89 ± 3	Aranda et al. (2024)
Republic of Korea	Fin whale *Balaenoptera physalus*	3.94	Park et al. (2023)
	Finless porpoises *Neophocaena asiaeorientalis*	4.25	
	Indo-Pacific bottlenose dolphin *Tursiops aduncus*	0.46	
	Dolphin *Delphinus delphis*	0.48	

4.4.3 Physical and Mechanical Impacts

Intestinal blockage, injury to the body, gastrointestinal histopathological changes, behavioural as well as lipid metabolism changes, and MPs being transferred to the liver are among the mechanical and physical effects of ingestion (Jovanović, 2017). MPs sourced from petroleum-derived polymer compounds, PE, and biopolymers-palatable food coating, EFC-were analyzed for possible effects on aerobic metabolic rate, escape effectiveness, and regular swimming regarding *Forsterygion capito* (the mottled triplefin) (Hawke et al., 2024). Compared to control fish the ones that were exposed to PE demonstrated longer latency to respond-about 25% slower to react-slower maximum speeds, and higher responsiveness in venting performance. Additionally, fish subjected to PE showed slower average speeds and a ~ 25% decrease in distance travelled. EFC contact, on the other hand, only had an adverse impact on the highest speed at an escape. The exposure to biopolymers had no effect on the duration of recovery, rate of metabolism, or directionality or average speed in escape effectiveness. Therefore, biodegradable polymers might prove advantageous for marine

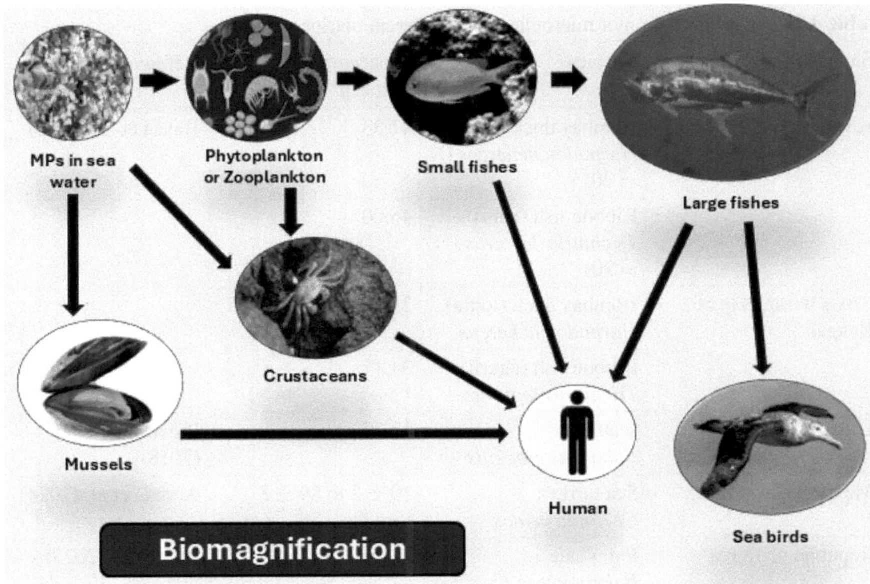

Fig. 4.3 Biomagnification

organisms owing to the lesser effect found when compared with petroleum-derived polymers in the present investigation (Hawke et al., 2024). By analysing individual, histopathological, and gastrointestinal variables, the effects of exposure to different MPs on the emergence and maturation of marine medaka larvae (*Oryzias melastigma*) were investigated (Wen et al., 2024). The results showed that although pathological harm to the tissues of the intestinal tract was noted, the composition of the microbiota in the GI tract did not significantly change while marine medaka larvae's body weight and size were unaffected by long-term contact with PE and PLA at a level of 200 μg/L. From physical and cellular points of view, (Isa et al., 2024) studied the cumulative effects of negative environmental conditions on *Pocillopora damicornis*. During a 14-day stress experiment, colonies were treated with three concentrations of PE MP beads, namely no MP beads (No MP), 1 mg/L (Low MP), and 10 mg/L (High MP), in combination with two separate temperatures, 25 °C and 30 °C, for 72 h. No stress signs, such as abnormal mucus production and polyp everted in corals were observed. Golden pompano (*Trachinotus blochii*), a marine aquaculture variety of fish of high profitable value, were exposed to 10 μg/L, 100 μg/L, and 1000 μg/L of 5 μm spherical PS MPs (Yao et al., 2024). According to the histopathological result, intestinal tract villi contracted. 16 s sequencing showed high impacts of MPs on gastrointestinal microbiota abundance and community structure that are likely to contribute to metabolic activities along the GI tracts. As stated by Yao et al., (2024) metabolomics sequencing on the intestinal content depicted that micro-plastics interfered with the colonic phase of biochemical reactions by causing disturbance on regular digestion and absorptions through an intestinal route disturbing its process concerning lipids, glucose, and amino acids.

4.5 Ecosystem-Level Impacts of Microplastics

4.5.1 Disruption of Food Webs

4.5.1.1 Marine Food Webs and Their Significance in Maintaining Ecosystem Stability

Food webs are the pathways of nutrient and energy transportation from microorganisms to predators. Marine food webs provide sufficient nutrient and energy consumption from phytoplankton, bacteria, and small shrimp-like organisms in the water to sponges and corals on the seabed, as well as fish, squid, whales, seals, and seabirds.

Primary producers, primary consumers, secondary and tertiary consumers, and detritivores and decomposers are the main key components of food webs. Each type of trophic level helps to balance ecosystem functions. Food webs also play a role in preserving the stability of marine ecosystems and adapting to environmental changes (Au et al., 2017; Ly, 2023). MPs are composed of different polymers, which affect the marine ecology functions. Polyurethane Foam (PUF) and Polylactic Acid (PLA) promote nitrification by increasing NO_3^- and NO^- abundance in sediments while decreasing NH_4^+ concentrations. Polyvinyl chloride (PVC) promotes NH_4^+ activation of remineralization. However, nitrification does not proceed immediately due to the abundance of NH_4^+ (Seeley et al., 2020). Tiny PS nanoparticles (less than 0.5 μm) get adverse effects on primary productivity by disrupting phytoplankton growth and function in marine ecosystems (Sjollema et al., 2016; Troost et al., 2018).

Biological pumps and microbial carbon pumps are the main sources of sequestered CO_2 in marine ecosystems. Phytoplankton use the ocean to manufacture particulate organic carbon (POC) through photosynthesis. Zooplankton feed phytoplankton and use their fecal pallets to transport carbon to ocean sediment. Fecal pallets are polluted with MP, decreasing the downward flow of sequestered carbon to the seafloor. As a result, reduces the amount of carbon sequestered. Reducing the intake of particulate organic carbon and contaminating the fecal pallet with MP substantially reduces carbon sequestration (Shen et al., 2020). Corals have a preference for ingesting MP, which can result in an excessive production of mucus due to a mutation in gene expression. This may lead to damage to coral's tissue, causing abrasions. Moreover, wound tissues are infected by opportunistic bacteria. Ultimately, coral reefs become dead and reduce the suspension feeder community that depends on corals (Corinaldesi et al., 2021). Baleen whales are more sensitive to MP pollution, and various types of plastic particles accumulate in baleen filters and oropharynx. MP pollution poses a high risk for whales due to disruption of feeding mechanisms and increasing filter porosity, diet, habitat and geographic distribution, and foraging ecology and

Table 4.2 Impacts of MPs on marine ecosystem functions

Ecosystem function	Role in marine ecosystems	Impact of microplastics	Case studies
Nutrient cycling	Recycling of essential nutrients (e.g., nitrogen, phosphorus)	Nitrification wasn't affected by MP while PUF and PLA increased NO_3^- and NO_3^- abundance	Green et al. (2016), Seeley et al. (2020)
Primary productivity	Production of organic compounds by photosynthetic organisms (e.g., phytoplankton)	High level of MP (25 mg/l) reduce algal growth. Southern part of the North Sea didn't reduce algal bio mass because it wasn't higher than 25 mg/l. Charged nano PS particles (0.2 μm) inhibit photosynthesis	Sjollema et al. (2016), Troost et al. (2018)
Carbon sequestration	Long-term storage of carbon in oceanic sinks	Reduction of sequestrated carbon in the ocean	Shen et al. (2020)
Habitat formation	Creation of habitats by organisms (e.g., coral reefs, seagrass beds)	Coral death and mass mortality of suspension feeders might happen	Corinaldesi et al. (2021)
Water filtration	Purification of water by filter-feeding organisms	Clogging and damaging of feeding apparatus, reduced filtration efficiency, and reduced feeding amount	Werth et al. (2024)

behavior (Werth et al., 2024). Table 4.2 demonstrates how different MP compounds disrupt marine ecosystem functions.

4.5.1.2 Trophic Transfer of Microplastics

MPs create biofilms in marine ecosystems and interact with marine life. Biofilms modify fish and seabird foraging behaviors by mimicking chemoreceptors used to find food. As a result of the degradation of MPs, dimethyl sulfide (DMS) is released into the marine ecosystem. Foraging shorebirds become attracted to dimethyl sulfide (DMS) and ingest more MP, altering the marine food chain (Carbery et al., 2018). Plastic fibers as well as larger plastics have been entangled in marine organisms. Studies found that swimming or feeding appendages of invertebrates and valve gapes of bivalves have been entangled with numerous plastic particles (Setälä et al., 2018). Even animals that were located in deep seas suffered from this issue (sea pens and hermit crabs). This may lead to entry into the marine food web by ingestion of MPs from animals at higher trophic levels (Taylor et al., 2016).

4.5.1.3 Microplastic Pollution and Its Effects on Non-Target Aquatic Life.

Zebrafish have been used as a model organism to demonstrate the negative impact of MPs on aquatic life (Bhagat et al., 2020; Mak et al., 2019). High concentration of PS-MP (20–200 pg/l) increased the formation of reactive oxygen species (ROS) due to reduced glutathione (GSH) and superoxide dismutase (SOD) activity, as well as inhibition of SOD1, SOD2, and nuclear factor erythroid-derived 2 (NFe212), all of which are associated with oxidative stress. Juvenile *Eriocheir sinesis* impaired glutathione (GSH), superoxide dismutase (SOD), Glutathione Peroxidase (GP), Glutathione S-Transferase (GST) and Catalase (CAT) activities after sub-chronic extended treatment with 40 to 40,000 pg/L concentrations of PS microspheres and raised hepatic and pancreatic malondialdehyde (MDA) levels. Arthropods have a protective chitinous covering that prevents particles larger than 300 nm in diameter from entering their bodies, which can be damaging to them (Pothiraj et al., 2023).

4.5.2 Effects on Biodiversity

Marine biodiversity consists of species, genetic, and ecosystem diversity. Species diversity ensures ecosystem stability, genetic diversity allows for adaptation, and ecosystem diversity supports overall ecosystem health. Biodiversity is essential for maintaining ecosystem resilience, functionality, and productivity (Worm et al., 2006). It improves ecological processes like nutrient cycling, primary production, habitat building, fisheries, coastal protection, and carbon sequestration (Cardinale et al., 2012; Hilborn et al., 2020).

4.5.2.1 Microplastics and Habitat Degradation

MPs are a developing thread in the marine aquatic system. Lamb et al., 2018 predicted that 11 billion pieces of plastic trash are detrimental to coral reefs in the Pacific Ocean alone, with an expected 40% increase by 2025. Plastic waste increases pathogen microbial colonization and promotes disease propagation. Corals exposed to plastic have a higher risk of disease, ranging from 4 to 89%. Plastic has an eight times higher impact on structurally complex corals. Case studies of habitat loss linked to MP pollution (Lamb et al., 2018). MPs mislead corals' feeding habitat by acting as prey. It may lead to enhanced consumption of MPs up to approximately 50 μg plastic $cm^2\ h^{-1}$. MPs were found on mesenteric tissue within the coral gut cavity, affecting badly coral health. In the presence of MPs, corals intend to secrete more mucus and need to spend more energy (Hall et al., 2015).

4.5.2.2 Impact on Keystone Species

Keystone species contribute to maintaining the structure and health of the marine ecosystem, directly affecting species abundance, biodiversity, food webs, and ecosystem functions. Sea otters, oysters, and sharks play a significant role in the sea as keystone species. Sea otters make a significant contribution to controlling sea urchin populations and maintaining kelp forests. Bioaccumulation of MPs toxic to sea otters via consuming contaminated prey leads to decreased population health and, consequently, kelp forest collapse and reduced biodiversity (Nopp-Mayr et al., 2024). Oysters are categorized as filter-feeding organisms that involve water filtration, creating habitat in marine aquatic systems. MPs that can disrupt the reproductive cycle and growth reduce water quality and biodiversity in marine ecosystems (Sussarellu et al., 2016). Pelagic predatory fish and sharks are apex predators that help to control prey populations in marine aquatic ecosystems. Accumulation of MPs in prey causes internal organ damage and toxicity, resulting in trophic cascade failure (Paul et al., 2024; Wang, 2023). MPs carry toxic chemicals that contribute adverse effects on marine ecosystems. MPs are uncharged particles. Organic molecules and contaminants bind to form an "Ecocorona", which increases the density and surface charge of the particles. This may lead to acting as carrier molecules that deplete feeding and energy stores, fecundity, and growth (Galloway et al., 2017).

4.5.3 Alteration of Ecosystem Functions

Persistent MP pollution impacts irreversible changes in ecosystem structure and function, indicating ecosystem regime shifts resulting in disruption of food webs, biodiversity loss, and reduced nutrient cycling (Cole et al., 2011; Hale et al., 2020). MPs can accumulate in the digestive tracts of marine organisms. It may lead to issues such as gut blockages and starvation due to a false sense of satiation (Paul et al., 2024). Seabirds such as the Northern Gannet and Great Shearwater were investigated to suffer from reduction of body weight and increased mortality due to plastic ingestion (Roy et al., 2022). Furthermore, the condition known as "plasticosis" has been identified, particularly affecting seabirds, where plastic consumption leads to significant gastrointestinal harm, including fibrosis and inflammation, thereby disrupting nutrient absorption and overall health (Bhardwaj et al., 2024). In addition to physical blockages, MPs can also induce reproductive toxicity. The ingestion of MPs can reduce the reproduction of certain organisms, such as the Japanese Quail and Flesh-footed Shearwater. Moreover, swimming behavior and prey avoidance in marine species were also reported due to the presence of plastic debris, and it can disrupt predator–prey dynamics within ecosystems (Bhardwaj et al., 2024; Roy et al., 2022). Those reductions in ecosystem functions provide significant consequences for the loss of ecosystem services, which are crucial for human communities, particularly those reliant on the ocean. Fish stocks for food, coastal protection, carbon sequestration, and water purification are significantly influenced by ecosystem functions, which may reduce marine resources for food income and protection, especially in coastal regions where livelihoods depend on fishing and tourism (Beaumont et al., 2019).

4.6 Global Hotspots of MPs Pollution in Marine Ecosystems

4.6.1 Identification of Marine MP Hotspots

The prevalence of MPs in the marine ecosystem underscores the necessity of addressing this rising global issue and the imperative of mitigating measures (Amelia et al., 2021). The widespread pollution of MP reached untouched marine environments such as deep waters in addition to coastal areas (Munhoz et al., 2022). The accumulation and dispersion of these contaminants can be attributed to several different causes of pollution. The construction of dense garbage patches is facilitated by several factors, including the development of demanding aquaculture sectors, oceanic currents and gyres, an increase in population, and maritime accidents (for example, the express pearl catastrophe that occurred in the coastal area of Sri Lanka). This significant issue has been spotted in specific hotspots around the world oceans (Bhuyan, 2022). The sizes (micro to nanoscale) and the dimensions (irregular) of plastic changed due to the weathering conditions, and physical and chemical degradations while posing the highest threats (Bonfanti et al., 2021; van Sebille et al., 2015).

Hotspots form alongside populated megacities, major ship routes and industrial zones. A study indicated that the predicted amount of plastic disposal in aquatic environments, including marine ecosystems, will potentially vary from 20 to 53 metric tons over the next six years (Borrelle et al., 2020). Moreover, recent studies (Borrelle et al., 2020; Yu & Singh, 2023) calculated that 66.7% of washed-up plastic sinks to the ocean floor sediments and hide from the surface region. Furthermore, the lifetime of plastics extends from pre-manufacturing to post-consumption, and debris made of plastic enters the ocean owing to deficient waste management (Yu & Singh, 2023).

4.6.2 Major Global Hotspots

The North Pacific gyre is home to the Great Pacific Garbage Patch (GPGP), a major MPs hotspot. The North Pacific Gyre has facilitated the accumulation of MPs and the growth of the GPGP over time, spreading over millions of square kilometers as floating particles. The MPs that make up the GPGP are fragmented into pieces that are about 5 mm in size. The size reduction of larger plastics may occur due to environmental factors and oceanic currents. However, complex processes in the hotspot, such as MP transportation, degradation, and environmental impacts, need to be explored. Additionally, biofouling helps further the distribution of MPs to the sea bottom and vertical transportation (Yu & Singh, 2023). GPGP's ability to accumulate and transport MPs needs to be understood since it could pose threats to other oceanic regions beyond the GPGP (Peng et al., 2020).

As major plastic suppliers in Southeast Asia, China, Indonesia, and the Philippines have increased MPs pollution in coastal areas (Bhuyan, 2022). The Bohai Sea and Yellow Sea, recognized as significant regions in China, indicate high concentrations of polymers, including PP, PE, and PS. In view of Bohai Sea, 46% of the fragmented polymers have primarily been associated with marine fishing activities and subsequently recovered. When considering the Yellow Sea, coastal currents are responsible for dispersing MPs close to the urban cities, thereby increasing the distribution (Bhuyan, 2022; Wang et al., 2019).

The high population density in South Asia has positioned the Indian Ocean as a significant hotspot. For instance, the coastlines of Asia are the greatest susceptible zone for heavy input of plastic into the marine environments than other coastal zones (Nakakuni et al., 2024). The maritime environment along the Bay of Bengal coast is turning into a reservoir for MPs. The Bay of Bengal as MP is impacted by air pollution, sewerage, fishing, and population increase. The Bay of Bengal has significantly greater concentrations of MPs especially within the Northeast region of it. The consequences of these effects could facilitate the movement of MPs within trophic levels, posing a significant threat to marine ecosystems (Mubin et al., 2023).

Even though it is only partially contained, the Mediterranean region as well the greatest concentrated MPs zone in the world when comparing the other hotspots. This is because of the restricted water circulation, high population density, and substantial tourism that occurs there. The coastal region of Lebanon, namely the Levantine Basin, is where 50% of MPs found in the marine water samples were PE and fibers. On the other side of the Mediterranean Sea, the prevalence of MP in Mersin Bay, Turkey, has been documented, highlighting significant discharge from wastewater treatment plants (Alomar et al., 2016; Amelia et al., 2021).

The extensive global impact of plastic contamination is underscored by the detection of MPs in the Arctic and Antarctic, which are transported by oceanic currents and atmospheric deposition. In the Antarctic, several studies investigated the occurrence of MP in sea ice while indicating reservoir for MP accumulation. The Arctic and East Greenland are distant destinations from major sources of MPs (Kelly et al., 2020). Nevertheless, MPs availability in this region has been confirmed by the discovery of MP ingestions in a variety of Arctic marine predators, including sea birds. The presence of MPs in East Greenland and the Arctic ice regions emphasizes their dispersion and transportation ability from southern parts of the world, the Eastern Arctic, or the Northern Pacific with the help of oceanic currents and wind (Baak et al., 2020). Additionally, areas in the North Atlantic and Pacific Oceans, the Barents Sea, and the Bering Sea are assessed as likely sources of plastic pollution in the Arctic Furthermore, the coastlines of North America and Europe, along with major shipping routes, have significant plastic-aggregated areas due to the impact of the North Atlantic gyre

(Kelly et al., 2020). The principal MP hotspots discussed have been threatened for decades, emphasizing the urgency of raising awareness and management of oceanic plastic pollution.

4.6.3 *Raising Awareness and Management Efforts*

The gyres, which are marine MP concentration zones in the open ocean, represent an accumulation of plastics from various geographic venues, their origin remains unknown. This indicates property differences of plastic waste in shoreline areas and oceanic areas. Thus, leveraging the features of partially closed regions including the Seto Inland Sea may provide valuable demonstrations that will help address marine plastic pollution challenges in coastal areas (Nakakuni et al., 2024). The need for regional approaches to detect and mitigate marine debris is urgent due to variations and abundance of MP content across countries. Future studies should focus on the mechanisms that facilitate convergence zone formation to enhance conservation (Desforges et al., 2015; Nakakuni et al., 2024).

Regional organizations' initiatives, such as the United Nations Environmental Program (UNEP), the National Oceanic and Atmospheric Administration (NOAA), and localized efforts including MPs dispersion monitoring, satellite data, oceanographic models, and concentration monitoring, are essential for mitigating impacts (Nielsen et al., 2023). Policy reforms, laws for effective oceanic protection, and ongoing public awareness are crucial for minimizing plastic influx into marine ecosystems and enhancing the sustainable use of this delicate environment (Chen et al., 2023).

References

Abeysinghe, K. M. S. N., Guruge, K. P. G. K. P., Bandara, T., & Kumara, P. B. T. P. (2024). Microplastic pollution status in the coral reef ecosystems on the Southern and Western coasts of Sri Lanka during the Southwest monsoon. *Marine Pollution Bulletin, 206*, Article 116713. https://doi.org/10.1016/j.marpolbul.2024.116713

Acarer Arat, S. (2024). An overview of microplastic in marine waters: Sources, abundance, characteristics and negative effects on various marine organisms. *Desalination and Water Treatment, 317*, Article 100138. https://doi.org/10.1016/j.dwt.2024.100138

Acharya, S., Rumi, S. S., Hu, Y., & Abidi, N. (2021). Microfibers from synthetic textiles as a major source of microplastics in the environment: A review. *Textile Research Journal, 91*(17–18), 2136–2156.

Alomar, C., Estarellas, F., & Deudero, S. (2016). Microplastics in the Mediterranean Sea: Deposition in coastal shallow sediments, spatial variation and preferential grain size. *Marine Environmental Research, 115*, 1–10. https://doi.org/10.1016/j.marenvres.2016.01.005

Amelia, T. S. M., Khalik, W. M. A. W. M., Ong, M. C., Shao, Y. T., Pan, H. J., & Bhubalan, K. (2021). Marine microplastics as vectors of major ocean pollutants and its hazards to the marine ecosystem and humans. *Progress in Earth and Planetary Science, 8*(1), 12.

Aranda, D. A., Sindou, P., Cauich Rodriguez, J. V., Saldaña, G. M., Coronado, R. F. V., González, W. D. N., Díaz, M. E., & Escalante, V. C. (2024). A non-invasive method of microplastics pollution quantification in green sea turtle Chelonia mydas of the Mexican Caribbean. *Marine Pollution Bulletin, 200*, Article 116092. https://doi.org/10.1016/j.marpolbul.2024.116092

Au, S. Y., Lee, C. M., Weinstein, J. E., van den Hurk, P., & Klaine, S. J. (2017). Trophic transfer of microplastics in aquatic ecosystems: Identifying critical research needs. *Integrated Environmental Assessment and Management, 13*(3), 505–509. https://doi.org/10.1002/ieam.1907

Auta, H. S., Emenike, C. U., & Fauziah, S. H. (2017). Distribution and importance of microplastics in the marine environment: A review of the sources, fate, effects, and potential solutions. *Environment International, 102*, 165–176. https://doi.org/10.1016/j.envint.2017.02.013

Baak, J. E., Linnebjerg, J. F., Barry, T., Gavrilo, M. V., Mallory, M. L., Price, C., & Provencher, J. F. (2020). Plastic ingestion by seabirds in the circumpolar Arctic: A review. *Environmental Reviews, 28*(4), 506–516. https://doi.org/10.1139/er-2020-0029

Bashir, S. M., Kimiko, S., Mak, C.-W., Fang, J.K.-H., & Gonçalves, D. (2021). Personal care and cosmetic products as a potential source of environmental contamination by microplastics in a densely populated Asian city. *Frontiers in Marine Science, 8*, Article 683482.

Beaumont, N. J., Aanesen, M., Austen, M. C., Börger, T., Clark, J. R., Cole, M., Hooper, T., Lindeque, P. K., Pascoe, C., & Wyles, K. J. (2019). Global ecological, social and economic impacts of marine plastic. *Marine Pollution Bulletin, 142*, 189–195. https://doi.org/10.1016/j.marpolbul.2019.03.022

Bell, J. D., Watson, R. A., & Ye, Y. (2017). Global fishing capacity and fishing effort from 1950 to 2012. *Fish and Fisheries, 18*(3), 489–505.

Bhagat, J., Zang, L., Nishimura, N., & Shimada, Y. (2020). Zebrafish: An emerging model to study microplastic and nanoplastic toxicity. *Science of the Total Environment, 728*, Article 138707. https://doi.org/10.1016/j.scitotenv.2020.138707

Bhardwaj, L. K., Rath, P., Yadav, P., & Gupta, U. (2024). Microplastic contamination, an emerging threat to the freshwater environment: A systematic review. *Environmental Systems Research, 13*(1), 8.

Bhuyan, M. S. (2022). Effects of microplastics on fish and in human health [Review]. *Frontiers in Environmental Science, 10*. https://doi.org/10.3389/fenvs.2022.827289

Bian, W., Zeng, Y., Li, Y., Na, G., Mu, J., Lv, S., & Liu, M. (2024). Microplastic pollution in tropical coral reef ecosystems from the coastal South China Sea and their impacts on corals in situ. *Journal of Hazardous Materials, 480*, Article 135898. https://doi.org/10.1016/j.jhazmat.2024.135898

Bonfanti, P., Colombo, A., Saibene, M., Motta, G., Saliu, F., Catelani, T., Mehn, D., La Spina, R., Ponti, J., Cella, C., Floris, P., & Mantecca, P. (2021). Microplastics from miscellaneous plastic wastes: Physico-chemical characterization and impact on fish and amphibian development. *Ecotoxicology and Environmental Safety, 225*, Article 112775. https://doi.org/10.1016/j.ecoenv.2021.112775

Borrelle, S. B., Ringma, J., Law, K. L., Monnahan, C. C., Lebreton, L., McGivern, A., Murphy, E., Jambeck, J., Leonard, G. H., Hilleary, M. A., Eriksen, M., Possingham, H. P., De Frond, H., Gerber, L. R., Polidoro, B., Tahir, A., Bernard, M., Mallos, N., Barnes, M., & Rochman, C. M. (2020). Predicted growth in plastic waste exceeds efforts to mitigate plastic pollution. *Science, 369*(6510), 1515–1518. https://doi.org/10.1126/science.aba3656

Carbery, M., O'Connor, W., & Palanisami, T. (2018). Trophic transfer of microplastics and mixed contaminants in the marine food web and implications for human health. *Environment International, 115*, 400–409. https://doi.org/10.1016/j.envint.2018.03.007

Cardinale, B. J., Duffy, J. E., Gonzalez, A., Hooper, D. U., Perrings, C., Venail, P., Narwani, A., Mace, G. M., Tilman, D., Wardle, D. A., Kinzig, A. P., Daily, G. C., Loreau, M., Grace, J. B., Larigauderie, A., Srivastava, D. S., & Naeem, S. (2012). Biodiversity loss and its impact on humanity. *Nature, 486*(7401), 59–67. https://doi.org/10.1038/nature11148

Chen, B., Zhang, Z., Wang, T., Hu, H., Qin, G., Lu, T., Hong, W., Hu, J., Penuelas, J., & Qian, H. (2023). Global distribution of marine microplastics and potential for biodegradation. *Journal of Hazardous Materials, 451*, Article 131198. https://doi.org/10.1016/j.jhazmat.2023.131198

Chen, R., Zhao, X., Wu, X., Wang, J., Wang, X., & Liang, W. (2024). Research progress on occurrence characteristics and source analysis of microfibers in the marine environment. *Marine Pollution Bulletin, 198*, Article 115834.

Chubarenko, I., Bagaev, A., Zobkov, M., & Esiukova, E. (2016). On some physical and dynamical properties of microplastic particles in marine environment. *Marine Pollution Bulletin, 108*(1–2), 105–112.

Chubarenko, I., Esiukova, E., Bagaev, A., Isachenko, I., Demchenko, N., Zobkov, M., Efimova, I., Bagaeva, M., & Khatmullina, L. (2018). Behavior of microplastics in coastal zones. In *Microplastic contamination in aquatic environments* (pp. 175–223). Elsevier.

Cole, M., Lindeque, P., Halsband, C., & Galloway, T. S. (2011). Microplastics as contaminants in the marine environment: A review. *Marine Pollution Bulletin, 62*(12), 2588–2597. https://doi.org/10.1016/j.marpolbul.2011.09.025

Corinaldesi, C., Canensi, S., Dell'Anno, A., Tangherlini, M., Di Capua, I., Varrella, S., Willis, T. J., Cerrano, C., & Danovaro, R. (2021). Multiple impacts of microplastics can threaten marine habitat-forming species. *Communications Biology, 4*(1), 431. https://doi.org/10.1038/s42003-021-01961-1

Dahl, M., Bergman, S., Björk, M., Diaz-Almela, E., Granberg, M., Gullström, M., Leiva-Dueñas, C., Magnusson, K., Marco-Méndez, C., Piñeiro-Juncal, N., & Mateo, M. Á. (2021). A temporal record of microplastic pollution in Mediterranean seagrass soils. *Environmental Pollution, 273*, Article 116451. https://doi.org/10.1016/j.envpol.2021.116451

Desforges, J.-P.W., Galbraith, M., & Ross, P. S. (2015). Ingestion of microplastics by zooplankton in the Northeast Pacific Ocean. *Archives of Environmental Contamination and Toxicology, 69*(3), 320–330. https://doi.org/10.1007/s00244-015-0172-5

Dey, U., Chell, S., Mondal, M., Das, K., Raj, D., Pandey, G., Meraj, G., Kumar, P., Almazroui, M., & Verma, S. (2024). Potential threat of microplastic pollution on coastal-marine ecosystem–An emerging economic setback and question to blue economy? *Earth Systems and Environment*, 1–29.

Emberson-Marl, H., Coppock, R. L., Cole, M., Godley, B. J., Mimpriss, N., Nelms, S. E., & Lindeque, P. K. (2023). Microplastics in the Arctic: a transect through the Barents Sea [Original Research]. 10. https://doi.org/10.3389/fmars.2023.1241829

Fernández, B., Vidal-Liñán, L., Bellas, J., Campillo, J. A., Chaves-Pozo, E., & Albentosa, M. (2024). The particle effect: Comparative toxicity of chlorpyrifos in combination with microplastics and phytoplankton particles in mussel. *Aquatic Toxicology, 275*, Article 107053. https://doi.org/10.1016/j.aquatox.2024.107053

Folbert, M. E., Corbin, C., & Löhr, A. J. (2022). Sources and leakages of microplastics in cruise ship wastewater. *Frontiers in Marine Science, 9*, Article 900047.

Galloway, T. S., Cole, M., & Lewis, C. (2017). Interactions of microplastic debris throughout the marine ecosystem. *Nature Ecology & Evolution, 1*(5), 0116. https://doi.org/10.1038/s41559-017-0116

Ghosh, G. C., Akter, S. M., Islam, R. M., Habib, A., Chakraborty, T. K., Zaman, S., Kabir, A. H. M. E., Shipin, O. V., & Wahid, M. A. (2021). Microplastics contamination in commercial marine fish from the Bay of Bengal. *Regional Studies in Marine Science, 44*, Article 101728. https://doi.org/10.1016/j.rsma.2021.101728

Gilman, E., Antonelis, K., Drinkwin, J., Gorgin, S., Suuronen, P., Thomas, S. N., & Wilson, J. (2023). Introduction to the Marine Policy special issue on abandoned, lost and discarded fishing gear: Causes, magnitude, impacts, mitigation methods and priorities for monitoring and evidence-informed management. *Marine Policy, 155*, Article 105738.

Green, D. S., Boots, B., Sigwart, J., Jiang, S., & Rocha, C. (2016). Effects of conventional and biodegradable microplastics on a marine ecosystem engineer (Arenicola marina) and sediment nutrient cycling. *Environmental Pollution, 208*, 426–434. https://doi.org/10.1016/j.envpol.2015.10.010

Guzzetti, E., Sureda, A., Tejada, S., & Faggio, C. (2018). Microplastic in marine organism: Environmental and toxicological effects. *Environmental Toxicology and Pharmacology, 64*, 164–171. https://doi.org/10.1016/j.etap.2018.10.009

Hale, R. C., Seeley, M. E., La Guardia, M. J., Mai, L., & Zeng, E. Y. (2020). A global perspective on microplastics. *Journal of Geophysical Research-Oceans, 125*(1). https://doi.org/10.1029/2018JC014719

Hall, N. M., Berry, K. L. E., Rintoul, L., & Hoogenboom, M. O. (2015). Microplastic ingestion by scleractinian corals. *Marine Biology, 162*(3), 725–732. https://doi.org/10.1007/s00227-015-2619-7

Hamman, M., van Schyff, V., Choong Kwet Yive, R. N. S., Iordachescu, L., Simon-Sánchez, L., & Bouwman, H. (2024). Microplastics in coral from three Mascarene islands, Western Indian Ocean. *Marine Pollution Bulletin, 208*, 116951. https://doi.org/10.1016/j.marpolbul.2024.116951

Hansani, K. U. D. N., Thilakarathne, E. P. D. N., Koongolla, J. B., Gunathilaka, W. G. I. T., Perera, B. G. D. O., Weerasingha, W. M. P. U., & Egodauyana, K. P. U. T. (2023). Contamination of microplastics in tropical coral reef ecosystems of Sri Lanka. *Marine Pollution Bulletin, 194*, Article 115299. https://doi.org/10.1016/j.marpolbul.2023.115299

Hasan, J., Islam, S. M. M., Alam, M. S., Johnson, D., Belton, B., Hossain, M. A. R., & Shahjahan, M. (2022). Presence of microplastics in two common dried marine fish species from Bangladesh. *Marine Pollution Bulletin, 176*, Article 113430. https://doi.org/10.1016/j.marpolbul.2022.113430

Hawke, A. M., Trujillo, J. E., Oey, I., Giteru, S. G., & Allan, B. J. M. (2024). Exposure to petroleum-derived and biopolymer microplastics affect fast start escape performance and aerobic metabolism in a marine fish. *Science of the Total Environment, 906*, Article 167423. https://doi.org/10.1016/j.scitotenv.2023.167423

Hilborn, R., Amoroso, R. O., Anderson, C. M., Baum, J. K., Branch, T. A., Costello, C., de Moor, C. L., Faraj, A., Hively, D., Jensen, O. P., Kurota, H., Little, L. R., Mace, P., McClanahan, T., Melnychuk, M. C., Minto, C., Osio, G. C., Parma, A. M., Pons, M., … Ye, Y. (2020). Effective fisheries management instrumental in improving fish stock status. *Proceedings of the National Academy of Sciences, 117*(4), 2218–2224. https://doi.org/10.1073/pnas.1909726116

Huang, L., Zhang, S., Li, L., Zhang, S., Wang, J., Liu, X., & Zhang, W. (2023). Research progress on microplastics pollution in polar oceans. *Polar Science, 36*, Article 100946. https://doi.org/10.1016/j.polar.2023.100946

Isa, V., Seveso, D., Diamante, L., Montalbetti, E., Montano, S., Gobbato, J., Lavorano, S., Galli, P., & Louis, Y. D. (2024). Physical and cellular impact of environmentally relevant microplastic exposure on thermally challenged Pocillopora damicornis (Cnidaria, Scleractinia). *Science of the Total Environment, 918*, Article 170651. https://doi.org/10.1016/j.scitotenv.2024.170651

Jovanović, B. (2017). Ingestion of microplastics by fish and its potential consequences from a physical perspective. *13*(3), 510–515.

Kaiser, D., Kowalski, N., & Waniek, J. J. (2017). Effects of biofouling on the sinking behavior of microplastics. *Environmental Research Letters, 12*(12), Article 124003.

Kelly, A., Lannuzel, D., Rodemann, T., Meiners, K. M., & Auman, H. J. (2020). Microplastic contamination in east Antarctic sea ice. *Marine Pollution Bulletin, 154*, Article 111130. https://doi.org/10.1016/j.marpolbul.2020.111130

Lamb, J. B., Willis, B. L., Fiorenza, E. A., Couch, C. S., Howard, R., Rader, D. N., True, J. D., Kelly, L. A., Ahmad, A., Jompa, J., & Harvell, C. D. (2018). Plastic waste associated with disease on coral reefs. *Science, 359*(6374), 460–462. https://doi.org/10.1126/science.aar3320

Lasee, S., Mauricio, J., Thompson, W. A., Karnjanapiboonwong, A., Kasumba, J., Subbiah, S., Morse, A. N., & Anderson, T. A. (2017). Microplastics in a freshwater environment receiving treated wastewater effluent. *Integrated Environmental Assessment and Management, 13*(3), 528–532.

Lebreton, L. C., Van Der Zwet, J., Damsteeg, J.-W., Slat, B., Andrady, A., & Reisser, J. (2017). River plastic emissions to the world's oceans. *Nature Communications, 8*(1), 15611.

Lei, K., Qiao, F., Liu, Q., Wei, Z., Qi, H., Cui, S., Yue, X., Deng, Y., & An, L. (2017). Microplastics releasing from personal care and cosmetic products in China. *Marine Pollution Bulletin, 123*(1–2), 122–126.

Lei, X., Cheng, H., Luo, Y., Zhang, Y., Jiang, L., Sun, Y., Zhou, G., & Huang, H. (2021). Abundance and characteristics of microplastics in seawater and corals from reef region of Sanya Bay, China [Original Research]. 8. https://doi.org/10.3389/fmars.2021.728745

Li, C., Zhu, L., Li, W.-T., & Li, D. (2023). Microplastics in the seagrass ecosystems: A critical review. *Science of the Total Environment, 902*, Article 166152. https://doi.org/10.1016/j.scitotenv.2023.166152

Li, Y., Sun, Y., Li, J., Tang, R., Miu, Y., & Ma, X. (2021). Research on the influence of microplastics on marine life. *IOP Conference Series: Earth and Environmental Science.*

Lim, K. P., Sun, C., Yusoff, S., Ding, J., Loh, K. H., Li, J., & Lim, P. E. (2024). Abundance of microplastics and its ecological risk assessment in coral reef regions of Peninsular Malaysia. *Marine Pollution Bulletin, 209*, Article 117112. https://doi.org/10.1016/j.marpolbul.2024.117112

Lloret, J., Pedrosa-Pamies, R., Vandal, N., Rorty, R., Ritchie, M., McGuire, C., Chenoweth, K., & Valiela, I. (2021). Salt marsh sediments act as sinks for microplastics and reveal effects of current and historical land use changes. *Environmental Advances, 4*, Article 100060. https://doi.org/10.1016/j.envadv.2021.100060

Ly, C. (2023). Exploring the structure of marine food webs: Identifying the complexity. *Journal of Oceanography and Marine Research, 11*, 277.

Mak, C. W., Yeung, K.C.-F., & Chan, K. M. (2019). Acute toxic effects of polyethylene microplastic on adult zebrafish. *Ecotoxicology and Environmental Safety, 182*, Article 109442.

Mao, Y., Ai, H., Chen, Y., Zhang, Z., Zeng, P., Kang, L., Li, W., Gu, W., He, Q., & Li, H. (2018). Phytoplankton response to polystyrene microplastics: Perspective from an entire growth period. *Chemosphere, 208*, 59–68. https://doi.org/10.1016/j.chemosphere.2018.05.170

Martinez, M., Minetti, R., La Marca, E. C., Montalto, V., Rinaldi, A., Costa, E., Badalamenti, F., Garaventa, F., Mirto, S., & Ape, F. (2024). The power of Posidonia oceanica meadows to retain microplastics and the consequences on associated macrofaunal benthic communities. *Environmental Pollution, 348*, Article 123814. https://doi.org/10.1016/j.envpol.2024.123814

Mendes, D. S., Silva, D. N. N., Silva, M. G., Beasley, C. R., & Fernandes, M. E. B. (2024). Microplastic distribution and risk assessment in estuarine systems influenced by traditional villages and artisanal fishery activities. *Scientific Reports, 14*(1), 29044. https://doi.org/10.1038/s41598-024-80468-1

Miller, M. E., Hamann, M., & Kroon, F. J. (2020). Bioaccumulation and biomagnification of microplastics in marine organisms: A review and meta-analysis of current data. *PLoS ONE, 15*(10), Article e0240792. https://doi.org/10.1371/journal.pone.0240792

Mubin, A.-N., Arefin, S., Mia, M. S., Islam, A. R. M. T., Bari, A. B. M. M., Islam, M. S., Ali, M. M., Siddique, M. A. B., Rahman, M. S., Senapathi, V., Idris, A. M., & Malafaia, G. (2023). Managing the invisible threat of microplastics in marine ecosystems: Lessons from coast of the Bay of Bengal. *Science of the Total Environment, 889*, Article 164224. https://doi.org/10.1016/j.scitotenv.2023.164224

Munhoz, D. R., Harkes, P., Beriot, N., Larreta, J., & Basurko, O. C. (2022). Microplastics: A review of policies and responses. *Microplastics, 2*(1), 1–26.

Nakakuni, M., Nishida, M., Nishibata, R., Kishimoto, K., Yamaguchi, H., Ichimi, K., Ishizuka, M., Suenaga, Y., & Tada, K. (2024). Convergence zones of coastal waters as hotspots for floating microplastic accumulation. *Marine Pollution Bulletin, 206*, Article 116691. https://doi.org/10.1016/j.marpolbul.2024.116691

Nanthini devi, K., Raju, P., Santhanam, P., & Perumal, P. (2022). Impacts of microplastics on marine organisms: Present perspectives and the way forward. *Egyptian Journal of Aquatic Research, 48*(3), 205–209. https://doi.org/10.1016/j.ejar.2022.03.001

Nielsen, M. B., Clausen, L. P. W., Cronin, R., Hansen, S. F., Oturai, N. G., & Syberg, K. (2023). Unfolding the science behind policy initiatives targeting plastic pollution. *Microplastics and Nanoplastics, 3*(1), 3. https://doi.org/10.1186/s43591-022-00046-y

Nopp-Mayr, U., Layendecker, S., Sittenthaler, M., Philipp, M., Kägi, R., & Weinberger, I. (2024). Microplastic loads in Eurasian otter (Lutra lutra) feces—Targeting a standardized protocol and first results from an alpine stream, the River Inn. *Environmental Monitoring and Assessment, 196*(8), 707. https://doi.org/10.1007/s10661-024-12791-z

Park, B., Kim, S.-K., Joo, S., Kim, J.-S., Jo, K., Song, N.-S., Im, J., Lee, H.-J., Kim, S. W., Lee, S. B., Kim, S., Lee, Y., Kim, B.-Y., & Kim, T. W. (2023). Microplastics in large marine animals stranded in the Republic of Korea. *Marine Pollution Bulletin, 189*, Article 114734. https://doi.org/10.1016/j.marpolbul.2023.114734

Park, J.-S., Yoo, J.-W., Lee, Y.-H., Park, C., & Lee, Y.-M. (2024a). Size- and shape-dependent ingestion and acute toxicity of fragmented and spherical microplastics in the absence and presence of prey on two marine zooplankton. *Marine Pollution Bulletin, 206*, Article 116768. https://doi.org/10.1016/j.marpolbul.2024.116768

Park, K., Kim, J.-H., Kim, W.-S., Park, J. W., & Kwak, I.-S. (2024b). Metabolic responses of the marine mussel Mytilus galloprovincialis after exposure to microplastics of different shapes and sizes. *Environmental Pollution, 360*, Article 124660. https://doi.org/10.1016/j.envpol.2024.124660

Paul, S., Nath, S., Bhattacharjee, S., & Mukherjee, S. (2024). Unveiling the effects of microplastics pollution on marine fauna. *Blue Biotechnology, 1*(1), 6.

Peng, G., Bellerby, R., Zhang, F., Sun, X., & Li, D. (2020). The ocean's ultimate trashcan: Hadal trenches as major depositories for plastic pollution. *Water Research, 168*, Article 115121. https://doi.org/10.1016/j.watres.2019.115121

Pinheiro, L. M., Britz, L. M. K., Agostini, V. O., Pérez-Parada, A., García-Rodríguez, F., Galloway, T. S., & Pinho, G. L. L. (2022). Salt marshes as the final watershed fate for meso- and microplastic contamination: A case study from Southern Brazil. *Science of the Total Environment, 838*, Article 156077. https://doi.org/10.1016/j.scitotenv.2022.156077

Pothiraj, C., Amutha Gokul, T., Ramesh Kumar, K., Ramasubramanian, A., Palanichamy, A., Venkatachalam, K., Pastorino, P., Barcelò, D., Balaji, P., & Faggio, C. (2023). Vulnerability of microplastics on marine environment: A review. *Ecological Indicators, 155*, Article 111058. https://doi.org/10.1016/j.ecolind.2023.111058

Provencher, J. F., Vermaire, J. C., Avery-Gomm, S., Braune, B. M., & Mallory, M. L. (2018). Garbage in guano? Microplastic debris found in faecal precursors of seabirds known to ingest plastics. *Science of the Total Environment, 644*, 1477–1484. https://doi.org/10.1016/j.scitotenv.2018.07.101

Raguso, C., Saliu, F., Lasagni, M., Galli, P., Clemenza, M., & Montano, S. (2022). First detection of microplastics in reef-building corals from a Maldivian atoll. *Marine Pollution Bulletin, 180*, Article 113773. https://doi.org/10.1016/j.marpolbul.2022.113773

Rahman, R. R., Baqee, A., Alam, M., Khan, M. W., Muhib, M. I., & Kabir, A. (2024). Organ-specific bioaccumulation of microplastics in market fish of Dhaka and size-dependent impacts of PVC microplastics on growth of Anabus testudineus. *Environmental Pollution, 361*, Article 124807. https://doi.org/10.1016/j.envpol.2024.124807

Ramili, Y., & Umasangaji, H. (2022). An assessment of microplastics (MPs) sedimentary accumulation in seagrass meadows of Mare Island conservation area, North Maluku, Indonesia. *Omni-Akuatika, 18*, 48. https://doi.org/10.20884/1.oa.2022.18.1.921

Richardson, K., Hardesty, B. D., & Wilcox, C. (2019). Estimates of fishing gear loss rates at a global scale: A literature review and meta-analysis. *Fish and Fisheries, 20*(6), 1218–1231.

Rifandi, R. A., & Ratnasari, A. V. (2023). Abundance of microplastics and hazard to the environment in estuary water in Pemalang, Central Java, Indonesia. *IOP Conference Series: Earth and Environmental Science, 1211*(1), Article 012012. https://doi.org/10.1088/1755-1315/1211/1/012012

Roy, P., Mohanty, A. K., & Misra, M. (2022). Microplastics in ecosystems: Their implications and mitigation pathways. *Environmental Science: Advances, 1*(1), 9–29.

Seeley, M. E., Song, B., Passie, R., & Hale, R. C. (2020). Microplastics affect sedimentary microbial communities and nitrogen cycling. *Nature Communications, 11*(1), 2372. https://doi.org/10.1038/s41467-020-16235-3

Setälä, O., Lehtiniemi, M., Coppock, R., & Cole, M. (2018). Chapter 11—Microplastics in marine food webs. In E. Y. Zeng (Ed.), *Microplastic contamination in aquatic environments* (pp. 339–363). Elsevier. https://doi.org/10.1016/B978-0-12-813747-5.00011-4

Sewwandi, M., Keerthanan, S., Perera, K. I., & Vithanage, M. (2023). Plastic nurdles in marine environments due to accidental spillage. *Microplastics in the Ecosphere: Air, Water, Soil, and Food*, 415–432.

Sharma, S., Sharma, V., & Chatterjee, S. (2021). Microplastics in the Mediterranean Sea: Sources, pollution intensity, sea health, and regulatory policies. *Frontiers in Marine Science, 8*, Article 634934.

Shen, M., Ye, S., Zeng, G., Zhang, Y., Xing, L., Tang, W., Wen, X., & Liu, S. (2020). Can microplastics pose a threat to ocean carbon sequestration? *Marine Pollution Bulletin, 150*, Article 110712. https://doi.org/10.1016/j.marpolbul.2019.110712

Singh, A., & Mishra, B. K. (2023). Microplastic in personal care products: An overlooked environmental concern. *Journal of Cleaner Production*, 139082.

Singh, A., & Upadhyay, J. (2025). Physiological and toxicological effects of nano/microplastics on marine birds. In *Global impacts of micro-and nano-plastic pollution* (pp. 257–288). IGI Global.

Sjollema, S. B., Redondo-Hasselerharm, P., Leslie, H. A., Kraak, M. H. S., & Vethaak, A. D. (2016). Do plastic particles affect microalgal photosynthesis and growth? *Aquatic Toxicology, 170*, 259–261. https://doi.org/10.1016/j.aquatox.2015.12.002

Soares, M. O., Rizzo, L., Ximenes Neto, A. R., Barros, Y., Martinelli Filho, J. E., Giarrizzo, T., & Rabelo, E. F. (2023). Do coral reefs act as sinks for microplastics? *Environmental Pollution, 337*, Article 122509. https://doi.org/10.1016/j.envpol.2023.122509

Srisiri, S., Haetrakul, T., Dunbar, S. G., & Chansue, N. (2024). Microplastic contamination in edible marine fishes from the upper Gulf of Thailand. *Marine Pollution Bulletin, 198*, Article 115785. https://doi.org/10.1016/j.marpolbul.2023.115785

Sui, Q., Yang, X., Sun, X., Zhu, L., Zhao, X., Feng, Z., Xia, B., & Qu, K. (2024). Bioaccumulation of polycyclic aromatic hydrocarbons and their human health risks depend on the characteristics of microplastics in marine organisms of Sanggou Bay, China. *Journal of Hazardous Materials, 473*, Article 134622. https://doi.org/10.1016/j.jhazmat.2024.134622

Sussarellu, R., Suquet, M., Thomas, Y., Lambert, C., Fabioux, C., Pernet, M. E. J., Le Goïc, N., Quillien, V., Mingant, C., Epelboin, Y., Corporeau, C., Guyomarch, J., Robbens, J., Paul-Pont, I., Soudant, P., & Huvet, A. (2016). Oyster reproduction is affected by exposure to polystyrene microplastics. *Proceedings of the National Academy of Sciences, 113*(9), 2430–2435. https://doi.org/10.1073/pnas.1519019113

Tan, E., & Mohd Zanuri, N. B. (2023). Abundance and distribution of microplastics in tropical estuarine mangrove areas around Penang, Malaysia [Original Research]. 10. https://doi.org/10.3389/fmars.2023.1148804

Taylor, M. L., Gwinnett, C., Robinson, L. F., & Woodall, L. C. (2016). Plastic microfibre ingestion by deep-sea organisms. *Scientific Reports, 6*(1), 33997. https://doi.org/10.1038/srep33997

Troost, T. A., Desclaux, T., Leslie, H. A., van Der Meulen, M. D., & Vethaak, A. D. (2018). Do microplastics affect marine ecosystem productivity? *Marine Pollution Bulletin, 135*, 17–29. https://doi.org/10.1016/j.marpolbul.2018.05.067

Trusler, M. M., Cook, S., Lomax, B. H., & Vane, C. H. (2025). Microplastic pollution in salt marsh and urban tributary sediment cores of the River Thames estuary, UK: Spatial and temporal accumulation trends. *Marine Pollution Bulletin, 211*, Article 117360. https://doi.org/10.1016/j.marpolbul.2024.117360

van Sebille, E., Wilcox, C., Lebreton, L., Maximenko, N., Hardesty, B. D., van Franeker, J. A., Eriksen, M., Siegel, D., Galgani, F., & Law, K. L. (2015). A global inventory of small floating

plastic debris. *Environmental Research Letters, 10*(12), Article 124006. https://doi.org/10.1088/1748-9326/10/12/124006
Waller, C. L., Griffiths, H. J., Waluda, C. M., Thorpe, S. E., Loaiza, I., Moreno, B., Pacherres, C. O., & Hughes, K. A. (2017). Microplastics in the Antarctic marine system: An emerging area of research. *Science of the Total Environment, 598*, 220–227. https://doi.org/10.1016/j.scitotenv.2017.03.283
Wang, J., Wang, M., Ru, S., & Liu, X. (2019). High levels of microplastic pollution in the sediments and benthic organisms of the South Yellow Sea, China. *Science of the Total Environment, 651*, 1661–1669. https://doi.org/10.1016/j.scitotenv.2018.10.007
Wang, J., Zheng, L., & Li, J. (2018). A critical review on the sources and instruments of marine microplastics and prospects on the relevant management in China. *Waste Management & Research, 36*(10), 898–911.
Wang, T., Li, B., Shi, H., Ding, Y., Chen, H., Yuan, F., Liu, R., & Zou, X. (2024). The processes and transport fluxes of land-based macroplastics and microplastics entering the ocean via rivers. *Journal of Hazardous Materials, 466*, Article 133623. https://doi.org/10.1016/j.jhazmat.2024.133623
Wang, W.-X. (2023). Environmental toxicology of marine microplastic pollution. *Cambridge Prisms: Plastics, 1*, e10, Article e10. https://doi.org/10.1017/plc.2023.9
Wen, S., Yin, X., Zhang, Y., & Diao, X. (2024). Chronic exposure to low concentrations of microplastics causing gut tissue damage but non-significant changes in the microbiota of marine medaka larvae (Oryzias melastigma). *Marine Environmental Research, 195*, Article 106381. https://doi.org/10.1016/j.marenvres.2024.106381
Werth, A., Kahane-Rapport, S., Potvin, J., Goldbogen, J., & Savoca, M. (2024). Baleen-plastic interactions reveal high risk to all filter-feeding whales from clogging, ingestion, and entanglement. *Oceans, 5*, 48–70. https://doi.org/10.3390/oceans5010004
Worm, B., Barbier, E. B., Beaumont, N., Duffy, J. E., Folke, C., Halpern, B. S., Jackson, J. B. C., Lotze, H. K., Micheli, F., Palumbi, S. R., Sala, E., Selkoe, K. A., Stachowicz, J. J., & Watson, R. (2006). Impacts of biodiversity loss on ocean ecosystem services. *Science, 314*(5800), 787–790. https://doi.org/10.1126/science.1132294
Yang, H., Chen, G., & Wang, J. (2021). Microplastics in the marine environment: Sources, fates, impacts and microbial degradation. *Toxics, 9*(2), 41.
Yao, F. C., Jin, C. X., Liang, H., Zhang, Y., Gu, Y., Song, F. B., Zhou, Z., Sun, J. L., & Luo, J. (2024). Microplastics weaken the digestion and absorption functions in the golden pompano (Trachinotus blochii) by affecting the intestinal structure, bacteria and metabolites. *Chemosphere, 362*, Article 142415. https://doi.org/10.1016/j.chemosphere.2024.142415
Yin, L., Liu, H., Cui, H., Chen, B., Li, L., & Wu, F. (2019). Impacts of polystyrene microplastics on the behavior and metabolism in a marine demersal teleost, black rockfish (Sebastes schlegelii). *Journal of Hazardous Materials, 380*, Article 120861. https://doi.org/10.1016/j.jhazmat.2019.120861
Yu, R.-S., & Singh, S. (2023). Microplastic pollution: Threats and impacts on global marine ecosystems. *Sustainability, 15*(17), 13252. https://www.mdpi.com/2071-1050/15/17/13252
Yu, X., Gutang, Q., Wang, Y., Wang, S., Li, Y., Li, Y., Liu, W., & Wang, X. (2024). Microplastic and associated emerging contaminants in marine fish from the South China Sea: Exposure and human risks. *Journal of Hazardous Materials, 480*, Article 136200. https://doi.org/10.1016/j.jhazmat.2024.136200

Chapter 5
Impact of Microplastics and Nanoplastics on Plant Health and Development

Mona F. A. Dawood, Hend A. Hamed, and Marwa T. El-Mahdy

Abstract In recent years, anthropogenic activities have emerged many persistent pollutants to the ecosystem as microplastics. Several studies has been explained the effect of microplastics on aquatic system, but scarcely applied on the agricultural systems. Microplastics reach the soils via supplementation of biosolids, compost, wastewater irrigation, mulching film, polymer-based fertilizers and pesticides, and atmospheric deposition. Terrestrial emissions of microplastics in industrialized countries predominantly originate from application of sewage sludge to farmlands. Sewage sludge is applied to improve and maintain productive soils and stimulate plant growth in many areas of the world. The pollution of MPs is worsening globally and the related hazards in agricultural soils need proper attention. This study will discuss the sources, fate, and dispersion of MPs to the agricultural soil. Further, the current chapter will includes the analytical techniques, ecological risks, and pollution characteristics of MPs in the soil environment. Also, the chapter also will discuss the adverse effects of microplastics on the soil biota and we will focus on the plant responses to microplastics contamination.

Keywords Microplastics · Nanoplastics · Ecosystem · Emerging pollutants · Metabolism · Antioxidants

M. F. A. Dawood (✉)
Botany and Microbiology Department, Faculty of Science, Assiut University, Assiut, Egypt
e-mail: mo_fa87@aun.edu.eg

H. A. Hamed
Department of Horticulture, Faculty of Agriculture, Sohag University, Sohag, Egypt

M. T. El-Mahdy
Department of Pomology, Faculty of Agriculture, Assiut University, Assiut, Egypt

Molecular Biology Research and Studies Institute, Assiut University, Assiut, Egypt

V. Kumar et al. (eds.), *Microplastics Pollution*,
https://doi.org/10.1007/978-3-031-99774-7_5

5.1 Introduction

Microplastics (MPs) and Nanoplastics (NPs), synthetic polymer compounds, are growing contaminants threatening ecosystems and biodiversity on a global scale (Li et al., 2016). These small plastic particles, typically less than 5 mm (Frias & Nash, 2019) in size pose a growing concern due to their non-biodegradable nature, their pervasive presence in marine, freshwater, soil, and atmospheric environment, and their long-term persistence (Bhat et al., 2024; Thacharodi et al., 2024b). Merely a part of this environmental problem is evident. With plastic daily generation reaching thousands of particles per cubic meter, the global production of plastics has doubled since the beginning of the century, to nearly 400 million metric tons per year in 2021, a 1.6% from the previous year (Statista, 2022). China remains the world's largest producer of plastic, accounting for 32% of the global production of plastic materials in 2022, with 19% and 17% coming from Asia and North America (Statista, 2022). Plastic waste and related chemicals and byproducts discharged into the environment throughout the plastic life cycle have both increased in tandem with advances in plastic production (Eriksen et al., 2023). By 2050, there will likely be 33 billion tons more plastic products on the planet (Phillips & Bonner, 2015).

In today's global economic system, environmental pollution and economic growth are closely related. As the demand for plastic materials increases, so does the risk of environmental degradation, promoting calls for shifts in economic strategies to address the plastic crisis (Andrée et al., 2019; Rao & Yan, 2020). The accumulation of plastics in the environment can directly impact the Environmental Performance Index (EPI), a crucial metric for assessing pollution levels in a country (Ave & Babolsar, 2010; Hsu et al., 2013). MPs are notorious for a slow biodegradation rate, with some types that can take up to 500 years to fully break down. Without adequate measures, and significant interventions, the amount of plastics is expected to double again in the coming few years (Isobe et al., 2019).

As MPs infiltrate the ecosystem, they can enter human food chains through several routes Fig. 5.1. Marine organisms including scallops, crabs, lobsters, mussels, oysters, shrimp, and clams are particularly vulnerable, as MPs are ingested through contaminated water or prey (Thacharodi et al., 2024a) These particles may carry harmful chemicals or pathogens, can accumulate in the bodies of these organisms, potentially leading to carcinogenic, and genotoxic damage Further, (Mato et al., 2001; Teuten et al., 2009). Furthermore, MPs act as vectors for environmental contaminants, which are then transferred up to the food chain (Bowmer & Kershaw, 2010; Wang et al., 2018).

Microplastics (MPs) can be categorized into two broad sources: primary and secondary Fig. 5.2. There are numerous ways that primary microplastics might directly enter the environment. They are made intentionally at small sizes for certain applications, such as in personal care goods products (e.g. microbeads), synthetic fibers in clothing, industrial manufacturing's tiny plastic pellets (nurdles), and plastic packaging. In contrast, secondary MPs, result from the breakdown of larger plastic debris due to environmental exposure like UV radiation, mechanical abrasion, and

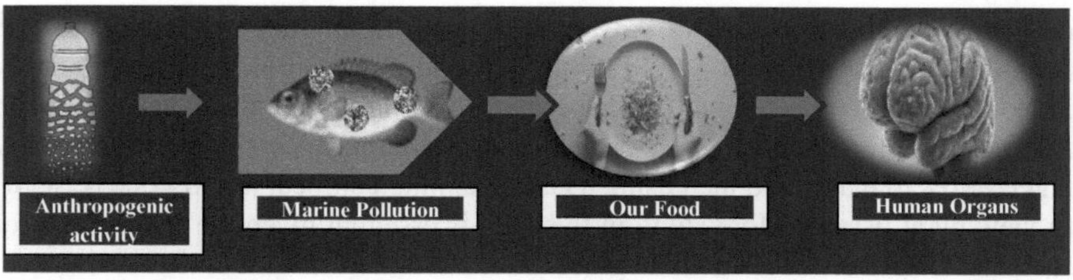

Fig. 5.1 A model illustrates how human activity leads to the entry of microplastics into human organs

microbiological activity. This process leads to the fragmentation of plastic items, such as bottles, bags, and packaging materials into smaller particles (Wei et al., 2021; Zhu et al., 2020).

The infiltration of MPs into plants is another emerging area of concern. Recent studies suggest that microplastics can be absorbed by plants through their root systems and build up in various plant tissues, including fruits and leaves. This raises significant questions regarding food safety and the potential risks to human health since microplastics may enter the food chain through agricultural products (Hua et al., 2023; Li et al., 2021b). Beyond its use by humans, microplastic presence in plants can interfere with nutrient dynamics in soil (Irshad et al., 2024; Saha et al., 2024) affecting the uptake of essential minerals such as nitrogen (N), potassium (K), and phosphorus (P) which are vital for plant growth and ecosystem productivity.

In strawberry plants for instance, MPs (polystyrene) affected the uptake of nutrients by strawberry plants, including NH_4^+–N, NO_3^-–N, K, P, Ca, and Mg Specifically, the pace at which N, K, and P were moved from the root to the shoot decreased

Fig. 5.2 Examples of plastic used in agricultural practices lead to the release of microplastics

considerably (Bai et al., 2024). Advanced imaging techniques like High-precision Scanning Electron Microscopes (SEM) allow for the observation of MPs morphology in plants. It is now possible to more successfully confirm that plants have absorbed MPs and identify the site of absorption and migration by applying fluorescently tagged MPs to plants and then using a laser scanning confocal microscope (LSCM) to view plant tissues. More study in this area is essential given the rising need for sustainable agriculture around the globe.

The toxicological effects of MPs add another layer of complexity to their environmental impact Studies have shown that MPs can absorb hazardous substances including persistent organic pollutants (POPs) and heavy metals (HVMs) which then pose risks to human health and wildlife. For example, research on mice exposed to polystyrene MPs (micro-PS) revealed significant reproductive toxicity including sperm malformation and reduced motility linked to oxidative stress and p38 MAPK signaling pathway activation (Xie et al., 2020). Beyond these effects, MPs are suspected of disturbing wildlife's hormonal balance with potential long-term effects on biodiversity and ecological stability.

Clarifying the processes of toxicity linked to microplastics and establishing safe exposure thresholds for both humans and wildlife are becoming more and more important as research advances. This growing database of research emphasizes the necessity of regulatory frameworks to address the potential risks related to microplastic pollution. To address MPs pollution, diverse strategies have been come forth and put into practice worldwide; actions such as improving waste disposal strategies and encouraging eco-friendly options are among the initiatives in this area.

Since 60–95% of marine plastic pollution worldwide is caused by single-use plastics (SUPs), legislative (top-down) action to eliminate SUPs is gaining traction. Additionally, non-legislative (bottom-up) measures are becoming more and more common (Schnurr et al., 2018). Additionally, new approaches such as efficient filtration systems and bioremediation techniques are being developed to identify and remove microplastics from ecosystems. To effectively implement these plans, cooperation between governments, organizations, academic institutions, and the public is required.

During this chapter, readers are provided a comprehensive overview of microplastics, shedding light on their sources, interactions with the environment, toxicological consequences, and viable remedies to lessen plastic pollution.

5.2 Microplastics Sources

Primary and secondary sources are the two broad categories into which microplastics (MPs) fall (Fig. 5.3).

(1) **Primary Sources**

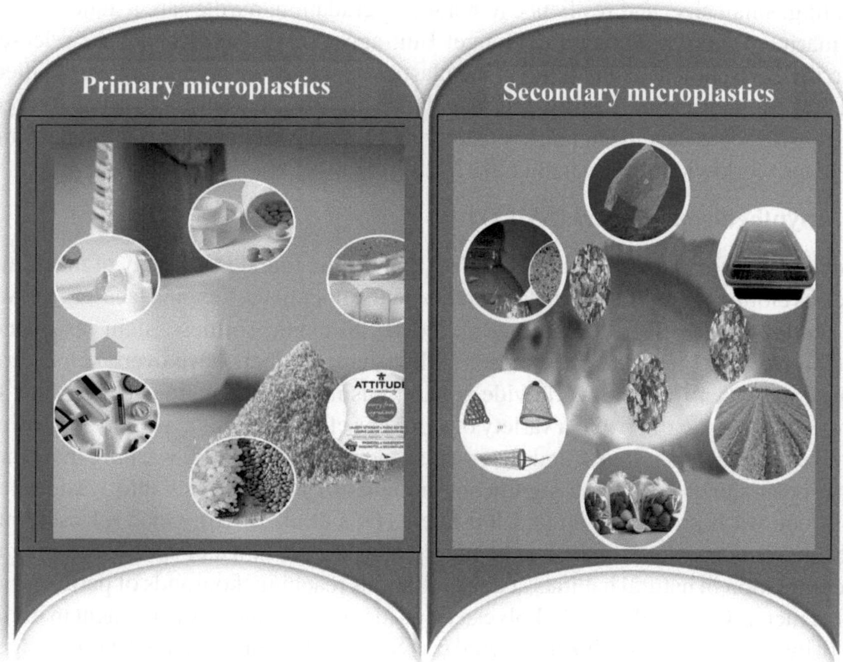

Fig. 5.3 Examples of primary and secondary microplastics

Primary microplastics are intentionally manufactured at small sizes for specific uses. These particles are used in various, commercial, industrial, and household products and are directly produced as microplastics. Common primary sources include:

I. **Microbeads**

The presence of tiny plastic beads, also known as microbeads, in personal care and cosmetics products (PCPs) like makeup cosmetics, body wash, hand sanitizer, soap, toothpaste, shaving cream, facial scraps, sunscreen, shower gel, shampoo conditioners, deodorants, lipsticks, eyeliners, nail polishes (Praveena et al., 2018) is one of the most well-known sources of primary microplastics (Nizzetto et al., 2016). The main plastic in PCPs is polyethylene and the MPs derived from these products typically have diameters smaller than 350 μm. The geometric averages for microplastic mass and abundance in PCPs are 0.04 g/g and 2162 particles/g, respectively. Based on the PCPs consumption and associated microplastic levels, the annual mass emission of microplastics from PCPs reaches up to 12.000 tons. (Sun et al., 2020).

II. **Pellets (nurdles)**

Nurdles, which are tiny plastic pellets utilized in industrialization, are the world's second-largest source of primary MPs pollution. An approximate 445,970 tons of nurdles are discarded into the environment worldwide annually. Because of their floating, small size, and resilience to natural degradation, nurdles are a major source of marine pollution (Galgani & Rangel-Buitrago, 2024). Nurdles may be released into the environment at any point in the plastics supply chain when they are handled, such as the fabrication facility, the manufacturing site, and during loading storage, or transit. There have been numerous reports of nurdle spillage caused by maritime mishaps worldwide (Sewwandi et al., 2023).

III. **Synthetic microfibers (SMFs)**

Tiny plastic threads, having diameters less than 5 mm, found in a variety of materials, such as clothing, tires, fishing nets, cigarette butts, floor mats, ties, and lace, are known as synthetic microfibers (Mishra et al., 2020). Strength, softness, stain resistance, and above all cost are the main reasons why synthetic fibers have overtaken natural fibers and taken over the worldwide apparel business. Without considering how to dispose of them, we all wear a variety of synthetic clothing items every day, including blankets, yoga pants, socks, undergarments, and jackets. Laundry washing machines have been shown to release significant amounts of microplastics into wastewater, with one wash producing up to 1900 fibers (Dris et al., 2016). Once released into the environment, these fibers can be ingested by a wide range of species, leading to accumulation in natural habitats at concentrations reaching thousands of particles per cubic meter (Cesa et al., 2017). Polyester was identified as the most prevalent material by (Tiwari et al., 2019), who discovered that SMFs in Indian coastal seashores ranged from 5 to 36 mm. In a related study, synthetic MFs with diameters ranging from 10 to 22 μm were extracted from the ocean surface water of Admiralty Bay, Antarctica, and their chemical compositions which primarily consisted of microplastics and SMFs were determined. According to (Sanchez-Vidal et al., 2018), the deep marine sediments of Southern Europe included 10–70 SMFs per 50 ml of microplastic polymers, including PE, PP, PA, and acrylic. There is growing concern for both human health and the atmosphere due to the pervasiveness of synthetic microfibers. Throughout the whole textile lifecycle, from the textile industry to consumer use, microfiber fleeces must be tackled.

IV. **Industrial Abrasives**

Industrial abrasives are an important source of primary microplastics (Sommer et al., 2018). These materials, often utilized in different industries for processes such as polishing, sandblasting, and cleaning, can participate in the generation of MPs when they outwear or break apart. MPs in the form of tiny abrasive particles are utilized in a wide range of applications, like cleaning metal surfaces, preparing concrete, or finishing products like plastic, wood, and glass (Vinay et al., 2023).

The abrasives are typically made from plastic materials such as polypropylene (PP), polyethylene (PE), HDPE, or other synthetic polymers, which are long-standing and can resist hard conditions (Savaş et al., 2019). However, as these abrasives are frequently used during industrial processes, they break down into smaller particles that are usually too small to be effectually seized by filtration systems or recycling processes.

In some cases, these microplastic particles are released directly into the environment, particularly in industrial settings near freshwater, where they can wash into lakes, rivers, and oceans. These microplastics then become part of the larger pollution problem, as they are often too small to be squared away effectively. The extensive use of plastic abrasives in industries like manufacturing, automotive repair, and construction makes industrial abrasives a noteworthy source of primary microplastics (Prasittisopin et al., 2023).

The effect of these MPs is concerning, as they can enter ecosystems, be ingested by wildlife, and participate in the expansion of plastics in marine environments, eventually impacting the food chain. The use of non-plastic abrasives or the development of more sustainable alternatives could help lessen the environmental impact of this source of microplastics.

(2) **Secondary Sources:**

The decomposition of bigger plastic debris produces secondary microplastics, which are not directly released into the environment in their microplastic form. Over time, weathering, UV rays, mechanical forces, and environmental factors cause these plastics to deteriorate.

I. **Plastic waste in the environment**

One of the major participants in MPs pollution is large plastic items that are inappropriately discarded encompassing waste from bottles, plastic bags, packaging, materials disposable plastic tableware (Ng et al., 2018).

- **Plastic bottles**: The primary purpose of plastic bottles is to serve as disposable containers for liquids or solids. They contain materials such as PE, PP, and PET. Less than half of the one million plastic bottles that are sold every minute worldwide get recycled. 480 billion plastic beverage bottles are sold globally each year. Because plastic bottles frequently decompose into microplastic particles and greatly contribute to ocean pollution, their continued use as ocean waste is a serious concern (Adetuyi et al., 2024).
- **Plastic bags**: Plastic bags are an essential part of an individual's everyday life. Their affordability, incredibly lightweight, vast capacity, and ease of storage make them popular. More than 5 trillion plastic bags are utilized annually all over the world. Many plastic bags have been discarded into the environment, especially on the land surrounding towns, riverbanks, and roadways. Despite their convenience, they have a long degradation cycle, and less than 10% are recycled. Several countries have used measures such as charging for bags or blocking them to lessen waste, even though a number of plastic bag prohibit have helped to reduce plastic bag pollution, many plastic bags remain in the environment and are likely a major source of environmental MPs, especially in oceans (Chen et al., 2021a, 2021b).
- **Disposable Plastic Tableware**: plastic tableware is produced by the thermoplastic shaping of resin or other thermoplastic materials. Widely utilized for meals and fast food, items made from PE, PP, and PS are lightweight and convenient but contribute significantly to waste. In the U.S. alone, hundreds of millions of

plastic utensils are used annually, most of which are not recycled. Inappropriate discarding results in plastic entering ecosystems, where it breaks down into MPS.

- **Plastic Packaging**: Used hugely in many industries, plastic packaging encompasses bags, films, and containers made from materials like PE, PP, and EVA. While it serves to shield products, it is a substantial source of waste. With the escalation of e-commerce and fast food, plastic packaging usage has ascended, helping to bring to vast amounts of plastic waste into oceans and waterways (Kershaw, 2016).
- **Agricultural practices**

The use of plastic in agriculture such as the use of plastic film in farming, irrigation pipes, the application of sewage sludge the use of compost made from contaminated organic waste, and plastic containers has grown significantly with modern agriculture (Kumar et al., 2020). Over time, these plastic materials break down due to environmental exposure and mechanical wear generating secondary microplastics. In addition to direct degradation, the improper disposal of plastic agricultural wastes, such as discarded plastic mulch can lead to the spread of microplastics in the soil and surrounding ecosystems. According to (Nizzetto et al., 2016), the density values of primary and secondary MPs released on land are equivalent to or higher than those of water. They can easily interact with soil organisms, impacting soil health and potentially entering the food chain through crop uptake.

III. **Road transport and tire wear**

According to estimates, road transport vehicle tires contribute 6200 tons of tire tread wear particles annually to soil and 1800 tons to surface water through pavement runoff, sewage system overflows, and effluents. Furthermore, 7400 tons of fine particulate matter are collected annually in open asphalt concrete roads, while an additional 900 tons are released into the atmosphere annually from tires. This relates to the tire wear of nine distinct vehicle categories, including various trucks, cars, buses, and motorbikes (Verschoor et al., 2016). According to Sommer et al. (2018), Tire wear makes up 30% of the microplastic particles that contaminate rivers, lakes, and oceans. This has an impact on aquatic life.

IV. **Marine activities**

Marine activities, such as shipping and fishing, are significant sources of secondary microplastics, and the oceans serve as the main sink for plastic trash. Between 0.13 and 135,000 tonnes of commercial fishing equipment are thought to be left annually worldwide. Boundaries, floating boxes, fishing poles, tanks, nets, fishing lines, and cables are examples of plastic fishing debris. Large amounts of polystyrene foam have been found in marine plastic trash (Lee et al., 2015). The primary causes of this waste are the extensive use of Styrofoam floating devices in the aquaculture sector, which has grown quickly in recent years, and the significant quantity of foam pontoons and abandoned fishing nets that end up in the ocean because of biological damage and natural wear after extended use.

Fishing waste, especially from commercial fishing operations, is a major participant in secondary MPs in marine habitats (An et al., 2020). The huge number of discarded equipment, like tanks, ropes, nets, fishing lines, and floating devices, all gradually break down into smaller plastic fragments over time, often because of the biological degradation, scraping, and natural weathering processes. Discarded, abandoned, or lost, or fishing gear (also known as "ghost gear") is a critical source of MPs contamination. Fishing nets, especially those made of plastic fibers, continue to catch marine life long after being abandoned. These nets not only entangle marine animals but also break down into MPs, compounding the issue. Studies show that fishing nets contribute a significant portion of the MPs contamination in specific areas, with an estimated 640,000 tonnes of fishing gear lost or discarded each year globally.

The breakdown of fishing equipment into MPs is increased by the resilience of the materials used, which can take decades to break down entirely. As the larger plastic debris fragments, it becomes more easily consumed by marine organisms, entering the food chain and causing damage to both aquatic life and humans (Lu et al., 2019). These secondary MPs, deriving from commercial fishing gear, consequently represent a substantial and persistent source of marine contamination, focusing the quest for stricter regulations, improved waste management practices, and the development of sustainable alternatives within the fishing industry (An et al., 2020).

V. **Others**

Along with the previously listed sectors, the construction, aerospace, and medical industries also use a lot of plastic. Hinged covers, hub caps and other car parts, touch screens, pipes and fittings, foundations, roofs, insulation, window frames, wall coverings, building insulation, valves, seals, and so forth are just a few of the many products that use plastic. Because they are resistant to climatic conditions, don't decay or rust, and require little upkeep, plastic products are frequently utilized outside (An et al., 2020).

Sources of microplastics according to the chemical composition

Primary and Secondary microplastics can be made from a variety of plastic materials, Some of the common materials used in the industry are listed in Table 5.1 along with their introduction and industrial applications.

5.3 Distribution of Microplastics

(1) **Environmental Media**

i. **Terrestrial soil**

Approximately 90% of global single-use plastic waste ends up in the terrestrial ecosystem (Ng et al., 2018), raising concerns about the potential contamination of agricultural end products and consumer health. This is further exacerbated by Terrestrial soil typically accumulates more MPs that are influenced by human activities,

Table 5.1 Synthetic plastics and their industrial uses

Synthetic plastics	Material introduction	Chemical composition	Industrial applications
Polyurethane (PUs)	The 6th most used polymer in the world, with an annual output of 18 million tons (Simón et al., 2018)	Polyol + Isocyanate	Automotive industry, textiles and fabrics, medical applications (Das & Mahanwar, 2020), marine applications, Coatings, building and construction applications, packaging (Akindoyo et al., 2016)
Polyvinyl chloride (PVC)	The most widely produced polymer with an annual output of 31 million tons per year (Lieberzeit et al., 2022)	Vinyl Chloride Monomer (VCM)	Agriculture, medicine, electrical engineering, pharmaceutical industries (Lieberzeit et al., 2022), packaging, military and aerospace industries, structural elements (Grosu, 2022)
Polypropylene (PP)	One of the most widely used thermoplastic material with global production over than 53 million tons (Gahleitner & Paulik, 2017)	Polymerization of propylene	Textile and clothes, scaffold in biomedical, nonwoven, nanomembrane for filtration, PP composite, wood-plastic composite, automobile, robes and Twines, packaging, geotextile (Hossain et al., 2024)
Polycarbonate (PC)	Amorphous thermoplastic with production estimated to be 3.3 million tons, growing at 7% per year (Antonakou & Achilias, 2013)	Bisphenol A + Phosgene	Optical/lighting, glass replacement, medical packaging, electrical products, household products (Gohil & Joshi, 2022), polycarbonate blend applications in biomedical, flame retardant, and membrane materials (Kausar, 2018)

(continued)

Table 5.1 (continued)

Synthetic plastics	Material introduction	Chemical composition	Industrial applications
Polystyrene (PS)	Polymer of styrene and it also releases a mixture of polyaromatic hydrocarbons (PAHs), (Chaukura et al., 2016)	Styrene monomer	Food packaging materials, medical applications, and crafts (Chaukura et al., 2016), Polymer foams (Ghoshal et al., 2023)
Polyethylene (PE)	Thermoplastic polymer made by polymerization of ethylene makes up about 1/4 of global plastic production (Chen & Lin, 2021)	Ethylene monomer	Packaging, agriculture pipes, plastic films for greenhouses, textiles, medical implants, 3D Printing, motor oil bottles (Paxton et al., 2019)
Nylon	Semi-crystalline polymer is part of the polyamides (PAs) family (Winnacker, 2017)	Hexamethylenediamine + Adipic acid	Shoes, clothing, rainwear, biomedical applications such as dental implants (Shakiba et al., 2021)
Polyethylene terephthalate (PET)	Widely used thermoplastic polymer, the 6th most important polymer in the market accounting for about 4 million tonnes (Nisticò, 2020)	Terephthalic acid + Ethylene glycol	Food packaging, bottles and containers, labels, and films (Nisticò, 2020) Textiles, construction, membrane filtration, and biomedical applications (Suhaimi et al., 2022)
Low-density polyethylene (LDPE)	One of the two types of polyethylene, characterized by a lower degree of crystallinity (Bonifazi et al., 2018)	Ethylene monomer	Medical, automotive, aerospace, electronics (Sanchis et al., 2008)

such as agroecosystems (Nizzetto et al., 2016). The United Nations Environment Program (UNEP) identified that large quantities of particulate plastics found within the marine environment globally result from land-based sources (UNEP & Assessment, 2016). Even though, The issue of terrestrial ecosystems has received little attention (Piehl et al., 2018).

For the first time, the amount of microplastic contamination on a farm in southeast Germany was measured by (Piehl et al., 2018). They discovered 0.34 ± 0.36 microplastic particles per kilogram dry weight of soil using Fourier transform infrared (FTIR) analysis. The most prevalent polymer type was often polyethylene, which was followed by polystyrene and polypropylene. The most common categories were films and fragments. Their findings reflect on plastic contamination in a location that solely receives traditional agricultural treatment since they purposefully selected a study site where agricultural plastic treatments and fertilizers containing microplastics were never employed. However, regions with agricultural plastic applications such as greenhouses, mulch, silage films, or fertilizers that contain plastic (sewage sludge, biowaste composts) are likely to have higher levels of contamination.

MPs in agroecosystems may negatively impact the physiology and behavior of soil organisms (Ju et al., 2019). They also affect the soil's biophysical function and structure (de Souza Machado et al., 2018), its physical characteristics, plant interactions (de Souza Machado et al., 2019), and the microbial community (Huang et al., 2019). Moreover, MPs can alter the transmission and occurrence of other pollutants (Zhou et al., 2019). The size, shape, and polymer composition of MPs are known to affect their uptake, retention, clearance, translocation, and toxicity within organism tissues in addition to their abundance (De Sá et al., 2018). The quantity and type of MPs can also influence their impact on the geochemistry, biophysical environment, and ecotoxicology of agricultural systems. Synthetic microplastic materials are highly prolific and recalcitrant posing significant risks. They are linked to organismal dysfunctions, disturbances in abiotic factors, and increased mortality and morbidity rates in both fauna and flora (Bosker et al., 2019).

ii. **Oceans, Marine and Freshwater Systems**

Marine environments, encompassing estuaries, coastal zones, and open oceans, serve as main sinks for microplastics. Rivers are a primary furrow for transferring MPs from land to the ocean, as they carry waste from urban, industrial, and agricultural activities. Once in the marine ecosystem, MPs may cumulate in a variety of sections, including the water column, sediments, and surface waters. The interconnection between ocean currents and MPs leads to the evolution of highly concentrated sites, like the Great Pacific Garbage Patch, which serves as a visual appearance of the scale of this issue.

Once in the marine ecosystem, MPs can undergo physical and chemical transformations because of subjection to seawater, sunlight, and biological processes. UV radiation causes the plastic particles to break down, leading to smaller pieces known as nanoplastics, which may have even more dangerous impacts because of their escalated reactivity and possible bioaccumulation. Moreover, these particles may adsorb and concentrate toxic materials like heavy metals, persistent organic pollutants (POPs), and dangerous chemicals from the adjacent environment, making them a vector for transporting these contaminants across environments (Xu et al., 2020a, 2020b).

The effects of MPs are not restricted to physical damage. They pose substantial risks to marine organisms, particularly plankton, filter feeders, and fish that fault these particles for food. Ingestion of MPs can result in digestive blockages, decreased feeding effectiveness, and even death in some species. Research has indicated that MPs can accumulate in the gastrointestinal zones of marine species, causing inflammation, metabolic disturbances, and harm to internal organs (Wang et al., 2021a, 2021b, 2021c). Moreover, MPs can enter the food web, resulting in possible risks to higher trophic concentrates, encompassing marine mammals and birds, which may ingest polluted prey.

The existence of MPs in marine sediments makes the issue more complex. Sediments act as both sinks and possible sources of MPs, as particles may cumulate in the seafloor and be resuspended into the water column during disturbances like storms or human activities such as dredging. This permits MPs to re-enter the environment, creating a cycle of pollution (Auta et al., 2017a, 2017b). The continuous accumulation of MPs in sediments also raises apprehensions for benthal organisms that live in these regions, possibly impacting species like worms, crabs, and other bottom-dwelling creatures.

Another noteworthy facet of MPs distribution in marine ecosystems is their potential to act as carriers for invasive pathogens, species, and microorganisms (Ali et al., 2023). As MPs provide a surface for biofouling organisms, including algae, bacteria, and barnacles, they can inadvertently ease the expansion of non-native species through vast oceanic distances. This may lead to disruptions in local environments, altering food webs and introducing pathogens to new regions. Moreover, the colonization of MPs by damaging bacteria and viruses can pose a threat to marine health, as these microorganisms can transfer diseases to marine organisms, including shellfish and fish.

The challenges of managing MPs contamination are made worse by the complex interconnections between human activities, ecological processes, and marine life. Decreasing the sources of plastic contamination is critical to reduce the spread of MPs in marine ecosystems. This includes both decreasing plastic generation and consumption, as well as improving waste-treating practices to inhibit plastics from entering the oceans. In addition, keeping track of and researching the fate, transfer, and effects of MPs are critical for informing policy decisions and advancing efficient alleviation strategies.

Only in the last fifteen years or so have the first studies on microplastics in freshwater habitats been published, making this a relatively new phenomenon. Microplastics are becoming a common pollutant; in fact, freshwater research frequently finds and reports microplastics at all sampling locations and frequently in all samples that are collected (Li et al., 2018).

iii. **Atmosphere**

Microplastics have been detected in the air, especially in urban and industrial regions, as well as in remote areas far from major human activity. These particles can be carried over long distances by wind currents, and the atmospheric build-up of MPs in terrestrial and aquatic ecosystems has become a concern (Nematollahi et al., 2022). Nanoplastics are particularly frequent in the air because of their tiny size and capability to be simply aerosolized.

Microplastics in the atmosphere represent an increasing and concerning source of contamination. Recent research has indicated that MPs can be found in the air at different locations, from urban centers to remote regions, even far from any known sources of plastic contamination. These airborne particles are tiny, often smaller than 10 μm in size, which permits them to travel long distances and be extended all over the world. The distribution of MPs in the atmosphere can have substantial ecological and health indications (Chen et al., 2020).

Mechanisms of Atmospheric Microplastic Distribution

Once microplastics are dumped into the air, various mechanisms contribute to their expansion:

Wind: Wind currents can carry these particles from urban areas to rural, remote, and even pristine ecosystems. This makes it not easy to spot the exact sources of contamination.

Precipitation: Microplastics in the atmosphere can settle onto land or water surfaces via precipitation, like rain or snow. This process, known as "wet deposition," can result in the deposition of MPs in both terrestrial and aquatic ecosystems (Belioka & Achilias, 2024).

Air Currents and Weather Patterns: Atmospheric conditions, such as storms and weather systems, can impact the movement and deposition of MPs (Belioka & Achilias, 2024). For instance, during acute weather events such as heavy winds, the particles may be lifted into the upper atmosphere and travel far distances, contributing to the worldwide widespread of MPs.

Impacts of Atmospheric Microplastics

1. **Human Health**: Intake of MPs can pose momentous health issues, particularly as they enter the respiratory system. These particles are tiny enough to reach the lungs (Tokunaga et al., 2023) and possibly enter the bloodstream, where they may cause inflammation, respiratory problems, or even more acute long-term health risks. Given that MPs have been found in urban air, workers in industrial settings, or even individuals living in contaminated regions, could be at sharpen risk.
2. **Ecological Risks**: Airborne MPs can have negative impacts on wildlife, particularly animals that depend on air or ground-based foraging (Enyoh et al., 2019). For example, birds that ingest or intake airborne MPs may suffer from internal wounds or poisoning. Moreover, MPs in the atmosphere can alter the composition of environments by impacting soil quality, water systems, and plant growth once they settle.
3. **Climate Impacts**: The presence of MPs in the atmosphere may also have indirect impacts on climate change (Li et al., 2024b). Some researchers recommend that microplastic particles could influence cloud formation, possibly altering precipitation patterns (Aeschlimann et al., 2022). Further, microplastics can serve as carriers for toxic chemicals, affecting air quality and the health of living organisms (Rafa et al., 2023).

(2) **Factors Affecting MPs Distribution**

Various factors impact microplastics and nano plastics distribution in the environment:

- **Size and density**: Due to changes in its physical characteristics (such as density, size, form, and hetero-aggregation with other particles), plastic waste will eventually sink to the seafloor. Larger microplastics (L-MPs; 1–5 mm) tend to sink, cumulating on the ocean floor or in freshwater sediments. Smaller microplastics (S-MPs; 0.02–1 mm), especially those in the nano plastic scope, have less density and are more likely to remain suspended in air and, water being transferred over long distances (Keisling et al., 2020).

- **Environmental conditions**: Wind, tide, water circulation, weather events like salinity, temperature, and chemical properties of water can affect the degradation rates and persistence of microplastics (Nel & Froneman, 2015). In warmer waters, plastics may degrade faster and decompose into smaller particles. On the contrary, colder waters or environments can preserve microplastics for expanded periods.
- **Anthropogenic factors**: Urbanization, manufacture, and growing plastic production collaborate higher concentrations of microplastics in regions with more human activity. Wastewater treatment plants WWTP, even though designed to filter pollutants, often dispense considerable volumes of microplastics into the environment, as these particles are not easy to filter entirely (Kim et al., 2015) found that the primary forces influencing the geographical distribution of microplastics on coastal beaches were wind and currents and that the high abundance was mostly caused by the enormous quantity of unrecycled plastic buoys originating from the aquaculture businesses.

5.4 Advanced Analytical Techniques for Detecting MPs/NPs in Plant

In the technology era, multiple tools are currently used to detect MPs/NPs in plant tissues and samples, although the accurate tracking inside the plant system is still challenging. Most of the modern analytical techniques have limitations due to NPs' size. Raman and Fourier transform infrared (FTIR) spectroscopies, for example, are not the most suitable techniques for measuring MPs/NPs in biological samples, however they are extensively used, because they are only providing measurements for plastic particles larger than 1 μm (Azeem et al., 2023). Microscopy techniques such as SEM and TEM are also used for identifying MPs/NPs in plant cells, and despite the accurate sample preparation, these samples became useless for continuous analysis (de Paiva Pinheiro et al., 2024). Therein, advanced applicable techniques combined with SEM and TEM are highly required to accurately determine the interaction level between MPs/NPs and plants and the toxicological endpoints. Detection of plastic pollution in the agricultural regions is of critical importance in assessing the degree of occurrence, impact, and toxicity in terrestrial ecosystems.

5.4.1 Pyrolysis Gas Chromatography/Mass Spectrometry

Pyrolysis gas chromatography-mass spectrometry (Py–GC–MS) is an informative and efficient analytical instrument at the laboratory scale that can characterize chemicals present in the milieu including MPs/NPs (Okoffo & Thomas, 2024). The workflow of Py–GC–MS depends on the thermal conversion of macromolecules to simpler vapor fragments through gas chromatography. In the Py–GC–MS, the sample is heated with GC carrier gas, and the high temperatures stimulate chemical changes in the MPs/NPs by bond breakdown, and the products are then returned to mass spectrometry for analysis (Kusch, 2017). Contrary to many analytical methods, Py–GC–MS is not hindered by particle sizes and at the same time displays high accuracy. Moreover, this technique has the advantages of using very small quantities of samples ranging (5 to 200 μg) with no specific pretreatments (Picó & Barceló, 2020). Recently, Li et al. (2021a) reported that Py–GC–MS has the capacity to precisely quantify and track cellular PS-NPs in different tissues of cucumber. Till the moment, Py–GC–MS is specifically deliberated to identify only PS in nanoscale, but the hope is to afford more choices through developing appropriate markers for this device in the near future.

5.4.2 Fluorescent Microscopy

The microscopy-based high-throughput approach is a powerful application to observe the imaging fluorescently labeled MPs/NPs in plant tissues. Plants are simply exposed to the targeted MPs/NPs in the experimental setting, and after fluorescent staining, slices from different organs are carefully gathered to be further examined. Numerous dyes are used for staining in this technique (Morgana et al., 2024). Red fluorescence is commonly utilized to probe plant roots, while green fluorescence is habitually employed to inspect leaves and aboveground materials. Recently, Luo et al. (2022) adopted a novel protocol, using lanthanide metal chelates as the principal fluorescent dyes in his study to visualize PS-NPs with fluorescence-free imaging to facilitate the next step detection by ICP-MS. LSCM at a definite wavelength is cooperatively used to measure the fluorescence in the plant slices (Azeem et al., 2023). The fluorescent microscopy has an optical $\sim$ 200 nm diffraction limitation on resolution (Nguyen & Tufenkji, 2022). Thus, allowing the fine mapping of MPs, whereas it is difficult to get clear images for localized NPs entire sample field; however, epifluorescence and confocal laser scanning microscopy could provide more clear images due to higher resolution and sensitivity, as well as eradication of out-of-spot light (Proença et al., 2022). The manipulation of these microscopes in plastic research has led to a rapid increase in the synthesis of fluorescent NPs, and many researchers have described protocols to synthesize labeled particles (Gao et al., 2023; Ilyas et al., 2024; Kim et al., 2024; Lu et al., 2023; Ouyang et al., 2024; Yin et al., 2024).

5.4.3 *Modern Techniques*

Besides the well-known methods to analyze Mps/NPs, more technological advances and analytical procedures have emerged in the last couple of years as reliable approaches for the rapid and sensitive characterization of MPS/NPs. As shown by Wu et al. (2024), an innovative photoluminescence-based detecting and imaging technique has been developed, presenting a simple and low-cost approach to identifying NPs in aqueous solutions and plant cells. The novelty of this technique, which successfully has been examined in *Arapidopsis thaliana* L., lies in designing a modified fluorescent probe with outstanding photostability properties; assists lighting up the NPs in variable environmental samples. Quantitative imaging by using laser ablation inductively coupled plasma mass spectrometry (LA-ICP-MS) has been implemented in an investigation to offer great spatial resolution and sensitivity for detecting traceable NPs in fresh samples (Wang et al., 2024). All these promising novelties could apply advantages over limitations of many techniques, and maybe by combining more than one technique, better characterization of these particles can be achieved with reliable results.

5.5 Atmospheric Transportation of MPs/NPs to Plant Body

Evidences have approved the possibility of airborne transportation of primary and secondary MPs (Allen et al., 2022; Sheraz et al., 2023). Importantly, MPs were found in atmospheric fallout, industrial emissions, dust, snow, and suspended microplastics, moving from air to ground, thus threatening the terrestrial ecosystem and plant health. For example, the quantity of atmospheric MPs in towns ranges from 2 to 925 items m^{-2} day^{-1}, while MPs in dust are 2.93–60 items g^{-1}, and 0.4–33.4 items m^{-2} (Wang et al., 2021a, 2021b, 2021c). Hence, MPs hold great perils, ending up with soil biota. Wind also lifts MPs, especially small particles, on the land up to the air; sometimes, they are transmitted long distances away from the original source. Boucher and Friot (2017) documented that 7% of terrestrial MPs are moved by wind and eventually leach into the ocean. Further, ocean spray can move some MPs/NPs from seawater back to their origin in the terrestrial system through atmospheric transport (Luo et al., 2024). This dynamic exchange between land, air, and water is likely to exacerbate pollution to plants.

5.6 Foliar and Root Uptake of MPs/NPs

Foliar and root uptake of MPs/NPs are potential pathways that facilitate the entry of plastic particles into plants. Several studies have shown the existence of MPs/NPs in plant leaves. MPs/NPs can stick on leaf surfaces and then be absorbed through the cuticle layer or enter when stomata pores are opening through atmospheric transition, precipitation, or polluted surfaces (Hua et al., 2023). Due to their tiny size, NPs > 200 nm can accumulate within the stomatal cavity and gradually cross the stomata into the biological membrane of leaf tissue. Further, accumulated NPs have the capability to penetrate leaves and migrate to plant roots. Sun et al. (2021) exhibited that maize leaves accumulated PS-COOH and PS-NH_2 and then moved them to roots through the stomatal pathway via the plant vasculature system. The authors showed that positively charged PS-NH_2 highly adhered with the leaf surfaces compared to negatively charged PS-COOH, suggesting that many factors such as surface charge and particle size could impact the accumulation and uptake of NPs. In addition, Ilyas et al. (2024) observed the different distribution of PS-NPs in leafy vegetables between the abaxial and adaxial sides of leaves, implying variable responses influenced by leaf structure. Moreover, studies demonstrated that plant leaves are exposed to hundreds of dangerous materials, and about 28% of stuck materials on leaf surfaces are MPs, proposing that plant leaves are active receptors for MPs/NPs.

The accumulation of MPs/NPs in root systems was detected in rice (Zhou et al., 2021), Pak choi, tomato, radish, and asparagus (Zytowski et al., 2024) in different developmental stages from germination till maturation. MPs/NPs move from soil to root cells via the root tips, the active apical meristems, and the primary and lateral root junction, followed by distribution to different plant organs. This process is controlled by the regulation of aquaporins in the solute transport system (Huang et al., 2024). The use of advanced tools such as confocal microscopy, scanning electron microscopy, and fluorescently tagged NPs facilitate the accurate detection of plastic particles entire the plant organs. Recently, Zhang et al. (2024) approved the translocation of fluorescently labeled PS-NPs (100 nm) from the root system to the stems, petioles, and leaves of grapevine. Moreover, some findings indicate that NPs can be transported to the edible fruits of some crops. Li et al. (2021c) observed the absorbance of PS-NPs (100–700 nm) by cucumber roots and the accumulation in leaves, flowers, and fruits. The take-up of NPs by root seems to happen through the crack-entry mode at lateral root appearance sites. This is the core tactic of NPs absorbance by plants. However, apoplast and symplast transports are additional pathways for NPs entry to plants. Through plant transpiration and root pressure, NPs can regulate their movements from the root epidermis to the extracellular channels and even reach the xylem vessel. Dong et al., (2021) pointed out that large MPs in carrot roots can break down into smaller pieces (0.2 μm) and deform the cell walls through transpiration, permitting SPS entry to cells. On the other hand, limited studies assume the possibility of transpiration of NPs by symplast transport route (Bandmann et al., 2012; Giorgetti et al., 2020). Only the study of Rong et al. (2024) confirmed for the first time the transportation of PS-MPs by the symplast pathway in oilseed rape roots, which may open the gate for more explorations supporting this theory.

5.7 Toxicity of MPs/NPs on Higher Plants

5.7.1 Germination

MPs/NPs as dangerous contaminants in agricultural fields, interact with plants and pose great risks to terrestrial vegetation, resulting in diverse biological interactions, physicochemical responses, and fates that threaten the food chain and human health. Toxicity metrics show that MPs/NPs can induce harm to plants at onset stages after sowing. In short bioassays, lentil plants exposed to 740–4990 nm PEMPs showed a rapid drop in germination percentages within the early 6 h (De Silva et al., 2022), while cress plants treated with 50, 500, and 4800 nm showed a similar decline in 8 h from exposure (Bosker et al., 2019). Yu et al. (2024) also documented that MPs/NPs significantly declined the germination rate of *Arabidopsis thaliana* L. seeds during the first 7 days of germination. Several MPs PS, PP, and PE markedly disturbed tomato germination, particularly at high concentrations (Shi et al., 2022). In contrast, Allium cepa exposed to 50 nm PS beads for three days showed no notable effect on germination, but localized PS beads entire roots caused severe oxidative stress and inhibition of root growth (Giorgetti et al., 2020). In most of these studies, all the phenological parameters of the shoot and root development were ominously repressed. NPs induce more toxic effects than MPs. NPs are readily absorbed through the seed surface and easily permeate bio-membranes, impeding seed capsule pores and affecting growth. Several research studies have evidenced that NPs have negative effects on the seed development and germination quality of some plants such as wheat (Lian et al., 2020), cabbage (Zhang et al., 2022), rice (Spanò et al., 2022; Zhang et al., 2021), and rapeseed (Li et al., 2023).

5.7.2 Oxidative Stress

One of the documented symptoms of toxicity induced by MPs/NPs is the rapid evoking of reactive oxygen species (ROS) such as hydrogen peroxide and hydroxyl radicals. Increased levels of ROS can instigate DNA destruction and protein misfolding at the cellular level (Abeed et al., 2023; El-Moneim et al., 2021). The intense ROS arouse beyond the antioxidant system's quenching ability triggers the peroxidation of lipid membranes and cell walls, which accelerates the programmed cell death and hostile menaces to the overall plant metabolism (Abdelmoneim et al., 2025; Dawood et al., 2025a, b). The anti-oxidative defense network is rapidly organized in different plant species to face challenged ROS in response to MPs/NPs. Rice (Wang et al., 2022), soybean (Qiu et al., 2023), garlic (Li et al., 2024a), maize (Li et al., 2024c), and grape (Zhang et al., 2024) modulated SOD, POD, and CAT activities under different concentrations of PS-NPs to scavenge ROS and reduce oxidative emergency. Moreover, Surgun-Acar (2022) reported that soybean copped with PS-NPs stress by regulating the level of proline, flavonoids, and phenolics, while Zhang et al. (2023) found that soluble proteins and soluble sugars were upregulated for active involvement in lessening PS-MPs toxicity in strawberry. In vitro analysis also showed the regulation of water-soluble vitamins and secondary metabolites to adapt with PS-NPs in micropropagated *Stevia rebaudiana* L. (Coman et al., 2023). These outcomes suggest that MPs/NPs could cause a high degree of oxidative stress and toxicity in plants; however, the interactions among MPs/NPs features (type, amount, and origin), soils, and plants greatly affect the plant response and mode of interaction.

5.7.3 Photosynthesis

Various stressors can prompt ROS-mediated chloroplast dysfunction and disturbance of central biochemical processes such as photosynthesis (Singh et al., 2022). Chloroplasts, as the primary locations of ROS provoke in stressed situations, are sensitive to external stresses. MPs/NPs were reported to jeopardize the photosynthesis cycle in many plant species (Ceccanti et al., 2024; Gao et al., 2024; Hasan & Jho, 2023; Liang et al., 2024; Li et al., 2024a; Tian et al., 2024; Zhang et al., 2022; Zhuang et al., 2023). The decline in photosynthetic pigments like chlorophyll a, chlorophyll b, and carotenoid contents is a reflector of negative impacts of MPs/NPs on photosynthesis performance, which may be ascribed to metabolic syndromes or to the aggressive accumulation of ROS in chloroplasts (Ren et al., 2021). Gao et al. (2019) stated that Rubisco activity in red lettuce may be abridged by NPs as a result of increased oxidative stress markers. In addition, due to the tiny size of NPs, they can block cell walls and reduce water uptake, thus affecting stomata function during photosynthesis. All these behaviors influence light use efficiency, exchange of gases and respiration, leading to physiological disorders and growth impartment.

5.7.4 Cytotoxicity and Genotoxicity

Vital plant compartments including nuclei can deleteriously be affected by MPs/NPs treatments. In response to PS-MPs in *broad beans*, the plants exhibited genotoxic impairment when the mitotic index was substantially reduced with enlarged micronucleus incidence in root cells (Jiang et al., 2019). Similar results were stated by Spanò et al. (2022). Some reports described a higher frequency of chromosomal and nuclear aberrations in root cells after exposure to MPs/NPs. An example is the cytotoxicy, genotoxicity, and oxidative damage that occurred to *Allium cepa* L. roots after exposure to PS-NPs (0.01–1.0 g/L) for 72 h (Giorgetti et al., 2020).

5.7.5 *Nutritional Quality*

MPs/NPs in soil–plant systems interrupt plant development by altering soil properties and root activities, decreasing nutrient availability. A report displayed that PS-NPs drastically influenced the total amount of K and Fe in wheat shoots, which suggests that NPs impact the transport and relocation of mineral elements in plant tissues (Lian et al., 2020). In another study, leaf C/N ratio declined considerably when boosting the foliar sprayed PS-NPs exposure dosages, leading to Fe, Mn, and Cu deficiency in lettuce (Lian et al., 2021). In *Cucurbita pepo* L., the impact of several MPs types involving PP, PE, PVC, and PET on element uptake in plant organs was examined (Colzi et al., 2022). The study revealed that the toxic influence of MPs on the nutritional value of fruits varied according to the type and dosages of MPs. The accumulation of MPs/NPs in plant tissues could disturb the essential physiological processes in plants. Lian et al. (2021) assumed that Fe deficiency directly correlated with the suppression in the photosynthesis capacity of Lactuca *sativa* L. when treated with PS-NPs. More recently, Ceccanti et al. (2024) found that PS-MP_S declined fruit size and anthocyanin content in strawberry. Bai et al., (2024) documented the reduced absorbance of essential macronutrients (N, P, and K) in strawberry seedlings by MPs, which are fundamental for plant growth and development. The authors found that the levels of these nutrients in roots were upper than the control, while they were less than the control in the plant leaves. This indicates that MPs perturb the translocation of N, P, and K and plant growth. This could be explained by the physical block of the nutrient uptake channel because the translocation by the xylem and plasma membrane of root cells is involved in the uptake and transport of ions in plant tissues. These outcomes demonstrate that plants exposed to MPs/NPs may show stunted growth, weakened metabolic activities, and nutritional deficiencies, threatening food security and human health.

5.8 Mitigation Strategies of Micro- or Nano-Plastics Pollution

5.8.1 *Remediation*

Phytoremediation is an eco-friendly method to reduce MPs and NPs in the environment. Certain plant species, including aquatic plants and grasses, absorb MPs through their roots, while NPs adhere to plant surfaces (Gong et al., 2023). Plants with extensive root systems, like Vetiver grass, trap MPs in the soil, limiting their mobility. Additionally, enhanced soil microbial activity around plant roots accelerates plastic degradation. Phytoremediation not only cleans the environment but also prevents plastics from entering the food chain, safeguarding ecosystems and human health. Plants like water hyacinth, duckweed, and common reed have shown significant potential in these processes (Ren et al., 2019). Optimizing this approach involves selecting suitable plant species and utilizing molecular tools to understand plant-plastic interactions (Kumar et al., 2024).

5.8.2 Biodegradation and Enzymatic Degradation

Enzymatic degradation provides a promising method to break down plastics. Microbial enzymes, such as lipases, proteases, and esterases, degrade polymers like polyethylene terephthalate (PET), polyethylene (PE), and polystyrene (PS) into smaller molecules (Mao et al., 2018). For instance, *Ideonella sakaiensis* produces PETase, an enzyme capable of hydrolyzing PET into its monomers. Fungal species like Aspergillus and Penicillium have also demonstrated plastic degradation capabilities. Scaling up these processes requires optimizing factors like temperature, pH, and nutrient availability, often in bioreactor systems. However, challenges such as efficiency, scalability, and environmental impacts of degradation byproducts persist (Arnal et al., 2023).

One of the most important enzymes used for plastics degradation, PETase. PETase enzymes derived from *I. sakaiensis* have been engineered for plastic degradation. Modifications to enhance enzyme efficiency enable the biodegradation of PET films and powders, achieving up to 98% conversion to terephthalic acid and ethylene glycol within 24 h (Jia et al., 2021; Mao et al., 2018). While enzymes like LCCICCG outperform others in large-scale applications, further advancements are needed to improve reaction rates and scalability (Arnal et al., 2023). Also, laccase enzymes, belonging to the oxidoreductase family, play a role in degrading MPs and NPs. They catalyze the oxidation of phenolic and non-phenolic compounds, creating reactive radicals that aid in breaking down complex polymers (Zhou et al., 2022). These enzymes modify the surface properties of plastics, making them more susceptible to further enzymatic or microbial degradation.

5.8.3 Nanotechnology-Based Remediation

Nanotechnology offers innovative solutions for mitigating MPs and NPs. Nanomaterials such as carbon nanotubes, graphene oxide, and magnetic nanoparticles show high adsorption capacities, capturing plastic particles effectively (Tang et al., 2021). Functionalized nanomaterials with tailored surface chemistries improve selectivity. Nanocatalysts, including titanium dioxide and iron oxide, leverage photocatalytic properties to degrade plastics into smaller molecules under light exposure (Domínguez-Jaimes et al., 2021). Enzymatic nano-catalysts enhance degradation efficiency for specific plastics (Ouda et al., 2023). Nanosensors incorporating carbon nanotubes, quantum dots, or gold nanoparticles enable real-time detection of plastic contamination, facilitating timely interventions (Jebril et al., 2024). Despite their potential, challenges such as toxicity, environmental persistence, and scalability require further research.

5.8.4 Advanced Filtration Technologies

Innovative filtration membranes with precise pore sizes are being developed to capture MPs and NPs. These membranes can be integrated into wastewater treatment systems, preventing plastic particles from contaminating water sources (Krishnan et al., 2023). One of these technologies is electrocoagulation and electrochemical methods. In this regard, electrocoagulation is an efficient way to remove MNPS from water by using electric current to destabilize and aggregate MNPs particles. The electrochemical methods like electrooxidation and electro-Fenton processes involves breaking down plastic particles (Malinović et al., 2022).

5.9 Phytoremediation of MPs/NPs

Phytoremediation is a remediation of toxic materials using plant-based technique with merits of low investment and less disturbance to the environment. Phytoremediation may be an alternative strategy for the decontamination of MPs/NPs.

There are possible mechanisms for phytoremediation that can be summarized as follows.

5.9.1 Phytostabilization

This method depends on the fact that plant leaves could act as anti-air pollution, by removing pollutants from the atmosphere. This occurs through large and uneven leaves surfaces serve as a sink of pollutants. In this context, the deposition of MPs/NPs on the aerial parts, such as leaves, may represent a potential mechanism for the phytostabilization of airborne MPs/NPs by plants that the air-borne MPs could deposit in the aerial parts of terrestrial plants (Liu et al., 2020). Mateos-Cárdenas et al. reported the settlement of PE MPs to the frond surface of *Lemna minor* (L.) (Mateos-Cárdenas et al., 2019). The aerial part surfaces of the plant allow hydrophobic attractions or formation of hydrogen bond, so MPs/NPs settle on their surfaces.

The absorption of MPs/NPs by plants is another phytostabilization way for reducing the contamination by these particles. Sundbæk et al. (2018) reported the sorption of fluorescent polystyrene (PS) MPs onto *Fucus vesiculosus*. They illustrated that the release of anionic polysaccharide substances on the plant surface plays a role in stabilizing MPs. Unlike nanoparticles (NPs), which plants can directly uptake, most MPs are too large to enter plant tissues. Gao et al. (2022), using scanning and transmission electron microscopy, demonstrated that PE MPs adhered to the root surface of *Lactuca sativa* but did not penetrate the roots. However, MPs captured by plant roots can reduce mobility, limit their spread to contaminated sites, and prevent interactions with other organisms, suggesting that phytostabilization could serve as an effective in situ restoration strategy for MPs/NPs in both aquatic and terrestrial ecosystems. Wu et al. (2016) found that pollutants could be immobilized by mycorrhizal associations, indicating that mycorrhizal fixation might also be a viable mechanism for the phytostabilization of MPs/NPs. Rozman et al. (2022) found that MPs/NPs settled on the surface of aquatic macrophytes, demonstrating their role in phytoremediation of MPs/NPs from aquatic environments. Rozman and co-workers (2022) observed that polyethylene (PE) MPs (average size of 149 ± 75 μm) adhered to the aquatic floating macrophyte *Lemna minor*, these adhered MPs represents 20% of the total particles. They stated that surface stickiness of aquatic plants may enable the agglomeration of MPs/NPs on roots and fronds, while electrostatic interactions between MPs/NPs and plant biomass could also contribute to this adhesion.

5.9.2 Phytoaccumulation and Translocation of MPs/NPs

Pollutants can enter plants through metabolic and non-metabolic absorption processes, with their movement facilitated by root pressure and transpiration. A key mechanism in phytoremediation involves the uptake and translocation of contaminants within plants. Recent studies have confirmed the phytoaccumulation of MPs/NPs in plants. For instance, Li et al. (2020) demonstrated that polystyrene (PS) plastic beads ranging from 0.2 to 2 μm were accumulated in plants. Similarly, PS MPs (100 nm) were shown to accumulate in plant roots at a concentration of 100 mg/L within 48 h (Jiang et al., 2019). Nanoparticles (NPs) have also been detected in specific plant tissues such as the cell walls of *Murraya exotica* (Zhang et al., 2019a, 2019b), and PS NPs were observed in the cytoplasm, vacuoles, and nuclei of root cells (Giorgetti et al., 2020). The mechanism by which MPs/NPs enter plant cells remains unclear. However, Bandmann et al., (2012) suggested that endocytosis could be involved, based on their observation of fluorescent nano-beads entering tobacco cells. Further research is needed to elucidate the exact pathways.

Beyond phytoaccumulation, the translocation of MPs/NPs within plants has been studied. Urbina et al. (2020) reported that MPs accumulated in roots without being transported to shoot. In contrast, other studies have confirmed the movement of MPs/NPs from root to aerial parts. In this regard, Lian et al. (2020) used scanning electron microscopy (SEM) and 3D laser confocal scanning microscopy (LCSM) to observe 100 nm PS NPs transported from roots to shoots and leaves. Li et al. (2020) found that PS (0.2–2 μm) and polymethyl methacrylate (PMMA, 0.2 μm) particles entered the root stele through cracks and were then pulled to shoots through transpirational pull.

Additionally, Azeem et al. (2021) observed a bidirectional movement of MPs/NPs. MPs/NPs entered leaves through stomata and were then transported down to stems and roots via vascular bundles. The confirmed phytoaccumulation and translocation of MPs/NPs in plants suggests that using plants to remediate MPs/NPs from contaminated environments is a viable strategy.

5.9.3 Rhizosphere Effects on MPs/NPs Phytoremediation

The effects of the rhizosphere on MPs/NPs can be categorized into direct and indirect impacts. Direct interactions influence the behavior of MPs/NPs, while indirect effects involve changes in soil properties, microbial communities, and plant stress responses, all of which play crucial roles in decontaminating MPs/NPs.

Direct Effects: Root Exudates and MPs/NPs Behavior

Plant root exudates, composed of organic acids, carbohydrates, amino acids, and enzymes, significantly affect phytoremediation. These exudates can alter the behavior of MPs/NPs in the environment. For instance, root exudates from *Zea mays* modified the aggregation and dissolution of nanosized pollutants (Shang et al., 2019). Additionally, the distribution and characteristics of MPs were shown to depend on rhizosphere interactions and organic matter content (Li et al., 2019). Citric acid, a common root exudate, enhanced the dispersion and uptake of nanosized pollutants (Bao et al., 2020). Conversely, Ye et al. (2018) found that amino acids such as aspartic acid and lysine promoted the agglomeration of pollutants when their electrical charge differed from the surface ligand of the pollutants, immobilizing MPs/NPs in the media. Root exudates, therefore, can either reduce MPs/NPs mobility to minimize adverse effects or enhance bioavailability for easier plant uptake and removal.

Indirect Effects: Environmental Changes, Microbial Interactions, and Stress Alleviation

1. Environmental Changes: Plants influence soil properties, indirectly affecting pollutant behavior. Xu et al., (2020a, 2020b) demonstrated that different plants altered total nitrogen, organic matter, and available potassium, correlating with pollutant immobilization and translocation. Luo et al. (2020) found that the immobilization of PS MPs was positively associated with soil organic carbon and Fe_2O_3 but negatively correlated with pH and clay content. Such plant-induced changes in soil physicochemical properties may enhance MPs/NPs immobilization.
2. Microbial Interactions: Root exudates modify microbial community composition, promoting microbes that assist in pollutant degradation. For example, exudates from *Festuca arundinacea* stimulated bacteria involved in degrading organic pollutants (Liu et al., 2014). Similarly, flavonoids in root exudates enhanced rhizobacterial degradation of pollutants (Pino et al., 2016). Specific microbes, such as *Bacillus cereus*, have been shown to degrade MPs like PE,

PET, PP, and PS (Auta et al., 2017a, 2017b). Thus, root exudates facilitate MPs/NPs degradation by fostering beneficial microbial communities.

3. Stress Alleviation: Root exudates also help plants mitigate abiotic stress. Organic acids like citric acid and glutaric acid reduced pyrene-lead-induced stress in *Scirpus triqueter*, promoting seedling growth and minimizing oxidative damage (Zhang et al., 2019a, 2019b). Additionally, root exudates shape the root microbiome and attract beneficial soil bacteria, enhancing plant growth and resistance to pollutants (Bora et al., 2016; Xiong et al., 2020). Enhanced plant growth increases pollutant accumulation, while improved resistance allows plants to tolerate higher pollutant concentrations, improving remediation efficiency.

Exposure to MPs/NPs has been shown to cause stress and toxicity in plants. For instance, MPs residues negatively affected the growth and PS MPs exhibited genotoxic effects in *Vicia faba* (Jiang et al., 2019). By alleviating these stress effects, root exudates contribute to the phytoremediation of MPs/NPs.

5.10 Key Determinants of Plant-Based Remediation for MPs/NPs Pollution

1. *Plant Factors*

Phytoremediation efficacy varies significantly across plant species due to differences in absorption, accumulation, and transportation of microplastics (MPs) and nanoplastics (NPs). For instance, Sun et al. (2020) reported that positively and negatively charged NPs were accumulated in the plants and the negatively charged NPs (PS-SO3H NPs (55 nm) and PS-NH2 NPs (71 nm)) mainly presented in the apoplast and xylem of *Arabidopsis thaliana.* Sun et al. (2021) reported that PS NPs of two types; PS-COOH NPs (22 nm) and PS-NH2 NPs (24 nm) were accumulated on the maize (*Zea mays* L.) leaves; PS NPs move to the vasculature through stomata opening, and PS NPs moved down to the roots through vascular bundle. Also, seagrasses accumulate higher MP densities than macroalgae (Seng et al., 2020), while *Arabidopsis thaliana* exhibited MPs/NPs accumulation only at root cap cells, without internal uptake (Taylor et al., 2020). The responses of plants to presence of MPs/NPs varied from positive to negative responses. Positive plant responses, such as increased root length and nutrient uptake under MP/NP stress, have been observed in *Triticum aestivum L.* (Lian et al., 2020). Conversely, MPs/NPs can cause toxicity, including growth inhibition, oxidative stress, and genotoxicity in plants like *Lolium perenne* and *Zea mays L.* (Wang et al., 2020). Plants also release root exudates and modify rhizosphere environments, facilitating MP/NP immobilization or promoting their uptake and removal (Li et al., 2023).

2. *MPs/NPs Characteristics*

The physical and chemical properties of MPs/NPs, such as size, shape, and charge, significantly influence their interaction with plants. Smaller NPs show higher bioavailability, as seen in *Raphanus sativus L.* (Wojcieszek et al., 2019). However, agglomeration of smaller particles can hinder their uptake (Wang et al., 2021a, 2021b, 2021c). The charge of MPs/NPs also matters; negatively charged particles are absorbed more readily (Sun et al., 2020). Chemical composition influences toxicity, with PVC MPs showing higher adverse effects than PE or PP MPs (Pignattelli et al., 2020).

3. *Environmental Factors*

Environmental conditions, including soil properties, pH, and climate, affect MPs/NPs mobility and plant uptake. For example, MPs exhibit stronger mobility in soils with high pH and low ionic strength, enhancing bioavailability (Wu et al., 2020). Additionally, combined pollution, such as the presence of heavy metals, may reduce MP/NP uptake. Interestingly, arsenic can increase MP uptake by altering surface properties, as observed in carrots (Dong et al., 2021).

4. *Challenges and Risks*

Despite its potential, phytoremediation of MPs/NPs faces challenges. Accumulated MPs/NPs in plants pose risks of bioaccumulation in the food chain, as seen with seaweed transferring MPs to herbivores (Gutow et al., 2016). Moreover, immobilized MPs/NPs can be re-released or degraded into smaller toxic fragments by root exudates or rhizosphere microorganisms, exacerbating their toxicity.

5.11 Sustainable Solutions for Reducing Microplastics and Nanoplastics in Ecosystems

Efforts to mitigate the harmful effects of plastic fragments, driven by the large-scale production and use of plastics, are gaining momentum. However, addressing the economic and environmental impacts of microplastics (MPs) and nanoplastics (NPs) in agriculture remains challenging due to significant knowledge gaps. Below is a concise overview of current strategies:

Addressing MPs/NPs in Agricultural Lands

The primary sources of MPs/NPs in cropping systems include plastic mulch. Replacing plastic mulches with organic materials like crop residues, rice straw, or biodegradable alternatives has shown promise in reducing contamination while promoting sustainability (Tan et al., 2016). Biodegradable films and natural fiber polymers also help minimize waste while enhancing soil nutrition when paired with biochar application (Palansooriya et al., 2022; Sintim et al., 2020).

Biosolids, such as sewage sludge and manure, often contain MPs/NPs. Hyper-thermophilic composting (HTC) is emerging as an effective method to process biosolids by reducing plastic and heavy metal contamination (Chen et al., 2021a, 2021b). Biodegradable plastics and microbial degradation are also gaining attention as environmentally safer solutions (Bhatt et al., 2021; Ju et al., 2021).

Advanced Remediation Techniques

1. Biodegradable Alternatives: Chito-beads in cosmetics have shown superior cleansing and soil degradation properties compared to traditional polyethylene microbeads, suggesting potential in other sectors (Ju et al., 2021).
2. Encapsulated Enzymes: Using enzymes near plant roots or seeds could degrade MPs/NPs directly in soils (Camenzuli et al., 2017).
3. Melatonin Application: Melatonin enhances plant tolerance to MPs/NPs by improving antioxidant activity and maintaining redox balance (Roy et al., 2023).

Planting windbreaks of trees near roads and industrial areas can reduce airborne MPs/NPs deposition, while harvested trees could be repurposed for furniture or bioenergy (Roy et al., 2023). Improving plastic recycling and promoting sustainable materials are critical steps in reducing plastic waste at the source.

Challenges in Wastewater Treatment

Traditional methods such as coagulation, filtration, and membrane bioreactors have shown potential for removing MPs/NPs from wastewater. For example, ferric chloride and polyaluminum chloride effectively coagulate MPs for removal (Rajala et al., 2020). Marine fungi like *Zalerion maritimum* have demonstrated the ability to degrade polyethylene MPs, though microbial remediation is limited by environmental factors and risks of secondary pollution (Paço et al., 2017).

While various technologies and approaches are being explored, the complexity of MPs/NPs pollution demands integrated, environmentally friendly, and widely applicable solutions. Continued research and innovation are essential to tackle this pressing environmental issue.

5.12 Conclusion, Gaps, and Future Recommendations

The widespread distribution of MPs/NPs in terrestrial ecosystems poses a significant threat to plant health, soil quality, food chain, and environmental sustainability. This chapter concluded the recent knowledge on distribution, sources, and the multifaceted impacts of MPs/NPs, from impairing plant growth and nutrient uptake to inducing oxidative stress and disrupting gene expression. These effects not only compromise plant biology but also destabilize soil structure, nutrient cycling, and microbial communities, creating a cascading impact on ecosystem dynamics. While challenges persist, promising strategies exist to mitigate MPs/NPs pollution. These include advancing biodegradable plastic alternatives, implementing strict plastic usage regulations, improving recycling and waste management systems, and adopting eco-friendly agricultural practices. However, gaps in policy frameworks and the lack of standardized monitoring protocols hinder the effective management of plastic pollution. Multidisciplinary collaboration, public awareness campaigns, and a shift toward a circular economy emphasizing waste reduction and responsible disposal are critical to solve this issue comprehensively.

Despite growing awareness of MPs/NPs pollution, several knowledge gaps remain such as limited understanding about how MPs/NPs are absorbed by plants and their toxicological effects at the physiological, molecular, and ecological levels. In addition, the environmental longevity and degradation pathways of MPs/NPs in soils and their interactions with various biotic and abiotic factors are poorly understood. Current remediation approaches are largely experimental, with limited scalability and region-specific applicability. There is also insufficient research on innovative technologies for the recovery and degradation of MPs/NPs. To address these gaps and mitigate MPs/NPs pollution, conducting interdisciplinary studies to elucidate MPs/NPs uptake mechanisms, toxicological impacts, and long-term environmental persistence. Advanced analytical techniques should be employed to improve detection and quantification in various environmental matrices. Innovative Remediation in terms of developing cost-effective and scalable technologies tailored to different ecosystems, such as enzyme encapsulation, engineered microorganisms, and hyperthermophilic composting. Importantly, establishing global standards for monitoring MPs/NPs pollution and harmonizing regulatory frameworks to promote the adoption of sustainable materials and practices. Public Awareness and Collaboration: Increase public education on plastic pollution and encourage collaboration among policymakers, industry leaders, researchers, and civil society to drive investment in eco-friendly solutions. Fostering a circular economic model that emphasizes waste reduction, recycling, biodegradable materials, and responsible disposal to curb the influx of plastics into ecosystems. By addressing these gaps and adopting holistic strategies, society can mitigate the adverse impacts of MPs/NPs and ensure the long-term sustainability of terrestrial ecosystems.

References

Abdelmoneim, M. S., Dawood, M. F., Hafez, E. E., Hammad, S. F., & Ghazy, M. A. (2025). Unveiling the role of cadaverine in mitigating salinity and/or Bisphenol A toxicity in tomato plants and reduced Bisphenol A accumulation in tomato roots. *Plant Physiology and Biochemistry, 224,* 109799.

Abeed, A. H., Al-Huqail, A. A., Albalawi, S., Alghamdi, S. A., Ali, B., Alghanem, S. M., Al-Haithloul, H. A. S., Amro, A., Tammam, S. A., & El-Mahdy, M. T. (2023). Calcium nanoparticles mitigate severe salt stress in Solanum lycopersicon by instigating the antioxidant defense system and renovating the protein profile. *South African Journal of Botany, 161,* 36–52.

Adetuyi, B. O., Mathew, J. T., Inobeme, A., Falana, Y. O., Adetunji, C. O., Shahnawaz, M., Oyewole, O. A., Eniola, K. I. T., Yerima, M. B., & Popoola, O. A. (2024). Tyres, bitumen wear and plastic bottles, other single used plastic as major sources of microplastic. In *Microplastic pollution* (pp. 141–160). Springer.

Aeschlimann, M., Li, G., Kanji, Z. A., & Mitrano, D. M. (2022). Potential impacts of atmospheric microplastics and nanoplastics on cloud formation processes. *Nature Geoscience, 15,* 967–975.

Akindoyo, J. O., Beg, M., Ghazali, S., Islam, M. R., Jeyaratnam, N., & Yuvaraj, A. R. (2016). Polyurethane types, synthesis and applications–A review. *RSC Advances, 6,* 114453–114482.

Ali, N., Khan, M. H., Ali, M., Ahmad, S., Khan, A., Nabi, G., Ali, F., Bououdina, M., & Kyzas, G. Z. (2023). Insight into microplastics in the aquatic ecosystem: Properties, sources, threats and mitigation strategies. *Science of the Total Environment,* 169489.

Allen, D., Allen, S., Abbasi, S., Baker, A., Bergmann, M., Brahney, J., Butler, T., Duce, R. A., Eckhardt, S., Evangeliou, N., Jickells, T., Kanakidou, M., Kershaw, P., Laj, P., Levermore, J., Li, D., Liss, P., Liu, K., Mahowald, N., & Masque, P. (2022). Microplastics and nanoplastics in the marine-atmosphere environment. *Nature Reviews Earth & Environment, 3*(6), 393–405.

An, L., Liu, Q., Deng, Y., Wu, W., Gao, Y., & Ling, W. (2020). Sources of microplastic in the environment. In *Microplastics in terrestrial environments: Emerging contaminants and major challenges* (pp. 143–159).

Andrée, B. P. J., Chamorro, A., Spencer, P., Koomen, E., & Dogo, H. (2019). Revisiting the relation between economic growth and the environment; a global assessment of deforestation, pollution and carbon emission. *Renewable and Sustainable Energy Reviews, 114*, Article 109221.

Antonakou, E. V., & Achilias, D. S. (2013). Recent advances in polycarbonate recycling: A review of degradation methods and their mechanisms. *Waste and Biomass Valorization, 4*, 9–21.

Arnal, G., Anglade, J., Gavalda, S., Tournier, V., Chabot, N., Bornscheuer, U. T., Weber, G., & Marty, A. (2023). Assessment of four engineered PET degrading enzymes considering large-scale industrial applications. *ACS Catalysis, 13*, 13156–13166.

Auta, H. S., Emenike, C. U., & Fauziah, S. H. (2017a). Screening of bacillus strains isolated from mangrove ecosystems in Peninsular Malaysia for microplastic degradation. *Environmental Pollution, 231*(Pt2), 1552–1559.

Auta, H. S., Emenike, C. U., & Fauziah, S. H. (2017b). Distribution and importance of microplastics in the marine environment: A review of the sources, fate, effects, and potential solutions. *Environment International, 102*, 165–176.

Ave, P., & Babolsar, I. (2010). Environmental performance index and economic growth: Evidence from some developing countries. *Australian Journal of Basic and Applied Sciences, 4*, 3098–3102.

Azeem, I., Adeel, M., Ahmad, M. A., Shakoor, N., Jiangcuo, G. D., Azeem, K., Ishfaq, M., Shakoor, A., Ayaz, M., Xu, M., & Rui, Y. (2021). Uptake and accumulation of nano/microplastics in plants: A critical review. *Nanomaterials, 11*(11), 2935.

Azeem, I., Shakoor, N., Chaudhary, S., Muhammad Adeel, Zain, M., Muhammad Arslan Ahmad, Li, Y., Zhu, G., Shah, A., Khan, K., Adnan Anwar Khan, Xu, M., & Rui, Y. (2023). Analytical challenges in detecting microplastics and nanoplastics in soil-plant systems. *Plant Physiology and Biochemistry*, 108132.

Bai, H., Yang, Y., & Huang, Y. (2024). Distribution of microplastics and their effects on nutrient absorption in strawberry plants. *Scientia Horticulturae, 332*, Article 113214.

Bandmann, V., Müller, J. D., Köhler, T., & Homann, U. (2012). Uptake of fluorescent nano beads into BY2-cells involves clathrin-dependent and clathrin-independent endocytosis. *FEBS Letters, 586*(20), 3626–3632.

Bao, Y., Ma, J., Pan, C., Guo, A., Li, Y., & Xing, B. (2020). Citric acid enhances Ce uptake and accumulation in rice seedlings exposed to CeO_2 nanoparticles and iron plaque attenuates the enhancement. *Chemosphere, 240*, 124897–124907.

Belioka, M.-P., & Achilias, D. S. (2024). The effect of weathering conditions in combination with natural phenomena/disasters on microplastics' transport from aquatic environments to agricultural soils. *Microplastics, 3*, 518–538.

Bhat, M. A., Gedik, K., & Gaga, E. O. (2024). A preliminary study on the natural aging behavior of microplastics in indoor and outdoor environments. *International Journal of Environmental Science and Technology, 21*, 1923–1936.

Bhatt, P., Pathak, V. M., Bagheri, A. R., & Bilal, M. (2021). Microplastic contaminants in the aqueous environment, fate, toxicity consequences, and remediation strategies. *Environmental Research, 200*, Article 111762. https://doi.org/10.1016/j.envres.2021.111762

Bonifazi, G., Capobianco, G., & Serranti, S. (2018). A hierarchical classification approach for recognition of low-density (LDPE) and high-density polyethylene (HDPE) in mixed plastic waste based on short-wave infrared (SWIR) hyperspectral imaging. *Spectrochimica Acta Part a: Molecular and Biomolecular Spectroscopy, 198*, 115–122.

Bora, K., Geun Cheol, S., & Choong-Min, R. (2016). Root exudation by aphid leaf infestation recruits root-associated *Paenibacillus* spp. to lead plant insect susceptibility. *Journal of Microbiology and Biotechnology, 26*, 549–557.

Bosker, T., Bouwman, L. J., Brun, N. R., Behrens, P., & Vijver, M. G. (2019). Microplastics accumulate on pores in seed capsule and delay germination and root growth of the terrestrial vascular plant Lepidium sativum. *Chemosphere, 226*, 774–781.

Boucher & Froit. (2017). Primary microplastics in the oceans: a global evaluation of sources.

Bowmer, T., & Kershaw, P. (2010). *Proceedings of the GESAMP international workshop on microplastic particles as a vector in transporting persistent, bio-accumulating and toxic substances in the ocean, 28–30th June 2010, UNESCO-IOC, Paris*. GESAMP.

Camenzuli, D., Wise, L. E., Stokes, A. J., & Gore, D. B. (2017). Treatment of soil cocontaminated with inorganics and petroleum hydrocarbons using silica: Implications for remediation in cold regions. *Cold Regions Science and Technology, 135*, 8–15. https://doi.org/10.1016/j.coldregions.2016.10.008

Ceccanti, C., Davini, A., Lo Piccolo, E., Lauria, G., Rossi, V., Ruffini Castiglione, M., Spanò, C., Bottega, S., Guidi, L., & Landi, M. (2024). Polyethylene microplastics alter root functionality and affect strawberry plant physiology and fruit quality traits. *Journal of Hazardous Materials, 470*, Article 134164.

Cesa, F. S., Turra, A., & Baruque-Ramos, J. (2017). Synthetic fibers as microplastics in the marine environment: A review from textile perspective with a focus on domestic washings. *Science of the Total Environment, 598*, 1116–1129.

Chaukura, N., Gwenzi, W., Bunhu, T., Ruziwa, D. T., & Pumure, I. (2016). Potential uses and value-added products derived from waste polystyrene in developing countries: A review. *Resources, Conservation and Recycling, 107*, 157–165.

Chen, G., Feng, Q., & Wang, J. (2020). Mini-review of microplastics in the atmosphere and their risks to humans. *Science of the Total Environment, 703*, Article 135504.

Chen, L., & Lin, Z. (2021). Polyethylene: properties, production and applications. In *2021 3rd International Academic Exchange Conference on Science and Technology Innovation (IAECST)* (pp. 1191–1196). IEEE.

Chen, Y., Awasthi, A. K., Wei, F., Tan, Q., & Li, J. (2021a). Single-use plastics: Production, usage, disposal, and adverse impacts. *Science of the Total Environment, 752*, Article 141772.

Chen, Z., Xing, R., Yang, X., Zhao, Z., Liao, H., & Zhou, S. (2021b). Enhanced in situ Pb (II) passivation by biotransformation into chloropyromorphite during sludge composting. *Journal of Hazardous Materials, 408*, Article 124973. https://doi.org/10.1016/j.jhazmat.2020.124973

Colzi, I., Renna, L., Bianchi, E., Castellani, M. B., Coppi, A., Pignattelli, S., Loppi, S., & Gonnelli, C. (2022). Impact of microplastics on growth, photosynthesis and essential elements in Cucurbita pepo L. *Journal of Hazardous Materials, 423*, Article 127238.

Coman, V., Scurtu, V.-F., Coman, C., Clapa, D., Iancu, ȘD., Leopold, N., & Leopold, L. F. (2023). Effects of polystyrene nanoplastics exposure on in vitro-grown Stevia rebaudiana plants. *Plant Physiology and Biochemistry, 197*, Article 107634.

Das, A., & Mahanwar, P. (2020). A brief discussion on advances in polyurethane applications. *Advanced Industrial and Engineering Polymer Research, 3*, 93–101.

Dawood, M. F., Tahjib-Ul-Arif, M., Shirazy, B. J., & Latef, A. A. H. A. (2025a). Unraveling the role of κ-carrageenan on the combined effect of drought and chromium stress in wheat (Triticum aestivium L.). *Plant Physiology and Biochemistry, 223*, 109781.

Dawood, M. F., Mazrou, Y. S., Sheteiwy, M. S., Nehela, Y., Makhlouf, A. H., Abd El, F. E. Z. A., & Hifney, A. F. (2025b). Exploring the role of Arthrospira platensis and earthworms (Lumbricus castaneus) biosystem in mitigating antimony toxicity and enhancing the biochemical properties of cucurbit plants. *Plant Physiology and Biochemistry*, 110096.

de Paiva Pinheiro, S. K., da Silva Pontes, M., Miguel, T. B. A. R., Grillo, R., Souza Filho, A. G., & de Castro Miguel, E. (2024). Nanoparticles and plants: a focus on analytical characterization techniques. *Plant Science*, 112225.

De Sá, L. C., Oliveira, M., Ribeiro, F., Rocha, T. L., & Futter, M. N. (2018). Studies of the effects of microplastics on aquatic organisms: What do we know and where should we focus our efforts in the future? *Science of the Total Environment, 645*, 1029–1039.

De Silva, Y. S. K., Rajagopalan, U. M., Kadono, H., & Li, D. (2022). Effects of microplastics on lentil (Lens culinaris) seed germination and seedling growth. *Chemosphere, 303*, Article 135162.

de Souza Machado, A. A., Kloas, W., Zarfl, C., Hempel, S., & Rillig, M. C. (2018). Microplastics as an emerging threat to terrestrial ecosystems. *Global Change Biology, 24*, 1405–1416.

de Souza Machado, A. A., Lau, C. W., Kloas, W., Bergmann, J., Bachelier, J. B., Faltin, E., Becker, R., Görlich, A. S., & Rillig, M. C. (2019). Microplastics can change soil properties and affect plant performance. *Environmental Science and Technology, 53*, 6044–6052.

Domínguez-Jaimes, L. P., Cedillo-González, E. I., Luévano-Hipólito, E., Acuña-Bedoya, J. D., & Hernández-López, J. M. (2021). Degradation of primary nanoplastics by photocatalysis using different anodized TiO_2 structures. *Journal of Hazardous Materials, 413*, 125452.

Dong, Y., Gao, M., Qiu, W., & Song, Z. (2021). Uptake of microplastics by carrots in presence of As (III): Combined toxic effects. *Journal of Hazardous Materials, 411*, Article 125055.

Dris, R., Gasperi, J., Saad, M., Mirande, C., & Tassin, B. (2016). Synthetic fibers in atmospheric fallout: A source of microplastics in the environment? *Marine Pollution Bulletin, 104*, 290–293.

El-Moneim, D. A., Dawood, M. F. A., Moursi, Y. S., Farghaly, A. A., Afifi, M., & Sallam, A. (2021). Positive and negative effects of nanoparticles on agricultural crops. *Nanotechnology for Environmental Engineering, 6*(2), 21.

Enyoh, C. E., Verla, A. W., Verla, E. N., Ibe, F. C., & Amaobi, C. E. (2019). Airborne microplastics: A review study on method for analysis, occurrence, movement and risks. *Environmental Monitoring and Assessment, 191*, 1–17.

Eriksen, M., Cowger, W., Erdle, L. M., Coffin, S., Villarrubia-Gómez, P., Moore, C. J., Carpenter, E. J., Day, R. H., Thiel, M., & Wilcox, C. (2023). A growing plastic smog, now estimated to be over 170 trillion plastic particles afloat in the world's oceans—Urgent solutions required. *PLoS ONE, 18*, Article e0281596.

Frias, J. P. G. L., & Nash, R. (2019). Microplastics: Finding a consensus on the definition. *Marine Pollution Bulletin, 138*, 145–147.

Gahleitner, M., & Paulik, C. (2017). Polypropylene and other polyolefins. In *Brydson's plastics materials* (pp. 279–309). Elsevier.

Galgani, F., & Rangel-Buitrago, N. (2024). White tides: The plastic nurdles problem. *Journal of Hazardous Materials, 470*, Article 134250.

Gao, H. H., Liu, Q., Yan, C. R., et al. (2022). Macro-and/or microplastics as an emerging threat effect crop growth and soil health[J]. *Resources, Conservation and Recycling, 186*, 106549.

Gao, M., Liu, Y., & Song, Z. (2019). Effects of polyethylene microplastic on the phytotoxicity of di-n-butyl phthalate in lettuce (Lactuca sativa L. var. ramosa Hort). *Chemosphere, 237*, 124482.

Gao, M., Wang, Z., Jia, Z., Zhang, H., & Wang, T. (2023). Brassinosteroids alleviate nanoplastic toxicity in edible plants by activating antioxidant defense systems and suppressing nanoplastic uptake. *Environment International, 174*, Article 107901.

Gao, W., Wu, D., Zhang, D., Geng, Z., Tong, M., Duan, Y., Xia, W., Chu, J., & Yao, X. (2024). Comparative analysis of the effects of microplastics and nitrogen on maize and wheat: Growth, redox homeostasis, photosynthesis, and AsA-GSH cycle. *Science of the Total Environment, 932*, Article 172555.

Ghoshal, T., Parmar, P. R., Bhuyan, T., & Bandyopadhyay, D. (2023). Polystyrene foams: Materials, technology, and applications. In *Polymeric foams: Fundamentals and types of foams* (Vol. 1, pp. 121–141). ACS Publications.

Giorgetti, L., Spanò, C., Muccifora, S., Bottega, S., Barbieri, F., Bellani, L., & Ruffini Castiglione, M. (2020). Exploring the interaction between polystyrene nanoplastics and Allium cepa during germination: Internalization in root cells, induction of toxicity and oxidative stress. *Plant Physiology and Biochemistry, 149*, 170–177.

Gohil, M., & Joshi, G. (2022). Perspective of polycarbonate composites and blends properties, applications, and future development: A review. In *Green sustainable process for chemical and environmental engineering and science* (pp. 393–424).

Gong, X., Shi, G., Zou, D., Wu, Z., Qin, P., Yang, Y., Hu, X., Zhou, L., & Zhou, Y. (2023). Micro- and nano-plastics pollution and its potential remediation pathway by phytoremediation. *Planta, 257*(2), 35.

Grosu, E. (2022). Applications of polyvinylchloride (PVC)/thermoplastic nano-, micro-and macroblends. In *Polyvinylchloride-based blends: Preparation, characterization and applications* (pp. 75–89).

Gutow, L., Eckerlebe, A., Gimenez, L., & Saborowski, R. (2016). Experimental evaluation of seaweeds as a vector for microplastics into marine food webs. *Environmental Science and Technology, 50*(2), 915–923.

Hasan, M. M., & Jho, E. H. (2023). Effect of different types and shapes of microplastics on the growth of lettuce. *Chemosphere, 339*, Article 139660.

Hossain, M. T., Shahid, M. A., Mahmud, N., Habib, A., Rana, M. M., Khan, S. A., & Hossain, M. D. (2024). Research and application of polypropylene: A review. *Discover Nano, 19*, 2.

Hsu, A., Lloyd, A., & Emerson, J. W. (2013). What progress have we made since Rio? Results from the 2012 Environmental Performance Index (EPI) and Pilot Trend EPI. *Environmental Science & Policy, 33*, 171–185.

Hua, Z., Ma, S., Ouyang, Z., Liu, P., Qiang, H., & Guo, X. (2023). The review of nanoplastics in plants: Detection, analysis, uptake, migration and risk. *TrAC Trends in Analytical Chemistry, 158*, Article 116889.

Huang, D., Ding, L., Wang, S., Ding, R., Qiu, X., Li, J., Hua, Z., Liu, S., Wu, R., Liang, X., & Guo, X. (2024). Metabolomics reveals how spinach plants reprogram metabolites to cope with intense stress responses induced by photoaged polystyrene nanoplastics (PSNPs). *Journal of Hazardous Materials, 466*, Article 133605.

Huang, Y., Zhao, Y., Wang, J., Zhang, M., Jia, W., & Qin, X. (2019). LDPE microplastic films alter microbial community composition and enzymatic activities in soil. *Environmental Pollution, 254*, Article 112983.

Ilyas, M., Liu, X., Yang, J., & Xu, G. (2024). Foliar implications of polystyrene nanoplastics on leafy vegetables and its ecological consequences. *Journal of Hazardous Materials*, 136346

Irshad, M. K., Aqeel, M., Saleem, S., Javed, W., Noman, A., Kang, M. W., Khalid, N., & Lee, S. S. (2024). Mechanistic insight into interactive effect of microplastics and arsenic on growth of rice (Oryza sativa L.) and soil health indicators. *Science of the Total Environment, 955*, 176875.

Isobe, A., Iwasaki, S., Uchida, K., & Tokai, T. (2019). Abundance of non-conservative microplastics in the upper ocean from 1957 to 2066. *Nature Communications, 10*, 417.

Jebril, S., Fredj, Z., Saeed, A. A., Gonçalves, A. M., Kaur, M., Kumar, A., & Singh, B. (2024). Nanomaterial-based electrochemical chemo (bio) sensors for the detection of nanoplastic residues: Trends and future prospects. *RSC Sustainability, 2*(4), 832–851.

Jia, Y., Samak, N. A., Hao, X., Chen, Z., Yang, G., Zhao, X., Mu, T., Yang, M., & Xing, J. (2021). Nano-immobilization of PETase enzyme for enhanced polyethylene terephthalate biodegradation. *Biochemical Engineering Journal, 176*, Article 108205.

Jiang, X., Chen, H., Liao, Y., Ye, Z., Li, M., & Klobučar, G. (2019). Ecotoxicity and genotoxicity of polystyrene microplastics on higher plant Vicia faba. *Environmental Pollution, 250*, 831–838.

Ju, H., Zhu, D., & Qiao, M. (2019). Effects of polyethylene microplastics on the gut microbial community, reproduction and avoidance behaviors of the soil springtail, Folsomia candida. *Environmental Pollution, 247*, 890–897.

Ju, S., Shin, G., Lee, M., Koo, J. M., Jeon, H., Ok, Y. S., et al. (2021). Biodegradable chito-beads replacing non-biodegradable microplastics for cosmetics. *Green Chemistry, 23*(8), 6953–6965. https://doi.org/10.1039/D1GC01588E

Kausar, A. (2018). A review of filled and pristine polycarbonate blends and their applications. *Journal of Plastic Film & Sheeting, 34*, 60–97.

Keisling, C., Harris, R. D., Blaze, J., Coffin, J., & Byers, J. E. (2020). Low concentrations and low spatial variability of marine microplastics in oysters (Crassostrea virginica) in a rural Georgia estuary. *Marine Pollution Bulletin, 150*, Article 110672.

Kershaw, P. J. (2016). Marine plastic debris and microplastics–Global lessons and research to inspire action and guide policy change.

Kim, D., Kim, H., Lee, J., Choi, M. J., Kweon, H. S., & An, Y. J. (2024). Evidence of parental transfer of nanoplastics in pea (Pisum sativum) plants. *Journal of Hazardous Materials, 465*, Article 133516.

Kim, K.-H., Kabir, E., & Kabir, S. (2015). A review on the human health impact of airborne particulate matter. *Environment International, 74*, 136–143.

Krishnan, R. Y., Manikandan, S., Subbaiya, R., Karmegam, N., Kim, W., & Govarthanan, M. (2023). Recent approaches and advanced wastewater treatment technologies for mitigating emerging microplastics contamination–A critical review. *Science of the Total Environment, 858*, Article 159681.

Kumar, D., Biswas, J. K., Mulla, S. I., Singh, R., Shukla, R., Ahanger, M. A., Shekhawat, G. S., Verma, K. K., Siddiqui, M. W., & Seth, C. S. (2024). Micro and nanoplastics pollution: sources, distribution, uptake in plants, toxicological effects, and innovative remediation strategies for environmental sustainability. *Plant Physiology and Biochemistry*, 108795.

Kumar, M., Xiong, X., He, M., Tsang, D. C. W., Gupta, J., Khan, E., Harrad, S., Hou, D., Ok, Y. S., & Bolan, N. S. (2020). Microplastics as pollutants in agricultural soils. *Environmental Pollution, 265*, Article 114980.

Kusch, P. (2017). Application of pyrolysis-gas chromatography/mass spectrometry (Py-GC/MS). In *Comprehensive analytical chemistry* (Vol. 75, pp. 169—207). Elsevier.

Lee, J., Lee, J. S., Jang, Y. C., Hong, S. Y., Shim, W. J., Song, Y. K., Hong, S. H., Jang, M., Han, G. M., & Kang, D. (2015). Distribution and size relationships of plastic marine debris on beaches in South Korea. *Archives of Environmental Contamination and Toxicology, 69*, 288–298.

Li, L., Luo, Y., Peijnenburg, W. J. G. M., Li, R., Yang, J., & Zhou, Q. (2020). Confocal measurement of microplastics uptake by plants. *MethodsX, 7*, 100750–100756.

Li, R., Zhang, L., Xue, B., & Wang, Y. (2019). Abundance and characteristics of microplastics in the mangrove sediment of the semienclosed Maowei Sea of the south China sea: New implications for location, rhizosphere, and sediment compositions. *Environmental Pollution, 244*, 685–692.

Li, C., Gao, Y., He, S., Chi, H. Y., Li, Z. C., Zhou, X. X., & Yan, B. (2021a). Quantification of nanoplastic uptake in cucumber plants by pyrolysis gas chromatography/mass spectrometry. *Environmental Science & Technology Letters, 8*(8), 633–638.

Li, G., Qiu, C., Zhang, D., Lv, M., Liao, X., Li, Q., & Wang, L. (2024a). Effects of polystyrene nanoplastics (PSNPs) on the physiology of Allium sativum L.: Photosynthetic pigments, antioxidant enzymes, phytohormones, and nutritional quality. *Environmental and Experimental Botany, 219*, 105654.

Li, J., Liu, H., & Chen, J. P. (2018). Microplastics in freshwater systems: A review on occurrence, environmental effects, and methods for microplastics detection. *Water Research, 137*, 362–374.

Li, J., Qu, X., Su, L., Zhang, W., Yang, D., Kolandhasamy, P., Li, D., & Shi, H. (2016). Microplastics in mussels along the coastal waters of China. *Environmental Pollution, 214*, 177–184.

Li, K., Du, L., Qin, C., Bolan, N., Wang, H., & Wang, H. (2024b). Microplastic pollution as an environmental risk exacerbating the greenhouse effect and climate change: A review. *Carbon Research, 3*, 9.

Li, Q., Zhang, B., Liu, W., & Zou, H. (2024c). Strigolactones alleviate the toxicity of polystyrene nanoplastics (PS-NPs) in maize (Zea mays L.). *Science of the Total Environment, 918*, 170626.

Li, S., Wang, T., Guo, J., Dong, Y., Wang, Z., Gong, L., & Li, X. (2021b). Polystyrene microplastics disturb the redox homeostasis, carbohydrate metabolism and phytohormone regulatory network in barley. *Journal of Hazardous Materials, 415*, Article 125614.

Li, T., Cao, X., Zhao, R., & Cui, Z. (2023). Stress response to nanoplastics with different charges in Brassica napus L. during seed germination and seedling growth stages. *Frontiers of Environmental Science & Engineering, 17*(4), 43.

Li, Z., Li, Q., Li, R., Zhou, J., & Wang, G. (2021c). The distribution and impact of polystyrene nanoplastics on cucumber plants. *Environmental Science and Pollution Research, 28*, 16042–16053.

Lian, J., Liu, W., Meng, L., Wu, J., Chao, L., Zeb, A., & Sun, Y. (2021). Foliar-applied polystyrene nanoplastics (PSNPs) reduce the growth and nutritional quality of lettuce (Lactuca sativa L.). *Environmental Pollution, 280*, 116978.

Lian, J., Wu, J., Xiong, H., Zeb, A., Yang, T., Su, X., & Liu, W. (2020). Impact of polystyrene nanoplastics (PSNPs) on seed germination and seedling growth of wheat (Triticum aestivum L.). *Journal of Hazardous Materials, 385*, 121620.

Liang, R., Sun, F., Yang, X., Liu, H., & Wang, X. X. (2024). The effects of diverse microplastics on adzuki bean (Vigna angularis) growth and physiologic properties. *Environmental Geochemistry and Health, 46*(10), 374.

Lieberzeit, P., Bekchanov, D., & Mukhamediev, M. (2022). Polyvinyl chloride modifications, properties, and applications. *Polymers for Advanced Technologies, 33*, 1809–1820.

Liu, K., Wang, X., Song, Z., Wei, N., & Li, D. (2020). Terrestrial plants as a potential temporary sink of atmospheric microplastics during transport. *Science of the Total Environment, 742*, 140523–140529.

Liu, W., Hou, J., Wang, Q., Yang, H., Luo, Y., & Christie, P. (2014). Collection and analysis of root exudates of *Festuca arundinacea* L. and their role in facilitating the phytoremediation of petroleum-contaminated soil. *Plant Soil, 389*(1–2), 109–119.

Lu, J., Liu, Y., Hu, Z., Xiang, H., Zhou, Z., Sun, B., Wu, Q., & Zhu, M. (2019). A study on the radial difference of PLA monofilament. In *Physics and techniques of ceramic and polymeric materials: Proceedings of Chinese materials conference 2018 19th* (pp. 141–153). Springer.

Lu, S., Huo, Z., Niu, T., Zhu, W., Wang, J., Wu, D., He, C., Wang, Y., Zou, L., & Sheng, L. (2023). Molecular mechanisms of toxicity and detoxification in rice (Oryza sativa L.) exposed to polystyrene nanoplastics. *Plant Physiology and Biochemistry, 199*, 107605.

Luo, Y., Zhang, Y., Xu, Y., Guo, X., & Zhu, L. (2020). Distribution characteristics and mechanism of microplastics mediated by soil physicochemical properties. *Science of the Total Environment, 726*, 138389–138395.

Luo, D., Chu, X., Wu, Y., Wang, Z., Liao, Z., Ji, X., Ju, J., Yang, B., Chen, Z., Dahlgren, R., Zhang, M., & Shang, X. (2024). Micro-and Nano-plastics in the atmosphere: A review of occurrence, properties and human health risks. *Journal of Hazardous Materials*, 133412.

Luo, Y., Li, L., Feng, Y., Li, R., Yang, J., Peijnenburg, W. J. G. M., & Tu, C. (2022). Quantitative tracing of uptake and transport of submicrometre plastics in crop plants using lanthanide chelates as a dual-functional tracer. *Nature Nanotechnology, 17*(4), 424–431.

Malinović, B.N., Markelj, J., Žgajnar Gotvajn, A., Kralj Cigić, I., & Prosen, H. (2022). Electrochemical treatment of wastewater to remove contaminants from the production and disposal of plastics: a review. *Environmental Chemistry Letters, 20*(6), 3765–3787.

Mao, Y., Ai, H., Chen, Y., Zhang, Z., Zeng, P., Kang, L., Li, W., Gu, W., He, Q., & Li, H. (2018). Phytoplankton response to polystyrene microplastics: Perspective from an entire growth period. *Chemosphere, 208*, 59–68.

Mateos-Cárdenas, A., Scott, D. T., Seitmaganbetova, G., van Pelt Frank, N. A. M., John, O. H., & Marcel, A. K. J. (2019). Polyethylene microplastics adhere to *Lemna minor* (L.), yet have no effects on plant growth or feeding by *Gammarus duebeni* (Lillj.). *Science of the Total Environment, 689*, 413–421.

Mato, Y., Isobe, T., Takada, H., Kanehiro, H., Ohtake, C., & Kaminuma, T. (2001). Plastic resin pellets as a transport medium for toxic chemicals in the marine environment. *Environmental Science and Technology, 35*, 318–324.

Mishra, S., Singh, R. P., Rath, C. C., & Das, A. P. (2020). Synthetic microfibers: Source, transport and their remediation. *Journal of Water Process Engineering, 38*, Article 101612.

Morgana, S., Casentini, B., Tirelli, V., Grasso, F., & Amalfitano, S. (2024). Fluorescence-based detection: A review of current and emerging techniques to unveil micro/nanoplastics in environmental samples. *TrAC Trends in Analytical Chemistry*, 117559.

Nel, H. A., & Froneman, P. W. (2015). A quantitative analysis of microplastic pollution along the south-eastern coastline of South Africa. *Marine Pollution Bulletin, 101*, 274–279.

Nematollahi, M. J., Keshavarzi, B., Mohit, F., Moore, F., & Busquets, R. (2022). Microplastic occurrence in urban and industrial soils of Ahvaz metropolis: A city with a sustained record of air pollution. *Science of the Total Environment, 819*, Article 152051.

Ng, E.-L., Lwanga, E. H., Eldridge, S. M., Johnston, P., Hu, H.-W., Geissen, V., & Chen, D. (2018). An overview of microplastic and nanoplastic pollution in agroecosystems. *Science of the Total Environment, 627*, 1377–1388.

Nguyen, B., & Tufenkji, N. (2022). Single-particle resolution fluorescence microscopy of nanoplastics. *Environmental Science & Technology, 56*(10), 6426–6435.

Nisticò, R. (2020). Polyethylene terephthalate (PET) in the packaging industry. *Polymer Testing, 90*, Article 106707.

Nizzetto, L., Bussi, G., Futter, M. N., Butterfield, D., & Whitehead, P. G. (2016). A theoretical assessment of microplastic transport in river catchments and their retention by soils and river sediments. *Environmental Science. Processes & Impacts, 18*, 1050–1059.

Okoffo, E. D., & Thomas, K. V. (2024). Mass quantification of nanoplastics at wastewater treatment plants by pyrolysis–gas chromatography–mass spectrometry. *Water Research, 254*, Article 121397.

Ouda, M., Banat, F., Hasan, S. W., & Karanikolos, G. N. (2023). Recent advances on nanotechnology-driven strategies for remediation of microplastics and nanoplastics from aqueous environments. *Journal of Water Process Engineering, 52*, Article 103543.

Ouyang, X., Ma, J., Feng, B., Liu, Y., Yin, P., Zhang, X., Li, P., Chen, Q., Zhao, Y., Weng, L., & Li, Y. (2024). Effects of nanoplastics on the growth, transcription, and metabolism of rice (Oryza sativa L.) and synergistic effects in the presence of iron plaque and humic acid. *Environmental Pollution*, 125246.

Paço, A., Duarte, K., Costa, J. P, Santos, P. S. M, Pereira, R., Pereira, M. E., Freitas, A. C., Duarte, A. C., & Rocha-Santos, T. A. P. (2017). Biodegradation of polyethylene microplastics by the marine fungus Zalerion maritimum. *Science of the Total Environment, 586*, 10–15.

Palansooriya, K. N., Sang, M. K., Igalavithana, A. D., Zhang, M., Hou, D., Oleszczuk, P., et al. (2022). Biochar alters chemical and microbial properties of microplasticcontaminated soil. *Environmental Research, 209*, Article 112807. https://doi.org/10.1016/j.envres.2022.112807

Paxton, N. C., Allenby, M. C., Lewis, P. M., & Woodruff, M. A. (2019). Biomedical applications of polyethylene. *European Polymer Journal, 118*, 412–428.

Phillips, M. B., & Bonner, T. H. (2015). Occurrence and amount of microplastic ingested by fishes in watersheds of the Gulf of Mexico. *Marine Pollution Bulletin, 100*, 264–269.

Picó, Y., & Barceló, D. (2020). Pyrolysis gas chromatography-mass spectrometry in environmental analysis: Focus on organic matter and microplastics. *TrAC Trends in Analytical Chemistry, 130*, Article 115964.

Piehl, S., Leibner, A., Löder, M. G. J., Dris, R., Bogner, C., & Laforsch, C. (2018). Identification and quantification of macro-and microplastics on an agricultural farmland. *Science and Reports, 8*, 17950.

Pignattelli, S., Broccoli, A., & Renzi, M. (2020). Physiological responses of garden cress (*L. sativum*) to different types of microplastics. *Science of the Total Environment, 727*, 138609–138616.

Pino, N. J., Muñera, L. M., & Peñuela, G. A. (2016). Root exudates and plant secondary metabolites of different plants enhance polychlorinated biphenyl degradation by rhizobacteria. *Bioremediation Journal, 20*(2), 108–116.

Prasittisopin, L., Ferdous, W., & Kamchoom, V. (2023). Microplastics in construction and built environment. *Developments in the Built Environment, 15*, Article 100188.

Praveena, S. M., Shaifuddin, S. N. M., & Akizuki, S. (2018). Exploration of microplastics from personal care and cosmetic products and its estimated emissions to marine environment: An evidence from Malaysia. *Marine Pollution Bulletin, 136*, 135–140.

Proença, P. L., Carvalho, L. B., Campos, E. V., & Fraceto, L. F. (2022). Fluorescent labeling as a strategy to evaluate uptake and transport of polymeric nanoparticles in plants. *Advances in Colloid and Interface Science, 305*, Article 102695.

Qiu, G., Han, Z., Wang, Q., Wang, T., Sun, Z., Yu, Y., Han, X., & Yu, H. (2023). Toxicity effects of nanoplastics on soybean (Glycine max L.): Mechanisms and transcriptomic analysis. *Chemosphere, 313*, 137571.

Rafa, N., Ahmed, B., Zohora, F., Bakya, J., Ahmed, S., Ahmed, S. F., Mofijur, M., Chowdhury, A. A., & Almomani, F. (2023). Microplastics as carriers of toxic pollutants: Source, transport, and toxicological effects. *Environmental Pollution,* 123190.

Rajala, K., Grönfors, O., Hesampour, M., & Mikola, A. (2020). Removal of microplastics from secondary wastewater treatment plant effluent by coagulation/flocculation with iron, aluminum and polyamine-based chemicals. *Water research, 183*, 116045.

Rao, C., & Yan, B. (2020). Study on the interactive influence between economic growth and environmental pollution. *Environmental Science and Pollution Research, 27*, 39442–39465.

Ren, C. G., Kong, C. C., Wang, S. X., & Xie, Z. H. (2019). Enhanced phytoremediation of uranium-contaminated soils by arbuscular mycorrhiza and rhizobium. *Chemosphere, 217*, 773–779.

Ren, X., Tang, J., Wang, L., & Liu, Q. (2021). Microplastics in soil-plant system: Effects of nano/microplastics on plant photosynthesis, rhizosphere microbes and soil properties in soil with different residues. *Plant and Soil, 462*, 561–576.

Rong, S., Wang, S., Liu, H., Li, Y., Huang, J., Wang, W., Han, B., Su, S., & Liu, W. (2024). Evidence for the transportation of aggregated microplastics in the symplast pathway of oilseed rape roots and their impact on plant growth. *Science of the Total Environment, 912*, Article 169419.

Roy, T., Dey, T. K., & Jamal, M. (2023). Microplastic/nanoplastic toxicity in plants: An imminent concern. *Environmental Monitoring and Assessment, 195*, 27. https://doi.org/10.1007/s10661-022-10654-z

Rozman, U., Jemec Kokalj, A., Dolar, A., Drobne, D., & Kalčíková, G. (2022). Long-term interactions between microplastics and floating macrophyte *Lemna minor*: The potential for phytoremediation of microplastics in the aquatic environment. *Science of the Total Environment, 831*, Article 154866.

Saha, A., Baruah, P., & Handique, S. (2024). Assessment of microplastic pollution on soil health and crop responses: Insights from dose-dependent pot experiments. *Applied Soil Ecology, 203*, Article 105648.

Sanchez-Vidal, A., Thompson, R. C., Canals, M., & De Haan, W. P. (2018). The imprint of microfibres in southern European deep seas. *PLoS ONE, 13*, Article e0207033.

Sanchis, R., Fenollar, O., García, D., Sánchez, L., & Balart, R. (2008). Improved adhesion of LDPE films to polyolefin foams for automotive industry using low-pressure plasma. *International Journal of Adhesion and Adhesives, 28*, 445–451.

Savaş, S., Gurbanov, N., & Doğan, M. (2019). Effect of fiber type, fiber content, and compatibilizer on two-body abrasive wear performance of HDPE matrix composites. *Journal of Composite Materials, 53*, 2743–2760.

Schnurr, R. E. J., Alboiu, V., Chaudhary, M., Corbett, R. A., Quanz, M. E., Sankar, K., Srain, H. S., Thavarajah, V., Xanthos, D., & Walker, T. R. (2018). Reducing marine pollution from single-use plastics (SUPs): A review. *Marine Pollution Bulletin, 137*, 157–171.

Seng, N., Lai, S., Fong, J., Saleh, M. F., Cheng, C., Cheok, Z. Y., & Todd, P. A. (2020). Early evidence of microplastics on seagrass and macroalgae. *Marine and Freshwater Research, 71*(8), 922–928.

Sewwandi, M., Keerthanan, S., Perera, K. I., & Vithanage, M. (2023). Plastic nurdles in marine environments due to accidental spillage. In *Microplastics in the ecosphere: Air, water, soil, food* (pp. 415–432).

Shakiba, M., Rezvani Ghomi, E., Khosravi, F., Jouybar, S., Bigham, A., Zare, M., Abdouss, M., Moaref, R., & Ramakrishna, S. (2021). Nylon—A material introduction and overview for biomedical applications. *Polymers for Advanced Technologies, 32*, 3368–3383.

Shang, H., Guo, H., Ma, C., Li, C., Chefetz, B., Polubesova, T., & Xing, B. (2019). Maize (*Zea mays* L.) root exudates modify the surface chemistry of CuO nanoparticles: Altered aggregation, dissolution and toxicity. *Science of the Total Environment, 690*, 502–510.

Sheraz, M., Kim, J., & Kim, J. (2023). Micro/Nanoplastic (MnPs) in indoor air: A critical review of synthesis routes for toxicity testing and preventative measure strategies. *Process Safety and Environmental Protection.*

Shi, R., Liu, W., Lian, Y., Wang, Q., Zeb, A., & Tang, J. (2022). Phytotoxicity of polystyrene, polyethylene and polypropylene microplastics on tomato (Lycopersicon esculentum L.). *Journal of Environmental Management, 317*, 115441.

Simón, D., Borreguero, A. M., De Lucas, A., & Rodríguez, J. F. (2018). Recycling of polyurethanes from laboratory to industry, a journey towards the sustainability. *Waste Management, 76*, 147–171.

Singh, A., Mehta, S., Yadav, S., Nagar, G. C., Ghosh, R., Roy, A., Chakraborty, A., & Singh, I. K. (2022). How to cope with the challenges of environmental stresses in the era of global climate change: An update on ROS stave off in plants. *International Journal of Molecular Sciences, 23*(4), 1995.

Sintim, H. Y., Bary, A. I., Hayes, D. G., Wadsworth, L. C., Anunciado, M. B., English, M. E., et al. (2020). In situ degradation of biodegradable plastic mulch films in compost and agricultural soils. *Science of the Total Environment, 727*, Article 138668. https://doi.org/10.1016/j.scitotenv.2020.138668

Sommer, F., Dietze, V., Baum, A., Sauer, J., Gilge, S., Maschowski, C., & Gieré, R. (2018). Tire abrasion as a major source of microplastics in the environment. *Aerosol and Air Quality Research, 18*, 2014–2028.

Spanò, C., Muccifora, S., Ruffini Castiglione, M., Bellani, L., Bottega, S., & Giorgetti, L. (2022). Polystyrene nanoplastics affect seed germination, cell biology and physiology of rice seedlings in-short term treatments: Evidence of their internalization and translocation. *Plant Physiology and Biochemistry, 172*, 158–166.

Statista. (2022). Annual production of plastics worldwide from 1950 to 2021.

Suhaimi, N. A. S., Muhamad, F., Abd Razak, N. A., & Zeimaran, E. (2022). Recycling of polyethylene terephthalate wastes: A review of technologies, routes, and applications. *Polymer Engineering & Science, 62*, 2355–2375.

Sun, H., Lei, C., Xu, J., & Li, R. (2021). Foliar uptake and leaf-to-root translocation of nanoplastics with different coating charge in maize plants. *Journal of Hazardous Materials, 416*, Article 125854.

Sun, Q., Ren, S.-Y., & Ni, H.-G. (2020). Incidence of microplastics in personal care products: An appreciable part of plastic pollution. *Science of the Total Environment, 742*, Article 140218.

Sundbæk, K. B., Koch, I. D. W., Villaro, C. G., Rasmussen, N. S., Holdt, S. L., & Hartmann, N. B. (2018). Sorption of fluorescent polystyrene microplastic particles to edible seaweed *Fucus vesiculosus*. *Journal of Applied Phycology, 30*(5), 2923–2927.

Surgun-Acar, Y. (2022). Response of soybean (Glycine max L.) seedlings to polystyrene nanoplastics: Physiological, biochemical, and molecular perspectives. *Environmental Pollution, 314*, 120262.

Tan, Z., Yi, Y., Wang, H., Zhou, W., Yang, Y., & Wang, C. (2016). Physical and degradable properties of mulching films prepared from natural fibers and biodegradable polymers. *Applied Sciences, 6*(5), 147. https://doi.org/10.3390/app6050147

Tang, Y., Zhang, S., Su, Y., Wu, D., Zhao, Y., & Xie, B. (2021). Removal of microplastics from aqueous solutions by magnetic carbon nanotubes. *Chemical Engineering Journal, 406*, Article 126804.

Taylor, S. E., Pearce, C. I., Sanguinet, K. A., Hu, D., Chrisler, W. B., Kim, Y. M., Wang, Z., & Flury, M. (2020). Polystyrene nano- and microplastic accumulation at *Arabidopsis* and wheat root cap cells, but no evidence for uptake into roots. *Environmental Science: Nano, 7*, 1942–1953.

Teuten, E. L., Saquing, J. M., Knappe, D. R. U., Barlaz, M. A., Jonsson, S., Björn, A., Rowland, S. J., Thompson, R. C., Galloway, T. S., & Yamashita, R. (2009). Transport and release of chemicals from plastics to the environment and to wildlife. *Philosophical Transactions of the Royal Society b: Biological Sciences, 364*, 2027–2045.

Thacharodi, A., Hassan, S., Meenatchi, R., Bhat, M. A., Hussain, N., Arockiaraj, J., Ngo, H. H., Sharma, A., Nguyen, H. T., & Pugazhendhi, A. (2024a). Mitigating microplastic pollution: A critical review on the effects, remediation, and utilization strategies of microplastics. *Journal of Environmental Management, 351*, Article 119988.

Thacharodi, A., Meenatchi, R., Hassan, S., Hussain, N., Bhat, M. A., Arockiaraj, J., Ngo, H. H., Le, Q. H., & Pugazhendhi, A. (2024b). Microplastics in the environment: A critical overview on its fate, toxicity, implications, management, and bioremediation strategies. *Journal of Environmental Management, 349*, Article 119433.

Tian, J., Xu, Z., Huang, W., Han, D., Dang, B., Xu, J., & Jia, W. (2024). Physiobiochemical and transcriptional responses of tobacco plants (Nicotiana tabacum L.) to different doses of polystyrene nanoplastics. *Industrial Crops and Products, 220*, 119218.

Tiwari, M., Rathod, T. D., Ajmal, P. Y., Bhangare, R. C., & Sahu, S. K. (2019). Distribution and characterization of microplastics in beach sand from three different Indian coastal environments. *Marine Pollution Bulletin, 140*, 262–273.

Tokunaga, Y., Okochi, H., Tani, Y., Niida, Y., Tachibana, T., Saigawa, K., Katayama, K., Moriguchi, S., Kato, T., & Hayama, S. (2023). Airborne microplastics detected in the lungs of wild birds in Japan. *Chemosphere, 321*, Article 138032.

Unep, A., Assessment, I.R.R. (2016). The rise of environmental crime. Nairobi UNEP.

Urbina, M. A., Correa, F., Aburto, F., & Ferrio, J. P. (2020). Adsorption of polyethylene microbeads and physiological effects on hydroponic maize. *Science of the Total Environment, 741*, 140216–140225.

Verschoor, A., De Poorter, L., Dröge, R., Kuenen, J., & de Valk, E. (2016). Emission of microplastics and potential mitigation measures: Abrasive cleaning agents, paints and tyre wear.

Vinay, Surana, D., Ghosh, P., Kumar, M., Varjani, S., Kumar, V., & Mannina, G. (2023). Contemporary drift in emerging micro (nano) plastics removal and upcycling technologies from municipal wastewater sludge: strategic innovations and prospects. *Current Pollution Reports, 9*, 174–197.

Wang, F., Zhang, X., Zhang, S., Zhang, S., & Sun, Y. (2020). Interactions of microplastics and cadmium on plant growth and arbuscular mycorrhizal fungal communities in an agricultural soil. *Chemosphere, 254*, 126791–126800.

Wang, X., Bolan, N., Tsang, D. C. W., Sarkar, B., Bradney, L., & Li, Y. (2021a). A review of microplastics aggregation in aquatic environment: Influence factors, analytical methods, and environmental implications. *Journal of Hazardous Materials, 402*, Article 123496.

Wang, F., Wong, C. S., Chen, D., Lu, X., Wang, F., & Zeng, E. Y. (2018). Interaction of toxic chemicals with microplastics: A critical review. *Water Research, 139*, 208–219.

Wang, G., Lu, J., Li, W., Ning, J., Zhou, L., Tong, Y., Liu, Z., Zhou, H., & Xiayihazi, N. (2021b). Seasonal variation and risk assessment of microplastics in surface water of the Manas River Basin, China. *Ecotoxicology and Environmental Safety, 208*, Article 111477.

Wang, J., Lu, S., Guo, L., Wang, P., He, C., Liu, D., Bian, H., & Sheng, L. (2022). Effects of polystyrene nanoplastics with different functional groups on rice (Oryza sativa L.) seedlings: Combined transcriptome, enzymology, and physiology. *Science of the Total Environment, 834*, 155092.

Wang, Y., Bai, J. J., Wei, Y. J., Zhao, C. X., Shao, Z., Chen, M. L., & Wang, J. H. (2024). Tracking and imaging nano-plastics in fresh plant using cryogenic laser ablation inductively coupled plasma mass spectrometry. *Journal of Hazardous Materials, 465*, Article 133029.

Wang, Y., Huang, J., Zhu, F., & Zhou, S. (2021c). Airborne microplastics: A review on the occurrence, migration and risks to humans. *Bulletin of Environmental Contamination and Toxicology, 107*(4), 657–664.

Wei, X.-F., Bohlén, M., Lindblad, C., Hedenqvist, M., & Hakonen, A. (2021). Microplastics generated from a biodegradable plastic in freshwater and seawater. *Water Research, 198*, Article 117123.

Winnacker, M. (2017). Polyamides and their functionalization: Recent concepts for their applications as biomaterials. *Biomater. Sci., 5*, 1230–1235.

Wojcieszek, J., Jimenez-Lamana, J., Bierla, K., Ruzik, L., Asztemborska, M., Jarosz, M., & Szpunar, J. (2019). Uptake, translocation, size characterization and localization of cerium oxide nanoparticles in radish (*Raphanus sativus* L.). *Science of the Total Environment, 683*, 284–292.

Wu, S., Zhang, X., Chen, B., Wu, Z., Li, T., Hu, Y., Sun, Y., & Wang, Y. (2016). Chromium immobilization by extraradical mycelium of arbuscular mycorrhiza contributes to plant chromium tolerance. *Environmental and Experimental Botany, 122*, 10–18.

Wu, X., Lyu, X., Li, Z., Gao, B., Zeng, X., Wu, J., & Sun, Y. (2020). Transport of polystyrene nanoplastics in natural soils: Effect of soil properties, ionic strength and cation type. *Science of the Total Environment, 707*, 136065–136071.

Wu, T., Hu, G., Ning, J., Yang, J., & Zhou, Y. (2024). A photoluminescence strategy for detection nanoplastics in water and biological imaging in cells and plants. *Journal of Hazardous Materials, 461*, Article 132695.

Xie, X., Deng, T., Duan, J., Xie, J., Yuan, J., & Chen, M. (2020). Exposure to polystyrene microplastics causes reproductive toxicity through oxidative stress and activation of the p38 MAPK signaling pathway. *Ecotoxicology and Environmental Safety, 190*, Article 110133.

Xiong, Y. W., Li, X. W., Wang, T. T., Gong, Y., Zhang, C. M., Xing, K., & Qin, S. (2020). Root exudates-driven rhizosphere recruitment of the plant growth-promoting rhizobacterium Bacillus flexus KLBMP 4941 and its growth promoting effect on the coastal halophyte *Limonium sinense* under salt stress. *Ecotoxicology and Environmental Safety, 194*, 110374–110383.

Xu, B., Liu, F., Cryder, Z., Huang, D., Lu, Z., He, Y., Wang, H., Lu, Z., Brookes, P. C., Tang, C., Gan, J., & Xu, J. (2020a). Microplastics in the soil environment: Occurrence, risks, interactions and fate—A review. *Crit Rev Env Sci Tec, 50*(21), 2175–2222.

Xu, S., Ma, J., Ji, R., Pan, K., & Miao, A.-J. (2020b). Microplastics in aquatic environments: Occurrence, accumulation, and biological effects. *Science of the Total Environment, 703*, Article 134699.

Ye, X., Li, H., Wang, Q., Chai, R., Ma, C., Gao, H., & Mao, J. (2018). Influence of aspartic acid and lysine on the uptake of gold nanoparticles in rice. *Ecotoxicology and Environmental Safety, 148*, 418–425.

Yin, J., Zhu, T., Li, X., Yin, X., Xu, J., & Xu, G. (2024). Polystyrene nanoplastics induce cell type-dependent secondary wall reinforcement in rice (Oryza sativa) roots and reduce root hydraulic conductivity. *Journal of Hazardous Materials, 477*, Article 135309.

Yu, Z., Xu, X., Guo, L., Yuzuak, S., & Lu, Y. (2024). Physiological and biochemical effects of polystyrene micro/nano plastics on Arabidopsis thaliana. *Journal of Hazardous Materials, 469*, Article 133861.

Zhang, C., Yue, N., Li, X., Shao, H., Wang, J., An, L., & Jin, F. (2023). Potential translocation process and effects of polystyrene microplastics on strawberry seedlings. *Journal of Hazardous Materials, 449*, Article 131019.

Zhang, H., Liang, J., Luo, Y., Tang, N., Li, X., Zhu, Z., & Guo, J. (2022). Comparative effects of polystyrene nanoplastics with different surface charge on seedling establishment of Chinese cabbage (Brassica rapa L.). *Chemosphere, 292*, 133403.

Zhang, Q., Zhao, M., Meng, F., Xiao, Y., Dai, W., & Luan, Y. (2021). Effect of polystyrene microplastics on rice seed germination and antioxidant enzyme activity. *Toxics, 9*(8), 179.

Zhang, S., Zhang, F., Cai, L., Xu, N., Zhang, C., Yadav, V., Zhou, X., Wu, X., & Zhong, H. (2024). Visual observation of polystyrene nano-plastics in grape seedlings of Thompson Seedless and assessing their effects via transcriptomics and metabolomics. *Journal of Hazardous Materials, 478*, Article 135550.

Zhang, T.-R., Wang, C.-X., Dong, F.-Q., Gao, Z.-Y., Zhang, C.-J., Zhang, X.-J., Fu, L.-M., Wang, Y., & Zhang, J.-P. (2019a). Uptake and translocation of styrene maleic anhydride nanoparticles in Murraya exotica plants as revealed by noninvasive, real-time optical bioimaging. *Environmental Science and Technology, 53*(3), 1471–1481.

Zhang, X., Chen, J., Liu, X., Chen, X., Liu, L., Niu, Y., & Wang, R. (2019b). The relief effects of organic acids on *Scirpus triqueter* L. under pyrene-lead stress. *Environmental Science and Pollution Research International, 26*(16):15828–15837.

Zhou, C. Q., Lu, C. H., Mai, L., Bao, L. J., Liu, L. Y., & Zeng, E. Y. (2021). Response of rice (Oryza sativa L.) roots to nanoplastic treatment at seedling stage. *Journal of Hazardous Materials, 401*, 123412.

Zhou, Y., Kumar, M., Sarsaiya, S., Sirohi, R., Awasthi, S. K., Sindhu, R., Binod, P., Pandey, A., Bolan, N. S., Zhang, Z., & Singh, L. (2022). Challenges and opportunities in bioremediation of micro-nano plastics: A review. *Science of the Total Environment, 802*, Article 149823.

Zhou, Y., Liu, X., & Wang, J. (2019). Characterization of microplastics and the association of heavy metals with microplastics in suburban soil of central China. *Science of the Total Environment, 694*, Article 133798.

Zhu, L., Zhao, S., Bittar, T. B., Stubbins, A., & Li, D. (2020). Photochemical dissolution of buoyant microplastics to dissolved organic carbon: Rates and microbial impacts. *Journal of Hazardous Materials, 383*, Article 121065.

Zhuang, H., Qin, M., Liu, B., Li, R., & Li, Z. (2023). Combination of transcriptomics, metabolomics and physiological traits reveals the effects of polystyrene microplastics on photosynthesis, carbon and nitrogen metabolism in cucumber (Cucumis sativus L.). *Plant Physiology and Biochemistry, 205*, 108201.

Zytowski, E., Mollavali, M., & Baldermann, S. (2024). Uptake and translocation of nanoplastics in mono and dicot vegetables. *Plant, Cell & Environment*.

Chapter 6
Integrating Remote Sensing and GIS for Effective Microplastics Pollution Management: A Pathway to Sustainable Development

Seylina Sathish Kumar, Shailender Kumar, and Monika Saroj

Abstract Microplastic pollution poses a profound and escalating threat to the biosphere, impacting ecosystems and human health on a global scale. In addition to assessing the potential of remote sensing and Geographic Information Systems (GIS) to enhance marine pollution control, this research investigates the multifaceted effects of microplastics. In an era of rapid technological advancements, remote sensing provides strategic tools for detecting, monitoring, mapping, and analysing microplastics, enabling more accurate insights into their distribution and movement in marine environments. Recent global estimates indicate that approximately 2.3 million tonnes of microplastics are now present in the world's seas, with concentrations rising sharply since 2005. By synthesising current literature, this paper offers a comprehensive perspective on using remote sensing to guide targeted mitigation strategies and inform policy development. The findings underscore the critical role of remote sensing technology in addressing microplastic pollution and driving progress toward the United Nations Sustainable Development Goals (SDGs), including SDG 15 (Life on Land), SDG 14 (Life Below Water), and SDG 6 (Clean Water and Sanitation). This technology offers innovative solutions to monitor, manage, and mitigate the environmental impacts of microplastics, fostering a healthier planet. This research highlights the urgency of integrating remote sensing into environmental monitoring and policy frameworks to mitigate the extensive ecological and health risks posed by microplastic pollution.

S. S. Kumar
University of Edinburgh, Edinburgh, Scotland
e-mail: seylinasathishkumar@gmail.com

S. Kumar (✉)
Centre for Environmental Studies and Disaster Management, Miranda House, University of Delhi, New Delhi, India
e-mail: shailukarela@gmail.com

M. Saroj
Regional Services Division, IGNOU Maidan Garhi, New Delhi 110068, India
e-mail: monikasaroj@ignou.ac.in

V. Kumar et al. (eds.), *Microplastics Pollution*,
https://doi.org/10.1007/978-3-031-99774-7_6

Keywords Microplastic pollution · Remote sensing · Geographical information systems · Marine ecology

6.1 Introduction

One of the most important environmental issues of our day is plastics. It is among the most durable materials humans have ever made. The fact that plastic can take millennia to disintegrate and that research indicates it might never do so completely is generally recognised today. Rather, it breaks up into particles known as microplastics (Fava, 2022) (maya0286, 2022). An estimated 8–10 million metric tonnes of plastic debris enter the ocean each year, making up an overwhelming 80% of marine pollution. According to estimates, plastic may surpass all marine life combined by 2050. Surprisingly, more plastic has been generated in the past ten years compared to the whole preceding century (Nations, 2017).

Astonishingly, more plastic has been manufactured in the last decade than was produced in the entire previous century. According to the Environmental Protection Agency (EPA), nearly all plastic ever produced still exists in some form, as it does not fully break down. Presently, our oceans are filled with an estimated 50–75 trillion pieces of plastic and microplastics, posing a persistent and severe threat to marine ecosystems (Fava, 2022).

Addressing plastic pollution requires comprehensive strategies, and policies like Extended Producer Responsibility (EPR) and the promotion Additional waste types, such as electronic waste (e-waste), play a significant role in the worsening issue of marine microplastic contamination. This category of waste includes obsolete electronic items like mobile phones, laptops, and televisions (Shaaban et al., 2024), often containing plastic components that, when improperly disposed of or inadequately recycled, can degrade into microplastics (Prata, 2024).

The pollution issue is exacerbated when these microplastics finally find their way into the oceans. Furthermore, hazardous compounds released by e-waste are detrimental to marine ecosystems. To lessen its harmful impacts on marine habitats, e-waste management laws must be strengthened, recycling procedures must be improved, and e-waste solutions must be incorporated into larger waste reduction plans (Plimmer, 2024).

A comprehensive, multidimensional strategy involving legislative actions, advancements in recycling techniques, technical innovation, and public awareness is needed to address the problem of e-waste and its role in marine contamination with micro plastics (Shaaban et al., 2024).

Over time, these plastics degrade into microplastics, contributing significantly to marine pollution. Furthermore, the chemicals released from e-waste, including heavy metals and toxic flame retardants, add to the environmental damage, posing risks to marine life.

Recent studies have demonstrated the presence of microplastics in human blood and breast milk, underscoring the pervasive nature of plastic pollution in the environment (Carrington, 2022). The research identified various potential sources of human exposure, including the use of plastic-containing personal hygiene products, the consumption of seafood, and the ingestion of food and beverages packaged in plastic materials.

Nevertheless, there was no clear link discovered between the identification of microplastics in human samples and these exposure pathways (Barceló et al., 2023). This absence of correlation raises the possibility that the human body contains microplastics as an inevitable byproduct of their pervasive environmental pollution. The authors propose that future studies using more thorough and focused approaches should clarify certain risk factors and exposure routes for microplastics, improving our comprehension of their possible health effects (Carrington, 2022).

Although the full impact on health is not yet understood, damage to human cells has been shown in laboratory studies. It is possible that these particles may become lodged in tissues, causing inflammation in a way similar to air pollution particles, or that harmful chemicals within the plastics could be released and cause adverse effects (Dutchen, 2023). "Additional threats come from chemicals present within and on microplastic particles, including plastic additives like BPA, phthalates, and heavy metals. These substances are known or suspected to disrupt the nervous, reproductive, and other bodily systems" (Dutchen, 2023) (Fig. 6.1).

In addition, in the 1990s, it was still legal to dispose of polymers in the water. Since it takes hundreds of years for these polymers to decompose, they have remained in the water for decades. Microplastics are created when plastics are broken down into microscopic particles by exposure to sunshine and ocean waves over time (Ashrafy et al., 2022). Over 300 animal species are thought to have consumed these microplastics (Ziani, 2023).

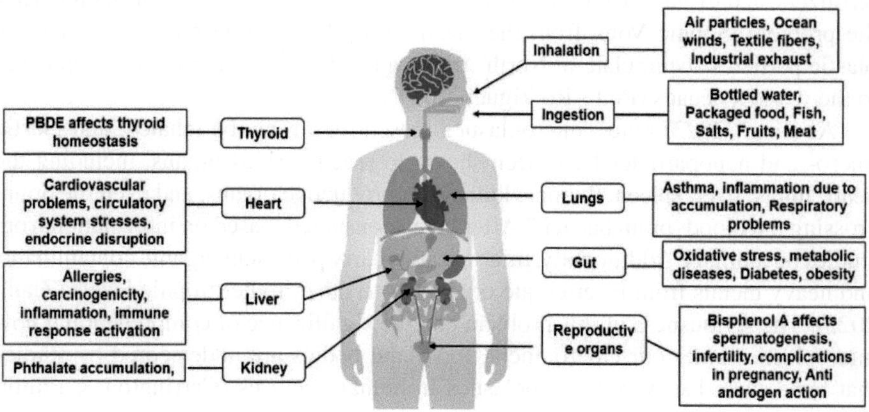

Fig. 6.1 Health effects of microplastics

Fish and other marine creatures ingest these microscopic particles, which are then eaten by larger fish, which are frequently thereafter eaten by humans (Alberghini et al., 2022). Microplastics and related substances enter animal cells in this manner, eventually making their way into human bodies.

While the detrimental effects of microplastics on marine species like seals (Hernandez-Milian et al., 2019), turtles—Female sea turtles may potentially transfer microplastics to their eggs during oogenesis (the process of egg formation). This possibility raises concerns about the early-life exposure of hatchlings to microplastics, which could have implications for their health and survival (Chemello et al., 2023), and seagulls (Wayman et al., 2024) are well-documented, raising concerns about human exposure, the primary question remains: how do these particles enter human systems?

Research also indicates that microplastics enter our bodies mainly through ingestion—via contaminated food and water sources—and inhalation, especially in environments with airborne plastic particles. These entry pathways are becoming increasingly evident, suggesting potential risks to human health due to the accumulation of microplastics in various organs (Giri et al., 2024).

According to research (Giri et al., 2024; Pironti et al., 2021), Ingestion and inhalation are the two major ways that microplastics enter the human body. There is evidence that the amount of microplastics in our food and water is rising. Evidence indicates that our food and water are increasingly contaminated with microplastics (Mamun et al., 2023); a 2020 Italian study detected these particles in everyday fruits and vegetables, likely due to uptake from soil (Oliveri Conti et al., 2020).

A study conducted (Harley-Nyang et al., 2022) in southwest England showed that applying treated sewage sludge to fields might release more microplastics than 20,000 plastic credit cards each month, indicating that sewage sludge used as fertiliser can introduce substantial amounts of microplastics into agricultural soils.

Agricultural practices, including the use of plastic matting sheets, plastic-coated fertilizers, and irrigation with water containing tiny plastic particles, further intensify the problem. Sophie Vonk from the Plastic Soup Foundation notes that far more plastic particles accumulate in North American and European farmlands compared to the world's oceans (Pinto-Rodrigues, 2023).

"As Ziani (2023) notes, microplastics present in air can be inhaled, and plastic micro- and nanoparticles have been shown to reach various organs, including the heart, lungs, liver, spleen, thymus, kidneys, reproductive organs, and even the brain, crossing the blood–brain barrier." After being consumed, based on information from studies on animals. Additionally, microplastics carry persistent organic contaminants and heavy metals from invertebrate organisms to other higher trophic levels (Ziani, 2023). The seriousness of this problem and the significance of comprehending how microplastics enter our environment, food, and bodies are underscored by studies that have even discovered microplastics in human foetuses (Carrington & Editor, 2024).

However, there are still a lot of obstacles to overcome in order to combat microplastic pollution. One of the main challenges is the unknowns—it is frequently difficult to create effective legislation and educational programs since the precise types of plastics used in different areas and their origins are not well-documented. Investing in technology for monitoring and controlling plastic pollution is crucial to overcoming these obstacles, particularly to protect our oceans for coming generations.

A crucial step in this process is understanding the entire lifecycle of microplastics: from their origins and formation to how they enter the environment, infiltrate food chains, and impact marine organisms. Microplastic pollution poses a unique threat, potentially even more concerning than large plastic debris, as it permeates marine ecosystems and food webs in ways that are difficult to control (Jain et al., 2023). Addressing this issue is essential, not only for the health of marine ecosystems but also to mitigate risks to human health. Through a better understanding of microplastics and the application of advanced monitoring technologies, we can develop more targeted solutions and policies to protect both our oceans and ourselves.

By enabling us to identify and track plastic pollution in marine areas that are challenging to reach, remote sensing technology provides a potent tool to better understand and address this issue. Modern technologies like remote sensing provide important chances to track and measure marine pollution, particularly in places that are inaccessible to human surveyors. Remote sensing can improve our capacity to locate pollution hotspots and monitor the flow of microplastic particles throughout seas by giving real-time data on concentration and distribution (Ma et al., 2023).

Ocean plastic pollution detection and mitigation have advanced significantly, with remote sensing technologies—especially satellite imagery—being essential in spotting huge trash and ghost nets. These are important markers of regions where marine plastic debris is prevalent.

Every year, approximately 10 million tons of plastic enter the oceans, accumulating in major ocean basins, including the Indian, Atlantic, and Pacific Oceans. These plastics form vast zones of plastic concentration known as oceanic gyres (Davaasuren et al., 2018; Waqas et al., 2023). One of the most well-known of these gyres is the Great Pacific Garbage Patch. Contrary to popular belief, however, these zones are not solid islands of waste but rather areas where plastic debris is dispersed, with a high prevalence of microplastics (Eriksen et al., 2014) (Fig. 6.2).

Because micro plastics penetrate the marine food web, their pervasiveness presents serious threats to the health of marine ecosystems (Marcharla et al., 2024). This can have potentially harmful effects on marine life and, ultimately, human health, though the full extent of these impacts is still uncertain (GESAMP, 2015).

One notable initiative tackling the global issue of ocean plastic pollution is *The Ocean Cleanup*, founded by Boyan Slat in the Netherlands (The Ocean Cleanup, 2023). This ambitious project aims to remove plastic waste from the oceans, particularly large debris, but the scale of the problem presents significant challenges (Sainte-Rose, 2018). Early research efforts, such as the Mega Expedition and aerial surveys, revealed far greater quantities of plastic than previously estimated, highlighting the vastness of the issue and the need for more effective solutions (Slat et al., 2014).

Traditional methods for identifying the most polluted regions, such as ship-based surveys, have proven inefficient due to their high costs and logistical complexities. As a result, there has been a growing shift toward leveraging advanced remote

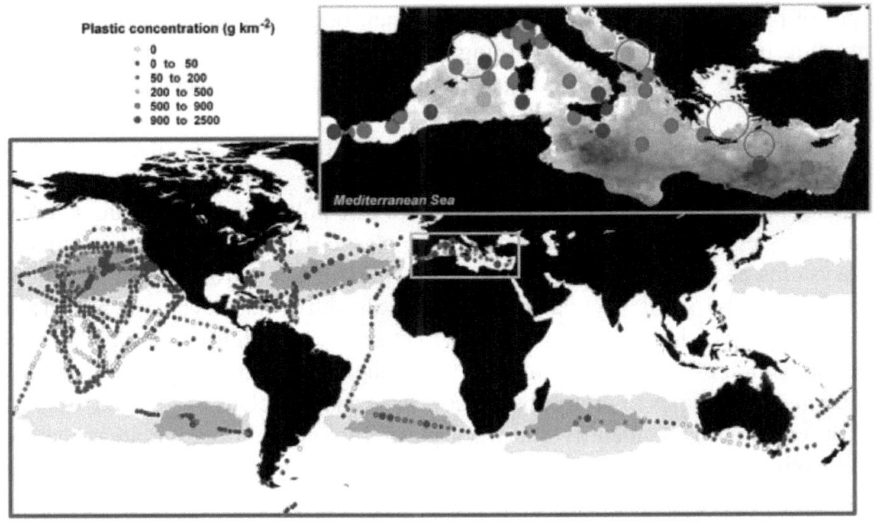

Fig. 6.2 Mapping of plastic litter hotspots in the global oceans. *Source* (Cózar et al., 2015), Windeck (2018)

sensing technologies to optimise the detection and mapping of plastic concentrations in oceanic environments (Lebreton, 2018). Remote sensing offers a cost-effective and scalable solution for monitoring vast oceanic areas, which would otherwise be difficult to access using traditional methods.

The integration of satellite-based technologies and artificial intelligence has significantly improved efforts to detect, track, and quantify marine plastic debris. These technologies allow for real-time monitoring of plastic accumulation, enabling a more targeted approach to combating pollution.

However, challenges remain, particularly in detecting The buildup of marine life on plastic surfaces and the requirement for high-resolution hyperspectral imaging. Unfortunately, the current capabilities of commercial satellites are still limited in this regard (Huang et al., 2020).

While technological advancements have greatly improved our ability to detect and monitor marine plastic pollution, the refinement of these methods and a thorough assessment of their long-term environmental impacts are still required. Going forward, the integration of remote sensing with machine learning algorithms, coupled with field validation, is expected to provide more efficient strategies for large-scale monitoring and intervention. These advancements are likely to play a key role in enhancing the accuracy and effectiveness of global efforts to address the growing problem of plastic pollution in the oceans (Astorayme et al., 2024).

Microplastic pollution in the ocean is a growing environmental concern, posing significant risks to marine ecosystems, wildlife, and human health (Amelia et al., 2021). Traditional methods of tracking microplastics, such as trawler-based sampling (Zheng et al., 2021), are often labour-intensive and limited in scope, providing only partial snapshots of the pollution spread. As a result, scientists have increasingly turned to remote sensing technologies, which offer the ability to monitor microplastics over large areas with higher accuracy and frequency (Goddijn-Murphy et al., 2024). This shift is essential in addressing the scale of the issue, as it allows for a more comprehensive and consistent approach to tracking pollution.

In this context, one of the key initiatives is *The Ocean Cleanup*, which, in collaboration with the European Space Agency (ESA) and Airbus, has pioneered the use of satellite imagery to enhance the detection of marine debris, particularly focusing on large debris and ghost nets (Sainte-Rose, 2018). While this approach shows significant potential for large-scale detection, several challenges remain. Notably, biofouling—the accumulation of marine organisms on plastic surfaces—hinders detection accuracy, and the need for high-resolution hyperspectral imaging, which is currently beyond commercially available satellite technologies, further limits its effectiveness. These challenges highlight the complexity of the problem and the need for ongoing innovation in monitoring methods.

However, despite these hurdles, such efforts have driven significant technological innovation (Santos et al., 2022). One notable development is the use of vessel-mounted cameras combined with deep learning algorithms, enabling real-time detection of floating debris.

These advancements represent a promising step forward, enhancing the identification and mitigation of marine plastic pollution (de Vries, 2021). In addition, the integration of these technologies into a broader monitoring system allows for more targeted and efficient interventions, making it possible to address the issue on a much larger scale.

These innovations mark a critical shift towards more sophisticated and scalable methods for monitoring and removing ocean plastics. The evolution from traditional manual expeditions to high-tech, satellite-based, and AI-driven solutions underscores the dynamic and evolving strategies being employed to address this global environmental crisis (Buczkowski, 2024). This transformation highlights the urgent need for advanced technologies that can keep pace with the growing scale of oceanic plastic pollution.

In order to address the rising problem of microplastic contamination, this paper seeks to highlight the significance of integrating remote sensing and GIS technologies. This work highlights the need for sophisticated geospatial technology for efficient monitoring and control by analysing the detrimental consequences of microplastics on marine ecosystems, the food chain, and ultimately human health.

Ultimately, the goal is to contribute to sustainable development and pollution mitigation efforts, ensuring that innovative technologies are applied to protect the health of both marine life and human populations (Ziani, 2023).

6.1.1 From Land to Sea: The Journey and Consequences of Plastic Waste

Plastic production began with the de novo synthesis of synthetic polymers, marking a revolutionary breakthrough in the early twentieth century (Worm et al., 2017). The invention of plastic revolutionised industries worldwide, offering a versatile, durable, and cost-effective material that transformed sectors from packaging and manufacturing to healthcare and construction (Pilapitiya & Ratnayake, 2024). However, what was once celebrated as an industrial marvel has now become a major environmental threat. The very properties that made plastic revolutionary—its durability and resistance to degradation—are now causing significant harm to ecosystems. Improper waste disposal has led to plastic accumulation in the environment, where it is transported via wind, rain, urban runoff, and drainage systems, eventually reaching waterways and oceans (Kye et al., 2023).

Recent research on oceanic plastic pollution reveals that approximately 80% of plastic waste in marine environments originates from land-based sources, with the remaining 20% attributed to marine activities (Schmaltz, 2020). A notable contributor to this pollution is synthetic fishing nets, which constitute nearly half of the mass of the Great Pacific Garbage Patch, largely due to ocean current dynamics and intensified fishing efforts. The prevalence of plastic in marine debris is primarily due to its durability, low cost, versatility, and inability to biodegrade, leading to its fragmentation into microplastics (Evers, 2024).

Photodegradation, driven by sunlight, further breaks down larger plastic items into smaller particles, including plastic bags, bottle caps, and Styrofoam containers. The impact of marine debris on ecosystems is profound, as evidenced by loggerhead sea turtles mistaking plastic bags for jellyfish and seabirds, such as albatrosses, misidentifying plastic pellets as fish eggs. These misidentifications can lead to starvation or organ damage in the affected species. Marine mammals, including seals, are particularly vulnerable to entanglement in abandoned fishing gear, leading to drowning in a phenomenon known as "ghost fishing" (Sigler, 2014).

Moreover, the accumulation of microplastics and debris in regions like the North Pacific Subtropical Gyre disrupts food webs by blocking sunlight, which inhibits the growth of essential autotrophs like algae and plankton. This decline in primary producers poses a threat to marine species that rely on them for sustenance, ultimately affecting apex predators such as tuna and sharks.

Plastics are not only a persistent environmental pollutant but also a significant vector for harmful chemicals. They are known to absorb contaminants like PCBs from seawater and leach dangerous compounds such as bisphenol A (BPA), raising serious concerns about their potential entry into the food chain and the subsequent effects on human health and marine life (Du et al., 2021). This issue becomes even more complex when considering microplastics, which are challenging to detect and monitor due to their minute size and widespread dispersion in aquatic environments.

6.2 Mapping the Invisible: Remote Sensing Solutions for Microplastic Analysis

Microplastics, defined as plastic particles smaller than 5 mm, present unique detection challenges. While larger microplastics (1–5 mm) are somewhat visible, smaller ones (1 μm–1 mm) typically require specialised equipment for accurate identification and analysis (Ziani, 2023). Their small size allows them to infiltrate marine ecosystems undetected, complicating efforts to quantify and assess their impact in large-scale studies of marine pollution. Addressing this challenge calls for innovative approaches, such as remote sensing and GIS technologies, to map and analyse these "invisible" pollutants effectively.

The small size of microplastics, which makes them difficult to detect, creates an opportunity for remote sensing to play a pivotal role in their monitoring and analysis. Lillesand et al., (2022) defined remote sensing as "the art and science of obtaining information, without physical contact, from the object under consideration". Remote sensing offers a robust, comprehensive, and efficient methodology for acquiring data on marine debris, enabling researchers to monitor and analyse the distribution and characteristics of microplastics in oceanic environments (Rußwurm et al., 2023; Salgado-Hernanz et al., 2021). By overcoming the limitations of traditional detection methods, remote sensing enhances the ability to track these contaminants on a larger scale.

Prior to the advent of remote sensing, conventional techniques primarily relied on methods such as manual sampling, filtering, and laboratory analysis for the detection and quantification of microplastics. The most commonly used method for separating microplastics from water samples is filtering or sieving, which is also applied to the supernatant containing plastics after density separation from sediment samples, Density differences can be utilised to separate plastics (0.8–1.6 g/cm^3) from sediment (2.7 g/cm^3) through techniques such as flotation and elutriation (Prata et al., 2019).These methods represent earlier approaches used for the detection of microplastics. These approaches, while effective on a localised scale, were often labour-intensive, time-consuming, and limited in their ability to provide large-scale or real-time data on microplastic pollution.

On the introduction of remote sensing which handled multitudes of benefits and advantage they are stated below (Skywatch, 2024) (Exploring the Pros and Cons: Advantages and Disadvantages of Remote Sensing | Spatial Post, 2023):

Wide Area Coverage: Remote sensing offers extensive coverage, allowing for the monitoring of microplastic pollution across large marine areas efficiently, even in remote or hard-to-reach regions like deep seas or isolated coastlines (Rauf et al., 2024).

Real-time Data: Delivering data in real-time or almost real-time facilitates the swift identification and monitoring of microplastic concentrations, allowing prompt reactions to pollution incidents or changing maritime conditions (Hafeez et al., 2018).

Access to Inaccessible Areas: In order to manage and mitigate microplastic pollution, remote sensing can gather data from maritime zones that are hazardous or impossible to access, such as open oceans or locations with high levels of contamination (Abdulhameed et al., 2024).

Multi-angle and Multi-spectral Imaging: By capturing data at different wavelengths and angles, remote sensing helps identify microplastic particles in water bodies, differentiating them from other materials like algae or sediments, supporting more accurate pollution assessments (Zhu et al., 2024).

Environmental Monitoring: Remote sensing is essential for tracking the distribution and movement of microplastics over time, helping inform marine conservation strategies, policy-making, and pollution management efforts (Ma et al., 2023).

To sum up, remote sensing offers significant benefits for detecting and monitoring microplastic pollution in marine ecosystems. Its wide coverage, real-time data collection, and ability to access remote areas make it essential for tracking microplastics, even in isolated ocean regions. Techniques like multi-spectral and multi-angle imaging enhance detection accuracy by distinguishing plastics from other materials. Continuous monitoring provides vital insights into the transport and accumulation of these pollutants (Ismaila Abimbola et al., 2024). The next section explores advanced techniques to further improve detection accuracy and efficiency.

Building on the central role of remote sensing in detecting microplastics, significant advancements have been made by harnessing the power of sensors that operate across different regions of the electromagnetic spectrum. Leveraging the power of sensors that operate across different regions of the electromagnetic spectrum (EMS) (Horning, 2008) is fundamental to the effectiveness of remote sensing in detecting and monitoring microplastics in marine environments.

The EMS encompasses various wavelengths, from ultraviolet to microwave (Horning, 2008), each of which interacts differently with materials found in the ocean. The unique properties of plastics allow sensors to capture their distinct spectral signatures, enabling the differentiation of microplastics from other oceanic elements—an essential step for accurately assessing plastic pollution (Schmidt et al., 2023). Optical sensors play a pivotal role in this process by capturing data in ultraviolet, visible, and infrared wavelengths. These sensors measure how surfaces reflect light, revealing the unique spectral reflectance properties of plastics. By analysing these properties, researchers can identify and quantify plastic debris, offering valuable insights into its distribution and impact on marine environments.

A particularly notable feature of plastics is their distinct reflectance signature in the near-infrared (NIR) region of the electromagnetic spectrum, typically ranging from 780 to 2500 nm. Plastics reflect significantly more light in this region compared to natural materials like water, soil, or vegetation, creating a noticeable reflection peak (Balsi et al., 2021). This characteristic makes it possible to detect and monitor plastic pollution using remote sensing technologies, especially in marine environments where visible light may be insufficient. By leveraging NIR reflectance patterns, scientists can enhance the accuracy of satellite imagery and other tools to uncover even hard-to-detect microplastics.

While optical sensors are highly effective for surface-level detection, microwave sensors add another critical dimension to remote sensing by penetrating through water. This capability enables the detection of submerged debris that optical methods might overlook, providing a more comprehensive understanding of microplastic pollution (VishnuRadhan et al., 2022). The integration of data from both optical and microwave sensors improves the ability to map and monitor microplastic distribution across diverse marine environments. This multi-faceted approach enhances detection accuracy and strengthens our capacity to address the global challenge of plastic pollution (Wang et al., 2020) (Fig. 6.3).

By leveraging the full spectrum of electromagnetic radiation in remote sensing, we can gain deeper insights into the scale and impact of microplastics on marine ecosystems. This knowledge not only enhances our understanding of the environmental threat posed by microplastics but also informs effective management strategies and raises awareness of this pressing issue.

Passive remote sensing techniques, particularly those utilising visible (VIS) to short-wave infrared (SWIR) bands, have demonstrated significant potential in identifying floating marine plastic debris. Plastics exhibit unique spectral absorption features that distinguish them from natural ocean components such as water, algae, or organic matter (Goddijn-Murphy et al., 2024). These spectral differences enable scientists to track the presence and movement of plastics across vast marine areas where in-situ observations would otherwise be infeasible or cost-prohibitive (Ismaila Abimbola et al., 2024) (Fig. 6.4).

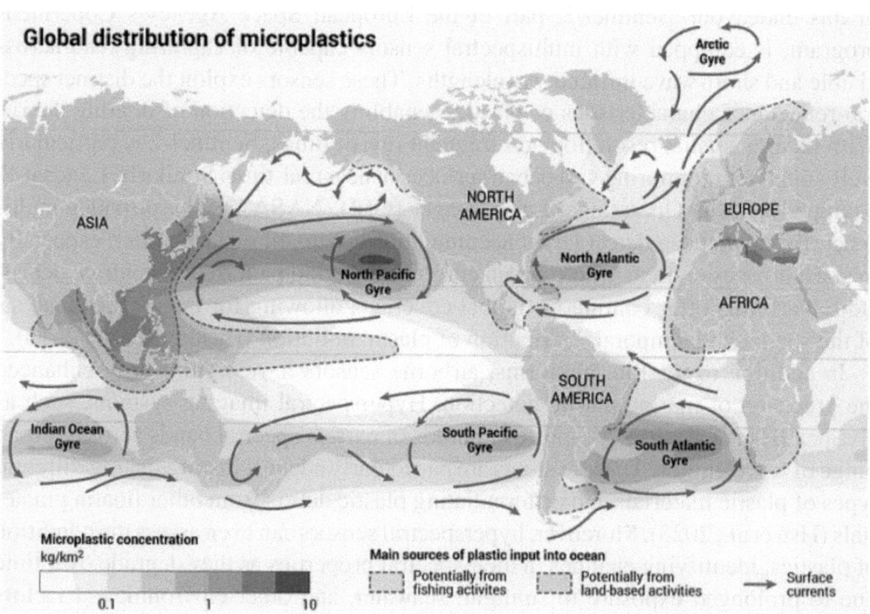

Fig. 6.3 Identifying plastic litter accumulation zones in the world's oceans (Pravettoni & Rekacewicz, 2019)

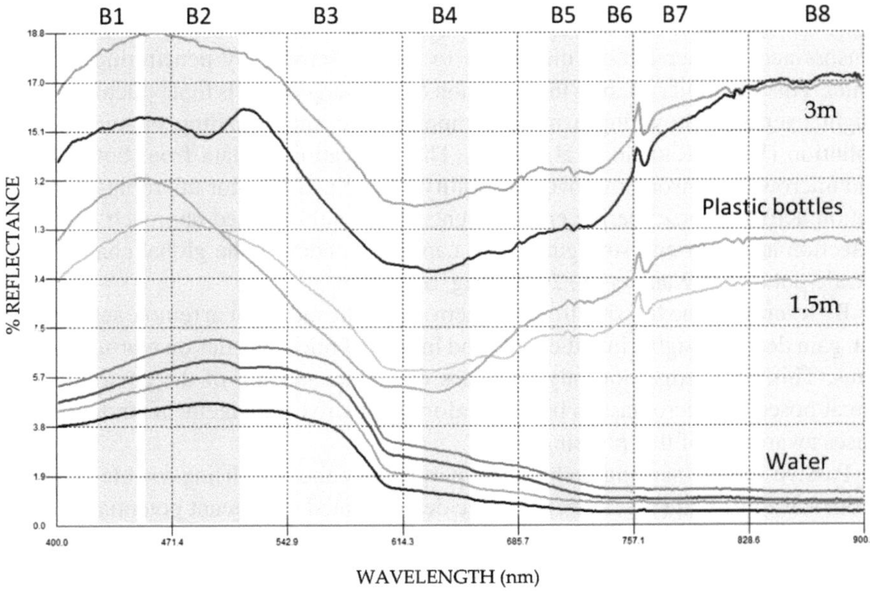

Fig. 6.4 Spectral reflectance curves for plastic bottles and water. *Source* Goddijn-Murphy and Dufaur (2018)

Satellites such as Sentinel-2 and Landsat-8 have become indispensable tools in this endeavour. Sentinel-2, part of the European Space Agency's Copernicus program, is equipped with multispectral sensors capable of capturing data across visible and short-wave infrared wavelengths. These sensors exploit the distinct spectral reflectance characteristics of plastics, enabling the detection of floating debris. With its high spatial resolution and frequent revisit times, Sentinel-2 is particularly well-suited for monitoring vast ocean surfaces in near real-time. Similarly, Landsat-8, equipped with its Operational Land Imager (OLI) (NASA, 2021), provides multispectral data that can detect larger accumulations of floating plastic litter, especially in coastal zones or areas where ocean currents and wind patterns concentrate debris. Both platforms offer continuous global coverage, allowing for large-scale analysis of the spatial and temporal distribution of plastic pollution (Hasituya et al., 2016).

In addition to satellite platforms, airborne sensors have significantly enhanced the precision of marine plastic detection. Hyperspectral imaging systems, such as HySpex (Hsu et al., 2023), capture hundreds of narrow spectral bands across a broad range of wavelengths. These systems are particularly adept at distinguishing different types of plastic materials and differentiating plastic debris from other floating materials (Hsu et al., 2023). Moreover, hyperspectral sensors can even assess the condition of plastics, identifying changes in their spectral properties as they degrade over time due to prolonged exposure to sunlight, seawater, and other environmental factors (Bonifazi et al., 2023).

Beyond satellite and airborne platforms, Unmanned Aerial Vehicles (UAVs) equipped with hyperspectral and multispectral sensors provide unmatched flexibility for monitoring localised regions, such as coastlines and estuaries. UAVs are particularly effective in accessing hard-to-reach areas and operating at lower altitudes, enabling the capture of ultra-high-resolution data (Mohsan et al., 2023). This capability makes UAVs indispensable for detailed mapping of plastic litter and tracking its movement in complex environments, especially where satellites encounter challenges such as cloud cover or limited spatial resolution (Abreo et al., 2023).

To further improve the accuracy of plastic detection, researchers are developing comprehensive spectral libraries that catalogue the unique spectral signatures of various plastic types and polymers. These libraries are vital for enhancing the classification capabilities of remote sensing systems, allowing them to better distinguish different plastic materials even when mixed with other marine debris or biomass (Gnann et al., 2022). By providing a detailed reference, these spectral libraries bolster the ability of Earth Observation (EO) technologies to identify and classify plastics with greater precision.

Plastic materials have distinct spectral reflectance properties, allowing EO data to effectively identify and monitor plastic pollution by leveraging their brightness and reflective qualities (Lavender, 2022). This capability aids in assessing and mitigating pollution across marine, coastal, and terrestrial environments, offering insights into the distribution and trends of plastic debris (Sivadas et al., 2022a). However, detection varies by environment: floating plastics are easier to identify due to their contrast against darker water surfaces, whereas on land, interference from reflective surfaces like vegetation and soil poses greater challenges (Lavender, 2022). Tailored strategies are essential to address these environmental differences effectively.

Water-Based Detection: Biermann et al. (2020) and Themistocleous et al. (2020) developed specialized indices to detect plastic waste on water surfaces using Sentinel-2 data. The Floating Debris Index and Plastic Index leverage the near-infrared (NIR) and red bands to identify floating plastics by utilizing the high brightness contrast of plastics compared to water surfaces, where plastics typically appear as bright features (Biermann et al., 2020; Themistocleous et al., 2020). This brightness difference enhances detection in open waters; however, caution is warranted, as not all bright patches in the ocean necessarily indicate plastic. For instance, convergence zones or windrows, where debris naturally accumulates due to ocean currents, may contain other reflective materials besides plastic (Biermann et al., 2020).

Land-Based Detection: Detecting plastic on land presents additional challenges due to the presence of bright background surfaces that can mimic plastic's spectral characteristics. Reflective roofs, synthetic turf, and greenhouses often contain plastic materials and other bright synthetic elements, which complicates distinguishing plastic waste from intended plastic use (Themistocleous et al., 2020). The spectral similarity between these surfaces and plastic waste reduces detection accuracy, especially in urban or agricultural areas where plastics are commonly incorporated into the built environment.

This overlap highlights the need for more refined indices and detection techniques to differentiate between plastic waste and functional plastic on land, as current EO methods struggle to isolate plastic pollution in complex terrestrial settings.

6.3 Types of Polymers and Their Spectral Reflectance in Different Environments

6.3.1 Polyethylene and Polypropylene in Coastal Sediments

Polyethylene and polypropylene, commonly linked to packaging waste, dominate plastic pollution in India's coastal environments, appearing in sediments, water, and biota (Sivadas et al., 2022a, 2022b). EO methods must target these polymers to monitor and mitigate their ecological impact in marine ecosystems effectively.

6.3.2 Plasticulture Detection Using Spectral Reflectance

The use of polyethylene in plasticulture—where it serves as a mulch covering farmland—illustrates a distinct application of EO for plastic detection. Plastic-mulched farmland has a unique spectral profile: it is brighter, smoother, and drier than typical soil due to reduced roughness and moisture retention, which results in higher reflectance values in the shortwave infrared (SWIR) bands (Hasituya et al., 2016). Levin et al. (2007) found that absorption features at **1218 nm and 1732 nm** are key for identifying plastic-covered agricultural fields (Levin et al., 2007). This spectral information aids in distinguishing plastic-mulched farmland from other agricultural surfaces, contributing to more accurate mapping and monitoring of agricultural plastic use.

6.3.3 Floating Plastic Detection *via* Absorption Peaks

Goddijn-Murphy and Dufaur (2018) identified absorption peaks around **1140 nm and 1680 nm** that are particularly valuable for detecting floating plastic waste in marine environments. These specific absorption features allow EO systems to distinguish floating plastics from other types of marine debris, such as wood or algae, that may not share the same spectral characteristics. By focusing on these peaks, researchers can improve the precision of remote sensing methods for detecting plastic in aquatic environments, where high-quality monitoring is critical to addressing the growing issue of oceanic plastic pollution (Goddijn-Murphy & Dufaur, 2018).

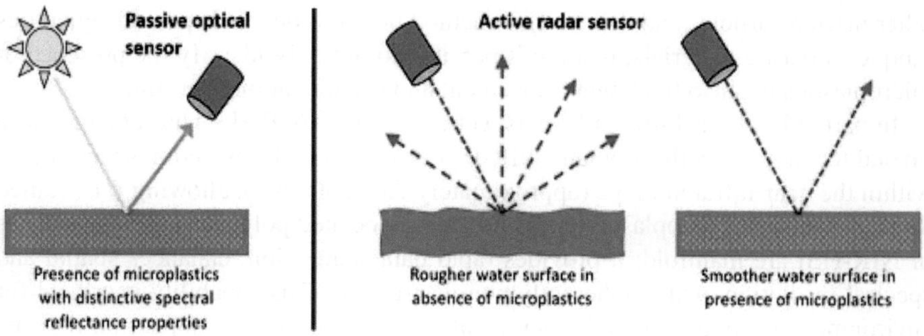

Fig. 6.5 Airborne remote sensing of aquatic microplastics primarily involves using passive optical sensors to measure solar radiation reflectance and active radar sensors to evaluate water surface roughness reduction. *Source* Chabot and Marteinson (2024)

Despite advancements in spectral indices and absorption features for plastic detection, several challenges remain. Separating plastic waste from natural and man-made materials in both terrestrial and aquatic environments requires continued refinement of EO techniques. The spectral similarities between plastic and other reflective surfaces, such as roofs or greenhouse coverings, complicate detection efforts, particularly on land.

Future research could enhance EO methodologies by developing more robust indices and conducting field-based validation studies to strengthen plastic detection accuracy across diverse environments. Additionally, advances in spectral libraries and the application of specific absorption peaks in EO could improve the monitoring of both floating and land-based plastic pollution, thereby aiding global environmental protection efforts (Sannigrahi et al., 2022). Moreover, the development of dedicated sensors specifically designed for mapping marine litter may provide even greater precision in identifying and characterising plastic pollution, creating a cohesive pathway for innovation and impact (Fig. 6.5).

The methods and applications utilised to improve microplastic monitoring and detection will be covered in depth in the next section. Advances in data processing and sensor technology are transforming our knowledge of marine plastic pollution, opening the door to workable mitigation plans and increasing awareness of this pressing problem on a worldwide scale.

Various studies have demonstrated the efficacy of spectral imaging in detecting microplastics within marine environments (Rußwurm et al., 2023). Hyperspectral imaging, a sophisticated remote sensing technique, captures and analyses electromagnetic radiation across multiple wavelengths, extending beyond the visible spectrum (Zahra et al., 2024; Wikipedia Contributors, 2024). This method enables researchers to discern material properties and composition based on unique spectral signatures, facilitating the identification of various substances, including microplastics (Faltynkova & Wagner, 2023).

Hyperspectral imaging has proven to be especially effective by analysing a broad spectrum of light wavelengths to detect and differentiate microplastics from other marine particles and debris. This method provides detailed spectral signatures unique to plastic materials, making it possible to not only identify the presence of microplastics but also track their movement and accumulation over time.

In particular, Near Infrared Hyperspectral Imaging (NIR-HSI) has emerged as a pivotal technology in this domain. NIR-HSI leverages high-resolution spectral data within the near-infrared range (approximately 750–2500 nm), allowing for detailed discrimination of microplastic types and their associated polymers. The advantages of NIR-HSI are manifold: it provides rapid data acquisition, enhances spatial and spectral resolution, and is inherently non-destructive. This capability is critical for environmental monitoring and assessment, as it minimises the impact on delicate marine ecosystems while ensuring accurate characterization of microplastic waste (Zhang et al., 2019). Consequently, NIR-HSI is increasingly recognized as an essential tool in the ongoing effort to quantify and understand microplastic pollution in marine systems, providing invaluable insights into the distribution, composition, and potential ecological impacts of microplastics (Waqas et al., 2023) (Fig. 6.6).

According to Davaasuren et al. (2018), surface wind speed and Langmuir cells, an ocean circulation pattern, work together to determine the characteristic look of surfactants, sea slicks, and microbial biofilms in Synthetic Aperture Radar (SAR) pictures. Preliminary conclusions suggest that SAR remote sensing may prove effective in detecting plastic pollution in open ocean environments. Furthermore, this approach could potentially be adapted and applied to other regions for broader monitoring of marine debris (Table 6.1).

6.4 Case Studies on Microplastic Detection in Marine Ecosystems

Remote sensing enables continuous tracking of microplastics across vast ocean regions, allowing researchers to identify key accumulation zones, understand seasonal and spatial variations, and better target mitigation efforts (Atwood et al., 2019). Case studies in this area highlight innovative applications of satellite and airborne sensors, demonstrating the power of remote sensing to transform our understanding of microplastic pollution. These studies underline how cutting-edge technology, combined with spatial data analysis, is opening new avenues for addressing this global change.

Researchers developed an inversion model to predict microplastic (MP) pollution in the Bohai Sea by integrating satellite remote sensing data with field measurements (Hong et al., 2023). Using a multiple regression approach, the team identified key variables for predicting MP levels and tested three selection methods: the successive projections algorithm (SPA), a band combination method, and a remote sensing index method. The SPA approach yielded the most accurate results, achieving a determination coefficient of 0.75 and a root mean square error of 0.38 items/m^3—well within an acceptable range (Hong et al., 2023).

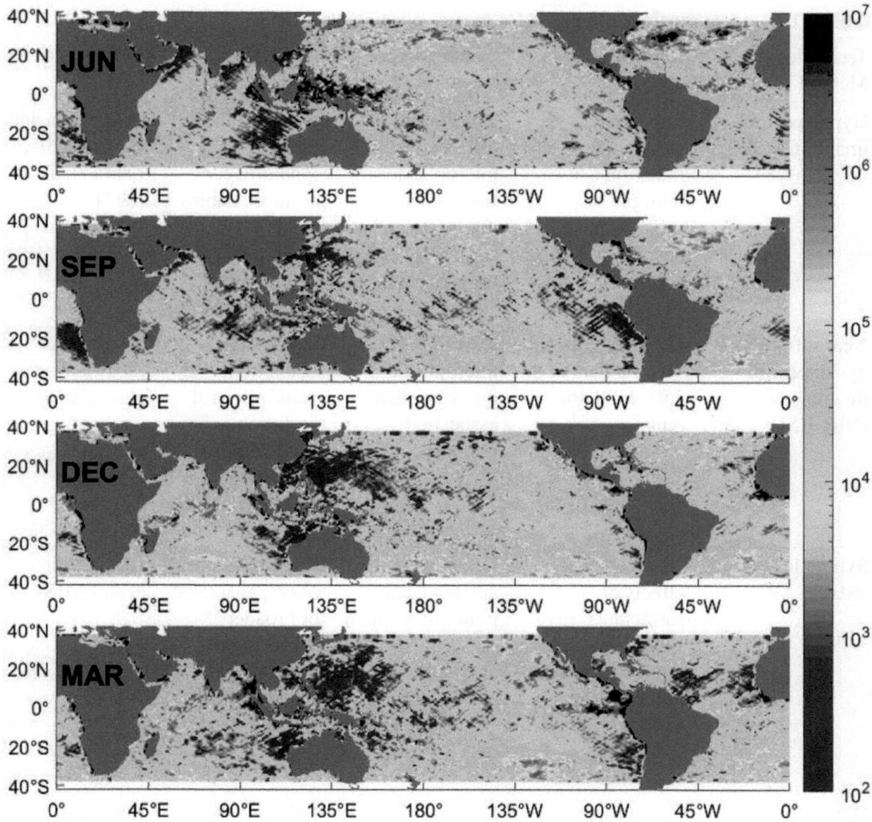

Fig. 6.6 Monthly average microplastic concentration for June–September-December 2017 and March 2018 shows higher concentrations in summer for both the Atlantic and Pacific basins, with the northern Indian Ocean peaking in spring, exhibiting repeating seasonal patterns. *Source* Evans and Ruf (2022)

Applying the SPA model, researchers analysed seasonal and regional variations in MP abundance across the Bohai Sea in 2022. They found significant fluctuations in microplastic concentrations across different areas and seasons, highlighting the dynamic nature of pollution in marine environments.

This study showcases the potential of satellite remote sensing to provide efficient, large-scale monitoring of microplastics, offering valuable insights for environmental management and pollution mitigation (Hong et al., 2023).

NASA's Cyclone Global Navigation Satellite System (CYGNSS) (Ruf et al., 2017) is revolutionising the way researchers track microplastic pollution across the oceans. Originally designed to monitor hurricanes, CYGNSS now offers a unique advantage for tracking microplastics (Sobhani et al., 2024), as it provides near-global coverage every one to two days. Unlike traditional methods, which rely on trawlers to collect

Table 6.1 Remote sensing techniques for marine plastic detection

Technology/ Method	Key features	Applications	Advantages	References
Hyperspectral imaging (HSI)	Captures electromagnetic radiation across multiple wavelengths beyond the visible spectrum	Detects and differentiates microplastics from other marine particles and debris based on spectral signatures	Provides unique spectral signatures for plastics, enables tracking movement and accumulation over time	Faltynkova and Wagner (2023), Zahra et al. (2024)
Near infrared hyperspectral imaging (NIR-HSI)	High-resolution spectral data in the 750–2500 nm range	Identifies microplastic types and their associated polymers	Rapid data acquisition, high spatial/spectral resolution, non-destructive, and minimal impact on marine ecosystems	Zhang et al. (2019), Waqas et al. (2023)
Synthetic aperture radar (SAR)	Analyzes the effects of surfactants, sea slicks, and microbial biofilms influenced by surface wind speed and Langmuir cells	Potential to detect plastic pollution in open ocean environments	Effective in open waters, adaptable for broader marine debris monitoring in other regions	Davaasuren et al. (2018)
Optical imaging	Uses visible and near-infrared light to observe and classify objects on the ocean surface	Detects floating plastics, particularly macroplastics, in coastal and marine environments	High-resolution imagery, easy integration with existing satellites, cost-effective for large-scale monitoring	Rußwurm et al. (2023), Wikipedia Contributors (2024)
Unmanned aerial **vehicles** (UAVs)	Equipped with high-resolution cameras and sensors, often integrating multispectral or hyperspectral capabilities	Monitors plastic pollution in remote or inaccessible areas, particularly coastal zones	Flexible deployment, high spatial resolution, effective for small-scale localized studies, and customizable for specific tasks	Zhang et al. (2019), Aitken et al. (2020)

limited samples, CYGNSS enables continuous monitoring of the ocean's surface, capturing real-time conditions (Bates, 2021).

CYGNSS identifies regions with high microplastic concentrations, providing valuable data on where plastics enter and accumulate in the ocean. This ability to pinpoint pollution hotspots is crucial for targeted clean-up efforts and for developing more effective strategies for tackling ocean pollution. As researcher Evans emphasises, the ultimate goal is to understand the sources and accumulation zones of microplastics to better inform mitigation efforts, promoting a cleaner and healthier ocean environment (Bates, 2021).

Marine plastic debris in satellite images, as the light reflection properties constantly change due to the dynamic nature of the ocean environment (Nivedita et al., 2024). Beyond these immediate technological limitations, additional challenges remain in accurately mapping marine plastics. The development of comprehensive spectrum libraries that incorporate material-level information for different kinds of marine plastic debris monitoring via remote sensing poses a number of difficulties, mostly because of spatial resolution constraints (Veettil et al., 2022).

Remote sensing images are captured at various spatial resolutions, which define the smallest object that can be detected (Waqas et al., 2023). Many current satellite sensors cannot capture microplastics (less than 5 mm), and even larger debris can be difficult to identify if the resolution is not high enough (Ismaila Abimbola et al., 2024).

Plastic debris, especially small or scattered pieces, often goes undetected in standard satellite imagery, particularly in open water environments where it is dispersed over vast areas (Nivedita et al., 2024).

Another significant challenge lies in the spectral resolution of remote sensing sensors. Spectral resolution refers to the number and width of wavelength bands that a sensor can detect, and it plays a key role in distinguishing between different materials. Plastics are made from a variety of polymers, such as polyethylene, polypropylene, and polystyrene, each of which has unique reflectance properties across the electromagnetic spectrum (Veettil et al., 2022). However, many satellite sensors do not have the spectral sensitivity required to differentiate between these subtle variations, making it difficult to distinguish plastic debris from other materials in the ocean, such as water, seaweed, or organic matter.

The surface reflectance properties of plastics also complicate detection. Different plastic types reflect sunlight in varying ways depending on their composition. Environmental factors like lighting conditions, ocean surface dynamics (waves and currents), and the degradation of plastic materials further alter reflectance patterns. These variables make it even harder to detect and classify plastics could significantly improve the accuracy of plastic detection.

Moreover, launching dedicated sensors specifically designed for marine litter mapping would enhance the ability to monitor plastic pollution more effectively in the future. Another major hurdle is the detection of submerged marine plastics, which are difficult to monitor with current remote sensing technology (de Fockert, 2024). Plastics mixed with biomass and marine litter that has been buried along coastlines due to accretion and erosion are also challenging to account for. Addressing these issues will require advancements in sensor technology and the development of new approaches to effectively map and monitor marine plastic debris on a global scale (Veettil et al., 2022).

6.5 Leveraging AI and ML for Effective Microplastic Pollution Detection

Artificial intelligence (AI) has revolutionised many fields, with some of the most significant advancements emerging in pollution prevention and environmental monitoring (Jin et al., 2024). A crucial application of AI and machine learning (ML) in this area is the monitoring and identification of microplastics, where these technologies enhance traditional detection methods by automating data analysis.

This automation allows for faster and more precise identification of microplastics in complex environments, processing large datasets from image processing, spectral analysis, and remote sensing to locate concentrations of microplastics and track their movement over time (Astorayme et al., 2024; Yu & Hu, 2022).

Pereira et al. (2024) highlighted that a major contribution of AI in environmental monitoring is its capability to analyze and interpret large volumes of data collected from various sources, such as satellite imagery, remote sensing devices, and weather stations. By processing this diverse information, AI can identify patterns, detect anomalies, and track environmental changes more efficiently and accurately than traditional methods (Jin et al., 2024).

Traditional ML methods, such as random forest and support vector machines, have shown reliability in detecting macroplastics (MPs) but face limitations with multiclass classification tasks. In contrast, deep learning (DL) approaches, particularly convolutional neural networks (CNNs) and pre-trained models like VGG16 (Simonyan & Zisserman, 2015) and YOLO (Reis et al., 2023), demonstrate higher accuracy for complex MP classification, although they require significant computational resources. Since 2020, there has been growing interest in transformer-based architectures for object detection, yet their application to MP detection remains largely unexplored (Astorayme et al., 2024).

Overall, AI has become essential to environmental monitoring, enhancing precision and efficiency in detecting contaminants like microplastics. Through automation in techniques such as image processing, Fourier transform infrared (FTIR) spectroscopy, Raman spectroscopy (Sunil et al., 2024) and hyperspectral imaging (HSI).

AI minimises manual labour and accurately distinguishes plastic types by analysing unique spectral signatures. Real-time data integration supports targeted monitoring and hotspot detection, marking AI's innovative application as a transformative strategy for safeguarding ecosystem health (Jin et al., 2024).

6.6 Microplastics in the Context of Sustainable Development

As we explored the technical aspects of detecting microplastics through remote sensing, it became clear that focusing on nature-based solutions is crucial for addressing this issue. Microplastics are pollutants that significantly affect the environment, and understanding their impact is essential for effective management.

The widespread presence of microplastics in our ecosystems highlights the urgent need to integrate sustainable practices into industrial development. These pollutants can disrupt the delicate balance of our natural environments, leading to negative consequences that go beyond short-term economic benefits. For example, microplastics can harm marine life, reduce biodiversity, and compromise ecosystem health.

By recognising the broader implications of microplastic pollution, we can better appreciate the importance of sustainable practices. This understanding emphasises the need for the Sustainable Development Goals (SDGs) (United Nations, 2024), which provide a framework for addressing environmental challenges while promoting economic growth.

In light of the urgent threats posed by environmental degradation, social inequality, and the decline of marine life and biodiversity, the United Nations launched the Sustainable Development Goals (SDGs) in 2015. These goals aim to create a world where sustainable practices and social fairness help protect our oceans, preserve ecosystems, and ensure a healthier, more prosperous future for everyone (Wikipedia Contributors, 2019a, 2019b).

No Poverty (SDG 1): Microplastics in water bodies and soil can affect local communities that rely on natural resources for their livelihoods, leading to poverty and limited access to clean, healthy environments (Kumar et al., 2021).

Zero Hunger (SDG 2): Contamination of marine and agricultural ecosystems by microplastics can disrupt food chains, impacting fish stocks and reducing the availability of safe, nutritious food (Okeke et al., 2022).

Good Health and Well-Being (SDG 3): Exposure to microplastics can lead to adverse health effects, including respiratory issues and toxicity, compromising the health of populations, especially vulnerable groups (Lee et al., 2023).

Quality Education (SDG 4): Educating communities about microplastic pollution and sustainable practices is essential for creating awareness and promoting solutions for environmental conservation (Jain et al., 2023).

Gender Equality (SDG 5): Women and children are often more vulnerable to the negative impacts of environmental degradation, including microplastic pollution, which affects their health and well-being disproportionately (Stoett et al., 2024).

Clean Water and Sanitation (SDG 6): Microplastics in water bodies threaten the availability of safe and clean water, making it critical to address their presence to ensure sustainable water management (Oyewo et al., 2024).

Affordable and Clean Energy (SDG 7): Reducing microplastics in marine environments can prevent pollution from impacting renewable energy sources, such as tidal and wind energy infrastructure (Matavos-Aramyan, 2024).

Decent Work and Economic Growth (SDG 8): Sustainable industries and economies depend on healthy ecosystems, which are threatened by microplastic pollution. Addressing this issue is essential for fostering economic growth (Stoett et al., 2024).

Industry, Innovation, and Infrastructure (SDG 9): Advances in AI and remote sensing enable innovative, scalable solutions for detecting and managing microplastic pollution, fostering sustainable industrialization and minimizing environmental impacts (Davaasuren et al., 2018).

Reduced Inequality (SDG 10): Microplastic pollution disproportionately impacts marginalized communities reliant on natural resources for their livelihoods, as they often lack the means to address environmental degradation. Tackling this issue promotes environmental justice and equitable access to clean resources, reducing inequalities (Sinha et al., 2024).

Sustainable Cities and Communities (SDG 11): Effective strategies to mitigate microplastic pollution include reducing plastic waste, improving urban waste management, raising public awareness, and integrating green infrastructure and pollution control measures into urban planning. Such efforts reduce overall microplastic exposure and promote inclusive, safe, and resilient communities (Ihenetu et al., 2024).

Responsible Consumption and Production (SDG 12): Minimizing microplastics through sustainable consumption and production patterns is essential for reducing waste and promoting resource efficiency (Copello de Souza, 2019).

Climate Action (SDG 13): Microplastics disrupt marine ecosystems, which play a vital role in carbon sequestration. Mitigating microplastic pollution enhances climate resilience and supports global climate action (Lincoln et al., 2022).

Life Below Water (SDG 14): Microplastics significantly harm marine life, ecosystems, and the sustainability of marine resources, necessitating urgent action to protect life below water (Vadrot, 2023).

Life on Land (SDG 15): Microplastics degrade soil and harm biodiversity, disrupting ecosystems. Sustainable land management and bioremediation, using microorganisms or plants to remove pollutants, can mitigate their impact and restore ecological balance (Thapliyal et al., 2024).

Peace, Justice, and Strong Institutions (SDG 16): Effective governance and collaboration are required to address microplastic pollution, ensuring that environmental laws and policies are enforced (Matavos-Aramyan, 2024).

Partnerships for the Goals (SDG 17): Combating microplastics pollution demands collective efforts from governments, industries, and civil society to implement sustainable solutions (Goh et al., 2024).

These goals aim to create a sustainable and equitable world for all by 2030. These 17 interconnected goals provide a comprehensive framework aimed at fostering sustainable development across social, economic, and environmental dimensions (Moallemi et al., 2020). Each goal addresses specific challenges while emphasising the need for collaboration and integrated approaches (United Nations Development Programme, 2024).

The escalating prevalence of microplastic pollution in marine environments necessitates the implementation of robust and scientifically informed strategies for sustainable management. The Sustainable Development Goals (SDGs) (United Nations, 2024) offer an essential framework for addressing this pervasive issue while supporting economic growth.

Specifically, SDG 14, which emphasises the conservation and sustainable use of oceans, seeks to mitigate marine pollution, including that from microplastics. This goal underscores the critical need for interdisciplinary collaboration among researchers, policymakers, and industry stakeholders to develop innovative solutions.

The 17 interconnected Sustainable Development Goals (SDGs) provide a comprehensive framework for addressing challenges related to social, economic, and environmental dimensions, with the aim of fostering a more equitable and sustainable world by 2030 (Moallemi et al., 2020). When addressing microplastic pollution in marine environments, these goals emphasise the importance of collaborative efforts and integrated strategies. For example, Goal 14 (Life Below Water) directly targets the need to conserve and sustainably use oceans and marine resources, while highlighting the interconnectedness of other goals like responsible consumption and production (Goal 12), which seeks to reduce waste and promote sustainable practices (United Nations, 2022).

This integrated approach underscores how progress in one area can support others. Reducing plastic waste (Goal 12) not only contributes to healthier oceans (Goal 14) but also has broader impacts on climate action (Goal 13) and improving human health and well-being (Goal 3), given the harmful effects of microplastics on ecosystems and communities (Lebreton, 2018; Nilsson et al., 2018). For instance, cutting down on microplastic pollution can improve marine biodiversity and fisheries, which, in turn, supports sustainable livelihoods and food security (Sachs et al., 2019). By addressing microplastic pollution through integrated SDG strategies, we can foster more holistic and sustainable solutions to the problem.

As discussed in earlier sections, the trailblazing advancements in remote sensing technologies, particularly the use of **hyperspectral imaging** and **Earth observation** systems, have revolutionised the way we gather data on microplastic pollution. These cutting-edge technologies offer a **multitude of valuable information** crucial for understanding the extent, distribution, and concentration of microplastics on a **global scale** (Faltynkova et al., 2021).

Integrating remote sensing technologies within the context of the SDGs can significantly enhance our ability to monitor and analyse microplastic distribution and impact. By employing these technologies, we can gather valuable data that informs sustainable practices, ensuring that marine ecosystems are protected from the detrimental effects of microplastic contamination while fostering responsible economic development.

6.7 Policies and Regulations

Policies and regulations play an essential role in managing plastic waste and mitigating microplastic pollution by establishing enforceable standards, fostering accountability, and promoting sustainable practices. Through bans on single-use plastics, recycling mandates, and Extended Producer Responsibility (EPR) programs, governments encourage industries to limit plastic use and innovate in waste management, ultimately helping to safeguard ecosystems, public health, and environmental sustainability (World Bank, 2022).

India, with its vast and diverse coastline, is particularly vulnerable to the impacts of plastic pollution. Stretching over 7500 kms along the Arabian Sea and the Bay of Bengal, the country has one of the longest coastlines in the world, which exposes it to significant levels of marine pollution, including microplastics. This extensive coastline encompasses a wide range of ecosystems, from pristine beaches to densely populated urban areas, amplifying the challenge of managing plastic waste (Veerasingam et al., 2020).

Our country, known for its diverse ecosystems and vibrant natural heritage, is grappling with a significant environmental challenge posed by plastic pollution. A recent study published in Nature highlights that India is a major contributor to global plastic waste, accounting for nearly 20% of the world's total. With approximately 9.3 million tonnes of plastic waste generated annually, around 3.5 million tonnes are mismanaged, often entering the environment and causing harm (Plastics for Change, 2024).

This figure surpasses the plastic waste contributions of major polluters like Nigeria, Indonesia, and China. Rapid urbanisation and population growth have significantly increased plastic waste in urban areas, where demand for single-use plastics remains high. Despite regulatory measures, single-use plastics still make up 43% of India's total plastic waste, mainly due to weak enforcement and the lack of affordable alternatives (Tables 6.2 and 6.3).

The table highlights the widespread and diverse nature of plastic pollution across India and surrounding regions, shedding light on the concentration, particle size, and types of polymers found in various aquatic ecosystems. India, with its vast and diverse coastline stretching over 7500 kms along the Arabian Sea and Bay of Bengal, is particularly vulnerable to the impacts of plastic pollution (Karthikeyan & Subagunasekar, 2023). The extensive coastline, encompassing both densely populated urban areas and pristine beaches, amplifies the challenge of managing plastic waste and underscores the need for targeted policy and action.

Key findings from the data can be stipulated as:

1. **Regional Pollution Variance**: Microplastics are notably abundant in both coastal and freshwater systems at locations such as Silver Beach in Southern India, Tuticorin, Marina Beach in Chennai, the Andaman Beaches, Nattika Beach on the Kerala Coast, and Vembanad Lake. Other impacted areas include South Juhu Creek in Mumbai, Pondicherry beaches, the Ganga River, Roche Park, and Port

Table 6.2 Microplastic concentrations and characteristics across locations in India

Location	Concentration	Size range	Polymer types	Identification method
Silver Beach, Southern India	204 items/kg; 100% < 0.5–1 mm	0.5–1 mm	PVC, polyethylene, nylon	ATR-FTIR
Tuticorin Beaches, Southern India	7 items/m^2 (macro); 9.5 items/m^2 (meso); 54 items/m^2	> 2.5 cm (macro), 5–2.5 cm (meso), < 5 mm (micro)	PE, PP, PET, NY, PS, PVC	ATR-FTIR
Marina Beach, Chennai, India	2275 items; 100% 5.5–25 mm	5.5–25 mm	LDPE, PE, PP, PA, PC	TGA–DSC, SEM
Andaman Beaches, India	414.35 items/kg; 100% 100–1000 pm	100–1000 pm	Polystyrene, polyamide, epoxy resin, PVC, nylon	Raman spectroscopy
Northeast Arabian Coast, India	22.4 kg dry weight/km^2; 40.6%	Variable	Styrofoam, beverage bottles, plastic ropes, etc.	Not specified
South Juhu Creek, Mumbai, India	2 g/kg	Variable	Not specified	Not specified
Coastal Tamil Nadu, India	385 items/kg; 100% 0.5–3 mm	0.5–3 mm	PE, PP, NY, PES, PS	SEM-EDAX, FTIR-ATR
Arabian Sea and Bay of Bengal, India	220 M Ps/kg, 181 M Ps/kg, 45 MPs/kg	> 1 mm	PET, PE, PVC, PP, PS, polyester, polyamides	SEM–EDS, fluorescence Microscopy, FTIR
Kerala Coast, India	0.55 kg/100 m^2; 73.8%	Variable	Plastic	Not specified
Nattika Beach, Kerala, India	191 items/kg; 70% 5–1 mm	5–1 mm	PE, PE + PP, PP, PS, PCU	FTIR, SEM

Blair Bay. These regions show varying pollution levels, influenced by factors such as human activity, environmental conditions, and proximity to urban centres.

2. **Polymer Types**: Commonly identified polymers in these environments include polyethylene (PE), polypropylene (PP), and polystyrene (PS). These polymers are frequently found in both coastal and freshwater ecosystems, signifying the widespread distribution of plastic waste.
3. **Particle Size**: The microplastic particles vary significantly in size, ranging from smaller than 1 mm to larger fragments. Notable occurrences of these varying sizes have been observed at locations like Chennai's East Coast and the Goa Coast. The variability in particle size is crucial for determining appropriate mitigation and

Table 6.3 Microplastic concentrations and characteristics across locations in India

Location	Concentration	Size range	Polymer types	Identification method
Puducherry, India	72.03 MPs/100 g; 56.32% 300 µm—1 mm	300 µm—1 mm	Polyurethane, HOPE, polypropylene, polystyrene, LDPE	Raman spectroscopy
Rameswaram Beach, SE Coast of India	403 pieces; 60.8% > 1.01–200 mm	> 1.01–200 mm	Polyethylene, polystyrene, nylon, PVC	FTIR spectroscopy
Nallathanni Island, SE of India	505 pieces	Variable	Polystyrene, polyethylene, polycarbonate, PVC, nylon	FTIR spectroscopy
Marine Beach, Chennai, India	8.96 kg/m; 56.42%	Variable	Food wrappers, cups, bottles, caps, thermocol	Not specified
Mumbai Beaches, India	7.49 g; 55.33% 1–5 mm	1–5 mm	Not specified	Not specified
Ganga River, India	134.29 items/kg (sediments), 19.87 items/L (water)	1–5 mm (sediment), 500 pm-1 mm (water)	PE, PP, PS, PET, PEST, PVC, RA, PVA	FTIR-ATR, SEM
Roche Park, Tuticorin, India	12.75 items/kg (sediment); 21.6 items/L (seawater)	1–3 mm (sediment); 0.5–1 mm (seawater)	PE, PP, polyester, polyamide, paint	FTIR-ATR
Vembanad Lake, Kerala, India	496 items/m^2; 91% < 5 mm	< 5 mm	Polypropylene, polyethylene, polystyrene	Raman spectra
Goa Coast, India	5500 pellets	Variable	PE, PP	FTIR-ATR
Vavvaru Island, Maldives	1029 items/m^2; 96% > 1 mm	> 1 mm	Polyethylene, polypropylene, polystyrene	ATR-FTIR spectroscop
Port Blair Bay, Andaman Islands	288 items/m^3 (water); 45.17 items/kg (sediments)	35.29–5010 pm (water); 46.72–5024 pm (sediments)	Lonomer surlyn, polyetherimide, acrylic, etc.	FTIR-ATR
Southernmost Coast of India	343 items/50 g; 100% < 5 mm	< 5 mm	Fiber, fragment	Not specified
Kerala, Southwest Coast of India	40.7 particles/m^2 (sediments); 1.25 particles/m^3 (water)	0.3–0.6 mm (sediments)j 1.18–2.36 mm (water)	PE, PP, PA, PS, PET, RY, PUR, PVC	FTIR-ATR, FP-XRF

cleanup strategies, as different-sized particles may require different approaches for effective removal and management.

4. **Analytical Techniques**: Advanced analytical methods such as Fourier-transform infrared spectroscopy (FTIR), Scanning Electron Microscopy (SEM), and Raman Spectroscopy were employed to identify and quantify these pollutants, providing reliable data on the types and distribution of microplastics across these regions.
5. **Location-Specific Data**: The data emphasises the importance of location-specific information to guide targeted policy and cleanup actions. For instance, regions with higher levels of specific polymer types may require tailored efforts in cleanup and prevention, while understanding the variation in debris types and sizes allows for a more focused approach to addressing plastic pollution.

Through a number of legislative measures, the government and policymakers have made important steps to combat microplastic contamination in order to effectively handle plastic waste in freshwater and coastal habitats. With an emphasis on lowering pollution in both freshwater and marine systems, these initiatives seek to address the expanding problem of plastic waste throughout India's varied ecosystems. The main laws and rules in place to address plastic pollution in the nation are listed in the section that follows.

This inflow of plastic has made it difficult for India's waste management system to keep up. 60% of plastic garbage is handled by the unorganised sector (Krishnan, 2022), which depends on ineffective recycling techniques, and about 77% of urban waste ends up in open landfills without adequate treatment. The open burning of plastic garbage in slums and rural areas generates toxic chemicals like dioxins, which contribute to air pollution and pose serious health hazards like cancer and respiratory disorders.

Along coastlines, plastic waste clogs water bodies, disrupts ecosystems, and significantly contributes to marine litter, with plastic making up 80% of marine pollution. This pollution not only threatens marine species but also introduces microplastics into the food chain, affecting human health and disrupting ecosystems (Government of India, 2024).

To combat these challenges, the Indian government has enacted several policies aimed at regulating plastic waste and mitigating microplastic pollution. The *Plastic Waste Management Rules* of 2016, amended in 2022, serve as a foundation for managing plastic waste by setting guidelines for reducing plastic production and promoting recycling. The 2021 amendment to these rules includes a ban on certain single-use plastic items, effective from July 2022, and mandates an increased thickness for plastic carry bags to reduce circulation and improve recyclability.

Key modifications also established EPR rules, transferring responsibility for plastic waste management to producers, minimising pollution at its source and the development of microplastics (Plastic Waste Management Rules, 2022).

The EPR guidelines, which were released in 2022 and focus on the sustainable handling of waste from plastic packaging, are another important component of India's waste reduction initiatives. EPR regulations encourage the circular economy by requiring businesses to meet predetermined recycling goals, reuse rigid plastic packaging, and add recycled materials to new goods. EPR helps restrict disposable and readily fragmentable plastics that contribute to microplastic pollution by providing incentives to industry to reduce plastic consumption and enhance waste management (CPCB, 2024).

In addition to regulatory frameworks, the National Centre for Coastal Research (NCCR), under the Ministry of Earth Sciences, leads the Pan India Coastal Clean-up Programs. These programs, conducted from 2017 to 2023, focus on cleaning beaches, raising awareness, and gathering data on marine litter. During this period, approximately 150 tons of litter were removed from 190 beaches. The data collected through these cleanups, including microplastic concentrations, is critical in shaping policies, identifying pollution hotspots, and engaging communities in sustainable waste management practices (Ministry of Earth Sciences, 2017–2023).

India is taking significant steps towards formulating a National Marine Litter Policy, spearheaded by the Ministry of Environment, Forest and Climate Change (MoEFCC), Ministry of Earth Sciences (MoES), and other relevant ministries. Key initiatives include the establishment of a National Task Force by the Ministry of Fisheries, Animal Husbandry and Dairying (MoFAH&D), which is preparing an actionable National Marine Litter Plan. The policy aims to track and monitor marine litter across riverine, coastal, and ocean ecosystems using advanced techniques like remote sensing and AI. In line with the United Nations Environment Programme's (UNEP) "Clean Seas Campaign," these efforts focus on source-to-sink tracking of plastic waste to inform effective cleanup strategies.

The government is also actively investing in research to understand microplastics' effects on health and ecosystems. The Food Safety and Standards Authority of India (FSSAI) has initiated studies to investigate microplastic contamination in food and water, highlighting the government's dedication to addressing the public health risks linked to plastic pollution (FSSAI Launches Project to Address Microplastic Contamination in Indian Food, 2024).

Public awareness and education are central to India's waste management strategy. The *Swachh Bharat Mission* emphasises reducing plastic waste in urban and rural areas through cleanliness and sanitation drives, while initiatives like the *Ganga Action Plan* and *Namami Gange Programme* focus on preventing plastic pollution in water bodies. These programs engage communities and raise awareness about the impact of plastic pollution on water quality, health, and ecosystem sustainability (Government of India, 2024; India, 2020).

With gyres of marine garbage upsetting marine life and introducing microplastics into coastal food chains, India's extensive coastline in the Indian Ocean increases its susceptibility to marine plastic pollution. The threats to the environment and human health from plastic pollution rise as the amount of plastic consumed each year rises by 16–20%. For India to lower plastic pollution and save its ecosystems, targeted regulations, effective waste management techniques, and public education are essential (Nature Research, 2024).

An increasing dedication to environmental conservation is shown in India's strategy for controlling plastic and microplastic pollution. A concerted strategy to lessen plastic pollution and its long-term effects on the environment and human health is made possible by these measures (Nature Research, 2024).Sustainable methods are supported by established regulations like the Plastic Waste Management Rules and EPR recommendations. To guarantee a cleaner and healthier environment for all, a specialised National Marine Litter Policy might improve India's response even more by combining scientific research, international best practices, and cross-sector cooperation (Chinglenthoiba et al., 2022).

6.8 Future Scope

One prominent trend in the study of microplastic pollution is the increasing awareness and research interest in this issue over the past few years. A significant portion of studies on microplastics have been published after 2022, indicating a growing recognition of the problem's urgency. This uptick in recent publications reflects a heightened global focus on understanding and managing microplastic pollution. Notably, in 2024, the first comprehensive review paper on microplastic pollution across various aquatic environments in India was authored by Perumal, marking a significant step for research in this region.

Notwithstanding India's vast and varied coasts, microplastic contamination is still comparatively under-researched, with only forty-four papers on the topic published in the Web of Science from 2013 to 2020 (Karthikeyan & Subagunasekar, 2023). This limited research presence highlights the need for continued focus and investment in studying microplastic pollution, especially within regions like India where the impacts on diverse ecosystems are yet to be fully understood.

Future research on microplastic pollution can benefit significantly from advances in remote sensing, artificial intelligence, policy innovations, and interdisciplinary approaches. Enhancing remote sensing capabilities through satellite advancements, such as hyperspectral and multispectral imaging, can improve the precision of microplastic detection, particularly in challenging marine environments. Miniaturised sensors for drones and underwater vehicles would further enable real-time monitoring, providing critical insights into pollution trends. Specialised sensors designed specifically to detect plastics, including microplastics, could enable even more detailed mapping of pollution, as these sensors could be optimised to identify different types of polymers.

Satellite remote sensing for detecting plastic litter in water is still in the research and development phase; however, it holds significant promise as a tool to overcome the limitations associated with other detection methods. The potential of satellite-based techniques lies in their ability to provide large-scale, high-resolution data, offering a comprehensive and efficient way to monitor plastic pollution across vast and otherwise inaccessible marine environments. While challenges remain, such as the differentiation of plastic from other materials and environmental conditions affecting data accuracy, advancements in sensor technology and image processing techniques continue to enhance the feasibility of satellite remote sensing as a vital tool in combating marine plastic pollution (Mukonza & Chiang, 2022).

The integration of machine learning (ML) and artificial intelligence (AI) into microplastic monitoring will also be transformative. Algorithms tailored to distinguish microplastics from organic debris could yield more accurate data on microplastic hotspots (Isaac Sajan et al., 2024). By automating the analysis of extensive remote sensing data, AI can help pinpoint problem areas on a large scale, improving both the effectiveness and efficiency of microplastic tracking.

In line with this, establishing a global database that compiles real-time data on microplastic distribution would support international collaboration. Citizen science initiatives, in which individuals contribute data through mobile apps, could also bolster this database, making it a valuable resource for researchers and policymakers. Policy innovation remains crucial in addressing plastic pollution. Policies based on real-time data will enable governments to respond promptly to pollution sources. Strengthening Extended Producer Responsibility (EPR) (UNEP, 2022) initiatives will motivate industries to reduce plastic use and prioritise sustainability, reinforcing the United Nations Sustainable Development Goals (SDGs).

Furthermore, developing regulations that address the life cycle of plastics, from production to disposal, will be essential in creating a circular economy, where plastic is reused and recycled rather than disposed of into marine ecosystems (Shamsuyeva & Endres, 2021). Addressing plastic pollution requires both innovative policy and technological advancements to foster long-term sustainability. Policies grounded in real-time data will enable governments to act swiftly in identifying and mitigating pollution sources, ensuring that responses are timely and effective. Strengthening Extended Producer Responsibility (EPR) initiatives will incentivize industries to reduce plastic production and adopt more sustainable practices, aligning with the United Nations Sustainable Development Goals (SDGs). Moreover, developing regulations that consider the entire life cycle of plastics—from production and use to disposal—will be critical in creating a circular economy. In such a system, plastics are reused, recycled, or safely biodegraded rather than accumulating in marine ecosystems (Shamsuyeva & Endres, 2021).

Alongside policy efforts, the development of biodegradable plastic alternatives, particularly those designed to degrade safely in marine environments, holds significant promise for reducing plastic waste. Research into advanced recycling methods, including techniques that address microplastics (Rabiu & Jaeger-Erben, 2024), can further mitigate environmental impact. Together, these efforts—policymaking, innovation in materials, and recycling technologies—will drive a more sustainable economy and a cleaner, healthier environment, ultimately supporting both ecological resilience and the long-term goals of the SDGs.

In tandem with stronger regulations, improving e-waste recycling practices is essential. Developing formal, safe recycling infrastructures and expanding them globally can help ensure that e-waste is disposed of properly. In many parts of the world, informal e-waste recycling often involves hazardous methods such as open burning or unsafe dismantling, which result in the breakdown of plastics into microplastics. Encouraging the establishment of certified e-waste recycling centers that follow environmentally sound practices will prevent microplastic leakage and allow for the safe recovery of valuable materials like metals

(Shamsuyeva & Endres, 2021). These measures can also mitigate the release of harmful substances like heavy metals and flame retardants into marine ecosystems.

Finally, tackling e-waste requires international cooperation. E-waste often crosses borders, with developed countries exporting discarded electronics to lower-income nations, where informal recycling practices are common (Alpizar et al., 2020). Strengthening international frameworks such as the Basel Convention, which aims to control the movement of hazardous waste across borders, is necessary to ensure that e-waste is managed in an environmentally responsible way (Liu et al., 2023). A coordinated global effort can ensure that e-waste is recycled safely, preventing the spread of microplastics to marine environments.

In addition to mitigating pollution, future efforts must prioritise addressing the impacts of microplastics through targeted ecosystem restoration. Projects focused on restoring marine environments, such as coral reefs and mangroves, are essential for enhancing ecosystem resilience specifically against microplastic contamination. These efforts can help mitigate the long-term ecological damage caused by microplastics, ensuring the health and functionality of these vital habitats (Ford et al., 2022). Furthermore, advancing interdisciplinary research that integrates environmental science, cutting-edge technology, policy innovation, and public health perspectives is crucial. Such an approach will provide a more comprehensive framework for understanding and combating microplastic pollution, paving the way for solutions that effectively address this pervasive issue across diverse sectors (Nasir et al., 2024).

A possible step towards lowering plastic pollution, especially in terrestrial and marine environments, is the introduction of biodegradable substitutes for conventional plastic straws, spoons, and other utensils. Compared to their plastic equivalents, biodegradable alternatives that are composed of materials like bamboo, cornflour, or plant-based polymers break down more quickly and have a far less environmental effect (Anders Ankarlid, 2023). This change not only contributes to a decrease in the amount of non-biodegradable plastic trash that accumulates, but it also fits with the rising trend towards natural alternatives. We can encourage a circular economy that reduces waste and maintains ecological balance by giving preference to environmentally friendly and sustainably supplied commodities. These substitutes provide a workable answer to the plastic dilemma and promote more harmony with nature when included into larger pollution reduction plans (Velenturf & Purnell, 2021).

To effectively tackle the pervasive issue of microplastic pollution, it is essential to promote sustainable practices at individual, community, and policy levels. One immediate step people can take is to reduce plastic usage in daily life by choosing natural, biodegradable alternatives. For instance, instead of relying on plastic carry bags, citizens can carry their own cloth bags, which are reusable and environmentally friendly. Swapping plastic containers and bottles for options like glass, copper, or stainless steel also helps limit plastic waste and microplastic contamination.

Beyond personal choices, we need a collective shift in societal practices and regulations. Retailers and vendors can play a crucial role by ceasing the sale of plastic carry bags and encouraging customers to bring reusable bags. Additionally, opting for clay pots to grow plants and using ceramic or steel containers to store water are other small but impactful steps that promote sustainable, non-plastic alternatives.

Spreading awareness is equally important to drive change. Schools, colleges, and community centres can be vital hubs for environmental education, teaching young people about the harms of plastic and encouraging them to choose sustainable options. Through awareness campaigns, workshops, and discussions, educational institutions can foster a culture of environmental responsibility among students and faculty.

Organising regular cleanup drives and establishing NGOs or environmental startups focused on plastic waste reduction can further mobilise communities. These initiatives not only help keep local areas clean but also raise awareness of plastic pollution's impacts. Citizens and volunteers can work together to address plastic waste in public spaces, waterways, and beaches, reinforcing the message that individual and community actions play a critical role in safeguarding the environment. Through these combined efforts—personal changes, awareness, and organised community actions—we can move toward a world with reduced plastic dependence, creating a healthier, cleaner future for generations to come.

6.9 Conclusion

Addressing the plastic pollution crisis requires a collective effort driven by strong government policies, innovative business initiatives, and societal engagement. India's National Marine Litter Policy, alongside international efforts, underscores the importance of coordinated action to mitigate plastic waste's adverse effects on marine ecosystems. By focusing on key shifts towards reuse, recycling, and product diversification, and adopting a circular economy, we can significantly reduce plastic waste and prevent it from entering marine environments.

The integration of Earth Observation (EO), Machine Learning (ML), and Artificial Intelligence (AI) into monitoring efforts offers transformative potential, enabling large-scale and real-time detection of pollution, the identification of hotspots, and automated analyses. EO technologies provide essential insights into plastic distribution patterns and accumulation zones, while ML and AI enhance these capabilities by processing complex datasets, predicting pollution trends, and classifying microplastics by type. Together, these technologies provide actionable insights that support effective cleanup, intervention, and prevention efforts.

Aligned with Sustainable Development Goals (SDGs), particularly SDG 14 on ocean conservation, SDG 12 on sustainable production, and SDG 3 on public health, EO, ML, and AI contribute significantly to global sustainability efforts. Arnold Kreilhuber, Director of UNEP's Europe Office, emphasises that plastic pollution is a pervasive issue affecting ecosystems everywhere, from cities to remote natural environments. However, through sustainable solutions, innovative waste management practices, and advanced technologies, it is possible to reduce plastic pollution by 80% by 2040, ensuring a healthier and cleaner environment for future generations.

In conclusion, the convergence of EO, ML, AI, and robust policy initiatives, combined with sustainable shifts in production and waste management, holds great promise in tackling the marine microplastic crisis. By addressing current challenges in data accessibility, computational resources, and standardisation, and fostering continued collaboration, we can build a sustainable blue economy, protect marine biodiversity, and create resilient ecosystems for the future.

Acknowledgements This chapter is dedicated to my beloved father, whose wisdom, strength, and unconditional support have been a guiding light in my life. Your belief in me has always been my greatest motivation, and I owe so much of who I am today to the values you instilled in me.

I also extend my heartfelt gratitude to my mother for her endless encouragement and love, which have been my constant source of comfort and strength.

A special thank you to Professor Shaliender for providing me the opportunity to contribute to this chapter and for trusting in my capabilities. Your mentorship has been invaluable.

Finally, I would like to thank myself for believing in my potential and persevering through challenges. This chapter is not only a reflection of my work but also of self-discovery and determination.

References

Abdulhameed, M., Mohammed, A., & Ans Ibrahim Mahameed. (2024). Application of remote sensing and GIS techniques for monitoring water volume variations in inaccessible reservoirs. *Hydrological Sciences Journal*, 1–11.https://doi.org/10.1080/02626667.2024.2402483

Abimbola, I., McAfee, M., Creedon, L., & Gharbia, S. (2024). In-situ detection of microplastics in the aquatic environment: A systematic literature review. *Science of the Total Environment*, 173111–173111.https://doi.org/10.1016/j.scitotenv.2024.173111

Abreo, N. A. S., Aurelio, R. M., Kobayashi, V. B., & Thompson, K. F. (2023). "Eye in the sky": Off-the-shelf unmanned aerial vehicle (UAV) highlights exposure of marine turtles to floating litter (FML) in nearshore waters of Mayo Bay, Philippines. *Marine Pollution Bulletin, 186*, Article 114489. https://doi.org/10.1016/j.marpolbul.2022.114489

Aerial Expedition|Milestones. (n.d.). The Ocean Cleanup. https://theoceancleanup.com/milestones/aerial-expedition/

Alberghini, L., Truant, A., Santonicola, S., Colavita, G., & Giaccone, V. (2022). Microplastics in fish and fishery products and risks for human health: A review. *International Journal of Environmental Research and Public Health, 20*(1), 789. https://doi.org/10.3390/ijerph20010789

Alpizar, F., Carlsson, F., Lanza, G., Carney, B., Daniels, R. C., Jaime, M., Ho, T., Nie, Z., Salazar, C., Tibesigwa, B., & Wahdera, S. (2020). A framework for selecting and designing policies to reduce marine plastic pollution in developing countries. *Environmental Science & Policy, 109*, 25–35. https://doi.org/10.1016/j.envsci.2020.04.007

Amelia, T. S. M., Khalik, W. M. A. W. M., Ong, M. C., Shao, Y. T., Pan, H.-J., & Bhubalan, K. (2021). Marine microplastics as vectors of major ocean pollutants and its hazards to the marine ecosystem and humans. *Progress in Earth and Planetary Science, 8*(1). https://doi.org/10.1186/s40645-020-00405-4

Anders Ankarlid. (2023, August 24). *4 eco-friendly alternatives to plastic straws.* Agood Company. https://www.agood.com/blogs/stories/4-eco-friendly-alternatives-to-plastic-straws?srsltid=AfmBOoppZBSFiEwCR_q3Kc8fCqbvtXqyMFpLaKKVWVS73rjVUkC-s9dp

Ashrafy, A., Liza, A. A., Islam, M. N., Billah, M. M., Arafat, S. T., Rahman, M. M., & Rahman, Sk. M. (2022). Microplastics pollution: A brief review of its source and abundance in different aquatic ecosystems. *Journal of Hazardous Materials Advances, 9*, 100215. https://doi.org/10.1016/j.hazadv.2022.100215

Astorayme, M. A., Vázquez-Rowe, I., & Kahhat, R. (2024). The use of artificial intelligence algorithms to detect macroplastics in aquatic environments: A critical review. *The Science of the Total Environment, 945*, 173843–173843. https://doi.org/10.1016/j.scitotenv.2024.173843

Atwood, E. C., Falcieri, F. M., Piehl, S., Bochow, M., Matthies, M., Franke, J., Carniel, S., Sclavo, M., Laforsch, C., & Siegert, F. (2019). Coastal accumulation of microplastic particles emitted from the Po River, Northern Italy: Comparing remote sensing and hydrodynamic modelling with in situ sample collections. *Marine Pollution Bulletin, 138*, 561–574. https://doi.org/10.1016/j.marpolbul.2018.11.045

Balsi, M., Moroni, M., Chiarabini, V., & Tanda, G. (2021). High-resolution aerial detection of marine plastic litter by hyperspectral sensing. *Remote Sensing, 13*(8), 1557. https://doi.org/10.3390/rs13081557

Barceló, D., Picó, Y., & Alfarhan, A. H. (2023). Microplastics: Detection in human samples, cell line studies, and health impacts. *Environmental Toxicology and Pharmacology, 101*, Article 104204. https://doi.org/10.1016/j.etap.2023.104204

Bates, S. (2021). *Scientists use NASA satellite data to track ocean microplastics from space—NASA.* https://www.nasa.gov/centers-and-facilities/goddard/scientists-use-nasa-satellite-data-to-track-ocean-microplastics-from-space/

Biermann, L., Clewley, D., Martinez-Vicente, V., & Topouzelis, K. (2020). Finding plastic patches in coastal waters using optical satellite data. *Scientific Reports, 10*(1), 1–10. https://doi.org/10.1038/s41598-020-62298-z

Bonifazi, G., Fiore, L., Pelosi, C., & Serranti, S. (2023). Evaluation of plastic packaging waste degradation in seawater and simulated solar radiation by spectroscopic techniques. *Polymer Degradation and Stability, 207*, Article 110215. https://doi.org/10.1016/j.polymdegradstab.2022.110215

Buczkowski, A. (2024, September 13). *How can remote sensing data be used to track plastic in the ocean?—Geoawesome.* Geoawesome. https://geoawesome.com/eo-hub/how-can-remote-sensing-data-be-used-to-track-plastic-in-the-ocean/

Carrington D (2022, October 7) Microplastics found in human breast milk for the first time. *The Guardian.* https://www.theguardian.com/environment/2022/oct/07/microplastics-human-breast-milk-first-time

Carrington, D., & Editor, D. C. E. (2024, February 27). Microplastics found in every human placenta tested in study. *The Guardian.* https://www.theguardian.com/environment/2024/feb/27/microplastics-found-every-human-placenta-tested-study-health-impact

Chabot, D., & Marteinson, S. C. (2024). Aerial remote sensing of aquatic microplastic pollution: The state of the science and how to move it forward. *Microplastics, 3*(4), 685–695. https://doi.org/10.3390/microplastics3040042

Chemello, G., Trotta, E., Notarstefano, V., Papetti, L., Di Renzo, L., Matiddi, M., Silvestri, C., Carnevali, O., & Gioacchini, G. (2023). Microplastics evidence in yolk and liver of loggerhead sea turtles (Caretta caretta), a pilot study. *Environmental Pollution, 337*, 122589–122589. https://doi.org/10.1016/j.envpol.2023.122589

Chinglenthoiba, C., Pukhrambam, B., Chanu, K. T., Devi, K. S., Meitei, N. J., Devika, Y., & Valiyaveettil, S. (2022). A review on microplastic pollution research in India. *58*, 102777–102777.https://doi.org/10.1016/j.rsma.2022.102777

Copello de Souza, L. (2019). *SDG 12—Initiatives to reduce the production and consumption of plastics.* https://www.2030spotlight.org/en/book/1883/chapter/sdg-12-initiatives-reduce-production-and-consumption-plastics

Cózar, A., Sanz-Martín, M., Martí, E., González-Gordillo, J. I., Ubeda, B., Gálvez, J. Á., Irigoien, X., & Duarte, C. M. (2015). Plastic accumulation in the Mediterranean Sea. *PLOS ONE, 10*(4), e0121762. https://doi.org/10.1371/journal.pone.0121762

CPCB. (2024). Centralized EPR Portal for Plastic Packaging. Cpcb.gov.in. https://eprplastic.cpcb.gov.in/

CYGNSS—NASA Science. (n.d.). Science.nasa.gov. https://science.nasa.gov/mission/cygnss/

Davaasuren, N., Marino, A., Boardman, C., Alparone, M., Nunziata, F., Ackermann, N., & Hajnsek, I. (2018, July 1). *Detecting microplastics pollution in world oceans using SAR remote sensing.* IEEE Xplore. https://doi.org/10.1109/IGARSS.2018.8517281

de Fockert, A. (2024, October 30). *Detecting plastic pollution in the ocean with advanced remote sensing technologies.* Deltares.nl. https://www.deltares.nl/en/news/detecting-plastic-pollution-remote-sensing

de Vries, R. (2021, April 9). *Towards spaceborne detection of ocean plastic.* The Ocean Cleanup. https://theoceancleanup.com/updates/towards-spaceborne-detection-of-ocean-plastic/

Du, S., Zhu, R., Cai, Y., Xu, N., Yap, P.-S., Zhang, Y., He, Y., & Zhang, Y. (2021). Environmental fate and impacts of microplastics in aquatic ecosystems: A review. *RSC Advances, 11*(26), 15762–15784. https://doi.org/10.1039/D1RA00880C

Dutchen, S. (2023). Microplastics everywhere|Harvard medicine magazine. *Magazine.hms.harvard.edu.* https://magazine.hms.harvard.edu/articles/microplastics-everywhere

Eriksen, M., Lebreton, L. C. M., Carson, H. S., Thiel, M., Moore, C. J., Borerro, J. C., Galgani, F., Ryan, P. G., & Reisser, J. (2014). Plastic pollution in the world's oceans: more than 5 trillion plastic pieces weighing over 250,000 tons afloat at sea. *PLoS ONE, 9*(12), Article e111913. https://doi.org/10.1371/journal.pone.0111913

Evans, M. C., & Ruf, C. S. (2022). Toward the detection and imaging of ocean microplastics with a spaceborne radar. *IEEE Transactions on Geoscience and Remote Sensing, 60*, 1–9. https://doi.org/10.1109/tgrs.2021.3081691

Evers, J. (2024, April 10). *Great Pacific garbage patch.* Education.nationalgeographic.org; National Geographic. https://education.nationalgeographic.org/resource/great-pacific-garbage-patch/

Exploring the Pros and Cons: Advantages and Disadvantages of Remote Sensing|Spatial Post. (2023, April 20). Spatialpost. https://www.spatialpost.com/advantages-and-disadvantages-of-remote-sensing/

Faltynkova, A., Johnsen, G., & Wagner, M. (2021). Hyperspectral imaging as an emerging tool to analyze microplastics: A systematic review and recommendations for future development. *Microplastics and Nanoplastics, 1*(1). https://doi.org/10.1186/s43591-021-00014-y

Faltynkova, A., & Wagner, M. (2023). Developing and testing a workflow to identify microplastics using near infrared hyperspectral imaging. *Chemosphere, 336*, Article 139186. https://doi.org/10.1016/j.chemosphere.2023.139186

Fava, M. (2022, May 9). *Ocean plastic pollution an overview: Data and statistics.* Ocean Literacy Portal; UNESCO. https://oceanliteracy.unesco.org/plastic-pollution-ocean/

Ford, H. V., Jones, N. H., Davies, A. J., Godley, B. J., Jambeck, J. R., Napper, I. E., Suckling, C. C., Williams, G. J., Woodall, L. C., & Koldewey, H. J. (2022). The fundamental links between climate change and marine plastic pollution. *Science of the Total Environment, 806*(1), Article 150392. https://doi.org/10.1016/j.scitotenv.2021.150392

FSSAI Launches Project to Address Microplastic Contamination in Indian Food. (2024). Pib.gov.in. https://www.pib.gov.in/PressReleasePage.aspx?PRID=2046403

GESAMP. (2015). *Sources, fate and effects of microplastics in the marine environment (Part 1)|GESAMP.* GESAMP. http://www.gesamp.org/publications/reports-and-studies-no-90

Giri, S., Lamichhane, G., Khadka, D., & Devkota, H. P. (2024). Microplastics contamination in food products: Occurrence, analytical techniques and potential impacts on human health. *Current Research in Biotechnology, 7*, 100190–100190. https://doi.org/10.1016/j.crbiot.2024.100190

Gnann, N., Baschek, B., & Ternes, T. A. (2022). *Close-range remote sensing-based detection and identification of macroplastics on water assisted by artificial intelligence: A review. 222*, 118902–118902. https://doi.org/10.1016/j.watres.2022.118902

Goddijn-Murphy, L., & Dufaur, J. (2018). Proof of concept for a model of light reflectance of plastics floating on natural waters. *Marine Pollution Bulletin, 135*, 1145–1157. https://doi.org/10.1016/j.marpolbul.2018.08.044

Goddijn-Murphy, L., Martínez-Vicente, V., Dierssen, H. M., Raimondi, V., Gandini, E., Foster, R., & Chirayath, V. (2024). Emerging technologies for remote sensing of floating and submerged plastic litter. *Remote Sensing, 16*(10), 1770–1770. https://doi.org/10.3390/rs16101770

Goh, K. C., Kurniawan, T. A., Sarpin, N., Masrom, M. A. N., Othman, M. H. D., Anouzla, A., Aziz, F., Ali, I., Casila, J. C., Khan, M. I., Zhang, D., Mahmoud, M., & Onn, C. W. (2024). Combating microplastic pollution in Malaysia's marine ecosystems using technological solutions, policy instruments, and public participation: A review. *Journal of Hazardous Materials Advances, 17*, Article 100542. https://doi.org/10.1016/j.hazadv.2024.100542

Government of India. (2024). Swachh Bharat Mission – Gramin. Department of Drinking Water and Sanitation. Ddws.gov.in. https://swachhbharatmission.ddws.gov.in/

Hafeez, S., Wong, M. S., Abbas, S., Kwok, C. Y. T., Nichol, J., Lee, K. H., Tang, D., & Pun, L. (2018). *Detection and monitoring of marine pollution using remote sensing technologies.* IntechOpen. https://www.intechopen.com/chapters/64603

Harley-Nyang, D., Memon, F. A., Jones, N., & Galloway, T. (2022). Investigation and analysis of microplastics in sewage sludge and biosolids: A case study from one wastewater treatment works in the UK. *Science of the Total Environment, 823*, Article 153735. https://doi.org/10.1016/j.scitotenv.2022.153735

Hasituya, Chen, Z., Wang, L., Wu, W., Jiang, Z., & Li, H. (2016). Monitoring plastic-mulched farmland by landsat-8 OLI imagery using spectral and textural features. *Remote Sensing, 8*(4), 353.https://doi.org/10.3390/rs8040353

Hernandez-Milian, G., Lusher, A., MacGabban, S., & Rogan, E. (2019). Microplastics in grey seal (Halichoerus grypus) intestines: Are they associated with parasite aggregations? *Marine Pollution Bulletin, 146*, 349–354. https://doi.org/10.1016/j.marpolbul.2019.06.014

Hong, P., Xiao, J., Liu, H., Niu, Z., Ma, Y., Wang, Q., Zhang, D., & Ma, Y. (2023). An inversion model of microplastics abundance based on satellite remote sensing: A case study in the Bohai Sea. *The Science of the Total Environment, 909*, 168537–168537. https://doi.org/10.1016/j.scitotenv.2023.168537

Horning, N. (2008). Remote sensing. *Encyclopedia of Ecology*, 2986–2994.https://doi.org/10.1016/b978-008045405-4.00237-8

Hsu, Y.-J., Huang, C., & Lee, M. (2023). Unveiling microplastic spectral signatures under weathering and digestive environments through shortwave infrared hyperspectral sensing. *Environmental Pollution, 342*, Article 123106. https://doi.org/10.1016/j.envpol.2023.123106

Huang, S., Ding, J., Liu, B., Ge, X., Jinjie, W., Zou, J., & Zhang, J. (2020). The capability of integrating optical and microwave data for detecting soil moisture in an Oasis Region. *Remote Sensing, 12*(9), 1358–1358. https://doi.org/10.3390/rs12091358

HySpex. (n.d.). https://www.hyspex.com/use-cases-application-notes/plastic/

Ihenetu, S. C., Enyoh, C. E., Wang, C., & Li, G. (2024). Sustainable urbanization and microplastic management: Implications for human health and the environment. *Urban Science, 8*(4), 252–252. https://doi.org/10.3390/urbansci8040252

India, N. M. for C. G. of J. S., Department of Water Resources, River Development & Ganga Rejuvenation, Government of. (2020, September). Namami Gange. National Mission for Clean Ganga (NMCG), Ministry of Jal Shakti, Department of Water Resources, River Development & Ganga Rejuvenation, Government of India. https://nmcg.nic.in/NamamiGanga.aspx

Isaac Sajan R, Manchu M, Felsy C, & Joselin Kavitha M (2024) Microplastic predictive modelling with the integration of Artificial Neural Networks and Hidden Markov Models (ANN-HMM). *Journal of Environmental Health Science and Engineering, 22*(2), 579–592.https://doi.org/10.1007/s40201-024-00920-2

Jain, R., Gaur, A., Suravajhala, R., Chauhan, U., Pant, M., Tripathi, V., & Pant, G. (2023). Microplastic pollution: Understanding microbial degradation and strategies for pollutant reduction. *Science of the Total Environment, 905*, Article 167098. https://doi.org/10.1016/j.scitotenv.2023.167098

Jin, H., Kong, F., Li, X., & Shen, J. (2024). Artificial intelligence in microplastic detection and pollution control. *Environmental Research, 262*, Article 119812. https://doi.org/10.1016/j.envres.2024.119812

Karthikeyan, P., & Subagunasekar, M. (2023). Microplastics pollution studies in India: A recent review of sources, abundances and research perspectives. *Regional Studies in Marine Science, 61*, Article 102863. https://doi.org/10.1016/j.rsma.2023.102863

Krishnan, A. (2022). *Like Japan and Singapore, why doesn't India burn plastic waste to produce energy?* Scroll.in. https://scroll.in/article/1014050/like-japan-and-singapore-why-doesnt-india-burn-plastic-waste-to-produce-energy

Kumar, R., Verma, A., Shome, A., Sinha, R., Sinha, S., Jha, P. K., Kumar, R., Kumar, P., Shubham, Das, S., Sharma, P., & Vara Prasad, P. V. (2021). Impacts of plastic pollution on ecosystem services, sustainable development goals, and need to focus on circular economy and policy interventions. *Sustainability, 13*(17), 9963. mdpi.

Kye, H., Kim, J., Ju, S., Lee, J., Lim, C., & Yoon, Y. (2023). Microplastics in water systems: A review of their impacts on the environment and their potential hazards. *Heliyon, 9*(3). https://doi.org/10.1016/j.heliyon.2023.e14359

Lavender, S. (2022). Detection of waste plastics in the environment: Application of Copernicus earth observation data. *Remote Sensing, 14*(19), 4772. https://doi.org/10.3390/rs14194772

Le Blanc, D., & Le, D. (2015). Towards integration at last? The sustainable development goals as a network of targets. https://www.un.org/esa/desa/papers/2015/wp141_2015.pdf

Lebreton, L. (2018). Evidence that the Great Pacific Garbage Patch is rapidly accumulating plastic. *Scientific Reports, 8*(1). https://doi.org/10.1038/s41598-018-22939-w

Lee, Y., Cho, J., Sohn, J., & Kim, C. (2023). Health effects of microplastic exposures: Current issues and perspectives in South Korea. *Yonsei Medical Journal, 64*(5), 301–308. https://doi.org/10.3349/ymj.2023.0048

Levin, N., Lugassi, R., Ramon, U., Braun, O., & Ben-Dor, E. (2007). Remote sensing as a tool for monitoring plasticulture in agricultural landscapes. *International Journal of Remote Sensing, 28*(1), 183–202. https://doi.org/10.1080/01431160600658156

Lillesand, T. M., Kiefer, R. W., & Kiefer, R. W. (2022). Lillesand, T. M., & Kiefer, R. W. (1987). Remote sensing and image interpretation (2nd ed.). John Wiley & Sons. References – Scientific Research Publishing. Scirp.org. https://www.scirp.org/reference/referencespapers?referenceid=3372690

Lincoln, S., Andrews, B., Birchenough, S. N. R., Chowdhury, P., Engelhard, G. H., Harrod, O., Pinnegar, J. K., & Townhill, B. L. (2022). Marine litter and climate change: Inextricably connected threats to the world's oceans. *Science of the Total Environment, 837*, Article 155709. https://doi.org/10.1016/j.scitotenv.2022.155709

Liu, K., Tan, Q., Yu, J., & Wang, M. (2023). A global perspective on e-waste recycling. *Circular Economy, 2*(1), Article 100028. https://doi.org/10.1016/j.cec.2023.100028

Ma, J., Ma, R., Pan, Q., Liang, X., Wang, J., & Ni, X. (2023). A global review of progress in remote sensing and monitoring of marine pollution. *Water, 15*(19), 3491–3491. https://doi.org/10.3390/w15193491

Mamun, A. A., Prasetya, T. A. E., Dewi, I. R., & Ahmad, M. (2023). Microplastics in human food chains: Food becoming a threat to health safety. *The Science of the Total Environment, 858*(Pt 1), Article 159834. https://doi.org/10.1016/j.scitotenv.2022.159834

Marcharla, E., Vinayagam, S., Gnanasekaran, L., Soto-Moscoso, M., Chen, W.-H., Thanigaivel, S., & Ganesan, S. (2024). Microplastics in marine ecosystems: a comprehensive review of biological and ecological implications and its mitigation approach using nanotechnology for the sustainable environment. *Environmental Research, 256*, 119181–119181. https://doi.org/10.1016/j.envres.2024.119181

Matavos-Aramyan, S. (2024). Addressing the microplastic crisis: A multifaceted approach to removal and regulation. *Environmental Advances, 17*, 100579–100579. https://doi.org/10.1016/j.envadv.2024.100579

maya0286. (2022, October 5). *What are microplastics and how are they formed?* Cruz Foam. https://www.cruzfoam.com/post/what-are-microplastics-and-how-are-they-formed#:~:text=Most%20microplastics%20originate%20from%20manufactured

Moallemi, E. A., Malekpour, S., Hadjikakou, M., Raven, R., Szetey, K., Ningrum, D., Dhiaulhaq, A., & Bryan, B. A. (2020). Achieving the sustainable development goals requires transdisciplinary innovation at the local scale. *One Earth, 3*(3), 300–313. https://doi.org/10.1016/j.oneear.2020.08.006

Mohsan, S. A. H., Othman, N. Q. H., Li, Y., Alsharif, M. H., & Khan, M. A. (2023). Unmanned aerial vehicles (UAVs): Practical aspects, applications, open challenges, security issues, and future trends. *Intelligent Service Robotics, 16*(1). https://doi.org/10.1007/s11370-022-00452-4

Mukonza, S. S., & Chiang, J.-L. (2022). Satellite sensors as an emerging technique for monitoring macro- and microplastics in aquatic ecosystems. *Water Emerging Contaminants & Nanoplastics, 1*(4), 17. https://doi.org/10.20517/wecn.2022.12

NASA. (2021, November 30). *Landsat 8—Landsat science*. Landsat Science. https://landsat.gsfc.nasa.gov/satellites/landsat-8/

Nasir, M. S., Tahir, I., Ali, A., Ayub, I., Nasir, A., Abbas, N., Sajjad, U., & Hamid, K. (2024). Innovative technologies for removal of micro plastic: A review of recent advances. *Heliyon, 10*(4), Article e25883. https://doi.org/10.1016/j.heliyon.2024.e25883

Nations, U. (2017). *Factsheet: Marine pollution.* https://sustainabledevelopment.un.org/content/documents/Ocean_Factsheet_Pollution.pdf

Nilsson, M., Chisholm, E., Griggs, D., Howden-Chapman, P., McCollum, D., Messerli, P., Neumann, B., Stevance, A.-S., Visbeck, M., & Stafford-Smith, M. (2018). Mapping interactions between the sustainable development goals: Lessons learned and ways forward. *Sustainability Science, 13*(6), 1489–1503. https://doi.org/10.1007/s11625-018-0604-z

Nivedita, V., Begum, S. S., Aldehim, G., Alashjaee, A. M., Arasi, M. A., Sikkandar, M. Y., Jayasankar, T., & Vivek, S. (2024). Plastic debris detection along coastal waters using Sentinel-2 satellite data and machine learning techniques. *Marine Pollution Bulletin, 209*, 117106–117106. https://doi.org/10.1016/j.marpolbul.2024.117106

Okeke, E. S., Okoye, C. O., Atakpa, E. O., Ita, R. E., Nyaruaba, R., Mgbechidinma, C. L., & Akan, O. D. (2022). Microplastics in agroecosystems-impacts on ecosystem functions and food chain. *Resources, Conservation and Recycling, 177*, Article 105961. https://doi.org/10.1016/j.resconrec.2021.105961

Oliveri Conti, G., Ferrante, M., Banni, M., Favara, C., Nicolosi, I., Cristaldi, A., Fiore, M., & Zuccarello, P. (2020). Micro- and nano-plastics in edible fruit and vegetables. The first diet risks assessment for the general population. *Environmental Research, 187*(187), 109677. https://doi.org/10.1016/j.envres.2020.109677

Oyewo, B., Tauringana, V., & Tingbani, I. (2024). Microplastics in aquatic bodies: Assessing the role of governance mechanisms in industrial wastewater management. *Journal of Environmental Management, 349*, 119563–119563. https://doi.org/10.1016/j.jenvman.2023.119563

Pereira, H. M., Martins, I. S., Isabel, M. D., Kim, H., Leadley, P., Popp, A., van Vuuren, D. P., Hurtt, G., & Quoss, L. (2024). Global trends and scenarios for terrestrial biodiversity and ecosystem services from 1900 to 2050. *Science, 384*(6694), 458–465.https://doi.org/10.1126/science.adn3441

Pinto-Rodrigues, A. (2023). Microplastics are in our bodies. Here's why we don't know the health risks. *Science News*. https://www.sciencenews.org/article/microplastics-human-bodies-health-risks

Pilapitiya, P. G. C. N. T., & Ratnayake, A. S. (2024). The world of plastic waste: A review. *Cleaner Materials, 11*(2772–3976), 100220. https://doi.org/10.1016/j.clema.2024.100220

Pironti, C., Ricciardi, M., Motta, O., Miele, Y., Proto, A., & Montano, L. (2021). Microplastics in the environment: Intake through the food web, human exposure and toxicological effects. *Toxics, 9*(9), 224. https://doi.org/10.3390/toxics9090224

Plastics for Change. (2024, September 25). *India emerges as the world's largest plastic polluter: What went wrong and what's next?* Plastics for Change. https://www.plasticsforchange.org/blog/india-emerges-as-the-worlds-largest-plastic-polluter-what-went-wrong-and-whats-next

Plimmer, G. (2024, November 25). Pressure builds to stem flow of plastic waste into the sea. *Financial Times*. https://www.ft.com/content/56418e4d-8af9-49ae-9ab8-ae01426321cf

Prata, J. C. (2024). The environmental impact of E-waste microplastics: A systematic review and analysis based on the driver–pressure–state–impact–response (DPSIR) framework. *Environments, 11*(2), 30–30. https://doi.org/10.3390/environments11020030

Prata, J. C., da Costa, J. P., Duarte, A. C., & Rocha-Santos, T. (2019). Methods for sampling and detection of microplastics in water and sediment: A critical review. *TrAC Trends in Analytical Chemistry, 110*, 150–159. https://doi.org/10.1016/j.trac.2018.10.029

Pravettoni, R., & Rekacewicz, P. (2019). Global distribution of microplastics | GRID-Arendal. Www.grida.no. https://www.grida.no/resources/13339

Rabiu, M. K., & Jaeger-Erben, M. (2024). Reducing single-use plastic in everyday social practices: Insights from a living lab experiment. *Resources, Conservation and Recycling, 200*, Article 107303. https://doi.org/10.1016/j.resconrec.2023.107303

Rauf, A. U., Sofiany, I. R., Fashani, Y., Al Naday, Q., & Sulistiowati, N. (2024). Application of Geographic Information System (GIS) and Remote Sensing (RS) in microplastic studies around Asia. *Emerging Contaminants and Associated Treatment Technologies*, 581–599.https://doi.org/10.1007/978-3-031-64253-1_25

Reis, D., Kupec, J., Hong, J., & Daoudi, A. (2023). *Real-time flying object detection with YOLOv8*. https://doi.org/10.48550/arxiv.2305.09972

Ruf, C., Gleason, S., Ridley, A., Rose, R., & Scherrer, J. (2017). *The nasa cygnss mission: Overview and status update*. 2641–2643. https://doi.org/10.1109/igarss.2017.8127537

Rußwurm, M., Venkatesa, S. J., & Tuia, D. (2023). Large-scale detection of marine debris in coastal areas with Sentinel-2. *Iscience, 26*(12), Article 108402. https://doi.org/10.1016/j.isci.2023.108402

Sachs, J. D., Schmidt-Traub, G., Mazzucato, M., Messner, D., Nakicenovic, N., & Rockström, J. (2019). Six transformations to achieve the sustainable development goals. *Nature Sustainability, 2*(9), 805–814. https://www.nature.com/articles/s41893-019-0352-9

Sainte Rose, B. (2018). *The Ocean Cleanup initiative with Pléiades imagery Maritime*. Airbus.com. https://intelligence.airbus.com/newsroom/case-studies/pleiades/pleiades-imagery-supports-ocean-cleanup-initiative/

Salgado-Hernanz, P. M., Bauzà, J., Alomar, C., Compa, M., Romero, L., & Deudero, S. (2021). Assessment of marine litter through remote sensing: Recent approaches and future goals. *Marine Pollution Bulletin, 168*, Article 112347. https://doi.org/10.1016/j.marpolbul.2021.112347

Sannigrahi, S., Basu, B., Basu, A. S., & Pilla, F. (2022). Development of automated marine floating plastic detection system using Sentinel-2 imagery and machine learning models. *Marine Pollution Bulletin, 178*, Article 113527. https://doi.org/10.1016/j.marpolbul.2022.113527

Santos, J., Pedersen, M. L., Ulusoy, B., Weinell, C. E., Pedersen, H. C., Petersen, P. M., Dam-Johansen, K., & Pedersen, C. (2022). A tunable hyperspectral imager for detection and quantification of marine biofouling on coated surfaces. *Sensors, 22*(18), 7074. https://doi.org/10.3390/s22187074

Schmaltz, E. (2020). Plastic pollution solutions: Emerging technologies to prevent and collect marine plastic pollution. *Environment International, 144*(144), Article 106067. https://doi.org/10.1016/j.envint.2020.106067

Schmidt, T., Kuester, T., Smith, T., & Bochow, M. (2023). Potential of optical Spaceborne sensors for the differentiation of plastics in the environment. *Remote Sensing, 15*(8), 2020–2020. https://doi.org/10.3390/rs15082020

Shaaban, M., Wang, X.-L., Song, P., Hou, X., & Wei, Z. (2024). Microplastic pollution and E-waste: Unraveling sources, mechanisms, and impacts across environments. *Current Opinion in Green and Sustainable Chemistry*, 100891–100891.https://doi.org/10.1016/j.cogsc.2024.100891

Shamsuyeva, M., & Endres, H.-J. (2021). Plastics in the context of the circular economy and sustainable plastics recycling: Comprehensive review on research development, standardization and market. *Composites Part C: Open Access, 6*, Article 100168. https://doi.org/10.1016/j.jcomc.2021.100168

Sigler, M. (2014). The effects of plastic pollution on aquatic wildlife: Current situations and future solutions. *Water, Air, & Soil Pollution, 225*(11). https://doi.org/10.1007/s11270-014-2184-6

Simonyan, K., & Zisserman, A. (2015, April 10). *Very deep convolutional networks for large-scale image recognition.* ArXiv.org. https://arxiv.org/abs/1409.1556

Sinha, R. K., Kumar, R., Phartyal, S. S., & Sharma, P. (2024). Interventions of citizen science for mitigation and management of plastic pollution: Understanding sustainable development goals, policies, and regulations. *Science of the Total Environment*, 176621.https://doi.org/10.1016/j.scitotenv.2024.176621

Sivadas, S. K., Mishra, P., Kaviarasan, T., & Hoehn, D. P. (2022a, March). *Litter and plastic monitoring in the Indian marine environment: A review of current research, policies.* ResearchGate; Elsevier BV. https://www.researchgate.net/publication/358607097_Litter_and_plastic_monitoring_in_the_Indian_marine_environment_A_review_of_current_research_policies_waste_management_and_a_roadmap_for_multidisciplinary_action

Sivadas, S. K., Mishra, P., Kaviarasan, T., Sambandam, M., Dhineka, K., Murthy, M. V. R., Nayak, S., Sivyer, D., & Hoehn, D. (2022b). Litter and plastic monitoring in the Indian marine environment: A review of current research, policies, waste management, and a roadmap for multidisciplinary action. *Marine Pollution Bulletin, 176*, Article 113424. https://doi.org/10.1016/j.marpolbul.2022.113424

Skywatch. (2024, January 9). *Advantages and disadvantages of remote sensing: Understanding the limits of a powerful technology—SkyWatch.* Skywatch.com. https://skywatch.com/advantages-and-disadvantages-of-remote-sensing-understanding-the-limits-of-a-powerful-technology/

Slat, B., Ardiyanti, A., Arens, E., Bolle, E., & Sonneville, J. D. (2014, March 1). Feasibility study – the ocean cleanup. https://doi.org/10.13140/RG.2.2.34283.54566

Sobhani, A., Siadatmousavi, S. M., & Mehdinia, A. (2024). Surface microplastics dynamics in the Persian Gulf and Arabian Sea using numerical modelling and CYGNSS satellite estimations. *Marine Environmental Research, 202*, Article 106749. https://doi.org/10.1016/j.marenvres.2024.106749

Stoett, P., Scrich, V. M., Elliff, C. I., Andrade, M. M., Natalia, & Turra, A. (2024). Global plastic pollution, sustainable development, and plastic justice. *World Development, 184*, 106756–106756.https://doi.org/10.1016/j.worlddev.2024.106756

Sunil, M., Pallikkavaliyaveetil, N., MIthun, N., Gopinath, A., Chidangil, S., Kumar, S., & Lukose, J. (2024). Machine learning assisted Raman spectroscopy: A viable approach for the detection of microplastics. *Journal of Water Process Engineering, 60*, 105150. https://doi.org/10.1016/j.jwpe.2024.105150

Thapliyal, C., Priya, A., Singh, S. B., Bahuguna, V., & Daverey, A. (2024). Potential strategies for bioremediation of microplastic contaminated soil. *Environmental Chemistry and Ecotoxicology, 6*, 117–131. https://doi.org/10.1016/j.enceco.2024.05.001

The Ocean Cleanup. (2023). *Boyan slat—The ocean cleanup.* The Ocean Cleanup. https://theoceancleanup.com/boyan-slat/

Themistocleous, K., Papoutsa, C., Michaelides, S., & Hadjimitsis, D. (2020). Investigating detection of floating plastic litter from space using sentinel-2 imagery. *Remote Sensing, 12*(16), 2648. https://doi.org/10.3390/rs12162648

UNEP. (2022, May 9). *Reducing plastic pollution through the extended producer responsibility.* UNEP—UN Environment Programme. https://www.unep.org/reducing-plastic-pollution-through-extended-producer-responsibility

United Nations. (2022). *Sustainable development goals: 17 goals to transform our world.* United Nations. https://www.un.org/en/exhibits/page/sdgs-17-goals-transform-world

United Nations. (2024). *The 17 sustainable development goals.* United Nations. https://sdgs.un.org/goals

United Nations Development Programme. (2024). *Sustainable development goals*. Sustainable Development Goals; United Nations. https://www.undp.org/sustainable-development-goals

Vadrot, A. B. M. (2023, November 23). *Protecting life below water: Competing normative, economic and epistemic orders (SDG 14)*. Bristoluniversitypressdigital.com; Bristol University Press. https://bristoluniversitypressdigital.com/display/book/9781529228021/ch006.xml

Veerasingam, S., Ranjani, M., Venkatachalapathy, R., Bagaev, A., Mukhanov, V., Litvinyuk, D., Verzhevskaia, L., Guganathan, L., & Vethamony, P. (2020). Microplastics in different environmental compartments in India: Analytical methods, distribution, associated contaminants and research needs. *TrAC Trends in Analytical Chemistry, 133*, Article 116071. https://doi.org/10.1016/j.trac.2020.116071

Veerendra Heggade, Dr. D. (2024). *Government of India Ministry of earth sciences Rajya Sabha*. https://moes.gov.in/sites/default/files/RS-English-662-08-02-2024.pdf

Veettil, B. K., Hong Quan, N., Hauser, L. T., Van Doan, D., & Quang, N. X. (2022). Coastal and marine plastic litter monitoring using remote sensing: A review. *Estuarine, Coastal and Shelf Science, 279*, Article 108160. https://doi.org/10.1016/j.ecss.2022.108160

Velenturf, A. P. M., & Purnell, P. (2021). Principles for a sustainable circular economy. *Sustainable Production and Consumption, 27*(1), 1437–1457. https://doi.org/10.1016/j.spc.2021.02.018

VishnuRadhan, R., Lonappan, A., & Eldho, T. I. (2022). A microwave-based technique as a feasible method to detect plastic pollutants in experimental samples. *Journal of Hazardous Materials, 428*, Article 128224. https://doi.org/10.1016/j.jhazmat.2022.128224

Wang, Z., Guo, P., Wan, H., Tian, F., & Wang, L. (2020). Integration of microwave and optical/infrared derived datasets from multi-satellite products for drought monitoring. *Water, 12*(5), 1504. https://doi.org/10.3390/w12051504

Waqas, M., Wong, M. S., Stocchino, A., Abbas, S., Hafeez, S., & Zhu, R. (2023). Marine plastic pollution detection and identification by using remote sensing-meta analysis. *Marine Pollution Bulletin, 197*, Article 115746. https://doi.org/10.1016/j.marpolbul.2023.115746

Wayman, C., González-Pleiter, M., Fernández-Piñas, F., Sorribes, E. L., Rocío Fernández-Valeriano, López-Márquez, I., González-González, F., & Rosal, R. (2024). Accumulation of microplastics in predatory birds near a densely populated urban area. *Science of the Total Environment*, 170604–170604.https://doi.org/10.1016/j.scitotenv.2024.170604

Wikipedia Contributors. (2019a, February 27). *Sustainable development goals*. Wikipedia; Wikimedia Foundation. https://en.wikipedia.org/wiki/Sustainable_Development_Goals

Wikipedia Contributors. (2019b, May 23). *Demographics of India*. Wikipedia; Wikimedia Foundation. https://en.wikipedia.org/wiki/Demographics_of_India

Wikipedia Contributors. (2024, August 24). *Spectral imaging*. Wikipedia; Wikimedia Foundation. https://en.wikipedia.org/wiki/Spectral_imaging

Windeck, C. (2018, May 15). *Monitoring and mapping microplastics in marine ecosystems|Geography Realm*. Geography Realm. https://www.geographyrealm.com/monitoring-mapping-microplastics-marine-ecosystems/

World Bank. (2022). *East Asia and Pacific Region: marine plastics series the role of extended producer responsibility schemes for packaging towards circular economies in APEC*. https://documents1.worldbank.org/curated/en/099640003102239957/pdf/P1709940b3dbd3092083b208e60bcd5719a.pdf

Worm, B., Lotze, H. K., Jubinville, I., Wilcox, C., & Jambeck, J. (2017). Plastic as a persistent marine pollutant. *Annual Review of Environment and Resources, 42*(1), 1–26. https://doi.org/10.1146/annurev-environ-102016-060700

Yu, F., & Hu, X. (2022). Machine learning may accelerate the recognition and control of microplastic pollution: Future prospects. *Journal of Hazardous Materials, 432*, Article 128730. https://doi.org/10.1016/j.jhazmat.2022.128730

Zahra, A., Qureshi, R., Sajjad, M., Sadak, F., Nawaz, M., Khan, H. A., & Uzair, M. (2024). Current advances in imaging spectroscopy and its state-of-the-art applications. *Expert Systems with Applications, 238*, 122172–122172. https://doi.org/10.1016/j.eswa.2023.122172

Zhang, Y., Wang, X., Shan, J., Zhao, J., Zhang, W., Liu, L., & Wu, F. (2019). Hyperspectral imaging based method for rapid detection of microplastics in the intestinal tracts of fish. *Environmental Science & Technology, 53*(9), 5151–5158. https://doi.org/10.1021/acs.est.8b07321

Zhao, B., Richardson, R. E., & You, F. (2024). Advancing microplastic analysis in the era of artificial intelligence: From current applications to the promise of generative AI. *Nexus*, 100043.https://doi.org/10.1016/j.ynexs.2024.100043

Zheng, Y., Li, J., Sun, C., Cao, W., Wang, M., Jiang, F., & Ju, P. (2021). Comparative study of three sampling methods for microplastics analysis in seawater. *Science of the Total Environment, 765*, Article 144495. https://doi.org/10.1016/j.scitotenv.2020.144495

Zhu, Y., Li, Y., Huang, J., Zhang, Y., Ho, Y., Fang, J. K., & Lam, E. Y. (2024). Advanced optical imaging technologies for microplastics identification: Progress and challenges. *Advanced Photonics Research*. https://doi.org/10.1002/adpr.202400038

Ziani, K. (2023). Microplastics: A real global threat for environment and food safety: A state of the art review. *Nutrients, 15*(3), 617. https://doi.org/10.3390/nu15030617

Chapter 7
Impact of Microplastic Contamination in Compost on Wheat Crop Growth: A Study of Bread and Durum Wheat Varieties

Guncha Sharma, Chirashree Ghosh, and Bhupinder Singh

Abstract Microplastics (MPs) are plastic particles less than 5 mm in size emerging as a significant human and health concern. In this study, we have inventoried microplastics content in market-available compost and subsequently examined their impact on wheat crop growth using the two most MP-contaminated composts. The crops selected for the study were bread wheat (*Triticum aestivum* L. cv.) and durum wheat (*Triticum durum* L. cv.). Market-available compost samples, categorized as organic and vermicompost, were analyzed for MP content and basic physicochemical properties. All compost samples were tested positive for microplastic contamination, corroborating that MPs can enter agroecosystems by applying organic fertilizers also such as vermicompost. The observed relationship between microplastic concentration and electrical conductivity could significantly affect compost fertility, soil health, and plant growth. Vermicompost samples exhibited higher MP concentrations than organic composts, potentially due to the activity of earthworms. In our laboratory-based study, no significant adverse effects were observed on shoot weight, root weight, vigor index I, and vigor index II. However, durum wheat showed greater sensitivity to microplastic-manure interactions than bread wheat. Specifically, bread wheat exhibited nearly 100% germination across all Polyvinyl chloride (PVC) and Polyethylene (PE) treatments, whereas durum wheat demonstrated approximately 60% germination under the same conditions. These findings suggest differential sensitivity of wheat varieties to microplastic contamination in compost. These results underscore the need for further research into the impacts of microplastic contamination in compost on different crop species and highlight the importance of mitigating microplastic pollution to ensure sustainable agricultural practices and to protect soil health.

G. Sharma · C. Ghosh (✉)
Environmental Pollution Laboratory, Department of Environmental Studies, University of Delhi, New Delhi, India
e-mail: chirashreeghosh.63@gmail.com

B. Singh
Division of Environment Science, Nuclear Research Laboratory, ICAR-Indian Agriculture Research Institute, New Delhi, India

V. Kumar et al. (eds.), *Microplastics Pollution*,
https://doi.org/10.1007/978-3-031-99774-7_7

Keyword Microplastics · Compost contamination · Wheat germination · Soil health · Agricultural sustainability

7.1 General Introduction

The accelerated global population and urbanization process causes unprecedented challenges for agricultural systems and environmental sustainability. As urban areas expand, they frequently encroach upon agricultural land, leading to significant reductions in arable soils and diminished capacity for food production. This phenomenon is well-documented, with urban sprawl contributing to the loss of productive agricultural land and disrupting local ecosystems (Johnson & Klemens, 2005). The conversion of land from agriculture to urban uses results in soil degradation and impairs ecosystem services that are vital for maintaining soil fertility and agricultural productivity (Kanianska, 2016). In addition to land conversion, agricultural practices contribute to environmental pollution through nutrient runoff, pesticide contamination, soil erosion, greenhouse gas emissions, and microplastic pollution. These pollutants degrade soil and water quality, and disrupt ecosystems, exacerbating the overall environmental impact on agriculture. Specifically, microplastics—emerging as a critical environmental threat—originate from sources such as degraded plastic products and synthetic fibers and can enter in both terrestrial and aquatic ecosystems.

7.1.1 Microplastic

Microplastics are a type of pollutant that has gained increasing attention due to their persistence and potential impacts on the environment. These particles, defined as plastic fragments smaller than 5 mm in diameter, originate from the breakdown of larger plastic debris (Halle et al., 2016). The fragmentation of plastics into microplastics occurs through various processes, including physical abrasion, weathering, and ultraviolet (UV) radiation (Andrady et al., 2022). Plastics exposed to environmental forces such as sunlight, wind, and water degrade into smaller particles and afterward distribute across different environmental compartments, including terrestrial and aquatic systems (Su et al., 2022). Microplastics are problematic because they are highly persistent in the environment and can accumulate in various ecological niches. They pose risks to wildlife and human health due to their potential toxicity along with associated additives or contaminants adhered on their surface.

7.1.2 Microplastics in Agriculture: Sources

Microplastics are increasingly recognized as a significant pollutant within agricultural systems. They can enter soils through multiple pathways as described below:

- **Plastic Residue**: Degraded plastic waste, including packaging materials and agricultural films, constitutes a major source of microplastics in soils. Plastics used in agricultural and other applications commonly fragment into microplastics through different processes (such as UV radiation, physical abrasion, and chemical weathering) (Qi et al., 2020; Hoang et al., 2024) This residue contributes to soil pollution, with studies indicating that plastic residues can persist in soils for decades and impact soil health and fertility (Zhang et al., 2022a, 2022b).
- **Fragmented Plastic Mulch Film**: Plastic mulches are commonly used in agriculture to enhance soil moisture retention and for weed control (Beltagi et al., 2022). These mulches, however, degrade over time due to mechanical wear and UV exposure, leading to the formation of microplastic fragments. Research had shown that the degradation of plastic mulches can result in significant microplastic contamination in soils, and potentially affecting soil structure and microbial communities (Huang et al., 2021a, 2021b).
- **Organic Fertilizers**: Organic fertilizers, particularly those derived from composted municipal waste, can introduce microplastics into agricultural soils (Zhang et al., 2024). Incomplete processing of waste materials during composting can lead to the presence of microplastic fragments in the final product. Studies have also documented the presence of microplastics in composts and the subsequent application of these contaminated fertilizers to agricultural fields (Scopetani et al., 2022).
- **Irrigation Water**: Microplastics can be transported into soils via irrigation practices. Surface runoff and wastewater irrigation are significant pathways through which microplastics are introduced into agricultural fields (Schell et al., 2022). Lemly et al. (1993) reported that microplastics can be mobilized by irrigation water, especially when sourced from contaminated reservoirs or runoff, thereby contributing to soil contamination.
- **Sludge Application**: Biosolids from wastewater treatment plants, frequently used as soil amendments, can be a substantial source of microplastics. These biosolids often contain microplastic fragments originating from municipal and industrial wastewater (Crossman et al., 2020). Hoang et al. (2024) stated that the application of biosolids to agricultural soils introduces these microplastics into the soil matrix, further exacerbating soil contamination.
- **Plastic-Based Agricultural Products**: Various plastic-based agricultural products, such as drip irrigation tubing and greenhouse coverings, can contribute to microplastic contamination. Over time, these products degrade due to exposure to environmental elements, leading to the release of microplastic particles into the soil. Studies have been highlighted the role of these products in increasing soil

microplastic concentrations and their potential impact on soil health (Zhou et. al., 2021; Chia et. al., 2022).

- **Manure and Animal Waste**: Animal manure, especially when sourced from confined animal feeding operations (CAFOs), can be a source of microplastics if the animals consume plastic-containing feed or bedding (Landrigan et al., 2020). Research has also shown that microplastics can accumulate in manure and be subsequently applied to fields as fertilizer, introducing additional microplastics into agricultural soils (Yang et al., 2021).
- **Atmospheric Deposition**: Atmospheric deposition of microplastics, carried by wind and atmospheric transport, can result in microplastics settling on agricultural fields (Belioka & Achilias, 2024). Rehm et al. (2021) have reported that microplastics can be deposited from the atmosphere and accumulate in the soil, contributing to the overall microplastic load in agricultural systems.

Once introduced into the agricultural systems, microplastics are distributed through various mechanisms influenced by soil texture and agricultural practices. Fine-textured soils, such as clays, retain microplastics more effectively, while coarse-textured soils, like sands, allow for greater mobility. Agricultural activities, such as soil preparation and planting, can redistribute microplastics within the soil profile (Wang et al., 2024). Irrigation methods also impact microplastic distribution, with surface irrigation generally leading to higher microplastic levels due to direct contact with contaminated water.

7.2 Effects of Microplastic

Microplastics in agriculture pose several significant risks to soil health, crop productivity, and microbial communities. Their presence in soils can alter various soil properties, disrupt nutrient cycling, and impact plant growth, creating a complex environmental issue.

One of the primary concerns is how microplastics affect soil texture and structure. Studies have shown that microplastics can disrupt soil porosity and bulk density, which influences water infiltration, soil stability, and erosion dynamics (Wang et al., 2022; Zhou et. al., 2021; Chia et. al., 2022). Microplastics can modify the distribution and connectivity of soil pores, which alters water movement and may impair root development, leading to reduced soil aggregation and increased vulnerability to erosion (Ren et al., 2021). Additionally, microplastics can influence soil's water-holding capacity. Depending on their size, shape, and the soil type, microplastics may either increase or decrease soil moisture retention (Guo et al., 2022). In some contexts, microplastics introduce additional voids into the soil, potentially improving water retention, whereas in other scenarios, they may disrupt the soil matrix and reduce its capacity to retain water (Li et al., 2024). This variability in water-holding capacity can impact the irrigation process especially soil moisture availability for crops.

The impact of microplastics on soil nutrient content is also complex. High concentrations of microplastics are linked to reduced soil nutrient levels, primarily through their interference with microbial activity and nutrient cycling (Aralappanavar et al., 2024). However, lower concentrations may enhance nutrient availability by altering microbial communities, which play a key role in nutrient mobilization (Fageria & Stone, 2006). Microplastics can disrupt microbial populations involved in nutrient mineralization, either improving or impairing nutrient availability based on concentration and soil context (Zhai et al., 2024). Similarly, microplastics can alter the chemical properties of the soil, including pH and adsorption properties. They may act as carriers for toxic substances, which could affect soil chemical interactions and plant health (Maity et al., 2021).

The presence of microplastics in soil can also have significant effects on plant growth. Disruption of soil structure and water dynamics impedes root development and water uptake, reducing overall plant growth and crop yields (Bodner et al., 2015). Microplastics can physically obstruct root growth and hinder penetration, compromising plant stability and nutrient absorption (Hasan & Tarannum, 2024). Moreover, microplastics can negatively affect nutrient availability by altering soil nutrient profiles and microbial functions, further impairing plant growth and productivity (Iqbal et al., 2023). Plants exposed to microplastics have been shown to exhibit reduced germination rates, impaired photosynthesis, and increased susceptibility to stress (Colzi et al., 2022). Furthermore, microplastics can act as vectors for toxic substances plants may absorb, exacerbating stress and toxicity, thus negatively impacting plant health and growth (Kumar et al., 2022).

Microplastics also influence soil microbial communities, disrupting their diversity and activity. Alterations in microbial composition can affect key soil functions such as nutrient cycling, organic matter decomposition, and soil fertility. Han et al. (2024) reported that microplastic contamination can lead to shifts in microbial community structure, which in turn may reduce the soil's ability to support plant growth. Microplastics can also affect microbial activity by creating physical barriers or altering soil porosity, which restricts microbial access to essential nutrients and oxygen (Sajjad et al., 2022). Additionally, microbes can colonize microplastics and form biofilms, which can further alter microbial community dynamics and interactions, affecting soil functions like organic matter decomposition and nutrient mobilization (Sooriyakumar et al., 2022). The disruption of microbial communities can impair important soil functions, including nitrogen fixation, phosphorus solubilization, and organic matter breakdown, which are essential for soil fertility and crop health (Sahu et al., 2019). Microplastics may also serve as vectors for the transport of pollutants and pathogens, further impacting microbial communities and overall soil health (Zhou et al., 2021).

7.2.1 Microplastic Effects on Food Safety and Public Health

The presence of microplastics in agricultural soils also raises concerns about food safety. As microplastics accumulate in soils, they can be absorbed by crops, potentially entering the food chain. Studies have demonstrated that microplastics can be taken up by plant roots and transferred to the edible parts of crops, including fruits and vegetables (Silva et al., 2021). While the direct impacts on human health remain unclear, the potential for microplastic ingestion through food consumption is a growing concern. The toxic chemicals associated with microplastics, such as additives and pollutants adsorbed onto their surfaces, could pose additional risks to human health if they are bioaccumulated in the food chain (Huang et al., 2021a, 2021b).

7.2.2 A Systematic Review of Microplastic in Agriculture

The growing body of research on microplastics in agricultural systems has provided valuable insights into their prevalence, distribution, and impacts on soil health, crop productivity, and ecosystems. Several international, national, and local studies have been conducted to assess the extent of microplastic contamination in agricultural ecosystems, and their findings reveal significant regional variations in concentrations, sources, and ecological effects. A summary of key international and national studies on the presence and impacts of microplastics in agricultural systems are listed below.

1. **Microplastic Concentrations in Agricultural Soils**

Studies conducted across various regions have revealed significant levels of microplastic contamination in agricultural soils, with concentrations varying depending on local agricultural practices and the extent of plastic use. In China, Yang et al. (2018) reported microplastic concentrations ranging from 0.1 to 8.6 g/kg, with areas using plastic mulches showing notably higher levels. The microplastics were primarily fibers and fragments, mostly composed of polyethylene. Similarly, in Europe, Heinze (2024) found that microplastic concentrations ranged from 0.4 to 1.5 g/kg, with agricultural soils in southern European countries like Germany and Italy having higher levels due to the widespread use of plastic mulches, biosolid applications, and plastic packaging for fertilizers and pesticides. Australia also exhibited high concentrations (1.2–5.5 g/kg) of microplastics in regions with intensive plastic mulch use (Sintim et al., 2021). In the United States, Carey (2024) found concentrations ranging from 0.3 to 3.0 g/kg in California and in Iowa, regions known for extensive plastic mulch use in row crop farming, particularly for tomatoes, strawberries, and lettuce. In South Africa, Beriot et al. (2023) observed microplastic levels between 0.5 and 2.0 g/kg, particularly in vegetable farming areas, and noted reductions in microbial biomass and changes in microbial diversity, which could impact soil fertility and crop productivity. In India, a study by Kumari et al. (2022) reported microplastic contamination between 0.1 to 1.0 g/kg in soils due to the application of treated sewage sludge, with polyethylene and polypropylene being the most common microplastic types found. These studies collectively highlight the global nature of microplastic pollution in agricultural soils, with varying concentrations depending on the region's farming practices and plastic use.

2. **Impact on Soil Properties and microflora Communities**

Recent studies conducted in various countries have highlighted the significant impacts of microplastics on soil health, microbial communities, and crop productivity. In Germany, Huang et al. (2019) found that even low levels of microplastic contamination could significantly alter soil microbial diversity. They observed shifts in the composition of soil bacteria and fungi, with microplastics reducing microbial activity and enzyme production essential for nutrient cycling. This disruption impaired nutrient mineralization and reduced organic matter decomposition, which are critical processes for soil fertility. Similarly, in Spain, Reverón et al. (2022) reported that microplastic contamination led to a reduction in soil nitrogen availability, negatively affecting crop yields. They also noted that microplastics altered soil aggregation and reduced water retention, making the soil more prone to erosion. Additionally, they found microplastics present in the rhizosphere, suggesting the potential for plant uptake and raising concerns about food safety. In Canada, Okeke et al. (2022) discovered that microplastics disrupted essential soil functions such as nutrient cycling. The study showed that microplastic contamination led to significant reductions in the metabolic activity of soil microbes, which hindered the breakdown of organic matter and the availability of nutrients, thereby negatively impacting plant growth. These findings underscore the complex and potentially detrimental effects of microplastic pollution on agricultural productivity.

3. **Microplastic Impacts on vegetation**

Worldwide, research on the effects of microplastics on plant growth and soil health has revealed significant negative impacts on crop development and soil properties. In Italy, Brenner et al. (2020) found that microplastic contamination reduced the germination rates of wheat and corn, with further observations of stunted plant growth and decreased photosynthetic activity. The study also noted that plants exposed to microplastics showed heightened stress levels and increased susceptibility to drought conditions, indicating the potential long-term effects on crop resilience. In China, Zhang et al. (2022a, 2022b) explored the uptake of microplastics by crops and found that wheat plants grown in contaminated soils exhibited microplastic particles in their roots and stems, primarily in the form of polyethylene fibers and fragments. This contamination led to reduced root growth, stunted plant development, and decreased nutrient uptake, which ultimately impaired overall plant health. Similarly, in South Korea, Sun et al. (2024) observed that exposure to microplastics reduced the photosynthetic rate and chlorophyll content in lettuce plants, which was attributed to physical obstructions of root growth, limiting water and nutrient uptake. In India, Khalid et al. (2023) studied the contamination of agricultural soils in Punjab from plastic mulches and irrigation pipes, finding microplastic concentrations between 0.2 and 3.0 g/kg. The presence of microplastics in the soil was linked to changes in soil physical properties, such as reduced porosity, increased bulk density, and impaired water infiltration, which led to soil compaction and decreased plant growth. These findings highlight the global scope of microplastic contamination in agriculture and its detrimental effects on plant growth and in crop productivity.

Global review has highlighted the widespread presence of microplastics (MPs) in agricultural ecosystems, this study aims to address the knowledge gap surrounding the interaction between crops and microplastics in agricultural systems. While numerous studies have identified microplastic contamination in soils, manures, and composts, the specific impact of microplastic-laden agricultural inputs on crop growth remains underexplored. The present study includes an assessment of microplastic contamination in commercially available manures to determine the extent of MPs in these products. Furthermore, the effects of microplastic contamination on crop growth were evaluated using the most commonly found microplastic-contaminated manure.

7.3 Material and Methods

- Selection of Market-Available Compost:

A total of four commercially available composts (two organic and two vermicomposts), which are widely used in the National Territory of India, Delhi, were selected for the evaluation of microplastics, as outlined in Table 7.1. Additionally, their basic physicochemical parameters were assessed. Since compost is widely used as a nutrient source for plant growth, it is produced through the natural process of recycling organic waste, such as food scraps, green waste, leaves, cow and buffalo dung, etc. Since microplastics can enter the composts via the organic waste itself or through atmospheric deposition and, such contaminated composts may act as direct sources of MPs for crops. For each compost type, three replicates were analyzed for basic physicochemical properties and microplastic content.

Table 7.1 List of compost samples

Compost type	Category name	Compost market brand name	Sample code
Type 1	Organic compost (Oc)	Simbhaoli organic	OC 1
		Sulpho-neem organic	OC 2
Type 2	Vermicompost (Vc)	Bhumi Kawach	VC 1
		Ecosmart home garden	VC 2

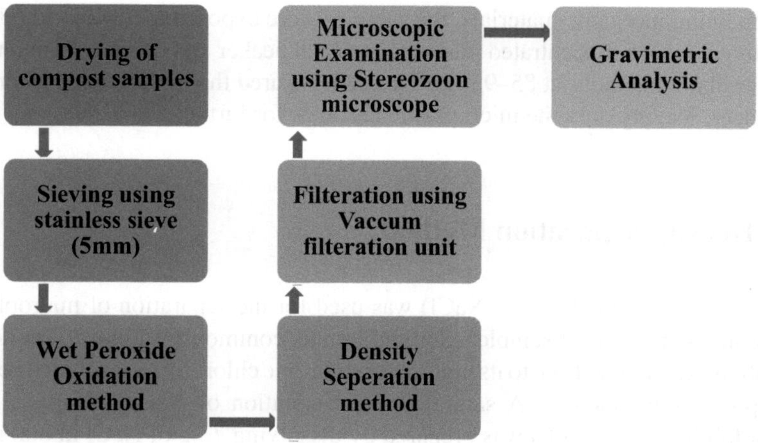

Fig. 7.1 Illustration of steps involved in microplastic estimation

7.4 Microplastic Analysis of Compost Samples

For Microplastic enumeration in compost samples, various steps were followed, as illustrated in Fig. 7.1.

For microplastic estimation, 10 g of the sample was sieved using 4BSS stainless steel sieve. They were first air-dried and then oven-dried at 90 °C until a constant weight was achieved. Followed by the analysis of basic physicochemical parameters such as pH, temperature, Electrical Conductivity (EC), Water Retention Capacity (WRC), and Moisture Content.

7.5 Sample Digestion

During microplastic enumeration to reduce interference from natural debris and organic matter, hydrogen peroxide (H_2O_2) was used for chemical digestion. As an oxidizing agent, hydrogen peroxide dissociates into water and produces highly reactive hydroxyl radicals ($OH^{\cdot}$), which react with and oxidize organic molecules in the sample, breaking them down into simpler compounds, such as carbon dioxide (CO_2) and water (H_2O). Glassware was first cleaned with 0.1N HCl and rinsed with filtered Milli-Q water, and the dried samples were transferred to well-labeled 250 mL glass beakers. For wet digestion, a Fe(II) catalyst solution was prepared using $FeSO_4{\cdot}7H_2O$ and concentrated sulfuric acid (H_2SO_4). This solution, along with hydrogen peroxide, was added to the samples and allowed to digest overnight at room temperature. Hot digestion was performed in a fume hood, with sample contents evaporated at temperatures below 75 °C. After evaporation, 30% hydrogen peroxide was added to the beakers until a bubble-free solution indicated complete digestion. To further remove any remaining inorganic materials, the samples were exposed to nitric acid (HNO_3) by adding 7 mL of concentrated nitric acid to each beaker and heating them on a hot plate for about one hour at 85–95 °C. This step ensured the dissolution of inorganic substances, leaving only the microplastic particles for further analysis.

7.6 Density Separation Method

In this study, sodium chloride (NaCl) was used for the separation of microplastics from water and sediment samples. Sodium iodide, commonly employed for density separation, was avoided due to its high cost, while zinc chloride was excluded because of disposal complications. A saturated saline solution of NaCl (density: 2.17 g/cm^3; solubility: 1.20 g/mL) was prepared by dissolving 73 g of NaCl in 800 mL of filtered deionized water (DDW) and adjusting the final volume to 1 L. The solution was heated and continuously agitated to ensure it remained clear, and then stored at room temperature for later use. The density separation process capitalizes on the different densities of various polymers, with some either floating or sinking when subjected to the saline solution. Polymers with a density greater than that of the saline solution, such as Polyvinyl Chloride (PVC) (up to 1.58 g/cm^3) and Polyethylene Terephthalate (PET) (up to 1.45 g/cm^3), settle to the bottom, while lighter polymer types remain suspended. The process involved rinsing the digested samples three times with the saturated saline solution. The rinsed solution was then poured into a labeled 250 mL volumetric cylinder, stirred for 2 h, and left overnight to allow proper density separation. This resulted in the formation of distinct layers: the top layer contained lighter particles, while the bottom layer contained denser sediment or polymer particles. Both layers were filtered individually using a vacuum filtration unit with 0.45 μm glass fiber filter paper, and the collected particles were transferred to closed glass Petri dishes for drying. The Petri dishes were kept at room temperature until the samples were fully dry and were then prepared for further microplastic enumeration. The same procedure was applied to the bottom layer of particles collected from the volumetric cylinder.

7.7 Microscopic Analysis

All preserved, exposed, and well-labeled filter papers were examined under a stereo-zoom microscope for the identification and enumeration of microplastic particles in the samples. The observations were conducted at magnifications ranging from 40× to 100×. Several criteria were applied to ensure accurate and reliable enumeration of microplastic particles, as outlined below:

- **Insolubility**: The particles on the filter paper were insoluble in water and resistant to all digestion steps conducted during sample preparation (Thiele et al., 2019).

- **Unnatural Appearance**: Microplastic particles exhibited unnatural characteristics, distinct from natural materials, such as unusual colors, textures, or shapes (Cowger et al., 2020).
- **Resistance to Pressure**: Microplastic particles were resistant to deformation or breakage when handled with tweezers (Fazio, 2022).
- **Scratch-Free Sound**: Microplastics produced a scratch-free sound when gently scratched on a Petri dish, indicating their solid, non-organic nature.
- **Absence of Cellular or Organic Structures**: Under the microscope, microplastics did not display any cellular or organic structures, appearing as synthetic materials without biological components (Lusher et al., 2020).

Based on these criteria, microplastic particles were classified into several types:

- **Fibers**: Thin, elongated particles resembling threads or filaments, characterized by a small diameter relative to their length, and uniform thickness throughout (Chubarenko et al., 2024).
- **Fragments**: Irregularly shaped particles with sub-angular, rounded, or sub-rounded edges, typically resulting from the fragmentation of larger plastic objects (Lusher et al., 2020).
- **Films**: Thin, smooth, semi-transparent sheets that are foldable but difficult to tear (Urgert, 2015).
- **Pellets**: Small, disk-shaped, flat, ovoid, or cylindrical particles with a smooth surface (Chubarenko et al., 2018).
- **Spheres**: Small, spherical particles with a smooth surface (Chubarenko et al., 2016).
- **Miscellaneous**: Particles with irregular, elongated, degraded, rough, or broken edges, exhibiting various shapes and surface textures.

7.8 Handpicking of Microplastic Particles

After observing the microplastic particles on the exposed filter paper, individual particles were handpicked using a sharp metal tweezer and transferred into pre-weighed 5 mL Eppendorf tubes. Only particles larger than 630 μm in size, which were visible to the naked eye, were considered for handpicking. Approximately 4 mL of ethanol solution was added to each tube to facilitate the transfer process. Microplastic particles were counted and recorded simultaneously in a record book. Once handpicking was completed, the ethanol in the Eppendorf tubes evaporated. After the ethanol had fully dried, the filled Eppendorf tubes were weighed using a precision balance, and the weight of each tube was recorded.

The numerical abundance of microplastic particles in compost samples was expressed as pieces per kilogram dry weight (particles/100 g dry weight) (Moller, 2022; Chaudhary & Suthar, 2023).

- **Lab-based exposure study**

After the inventory of microplastics in market-available compost, lab-based experiments were conducted to understand the interaction between crops and microplastics when grown in cocopeat as a substrate. For this, the following crops and polymers were selected:

- **Experimental crop**: Hexaploid and tetraploid wheat i.e. Bread wheat *(Triticum aestivum L)* cv. HD 3086 and Durum wheat cv. HD 4728 *(Triticum durum L)* respectively was used for the study. Seeds of both species were procured from the Indian Agricultural and Research Institute (IARI), New Delhi, India
- **Experimental plastic polymer**: The study used Polyethylene (PE), Polypropylene (PP), and Polyvinyl Chloride (PVC), which were selected based on the most abundant polymers found in the market-available compost analysis. The experimental pellets and powders were procured from G-Lex, GAIL (India) Limited.

To study the impact of MP-associated compost i.e. Organic Compost (OC) and Vermicompost (VC) natural and enriched were selected using seed germination and seed vigour as Index in bread and durum wheat for the exposure study.

Cocopeat was used as a basic substrate where different plastic polymers were added to design different experimental treatments. The mixture was added to germination/planting trays and two seeds/pits were grown in triplicates. Before sowing, seeds were surface sterilized by soaking seeds in a 3% Hydrogen peroxide (H_2O_2) solution, followed by rinsing the seeds thrice with distilled water (Bakhsh et al., 2016). Barampuram et al. (2014) reported, H_2O_2 solution is a superior surface sterilization agent, in comparison to chlorine gas and bleach resulting in lower contamination levels and higher germination rates. The trays were maintained in a controlled growth chamber at following conditions (Table 7.2).

At the end of the experiment, after 11 days, the seedlings were removed and separated into roots and shoots for further analysis. Germination Index, Vigour Index I, and II were calculated using the following formulas.

$$\text{Germination index (TGI)} = \sum \text{G/T}, \tag{7.1}$$

where G is the percentage of seed germinated per day; T is the germination period

$$\text{Vigour Index I} = \text{MeanSeedling length} \times \text{percent Germination} \tag{7.2}$$

$$\text{Vigour Index II} = \text{Mean Seedling mass} \times \text{percent Germination} \tag{7.3}$$

Table 7.2 Diurnal variation of growth chamber

Components	Day	Night
Light	16 h	8 h
Temperature	25 °C	16 °C
Relative humidity	65%	65%

7.9 Result and Discussion

The physicochemical characteristics of organic compost (OC) and vermicompost (VC), highlight key parameters such as pH, water retention capacity (WRC), moisture content, electrical conductivity (EC), and microplastic (MP) count as shown in Table 7.3. These parameters are critical in determining the quality, fertility, and environmental impact of composts and vermicomposts. A detailed analysis of each of these properties, offering insights into the differences between Organic compost (OC) and Vermicompost (VC) and the potential implications for soil health and plant growth are documented below.

7.9.1 Quantitative Analysis of Market Available Compost

The basic physicochemical characteristics of both the compost samples were observed (Table 7.3), **pH** of the compost samples ranged between 8.13 and 8.69, reflecting slightly alkaline nature. This alkalinity can be attributed to microbial degradation of organic matter, which produces basic compounds like ammonium and other alkaline substances. The presence of mineral-based materials, such as lime or calcium carbonate, likely contributed to elevated pH levels. These values align with the typical pH range (7.0–8.5), considered optimal for nutrient availability, including nitrogen, phosphorus, and potassium (Asaye et al., 2022). In the case of **Water Retention Capacity (WRC), it** varied between samples, with Organic compost-2 showing the highest value (43%) and Vermicompost-1 the lowest (30%). This difference is linked to the structural characteristics of organic compost and vermicompost. Organic compost, composed of coarser materials like decomposed plant matter which retains more water effectively due to larger particle size and greater surface area. In contrast, the finer texture of vermicompost, primarily worm castings, promotes faster drainage, thereby reducing WRC. These observations are consistent with previous studies conducted (Banda et al., 2023).

A similar trend was observed in **Moisture Content**, with a range 30–42.3%, where vermicompost (1) has the highest, which could be attributed to vermicompost's finer texture and higher microbial activity. The compact structure of worm castings retains moisture effectively, whereas the coarser particles in organic compost allow for faster evaporation, leading to reduced moisture content (Ali, 2007). In the case of **Electrical Conductivity (EC),** it was observed highest in Organic compost-1 (9.79 μS/cm) and lowest in Vermicompost-1 (2.02 μS/cm). EC reflects ion concentration, particularly soluble salts, which influence soil fertility. The higher EC in OC 1 suggests the presence of mineral-rich materials, such as animal manures or lime, incorporated during composting. Vermicompost, by contrast, benefits from earthworm digestion, which reduces salinity by filtering and breaking down salts, resulting in a stable, plant-friendly product with lower EC (Bhat et al., 2017; Kameswaran et al., 2024).

Table 7.3 Microplastic concentration and physicochemical trends in compost

Category name	Sample code	Avg. pH	Avg. WRC (%)	Avg. moisture content (%)	Avg. EC (µS/cm)	MP count (particles/Kg)
Organic compost (OC)	OC 1	**8.56 ± 0.06**	41.3 ± 2.08	**36. 4 ± 0.95**	**9.79 ± 0.09**	233.33 ± 577.35
	OC 2	8.13 ± 0.02	**43 ± 2.0**	34.2 ± 0.61	4.38 ± 0.16	**400 ± 111.55**
Vermicompost (VC)	VC 1	8.23 ± 0.07	30 ± 3.61	**43.2 ± 1.18**	2.02 ± 0.19	866.67 ± 141.92
	VC 2	**8.69 ± 0.04**	**42.3 ± 3.21**	23.9 ± 0.25	**2.92 ± 0.16**	**926.67 ± 155.49**

Maximum values highlighted; *EC* Electrical Conductivity; *WRC* Water Retention Capacity; *MP* Microplastic content

The **microplastic (MP)** count was significantly higher in vermicompost (VC 2; 926.67 particles/kg) compared to organic compost (OC 2; 400 particles/kg). This marked difference underscores the potential vulnerability of vermicomposting processes to microplastic contamination due to differences in raw materials and processing methodologies. Microplastics predominantly enter composting systems through contaminated feedstocks, which may include organic waste with plastic residues, synthetic fibers, or other debris (Zhang et al., 2023a, 2023b). Vermicomposting, in particular, is prone to higher MP counts due to the finer and varied organic materials used as feedstocks, including bedding materials that often harbor small plastic particles (Oyege & Bhaskar, 2023). Additionally, earthworm activity within vermiculture plays a significant role in amplifying microplastic contamination. Earthworms ingest ambient microplastic particles, which are fragmented in their gut and subsequently excreted as part of their castings, concentrating microplastics in the final vermicompost product (Meng, 2024). In contrast, traditional composting involves coarser organic materials and typically operates at higher temperatures. These elevated temperatures, coupled with prolonged microbial activity, can aid in the partial degradation or fragmentation of larger plastic debris, potentially reducing the overall MP count in the final compost (Kurniawan et al., 2023). The observed lower microplastic concentration in organic compost may also reflect a reduced ability of larger particles to integrate into coarser compost matrices compared to the finer particle composition of vermicompost (Zhang et al., 2024). The analysis of market-available compost revealed microplastic concentrations ranging from 400 $\pm$ 111.55 to 926.67 $\pm$ 155.49 particles/kg, with vermicompost samples exhibiting two to three times higher MP content (mean 896.67 particles/kg) compared to organic compost (316.67 particles/kg) (Table 7.3). This disparity highlights the significant role of biotranslocation in vermiculture, where microplastics ingested by earthworms are concentrated and excreted as castings. Furthermore, the finer texture of vermicompost facilitates the retention of smaller plastic particles, enhancing MP accumulation.

- ***Establishing correlation between physicochemical parameters and microplastic concentration***

No significant correlation was observed between microplastic (MP) concentration and most physico-chemical parameters of the compost samples, pH, Electrical Conductivity (EC), water retention capacity (WRC), and moisture content (Table 7.4). This indicates that microplastic contamination may not directly influence these parameters in compost or vermicompost. In contrast, a strong and highly significant positive correlation ($r = 0.98$) was found between MP concentration and electrical conductivity (EC). This suggests that higher MP levels are associated with elevated EC values, potentially due to the co-occurrence of microplastics with soluble salts and minerals. Such association could arise from feedstock materials containing plastic residues, which may increase the mineral content and salinity of the compost (Rynk et al., 2022). Additionally, the interaction between microplastic particles and ionic compounds likely contributes to this relationship, as microplastics are known

Table 7.4 Correlation between physicochemical parameters and microplastic concentration

Correlation	pH	Water retention capacity	Moisture content	Electrical conductivity	MP concentration
pH	1				
Water retention capacity	0.794	1			
Moisture content	0.106	− 0.443	1		
Elec. conductivity	0.271	0.737	− 0.510	1	
Microplastic concentration	0.134	0.595	− 0.408	**0.978**	1

Significant at 5%; significant at 1% level

to adsorb and retain salts and other solutes on their surfaces, thereby increasing the ionic strength and conductivity of the compost (Hüffer et al., 2019).

7.9.2 Mechanisms Linking Microplastics and Electrical Conductivity

Microplastics have a high surface area and the ability to adsorb pollutants from their environment, including heavy metals and ionic species (Khalid et al., 2021). This phenomenon can lead to the accumulation of heavy metals and other ions in composts containing microplastics, elevating EC values. As these ions are released from the microplastic matrix, they increase salinity and ionic conductivity. This aligns with previous studies reporting that microplastics can act as vectors for heavy metal ions, enhancing the overall ionic strength of soils and composts (Liu et al., 2024). Furthermore, the decomposition of organic materials with high mineral content in microplastic-laden feedstock may also contribute to the observed correlation between MP concentration and EC. These findings highlight the potential for microplastics to alter the chemical properties of compost, with implications for soil fertility and plant health.

A. ***Lab-Based Controlled Studies: Microplastics in the agricultural system***

Physico-chemical characteristics of compost for crop germination

The physico-chemical characteristics of the substrate, OC2 (Sulpho-Neem Organic), and VC2 (Ecosmart Home Garden) reveal notable differences, as summarized in Table 7.5. The substrate exhibited a neutral pH (7.50), the highest organic content (90%), and the lowest electrical conductivity (2.48 mS/cm), indicating minimal soluble salts. These properties make the substrate ideal for improving soil structure and fertility while minimizing the risk of salinity stress. Additionally, it demonstrated moderate water retention capacity (67%), supporting balanced moisture availability for plant growth. In contrast, OC2 showed a more alkaline pH (8.24), the highest electrical conductivity (6.05 mS/cm), and the best water retention capacity (72%). However, its lower organic content (33.7%) suggests a composition rich in minerals and salts, which may enhance nutrient availability and increase the risk of soil salinity stress. VC2 exhibited properties intermediate between the substrate and OC2, with an alkaline pH (8.31), moderate electrical conductivity (2.83 mS/cm), and organic content (57.6%). While it provides a balanced option for general gardening, its relatively low water retention capacity (59%) may limit its effectiveness in moisture-intensive applications.

(a) ***Microplastic exposure study: Seed germination as a manure***

Under polyethylene (PE) and polyvinyl chloride (PVC) treatments near 100% germination rate was observed in bread wheat (*Triticum aestivum*) which can be attributed to their beneficial effects on the germination environment (Fig. 7.2). PE and PVC are commonly used in agricultural practices, particularly as mulching materials, due to their ability to retain soil moisture. This moisture retention ensures a consistent water supply during the germination phase, facilitating the enzymatic activity required to mobilize seed reserves and support early seedling growth. Additionally, these plastics help to regulate soil temperature, maintaining an optimal thermal environment for wheat germination by preventing extreme fluctuations. Furthermore, PE and PVC create a protective barrier against environmental stressors, such as wind, heavy rainfall, and soil compaction, all of which can impede seedling establishment. These plastics also stabilize soil structure, creating conditions conducive to seed emergence. While concerns about the long-term environmental impacts of plastics remain, PE and PVC are chemically inert under typical agricultural conditions and do not release harmful substances during the short-term germination phase. The combined benefits of moisture retention, temperature regulation, and environmental protection provided by PE and PVC likely contributed to this study's exceptionally high germination rates.

Durum wheat (*Triticum durum*) demonstrated significantly lower germination rates (~ 60%) compared to bread wheat (*Triticum aestivum*) under all treatment

Table 7.5 Physicochemical characteristics of substrate and compost

Compost	Avg. pH	Avg. EC (ms/cm)	Avg. WRC (%)	Avg. OC (%)
Substrate	7.50	2.48	67	90.00
OC2	8.24	6.05	72	33.70
VC2	8.31	2.83	59	57.60

OC2 Sulpho-Neem Organic; *VC2* Ecosmart home garden; *EC* Electrical Conductivity; *WRC* Water Retention Capacity; *OC* Organic Content

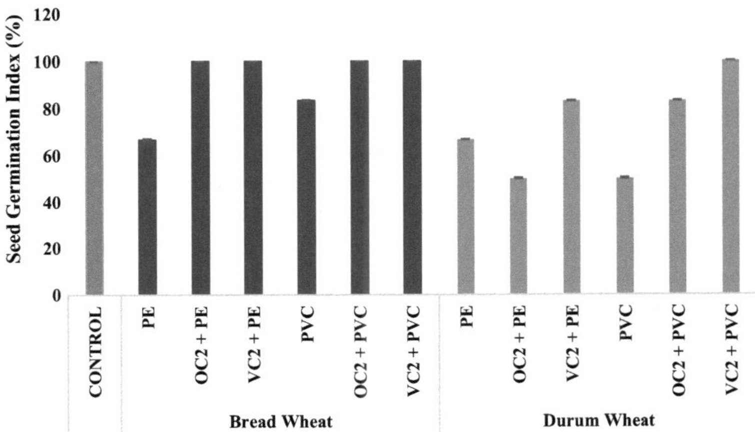

Fig. 7.2 Germination index of (Bread & Durum) wheat crops (PE: polyethylene; PVC: polyvinyl chloride; OC2: organic compost (Sulpho-neem organic); VC2 vermicompost (Ecosmart home garden)

combinations involving polyethylene (PE) and polyvinyl chloride (PVC) microplastics in organic and vermicompost amendments. This disparity highlights species-specific sensitivities to environmental changes induced by microplastic contamination. Durum wheat's reduced germination can be attributed to its higher sensitivity to soil modifications caused by microplastics. PE and PVC microplastics reduce soil porosity and permeability, potentially causing waterlogging or inadequate water availability, both of which are critical during germination (Zhang et al., 2023a, 2023b). Additionally, microplastics can contribute to soil compaction, restricting root penetration and limiting oxygen diffusion, adversely affecting seedling development (Waghamare, 2025). Microplastics also act as absorbents for essential nutrients and heavy metals, reducing the bioavailability of nitrogen, phosphorus, and other ions necessary for seedling establishment (Roy et al., 2024). This nutrient sequestration disrupts the nutrient balance in the rhizosphere, compounding stress on germinating seeds. Compost amendments typically enhance microbial diversity and nutrient cycling. However, microplastics can interfere with microbial communities, reducing compost efficacy and further hindering germination (Luo et al., 2024). Vermicompost's fine particulate nature may intensify this effect, potentially allowing higher adsorption of microplastics and further impairing nutrient availability (Sarwar et al., 2024). These combined factors underscore durum wheat's heightened sensitivity to soil and environmental stressors. Bread wheat's near 100% germination across similar conditions suggests a greater adaptive capacity to environmental stressors. These traits likely buffer the impact of soil compaction, nutrient sequestration, and microbial disruption, allowing consistent germination rates. The differential responses of bread and durum wheat highlight the complex interactions between plant physiology, soil amendments, and microplastic pollution. Microplastics, while inert in the short term, have the potential to disrupt soil structure, nutrient dynamics, and microbial activity, affecting plant health and soil fertility in course of time (Zhou et al., 2021). These findings emphasize the importance of incorporating species-specific sensitivities when evaluating the ecological and agricultural implications of microplastic contamination. The germination results underscore the role of environmental conditions and inherent species characteristics in determining plant response to microplastic-contaminated soils. Durum wheat's lower germination rates reflect its sensitivity to changes in soil structure and nutrient dynamics caused by microplastics, while bread wheat's resilience demonstrates its ability to withstand these altered conditions. Further research is essential to elucidate the long-term impacts of microplastics on soil–plant systems and to develop strategies to mitigate these effects in agricultural practices.

(b) ***Microplastic exposure study: Vigour Index I***

The positive effect on vigour index 1 (VI1) in bread wheat (*Triticum aestivum*) treated with polyethylene (PE) and polyvinyl chloride (PVC) in combination with organic compost (OC2) and vermicompost (VC2) in a cocopeat substrate can be attributed to the synergistic impact of the materials and amendments (Fig. 7.3). PE and PVC are extensively used in agricultural practices for their moisture retention and temperature stabilization properties. Their ability to moderate temperature fluctuations creates a thermally stable environment conducive to consistent seedling development, thus fostering better root establishment and overall plant vigour. The addition of organic composts and vermicompost further enhanced the substrate's properties, including water-holding capacity, organic matter content, and microbial diversity. Organic amendments improve soil structure and aeration, facilitating root proliferation and nutrient uptake (Rastogi et al., 2023). Vermicompost, rich in microbial and enzymatic activity, promotes the availability of essential nutrients like nitrogen and phosphorus, crucial for seedling metabolism and growth (Rehman et al., 2023). This enhanced nutrient availability synergized with the benefits of PE and PVC, optimizing conditions for seedling development. Additionally, vermicompost's granular texture and high cation exchange capacity improve substrate porosity and ion retention, ensuring sustained nutrient supply during critical growth phases. The combination of physical enhancements provided by PE and PVC and the biological benefits of organic amendments resulted in a favorable microenvironment, promoting enhanced VI1 values in bread wheat. This study highlights the importance of understanding the interactions between synthetic materials and organic amendments in agricultural systems. By combining the water-retention and thermal stabilization properties of plastics with the nutrient-enrichment and microbial benefits of organic composts, optimal growth conditions for crops can be achieved. Future research should focus on the long-term implications of such interactions, particularly regarding microplastic accumulation and its potential environmental impacts.

The greater sensitivity of durum wheat (*Triticum durum*) to the interaction between microplastics and organic amendments, as indicated by significant variations in vigour index 1 (VI1), highlights the complex interplay between crop-specific physiological traits and environmental stressors. The vigour index, a comprehensive measure of seedling health incorporating parameters such as seedling length, dry

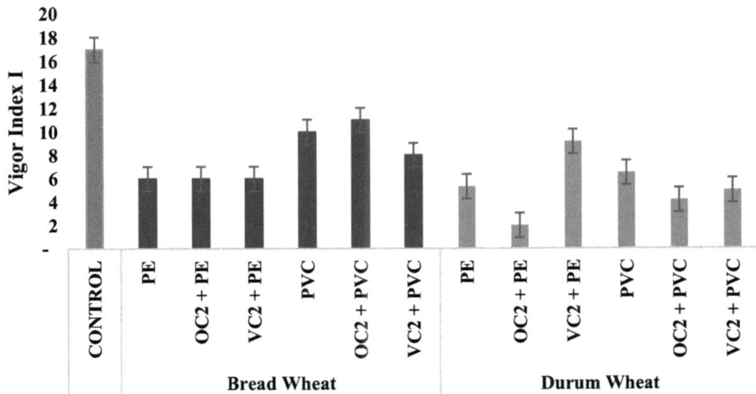

Fig. 7.3 Vigour index I of seedlings of Bread & Durum wheat as affected by MP in composting material (PE: polyethylene; PVC: polyvinyl chloride; OC2: organic compost (Sulpho-neem organic); VC2 vermicompost (Ecosmart home garden)

weight, and establishment success, serves as a critical indicator of the seedling's ability to thrive in its growing environment (Mahender et al., 2015). Durum wheat's reduced and more variable VI1 in the presence of microplastics and organic manures underscores its heightened vulnerability to soil disruptions. Microplastics such as polyethylene (PE) and polyvinyl chloride (PVC) are known to alter soil properties by reducing porosity and permeability, which adversely impacts water retention and aeration (Wang et al., 2022). These changes disrupt the delicate balance of soil conditions required for optimal germination and growth, particularly in durum wheat, which is less tolerant of such fluctuations compared to bread wheat. The physiological demands of durum wheat for stable soil and nutrient conditions make it less equipped to cope with these disruptions, resulting in a lower VI1.

The ability of bread wheat (*Triticum aestivum*) to maintain a higher and more stable VI1 under similar conditions reflects its adaptability to environmental stressors and its ability to leverage the beneficial properties of organic amendments. Organic composts and vermicompost can enhance soil structure, aeration, and microbial diversity, thereby mitigating some of the negative impacts of microplastic contamination (Poornima et al., 2024). These amendments may have partially offset the adverse effects of microplastics by improving nutrient cycling and water availability, but the benefits were more pronounced in bread wheat due to its inherently higher resilience to environmental variability. Durum wheat's dependence on stable soil conditions is further complicated by the potential disruption of microbial communities by microplastics, which can impair nutrient cycling and reduce soil fertility (Corato et al., 2024). Vermicompost and organic compost, while beneficial in many respects, may not have been sufficient to counteract these disruptions in the context of durum wheat's physiological requirements. In contrast, bread wheat's broader tolerance range allowed it to thrive under these conditions, benefiting from the synergistic effects of the organic amendments and plastic treatments.

(c) ***Microplastic exposure study: Vigour Index II***

The variation in vigour index 2 (VI2) between bread wheat (*Triticum aestivum*) and durum wheat (*Triticum durum*) under microplastic (PVC and PE) and organic manure treatments reflects the differential adaptability of these two wheat varieties to environmental stressors and in soil conditions. VI2, (which accounts for seedling root and shoot growth along with biomass accumulation), serves as a key indicator of overall seedling health and development potential (Shi et al., 2020). A higher VI2 suggests enhanced seedling vigour, faster establishment, and superior growth dynamics. In bread wheat, the observed positive response in VI2 under treatments involving PVC and PE combined with organic compost (OC2) and vermicompost (VC2) indicates a favorable interaction between these microplastics and organic amendments. Plastics reduce water loss through evaporation and moderate temperature fluctuations, ensuring consistent soil moisture and thermal stability. Such conditions are conducive to robust root and shoot growth, contributing to the improved VI2 values observed in bread wheat. The synergistic effect was further enhanced by the addition of organic amendments such as OC2 and VC2, which are known to improve soil structure, increase water-holding capacity, and support microbial diversity. These amendments facilitate better aeration and nutrient availability, enhancing root access to essential resources and promoting efficient nutrient uptake. The combined effect of these organic amendments and microplastics likely provided bread wheat seedlings with the ideal conditions for growth, as evidenced by the higher VI2 values. In contrast, durum wheat demonstrated a comparatively lower and more variable VI2 under similar conditions, suggesting its greater sensitivity to the combined effects of microplastic and organic amendments. Microplastics can alter soil properties, including porosity and permeability, potentially creating conditions of waterlogging or localized drought stress (Li et al., 2023). These disruptions are particularly detrimental to durum wheat, which requires stable soil conditions and consistent nutrient availability for optimal growth. The nutrient cycling benefits provided by OC2 and VC2 may not have been sufficient to offset these disruptions for durum wheat, given its physiological sensitivity to environmental stressors. The contrasting responses in VI2 between the two wheat varieties underscore the importance of species-specific traits in determining their resilience to microplastic contamination and their ability to benefit from organic amendments (Fig. 7.4).

VI2 serves as a critical indicator of seedling health, integrating growth and biomass accumulation to reflect the seedling's ability to establish and thrive in specific environmental conditions (Shandu, 2021). In bread wheat, while treatments with PVC and PE generally exhibited positive effects on VI2 due to enhanced water retention and temperature stabilization, a notable exception was observed with OC2 + PE and VC2 + PE treatments, where VI2 values were reduced. This indicates that PE, under these specific conditions, negatively impacted soil aeration and porosity, which are key factors necessary for oxygen availability and root penetration (Khan et al., 2000). PE's capacity to retain water, although beneficial in reducing evaporation, may have led to localized waterlogging, creating hypoxic conditions that impaired root respiration and nutrient uptake. Furthermore, the microplastic particles may have

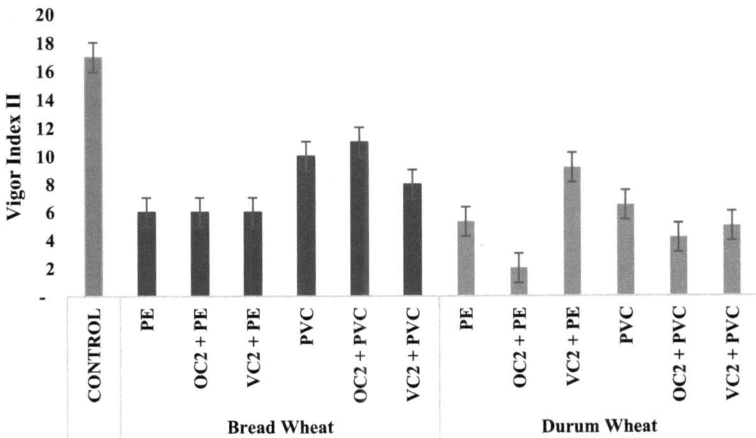

Fig. 7.4 Vigour index II of seedlings of Bread & Durum wheat as affected by MP in composting material (PE: polyethylene; PVC: polyvinyl chloride; OC2: organic compost (Sulpho-neem organic); VC2 vermicompost (Ecosmart home garden)

contributed to nutrient imbalances by adsorbing essential nutrients, making them less bioavailable for seedling development (Roy et al., 2024).

Durum wheat, known for its lower resilience to abiotic stressors, showed a greater decline in VI2 across treatments, particularly those involving PE. The adverse effects of PE were consistent across organic amendments (OC2 + PE, VC2 + PE), suggesting that PE significantly disrupted the soil's physical structure and nutrient dynamics, which are critical for durum wheat's growth (Melash et al., 2023). The heightened sensitivity of durum wheat to microplastic-induced stress can be attributed to its physiological reliance on stable soil conditions and its vulnerability to alterations in water retention and aeration. This disruption is particularly detrimental for durum wheat, which lacks the adaptive flexibility exhibited by bread wheat in response to environmental variability. The combined effects of reduced aeration, impaired root development, and nutrient sequestration by PE likely explain the significant reductions in VI2 observed in durum wheat, regardless of the organic amendment used. In conclusion, the contrasting responses of bread and durum wheat to microplastics and organic amendments underscore the complex interplay between soil properties, plant physiological traits, and environmental stressors.

7.10 Conclusion

The findings from this study provide valuable insights into the complex interactions between microplastics, organic manures, and wheat seedling performance, influenced by key physico-chemical parameters of the growth substrate. The presence of microplastics (PE and PVC) in organic manures (OC2 and VC2) significantly altered substrate properties, including pH, electrical conductivity, and water retention capacity. While no significant correlation was observed between microplastic concentration and pH, moisture content, or water retention capacity, a strong positive correlation with electrical conductivity suggests that microplastics may influence the ionic composition of the substrate, which in turn could affect seedling growth and nutrient uptake. These results indicate that microplastics, particularly in combination with organic amendments, can alter the substrate's physical and chemical environment, influencing seedling development.

Germination rates and vigour indices varied significantly between wheat varieties, with bread wheat (*Triticum aestivum*) showing greater tolerance to microplastic contamination and organic manure interactions compared to durum wheat (*Triticum durum*). The positive effects observed on vigour index 1 (VI1) in bread wheat, particularly under PVC and PE treatments combined with organic composts and vermicompost, suggest enhanced moisture retention, temperature regulation, and improved substrate structure. These factors likely created a more favorable environment for seedling growth. In contrast, durum wheat exhibited greater sensitivity to these interactions, especially in vigour index 2 (VI2), with significant reductions in seedling performance. In conclusion, a better understanding is needed to explain microplastic-organic manure interactions and their implications for sustainable farming practices. Further research is required to explore the long-term effects of microplastic contamination on soil health, nutrient cycling, and crop productivity, as well as strategies to mitigate these impacts in agricultural systems.

Acknowledgements The author would like to acknowledge the Indian Agricultural Research Institute (IARI), Pusa Road, Delhi for providing the necessary facilities and resources for conducting this study. We would like to acknowledge the Institution of Eminence and University R & D grant for providing financial support for the study.

References

Ali, M. (2007). *Quality enhancement of compost using vermicomposting and air separation*. Cardiff University.

Andrady, A. L., Barnes, P. W., Bornman, J. F., Gouin, T., Madronich, S., White, C. C., Zepp, R. G., & Jansen, M. A. K. (2022). Oxidation and fragmentation of plastics in a changing environment; from UV-radiation to biological degradation. *Science of the Total Environment, 851*, Article 158022.

Aralappanavar, V. K., Mukhopadhyay, R., Yu, Y., Liu, J., Bhatnagar, A., Praveena, S. M., Li, Y., Paller, M., Adyel, T. M., Rinklebe, J., Bolan, N. S., & Sarkar, B. (2024). Effects of microplastics on soil microorganisms and microbial functions in nutrients and carbon cycling—A review. *Science of the Total Environment*, 171435.

Asaye, Z., Kim, D. G., Yimer, F., Prost, K., Obsa, O., Tadesse, M., Gebrehiwot, M., & Brüggemann, N. (2022). Effects of combined application of compost and mineral fertilizer on soil carbon and nutrient content, yield, and agronomic nitrogen use efficiency in maize-potato cropping systems in southern Ethiopia. *Land, 11*(6), 784.

Bakhsh, A., Anayol, E., Sancak, C. E. N. G. İ. Z., & Özcan, S. (2016). An efficient and cost effective sterilizing method with least microbial contamination and maximum germination ratio for in vitro cotton (Gossypium hirsutum L.) culture. *JAPS: Journal of Animal & Plant Sciences, 26*(3).

Banda, B., Habtu, N. G., Gebreeyessus, G. D., & Meshesha, B. T. (2023). Vermicomposting as an effective approach to municipal sewage sludge management through optimization of the selected process variables. *Water Science & Technology, 88*(8), 1957–1973.

Barampuram, S., Allen, G., & Krasnyanski, S. (2014). Effect of various sterilization procedures on the in vitro germination of cotton seeds. *Plant Cell Tissue and Organ Culture (PCTOC), 118*, 179–185.

Belioka, M. P., & Achilias, D. S. (2024). The effect of weathering conditions in combination with natural phenomena/disasters on microplastics' transport from aquatic environments to agricultural soils. *Microplastics, 3*(3), 518–538.

Beriot, N., Zornoza, R., Lwanga, E. H., Zomer, P., van Schothorst, B., Ozbolat, O., Lloret, E., Ortega, R., Miralles, I., Harkes, P., van Steenbrugge, J., & Geissen, V. (2023). Intensive vegetable production under plastic mulch: A field study on soil plastic and pesticide residues and their effects on the soil microbiome. *Science of the Total Environment, 900*, Article 165179.

Bhat, S. A., Singh, J., & Vig, A. P. (2017). Instrumental characterization of organic wastes for evaluation of vermicompost maturity. *Journal of Analytical Science and Technology, 8*, 1–12.

Bodner, G., Nakhforoosh, A., & Kaul, H. P. (2015). Management of crop water under drought: A review. *Agronomy for Sustainable Development, 35*, 401–442.

Carey, A. M. (2024). *Strategies to enhance organic vegetable production systems: Soil blocking, cover crop mixtures, and poultry integration* (Doctoral dissertation, Iowa State University).

Chaudhary, M., & Suthar, S. (2023). Microplastic abundance and characterization in municipal sewage sludge from cities across upper Ganga River, India: Apprising microplastic uptakes and their toxicity in the plant during sludge application in agriculture. *Physics and Chemistry of the Earth, Parts A/B/C, 132*, Article 103468.

Chia, R. W., Lee, J. Y., Jang, J., Kim, H., & Kwon, K. D. (2022). Soil health and microplastics: a review of the impacts of microplastic contamination on soil properties. *Journal of Soils and Sediments, 22*(10), 2690–2705.

Chubarenko, I., Bagaev, A., Zobkov, M., & Esiukova, E. (2016). On some physical and dynamical properties of microplastic particles in marine environment. *Marine Pollution Bulletin, 108*(1–2), 105–112.

Chubarenko, I., Esiukova, E., Bagaev, A., Isachenko, I., Demchenko, N., Zobkov, M., Efimova, I., Bagaeva, M., & Khatmullina, L. (2018). Behavior of microplastics in coastal zones. In *Microplastic contamination in aquatic environments* (pp. 175–223). Elsevier.

Chubarenko, I., Esiukova, E., Bagaev, A., Isachenko, I., Zobkov, M., Bagaeva, M., Khatmullina, L., & Fetisov, S. (2024). Microplastics particles in coastal zone: Approach of physical oceanography. In *Microplastic Contamination in Aquatic Environments* (pp. 249–310). Elsevier.

Colzi, I., Renna, L., Bianchi, E., Castellani, M. B., Coppi, A., Pignattelli, S., Loppi, S., & Gonnelli, C. (2022). Impact of microplastics on growth, photosynthesis and essential elements in Cucurbita pepo L. *Journal of Hazardous Materials, 423*, Article 127238.

Cowger, W., Gray, A., Christiansen, S. H., DeFrond, H., Deshpande, A. D., Hemabessiere, L., Lee, E., Mill, L., Munno, K., Ossmann, B. E., Pittroff, M., Rochman, C., Sarau, G., Tarby, S., & Primpke, S. (2020). Critical review of processing and classification techniques for images and spectra in microplastic research. *Applied Spectroscopy, 74*(9), 989–1010.

Crossman, J., Hurley, R. R., Futter, M., & Nizzetto, L. (2020). Transfer and transport of microplastics from biosolids to agricultural soils and the wider environment. *Science of the Total Environment, 724*, Article 138334.

De Corato, U., Viola, E., Keswani, C., & Minkina, T. (2024). Impact of the sustainable agricultural practices for governing soil health from the perspective of a rising agri-based circular bioeconomy. *Applied Soil Ecology, 194*, Article 105199.

El-Beltagi, H. S., Basit, A., Mohamed, H. I., Ali, I., Ullah, S., Kamel, E. A., Shalaby, T. A., Ramadan, K. M., Alkhateeb, A. A., & Ghazzawy, H. S. (2022). Mulching as a sustainable water and soil saving practice in agriculture: A review. *Agronomy, 12*(8), 1881.

Fageria, N. K., & Stone, L. F. (2006). Physical, chemical, and biological changes in the rhizosphere and nutrient availability. *Journal of Plant Nutrition, 29*(7), 1327–1356.

Fazio, A. (2022). The effects of biofilms on the settling and transport of microplastic particles.

Guo, Z., Li, P., Yang, X., Wang, Z., Lu, B., Chen, W., Wu, Y., Li, G., Zhao, Z., Liu, G., Ritsema, C., Geissen, V., & Xue, S. (2022). Soil texture is an important factor determining how microplastics affect soil hydraulic characteristics. *Environment International, 165*, Article 107293.

Han, L., Chen, L., Feng, Y., Kuzyakov, Y., Zhang, S., Chao, L., Cai, Y., Ma, C., Sun, K., & Rillig, M. C. (2024). Microplastics alter soil structure and microbial community composition. *Environment International, 185*, Article 108508.

Hasan, M. M., & Tarannum, M. N. (2024). Adverse impacts of microplastics on soil physicochemical properties and crop health in agricultural systems. *Journal of Hazardous Materials Advances*, 100528.

Heinze, W. M. (2024). Microplastic distribution and transport in agricultural soils. *Acta Universitatis Agriculturae Sueciae*, (2024: 80).

Hoang, V. H., Nguyen, M. K., Hoang, T. D., Ha, M. C., Huyen, N. T. T., Bui, V. K. H., Pham, M.-T., Nguyen, C.-M., Woong Chang, S., & Nguyen, D. D. (2024). Sources, environmental fate, and impacts of microplastic contamination in agricultural soils: A comprehensive review. *Science of the Total Environment*, 175276.

Huang, D., Xu, Y., Lei, F., Yu, X., Ouyang, Z., Chen, Y., Jia, H., & Guo, X. (2021a). Degradation of polyethylene plastic in soil and effects on microbial community composition. *Journal of Hazardous Materials, 416*, Article 126173.

Huang, W., Song, B., Liang, J., Niu, Q., Zeng, G., Shen, M., Deng, J., Luo, Y., Wen, X., & Zhang, Y. (2021b). Microplastics and associated contaminants in the aquatic environment: A review on their ecotoxicological effects, trophic transfer, and potential impacts to human health. *Journal of Hazardous Materials, 405*, Article 124187.

Huang, Y., Zhao, Y., Wang, J., Zhang, M., Jia, W., & Qin, X. (2019). LDPE microplastic films alter microbial community composition and enzymatic activities in soil. *Environmental Pollution, 254*, Article 112983.

Hüffer, T., Metzelder, F., Sigmund, G., Slawek, S., Schmidt, T. C., & Hofmann, T. (2019). Polyethylene microplastics influence the transport of organic contaminants in soil. *Science of the Total Environment, 657*, 242–247.

Iqbal, B., Zhao, T., Yin, W., Zhao, X., Xie, Q., Khan, K. Y., Zhao, X., Nazar, M., Li, G., & Du, D. (2023). Impacts of soil microplastics on crops: A review. *Applied Soil Ecology, 181*, Article 104680.

Johnson, E. A., & Klemens, M. W. (2005). The impacts of sprawl on biodiversity. In *Nature in fragments: The legacy of sprawl* (pp. 18–54). Columbia University Press.

Kanianska, R. (2016). Agriculture and its impact on land-use, environment, and ecosystem services. In *Landscape ecology-The influences of land use and anthropogenic impacts of landscape creation* (pp. 1–26).

Khalid, A. R., Shah, T., Asad, M., Ali, A., Samee, E., Adnan, F., Bhatti, M. F., Marhan, S., Kammann, C. I., Haider, G., & Haider, G. (2023). Biochar alleviated the toxic effects of PVC microplastic in a soil-plant system by upregulating soil enzyme activities and microbial abundance. *Environmental Pollution, 332*, Article 121810.

Khalid, N., Aqeel, M., Noman, A., Khan, S. M., & Akhter, N. (2021). Interactions and effects of microplastics with heavy metals in aquatic and terrestrial environments. *Environmental Pollution, 290*, Article 118104.

Khan, A. R., Chandra, D., Quraishi, S., & Sinha, R. K. (2000). Soil aeration under different soil surface conditions. *Journal of Agronomy and Crop Science, 185*(2), 105–112.

Kumar, R., Ivy, N., Bhattacharya, S., Dey, A., & Sharma, P. (2022). Coupled effects of microplastics and heavy metals on plants: Uptake, bioaccumulation, and environmental health perspectives. *Science of the Total Environment, 836*, Article 155619.

Kumari, A., Rajput, V. D., Mandzhieva, S. S., Rajput, S., Minkina, T., Kaur, R., Sushkova, S., Kumari, P., Ranjan, A., Kalinitchenko, V. P., & Glinushkin, A. P. (2022). Microplastic pollution: An emerging threat to terrestrial plants and insights into its remediation strategies. *Plants, 11*(3), 340.

Kurniawan, T. A., Haider, A., Ahmad, H. M., Mohyuddin, A., Aslam, H. M. U., Nadeem, S., Javed, M., Othman, M. H. D., Goh, H. H., & Chew, K. W. (2023). Source, occurrence, distribution, fate, and implications of microplastic pollutants in freshwater on environment: A critical review and way forward. *Chemosphere, 325*, Article 138367.

Lemly, A. D., Finger, S. E., & Nelson, M. K. (1993). Sources and impacts of irrigation drainwater contaminants in arid wetlands. *Environmental Toxicology and Chemistry: An International Journal, 12*(12), 2265–2279.

Landrigan, P. J., Stegeman, J. J., Fleming, L. E., Allemand, D., Anderson, D. M., Backer, L. C., Brucker-Davis, F., Chevalier, N., Corra, L., Czerucka, D., Dechraoui Bottein, M.-Y., Demeneix, B., Depledge, M., Deheyn, D. D., Dorman, C. J., Fénichel, P., Fisher, S., Gaill, F., Galgani, F., ... Rampal, P. (2020). Human health and ocean pollution. *Annals of global health, 86*(1).

Li, S., Lu, S., Wang, J., Chen, Z., Zhang, Y., Duan, J., Liu, P., Wang, X., & Guo, J. (2023). Responses of physiological, morphological and anatomical traits to abiotic stress in woody plants. *Forests, 14*(9), 1784.

Li, Y., Ding, B. H., & Geng, X. (2024). Effect of biochar on microplastics penetration treatment within soil porous medium under the wetting-drying cycles and optimisation of soil-biochar mixing format. *Science of the Total Environment, 935*, Article 173194.

Liu, B., Zhao, S., Qiu, T., Cui, Q., Yang, Y., Li, L., Chen, J., Huang, M., Zhan, A., & Fang, L. (2024). Interaction of microplastics with heavy metals in soil: Mechanisms, influencing factors and biological effects. *Science of The Total Environment*, 170281.

Luo, C., Zhang, L., Ali, M. M., Xu, Y., & Liu, Z. (2024). Environmental risk substances in soil on seed germination: Chemical species, inhibition performance, and mechanisms. *Journal of Hazardous Materials*, 134518.

Lusher, A. L., Bråte, I. L. N., Munno, K., Hurley, R. R., & Welden, N. A. (2020). Is it or isn't it: The importance of visual classification in microplastic characterization. *Applied Spectroscopy, 74*(9), 1139–1153.

Mahender, A., Anandan, A., & Pradhan, S. K. (2015). Early seedling vigour, an imperative trait for direct-seeded rice: An overview on physio-morphological parameters and molecular markers. *Planta, 241*, 1027–1050.

Maity, S., Guchhait, R., Chatterjee, A., & Pramanick, K. (2021). Co-occurrence of co-contaminants: Cyanotoxins and microplastics, in soil system and their health impacts on plant—A comprehensive review. *Science of the Total Environment, 794*, Article 148752.

Melash, A. A., Bogale, A. A., Bytyqi, B., Nyandi, M. S., & Ábrahám, É. B. (2023). Nutrient management: bas a panacea to improve the caryopsis quality and yield potential of durum wheat (Triticum turgidum L.) under the changing climatic conditions.

Meng, K. (2024). *Exploring the potential of earthworms to reduce microplastic pollution in soils* (Doctoral dissertation, Wageningen University and Research).

Möller, J. N. (2022). *Microplastic pollution in composts, digestates and soils* (Doctoral dissertation, University of Bayreuth).

Okeke, E. S., Okoye, C. O., Atakpa, E. O., Ita, R. E., Nyaruaba, R., Mgbechidinma, C. L., & Akan, O. D. (2022). Microplastics in agroecosystems-impacts on ecosystem functions and food chain. *Resources, Conservation and Recycling, 177*, Article 105961.

Oyege, I., & Balaji Bhaskar, M. S. (2023). Effects of vermicompost on soil and plant health and promoting sustainable agriculture. *Soil Systems, 7*(4), 101.

Pérez-Reverón, R., Álvarez-Méndez, S. J., Kropp, R. M., Perdomo-González, A., Hernández-Borges, J., & Díaz-Peña, F. J. (2022). Microplastics in agricultural systems: Analytical methodologies and effects on soil quality and crop yield. *Agriculture, 12*(8), 1162.

Poornima, S., Dadi, M., Subash, S., Manikandan, S., Karthik, V., Deena, S. R., Balachandar, R., Kumaran, S. K. N., & Subbaiya, R. (2024). Review on advances in toxic pollutants remediation by solid waste composting and vermicomposting. *Scientific African*, e02100.

Qi, R., Jones, D. L., Li, Z., Liu, Q., & Yan, C. (2020). Behavior of microplastics and plastic film residues in the soil environment: A critical review. *Science of the Total Environment, 703*, Article 134722.

Rastogi, M., Verma, S., Kumar, S., Bharti, S., Kumar, G., Azam, K., & Singh, V. (2023). Soil health and sustainability in the age of organic amendments: A review. *International Journal of Environment and Climate Change, 13*(10), 2088–2102.

Rehm, R., Zeyer, T., Schmidt, A., & Fiener, P. (2021). Soil erosion as transport pathway of microplastic from agriculture soils to aquatic ecosystems. *Science of the Total Environment, 795*, Article 148774.

Rehman, S. U., De Castro, F., Aprile, A., Benedetti, M., & Fanizzi, F. P. (2023). Vermicompost: Enhancing plant growth and combating abiotic and biotic stress. *Agronomy, 13*(4), 1134.

Ren, Z., Gui, X., Xu, X., Zhao, L., Qiu, H., & Cao, X. (2021). Microplastics in the soil-groundwater environment: Aging, migration, and co-transport of contaminants—A critical review. *Journal of Hazardous Materials, 419*, Article 126455.

Roy, R., Hossain, A., Sultana, S., Deb, B., Ahmod, M. M., & Sarker, T. (2024). Microplastics increase cadmium absorption and impair nutrient uptake and growth in red amaranth (Amaranthus tricolor L.) in the presence of cadmium and biochar. *BMC Plant Biology, 24*(1), 608.

Rynk, R., Schwarz, M., Richard, T. L., Cotton, M., Halbach, T., & Siebert, S. (2022). Compost feedstocks. In *The composting handbook* (pp. 103–157). Academic Press.

Sahu, P. K., Singh, D. P., Prabha, R., Meena, K. K., & Abhilash, P. C. (2019). Connecting microbial capabilities with the soil and plant health: Options for agricultural sustainability. *Ecological Indicators, 105*, 601–612.

Sajjad, M., Huang, Q., Khan, S., Khan, M. A., Liu, Y., Wang, J., Lian, F., Wang, Q., & Guo, G. (2022). Microplastics in the soil environment: A critical review. *Environmental Technology & Innovation, 27*, Article 102408.

Sarwar, M. J., Shabaan, M., Ayyub, M., Zahir, Z. A., Asghar, H. N., Asif, K., Shareef, A., & Zulfiqar, U. (2024). Harnessing the duo effect of plant growth-promoting rhizobacteria and organic amendments in the phytoremediation of cadmium: A review. *Journal of Soil Science and Plant Nutrition*, 1–17.

Schell, T., Hurley, R., Buenaventura, N. T., Mauri, P. V., Nizzetto, L., Rico, A., & Vighi, M. (2022). Fate of microplastics in agricultural soils amended with sewage sludge: Is surface water runoff a relevant environmental pathway? *Environmental Pollution, 293*, Article 118520.

Scopetani, C., Chelazzi, D., Cincinelli, A., Martellini, T., Leiniö, V., & Pellinen, J. (2022). Hazardous contaminants in plastics contained in compost and agricultural soil. *Chemosphere, 293*, Article 133645.

Shandu, M. (2021). *The influence of zinc content on wheat seed quality and vigour* (Doctoral dissertation, Stellenbosch: Stellenbosch University).

Shi, Z., Chang, T. G., Chen, F., Zhao, H., Song, Q., Wang, M., Wang, Y., Zhou, Z., Wang, C., Zhou, S.-C., Wang, B., Chen, G., & Zhu, X. G. (2020). Morphological and physiological factors contributing to early vigor in the elite rice cultivar 9,311. *Scientific Reports, 10*(1), 14813.

Silva, G. C., Galleguillos Madrid, F. M., Hernández, D., Pincheira, G., Peralta, A. K., Urrestarazu Gavilán, M., Vergara-Carmona, V., & Fuentes-Peñailillo, F. (2021). Microplastics and their effect in horticultural crops: Food safety and plant stress. *Agronomy, 11*(8), 1528.

Sintim, H. Y., Bandopadhyay, S., English, M. E., Bary, A., y González, J. E. L., DeBruyn, J. M., Schaeffer, S. M., Miles, C. A., & Flury, M. (2021). Four years of continuous use of soil-biodegradable plastic mulch: impact on soil and groundwater quality. *Geoderma, 381*, 114665.

Sooriyakumar, P., Bolan, N., Kumar, M., Singh, L., Yu, Y., Li, Y., Weralupitiya, C., Vithanage, M., Ramanayaka, S., Sarkar, B., Wang, F., Gleeson, D. B., Zhang, D., Kirkham, M. B., Rinklebe, J., & Siddique, K. H. (2022). Biofilm formation and its implications on the properties and fate of microplastics in aquatic environments: A review. *Journal of Hazardous Materials Advances, 6*, Article 100077.

Srinivasan Kameswaran, K., Bellamkonda, R., & Bangeppagari, M. (2024). Recycling resources of soil and agroecosystem. In *Prospects for soil regeneration and its impact on environmental protection* (pp. 173–203). Springer Nature Switzerland.

Su, L., Xiong, X., Zhang, Y., Wu, C., Xu, X., Sun, C., & Shi, H. (2022). Global transportation of plastics and microplastics: A critical review of pathways and influences. *Science of the Total Environment, 831*, Article 154884.

Sun, X., Withana, P. A., Palansooriya, K. N., Vithanage, M., Yang, X., Lee, S. R., Bank, M. S., You, S., & Ok, Y. S. (2024). Impacts of microplastics on terrestrial plants: A critical review. *Land Degradation & Development, 35*(5), 1629–1643.

Ter Halle, A., Ladirat, L., Gendre, X., Goudouneche, D., Pusineri, C., Routaboul, C., Tenailleau, C., Duployer, B., & Perez, E. (2016). Understanding the fragmentation pattern of marine plastic debris. *Environmental Science & Technology, 50*(11), 5668–5675.

Thiele, C. J., Hudson, M. D., & Russell, A. E. (2019). Evaluation of existing methods to extract microplastics from bivalve tissue: Adapted KOH digestion protocol improves filtration at single-digit pore size. *Marine Pollution Bulletin, 142*, 384–393.

Urgert, W. (2015). *Microplastics in the rivers Meuse and Rhine*. Open University of the Netherlands.

Waghamare, S. (2025). Soil physio-biochemical alterations due to micro/nano-plastic pollution. In *Societal and environmental ramifications of plastic pollution* (pp. 89–114). IGI Global.

Wang, F., Wang, Q., Adams, C. A., Sun, Y., & Zhang, S. (2022). Effects of microplastics on soil properties: Current knowledge and future perspectives. *Journal of Hazardous Materials, 424*, Article 127531.

Wang, Y., Xv, X., Shao, T., He, Q., Guo, Z., Wang, Y., Guo, Q., & Xing, B. (2024). A case on source to soil to solutions: Distribution characteristics of microplastics in farmland soil of the largest vegetable base in Northwest China. *Science of the Total Environment, 907*, Article 167910.

Yang, J., Li, L., Li, R., Xu, L., Shen, Y., Li, S., Tu, C., Wu, L., Christie, P., & Luo, Y. (2021). Microplastics in an agricultural soil following repeated application of three types of sewage sludge: A field study. *Environmental Pollution, 289*, Article 117943.

Yang, X., Bento, C. P., Chen, H., Zhang, H., Xue, S., Lwanga, E. H., Zomer, P., Ritsema, C. J., & Geissen, V. (2018). Influence of microplastic addition on glyphosate decay and soil microbial activities in Chinese loess soil. *Environmental Pollution, 242*, 338–347.

Zhai, Y., Bai, J., Chang, P., Liu, Z., Wang, Y., Liu, G., Cui, B., Peijnenburg, W., & Vijver, M. G. (2024). Microplastics in terrestrial ecosystem: Exploring the menace to the soil-plant-microbe interactions. *TrAC Trends in Analytical Chemistry*, 117667.

Zhang, J., Guo, N., Ding, W., Han, B., Zhao, M., Wang, X., Wang, J., Cao, B., Zou, G., & Chen, Y. (2023a). Microplastic pollution and the related ecological risks of organic composts from different raw materials. *Journal of Hazardous Materials, 458*, Article 131911.

Zhang, J., Ren, S., Xu, W., Liang, C., Li, J., Zhang, H., Li, Y., Liu, X., Jones, D. L., Chadwick, D. R., Zhang, F., & Wang, K. (2022a). Effects of plastic residues and microplastics on soil ecosystems: A global meta-analysis. *Journal of Hazardous Materials, 435*, Article 129065.

Zhang, K., Gao, N., Li, Y., Dou, S., Liu, Z., Chen, Y., Ma, C., & Zhang, H. (2023b). Responses of maize (Zea mays L.) seedlings growth and physiological traits triggered by polyvinyl chloride microplastics is dominated by soil available nitrogen. *Ecotoxicology and Environmental Safety, 252*, 114618.

Zhang, S., Li, Y., Jiang, L., Chen, X., Zhao, Y., Shi, W., & Xing, Z. (2024). From organic fertilizer to the soils: What happens to the microplastics? A critical review. *Science of The Total Environment*, 170217.

Zhang, Z., Cui, Q., Chen, L., Zhu, X., Zhao, S., Duan, C., Zhang, X., Song, D., & Fang, L. (2022b). A critical review of microplastics in the soil-plant system: Distribution, uptake, phytotoxicity and prevention. *Journal of Hazardous Materials, 424*, Article 127750.

Zhou, J., Wen, Y., Marshall, M. R., Zhao, J., Gui, H., Yang, Y., Zeng, Z., Jones, D. L., & Zang, H. (2021). Microplastics as an emerging threat to plant and soil health in agroecosystems. *Science of the Total Environment, 787*, 147444.

Chapter 8
Impact of Microplastic Pollution on Soil Health

Vaibhav Bhagwan Pandit, Ch. Ravali, and Shabir H. Wani

Abstract Microplastics are small particles of plastic, having a diameter of less than 5 mm, emerging as a new form of pollution in soil, which poses significant threats to soil health. The primary sources of MPs in agricultural soil include residual plastic fragments from mulching, biosolids (sewage sludge), atmospheric deposition, municipal waste (compost) and plastic coated fertilizers. A soil is considered polluted when the concentration of MPs reaches 0.1% by weight. The significant microplastic concentrations in soil is concerning due to their potential to disrupt ecosystems environment, including altering soil characteristics, disrupting the microbial activities, harming soil biota and negatively impacting the growth of plants. Microplastics create a significant menace to soil ecology and ecosystems due to their potential toxicity, long life span and persistence. In soil, they contribute to degradation of soil structure, reduce the infiltration capacity of rain as well as irrigation water and inturn negatively impact soil water retention. Microplastics in the soil environment greatly impact soil structure, particularly by altering its porosity. Bioturbation causes microplastics to move downward through the soil, and when accumulated in large quantities, these particles can clog soil pores, reducing infiltration capacity. This disruption can interfere with the nutrient cycles of soil, alter the composition of microbial communities and eventually hinder crop growth.

Keywords Microplastics · Sewage · Sludge · Soil pollution · Microbial activity

V. B. Pandit · Ch. Ravali
SR University, Warangal, Telangana, India

S. H. Wani (✉)
Mountain Research Centre for Field Crops, Sher-e-Kashmir University of Agricultural Sciences and Technology of Kashmir, Khudwani, Srinagar, Jammu and Kashmir, India
e-mail: shabirhwani@skuastkashmir.ac.in

V. Kumar et al. (eds.), *Microplastics Pollution*,
https://doi.org/10.1007/978-3-031-99774-7_8

8.1 Introduction

The role of plastic materials in everyday life has become irreplaceable. Due to their inexpensive nature, exceptional portability and durability, they are widely used. By 2015, the global production of plastic waste has reached an estimated 6.3 billion tons, with roughly 79% of it ultimately find its way to garbage dumps or the natural surroundings (Geyer et al., 2017). Statistics indicate that the total world wide plastic production has amounted to 8.3 billion tons (Plastics Europe, 2019). Of this, around 20% is recycled, leaving the rest 80% to accumulate in rivers, soil, and oceans. Plastic waste that builds up in the environment undergoes breakdown through physical, chemical, or biological processes, gradually fragmenting into smaller particles, which eventually become microplastics. The large fragments of plastic, when exposed to external environment are exposed to UV rays, collision, hydrolysis, soil erosion or biological agents to form solid particles which are < 5 mm size and these small particles are called as "microplastics (MPs)" (Mao et al., 2020).The issue of microplastic pollution in the biosphere has gained enormous attention and presents an intriguing area of research concerning its origin, possible toxicity, ecological pathways and sinks within the ecosystem.

Agricultural lands and soil ecosystems have emerged as key repositories for microplastics (MPs). Owing to their very tiny particle size, they diffuse very quickly into the soil environment. Apart from this, some of the human activities viz., plastic mulching, littering, sewage, fertilizer coatings and factors like runoff and air transmission also favour the accumulation of MPs in soil. There are two distinct types of MPs in the environment. Primary MPs, which are produced directly and used as raw materials in industrial manufacturing or cosmetics like plastic microbeads, which are predominantly used in personal care products (PCPs). The second type of MPs are secondary MPs. Secondary microplastics are generally produced by disintegration and fragmentation of large sized plastic particles. These secondary MPs are the main source of microplastic pollution (Prata et al., 2020).

MPs are commonly reported cause harm to animals, plants, microbes and also humans. They serve as a potential threat to soil biota. MPs cause inhibition of growth, cause gut damage, weight reduction, increase in mortality rate of earth worms. Brennecke et al. (2016) discovered that copper and zinc can leach from polystyrene (PS) and polyvinyl chloride (PVC) microplastics when encountered with seawater. Additionally, MPs in the environment are frequently associated with hazardous organic compounds (HOCs) such as hexachlorocyclohexane (HCH), polycyclic aromatic hydrocarbons (PAHs), polychlorinated biphenyls (PCBs), and polybrominated diphenyl ethers (PBDEs).

Once MPs enter into soil, they become dispersed throughout the soil matrix as a result of alternate dry and wet cycles, as well as soil management practices and biological activity. This spread can alter the physical and chemical characteristics of the soil, inturn affecting its carbon, nitrogen and phosphorus content, as well as its pH. Due to MPs accumulation, the soil enzyme activity, microbial activity are highly affected. Furthermore, because of their extensive surface area, microplastics can adsorb persistent organic pollutants from the soil. These contaminants are subsequently be passed on to both aquatic and terrestrial organisms, leading to concerns about potential accumulation of toxins within the food chain. This chapter investigates the role of soils as potential reservoirs for microplastics, focusing on their distribution, persistence and effects on soil biota. It provides an in-depth knowledge of the ecological risks linked to microplastics in soil ecosystems.

8.1.1 Potential Sources of MPs in the Soil

Many contributors to microplastics are well known. Among them, polyethylene film mulching and organic fertilizers are regarded as the predominant suppliers of microplastics in soil.

8.1.2 Plastic Mulching—A Possible Contributor to Microplastic Contamination

Plastic mulching was introduced during 1970s and quickly gained widespread use across the world. Plastic film mulching is the dominant source of soil microplastic pollution, due to its high usage and poor rate of disposal. After being used, discarded plastic residue from plastic mulching practices in agriculture are broken down into minute fragments by multiple transformations in the soil (Ryan et al., 2009). These particles ultimately break to be smaller than 5 mm size and are labelled as microplastics (MPs). The resdues left after the removal of plastic mulch are subjected to physical, chemical and biological processes like UV radiations, wind/water erosion or by the action of earth worms. Furthermore they undergo photo-oxidative degradation, as a result they are broken down into smaller particles (Gasperiet al., 2018).

Huang et al. (2020) reported that, continuous mulching of fields with 5, 15 and 24 years, the concentrations of microplastic particles were 80.3 ± 49.3, 308 ± 138.1 and 1075.6 ± 346.8 pieces/kg soil, respectively. These concentrations of microplastic particles has increased over time in the sites where plastic mulch was constantly used. Plastic mulching serves as a direct source of microplastics in soil. The quantity of microplastics in soil are very high in soils with plastic mulch compared to the soils where there was no plastic mulch.

There are some alternatives to replace plastic mulch, which could serve as a possible solution for this problem. One of the alternatives is replacement of plastic mulch with organic or live mulch. Growing cover crops to cover the soil surface, or use of crop residues help in releasing nutrients as they decompose. But, the main crop and the live mulch (cover crop) could compete for nutrients and water. Another important alternative is the replacement with biodegradable plastics. Usually bioplastics are synthesized from petroleum based products, microbes, renewable resources

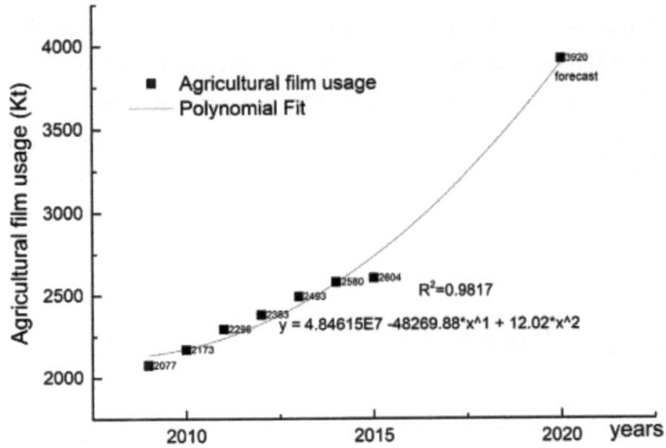

Fig. 8.1 Annual production of agricultural plastic film

Fig. 8.2 MPs from plastic mulch

or combination of all three of them. Biodegradable plastics can be degraded by the action of microbes and which consequently emit CO_2, so there is no problem of accumulation in soil. Also, they are non-toxic to living organisms in soil. They offer a potential replacement for synthetic plastics. However, due to their high cost and low durability, their use in agriculture is limited (Figs. 8.1 and 8.2).

8.1.3 Migration of MPs in Soil

The existence of microplastics in the aquatic ecosystems and their harmful impacts are well noticed in many studies. In recent days, the prevalence of microplastics in soil environment is gaining much importance. Evaluating the effects of microplastics on the soil environment is crucial for understanding their toxicological properties. It is impossible to ignore the high concentration of MPs in soil because of their ability to change various soil properties, alter microbial activity in the soil, damage soil biota and hinder the growth of plant. The dynamics of migration of MPs in soil involves physical, chemical and biological mechanisms.

8.1.4 Physical Migration

Physical migration signifies vertical and horizontal migration of microplastics into atmosphere. Water bodies and soil environments are influenced by external agents such as wind and water flow. The process of migration is primarily driven by various environmental factors.

8.1.5 Migration by Wind

Wind currents carry microplastics (low-density MPs) and release into atmosphere my multiple ways like emissions can result from the impact during saltation (the displacement of sand particles over the surface), direct release, or the breakdown of large aggregate particles. After entering into the atmosphere, due to the action of wind, MPs then migrate to remote areas and settle down. Wind speed is one of the important factors which have positive correlation with the dispersal of MPs. The minimum wind speed required for the transportation and deposition of MPs is known as wind speed threshold, which is crucial for understanding how wind erosion affects the microplastic movement. The wind speed threshold needed for polyethylene (PE) particles and polyethylene terephthalate (PET) fibers to move out of soil was observed to be 6.9 ms^{-1} in quartz sand and 10.8 ms^{-1} in loam soil, according to wind tunnel experiments (Bullard et al., 2020). Vertical transport of MPs also plays a significant role in transport of MPs into atmosphere. Wind flow and turbulence are the important characteristics which play vital role in vertical transport of MPs into air (Esders et al., 2022). These particles hang out in the atmosphere for longer time, they have a huge impact on the quality of the air. Consequently, MP abundance showed a decreasing pattern as height increased, although AMP (Amine Modified Polysterene) abundance peaked locally at 168 m (Zhang et al., 2020). Precipitation serves as one of the efficient ways for removal of airborne particulate matters (PMs), which may enter into urban soils as well as water bodies.

Apart from these, land surfaces and characteristics of microplastics also affect migration and dispersal of microplastics into atmosphere. Additionally, MPs in water may travel into the atmosphere due to wind. According to research, surface circulation caused by wind in inland surface waters might cause MPs to migrate downward and accumulate there (Imhof et al., 2013). Nevertheless, it is still unknown if MPs in inland freshwater habitats migrate and escape into the atmosphere when wind blows across them.

Characteristics of MPs also plays a significant role in their transport by wind. The low density MPs with particle size < 100 μm can be easily carried by wind (Rezaei

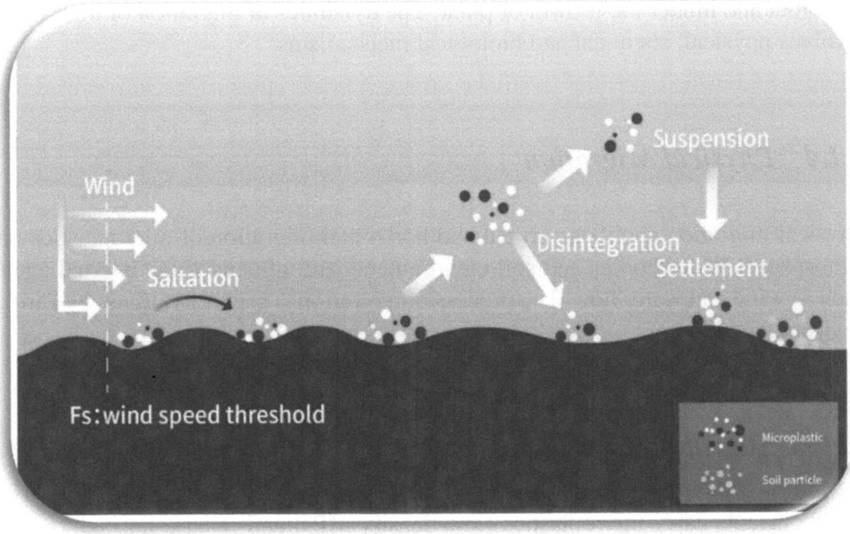

Fig. 8.3 Transport of microplastics by wind

et al., 2019). The particle drift is also influenced by shape of MPs, where the horizontal motion is highest for fibers and films. Compared to spherical shaped particles and cylindrical shaped particles, fibres and films are more susceptible for wind transport. Flat fibers are far more effective for transmission of long distances than cylindrical fibers because they have a higher cross-sectional area, which increases the average residence duration in the atmosphere increased by over 450%. (Xiao et al., 2023) (Fig. 8.3).

8.1.6 Migration by Water

Water serves as potential agent for the transport of microplastics through water primarily happens through vertical sedimentation or horizontal displacement or by suspension and floating. Surface runoff induced by rainfall is one of the important pathways for transport of MPs. Rainfall also flushes the dispersed microplastics from the atmosphere into aquatic environments. The rainfall intensity plays a vital role in migration of microplastics. High flow velocities in sediments show negative correlation with the concentration of microplastics in water bodies. Intense hydrodynamic conditions enhance the horizontal movement of microplastics in water bodies, but those conditions are not suitable for deposition of MPs (Zhang et al., 2017). From river streams, they also enter into inland water bodies. Later from these inland waters, these MPs infiltrate into ground water or may enter into urban water supply systems via ground water exploitation. Later these MPs enter into human bodies through

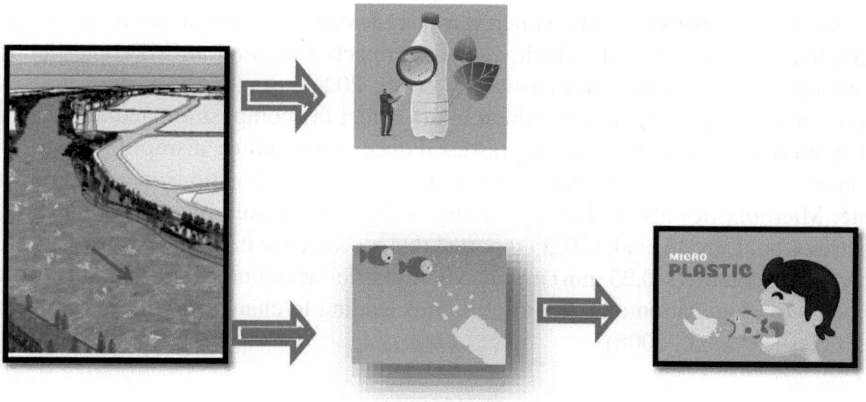

Fig. 8.4 Schematic diagram of migration of MPs and their entry into food chain

drinking water (Pannoet al., 2019), also due to enrichment of MPs in water bodies, the aquatic organisms may consume them (Fig. 8.4).

8.1.7 Impact of MPs on Soil Properties

Owing to their small size, high dispersal rate, MPs get easily mixed up with soil profile via infiltration and other agricultural activities. This may disturb soil functions like water retention, water flow, nutrient cycling etc. especially in agricultural soils. Many reports found that, microplastics mainly affect soil physical, chemical and biological properties.

8.1.8 Physical Properties

When microplastics are present in larger quantities, they block the soil pores and ultimately reduce the infiltration. Due to their small size, they integrate into soil aggregates and get adhered to these aggregates so tightly (Zhang & Liu, 2018). Each microplastic type differs in its impact on soil properties. Presence of MPs initiates the microbial processing of organic matter and increased aggregation (Zhang et al., 2019). This leads to accommodation of more microplastic fibers into newly formed soil aggregates. The increased inclusion of microplastic fibers in aggregates creates fracture points, which weaken the stability of the aggregates (Liang et al., 2021). Microplastics block the soil pores, thus decreasing infiltration and water holding capacity and creates anoxic conditions (Liu et al., 2014). Due to bioturbation, MPs move towards bottom layers of soil and pollute the inner layers of soil.

Numerous reports found that microplastics can change the permeability and water retention capacity of soil, which, in turn, impacts evaporation. Microplastics in soils decrease the soil bulk density (Qi et al., 2020). Smaller MPs clog the soil pores and reduce soil aeration, which inturn affect the composition and abundance of microbial communities. Microplastics reduce soil stability, disrupt the distribution of water-stable aggregates, and impair essential soil functions viz., infiltration rate. Microplastics also reduce hydraulic conductivity of soil, by altering pore-size distribution. Lozano et al. (2021) reported that, occurrence of microplastics reduced macro-aggregates (> 0.25 mm) in soil. Microplastics accumulation in soil affects the process of aggregation of soil particles, thus leading to changes in the sizes of soil particles (Seyet al., 2008).

8.1.9 Chemical Properties

Microplastics adversely affect many soil functions. Among them, nutrient availability is the major one. Microplastics influence soil chemical properties by altering soil pH, electrical conductivity, nutrient availability and enzyme activities. Microplastic pollution can also lead to heavy metal accumulation in plants. Microplastics affect the CEC of soil. Owing to their small size, large surface area and negative charge, they mimic the soil colloids (Sharma et al., 2021). So, they absorb the positively charged ions from the soil, hence they influence CEC directly. Also the functional groups present on the surface of MPs react with different anions and cations in the soil solution, hence in this way they also affect CEC (Zhang et al., 2015). Microplastics also influence the organic matter content in soil by altering the microbial biomass and enzymes (Liu et al., 2021). Materials like high density polyethylene (HDPE) and polylactic acid (PLA) at 1% (w/w) and 10% (w/w) enhanced the soil pH.

Accumulation of MPs in soil affects the prevalence of functional genes associated to transformations of nitrogen, within soil aggregates. In the soils polluted with MPs, when there is elevated pH, the rates of mineralization, nitrification and reduction of nitrate to ammonium has been significantly increased (Zhao et al., 2018). MPs can indirectly affect the transformations of nitrogen by altering the concentrations of dissolved organic carbon. It is well known fact that, changes in the ratio of nitrate nitrogen to dissolved organic carbon affects the nitrogen transformations in soil. If the ratio is in between 3.00 to 6.00, it leads to nitrification process and if the ratio is close to 1.00, then accumulation of nitrate nitrogen is limited. Contamination of MPs in soil significantly alter C:N ratio, which inturn affects nitrogen transformations (Zhou et al., 2024). Plastisphere microbes play key role in the nitrogen transformations in soil. The surface of biodegradable MPs have abundant microbes which are involved in nitrogen transformations. The surface of most of biodegradable MPs is enriched with *Ramlibacter* spp. When the soil has accumulated biodegradable MPs, it leads to enrichment of *Ramlibacter* spp. into the surrounding soil from surface of MPs, resulting in changes in the cycling processes of nitrogen between plastisphere and soil (Hu et al., 2023).

The occurrence of MPs in soil leads to changes in phosphorous in plant–microbe-soil systems. Many studies have reported that MPs accumulation in soil have direct or indirect impact on phosphorous concentration in soil. MPs alter soil phosphorous availability by altering the microbial processes which are involved in making phosphorous available. The additives in MPs like plasticizers, flame retardants, antioxidants, photostabilizers etc., potentially reduce the soil microbial activity, which inturn leads to decrease in soil microbial biomass phosphorous, phosphatase activity and finally reduce the concentration of available phosphorous in soil. However, the phosphorous dynamics in soil are influenced by shape, size and type of MPs. MPs have ability to inhibit the neutral phosphatase activity in soil and also increase phosphorous leaching (Zhou et al., 2024).

Generally, MPs are composed of more than 90% of carbon (Riling & Lehmann, 2020). Microplastics affect the decomposition process of soil organic matter, thus directly affects the soil carbon pools. According to available data, microplastics have an effect on the rhizospheric environment, which includes soil aggregates and microbial communities (Rong et al., 2023). This has an impact on the stability of soil organic carbon (SOC) and the transfer of rhizodeposits into soil carbon pools. Microplastic carbon is resistant to breaking down, therefore it tends to build up in the soil and, after a considerable amount of time, is immobilized by microorganisms. This could potentially have an impact on SOM turnover and the carbon cycle (Chen et al., 2022). The microplastic leachates (weathered products of MPs viz., benzoic ethers, long chain alkanes, high molecular weight acids) have the ability to separate from microplastics and infiltrate into the soil. These leachates participate in the soil carbon cycle and act as an active carbon pool (Cristina et al., 2023). Due to microbial degradation of carbon derived from PBAT and consumption of dissolved organic nitrogen, polybutylene adipate terephthalate (PBAT) microplastics mimic soil carbon, resulting in the formation of particulate organic carbon and thus promoting the creation of dissolved organic carbon and mineral-associated organic carbon. However, to fully understand how microplastics affect soil carbon fractions, further investigation is to be done to determine the ways in which conventional and biodegradable microplastics affect the transformations that lead to the accumulation and stabilization of various carbon fractions in soil (Chen et al., 2024).

8.1.10 Biological Properties

Microplastic accumulation in soil can influence biological properties especially like soil enzyme activities and soil microbial activity. Huang et al. (2020) found that, the presence of MPs in soil significantly altered enzyme activities like urease and catalase. MPs will change the microbial activities which ultimately changes the enzyme activities. Soil MPs have negative impact on soil enzyme activity. MPs when applied to soil along with organic matter, caused decrease in the enzyme activities like β-glucosidase, cellobiohydrolase, phosphotase and N-acetyl glucosamine (Liang et al., 2021). Microplastic fibers may have shifted the composition of the microbial community by altering the physical properties of the soil, releasing non-water-soluble additives, or acting as dwellings for microbes. Such community changes may have resulted in modifications to the enzymatic activity.

Occurrence of elevated levels of polypropylene (28%) led to increase in Fluorescein diacetate hydrolase (FDAse) activities (Liu et al., 2021). According to Zhang and Liu (2018), microplastics alter the porosity and structure of the soil, which ultimately affects the soil enzymatic activities. Polystyrene, polyethylene and polypropylene MPs have significantly reduced phytase activity and enhanced dehydrogenase activity (Lian et al., 2024). Soil containing low density polyethylene (LDPE) have significantly decreased the activities of urease, dehydrogenase and β-glucosidase. MPs have negative correlation with these enzyme activities (Khan & Sikder, 2024).

Most of the researchers reported that, microplastic accumulation in soil has negative impact on soil microbes. The role that MPs play as both a source and a sink for harmful chemical pollutants and pathogenic bacteria, as well as their potential entry and accumulation into the food chain, are now the most crucial issues concerning MPs in the environment. Microorganisms interact with MPs in soil environment. Many studies have reported that, there are several genera of bacteria, algae and fungi attached to the surface of microplastics (De-Tender et al., 2017). Different types of MPs promote the growth and multiplication of specific genera of soil microbes. The environment immediately surrounding the plastic (including the microbes in that environment) is known as 'plastisphere' and the microbial community in that environment is called 'plastisphere microbial community' (Fig. 8.5).

The microbiome in the plastisphere differs from the microbial communities in the soil rhizosphere, bulk soil and drilosphere. The microbial community in plastisphere has lower diversity. However some bacterial genera are abundant in plastisphere compared to that of bulk soil and rhizosphere. For example, the bacterial genera like *Acidobacteria, Alphaproteobacteria* and *Firmicutes* were abundant in polyethylene plastisphere compared to bulk soil in maize field, under continental monsoon climatic

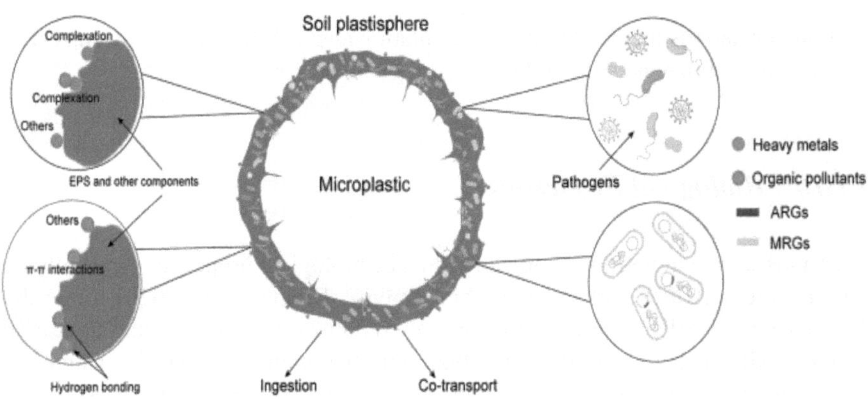

Fig. 8.5 Soil plastisphere

conditions in China (Shi et al., 2022). Results from various incubation studies indicated that, *Proteobacteria* is the dominant bacterial genera in plastisphere. Microplastics have antagonistic effect on soil microbes. The low density polyethylene (LDPE) have decreased the bacterial count in the incubated soil (Khan & Sikder, 2024), due to antagonistic effect. This antagonistic effect resulted from increased nutrient availability and decreased enzyme activity. MPs cause mortality to all those microbes which are involved in nutrient cycling. MPs also decrease soil respiration which inturn lead to reduced microbial counts. Number of studies revealed that presence of MPs in soil modifies structure of gut microbes, which leads to irritation symptoms in the gut of zebrafish and intestinal failure in mice (Jin et al., 2019).

According to studies, plasticizing chemicals like phthalates that are produced from plastic can harm soil-dwelling species. Such extra chemicals are easily washable out and are only weakly incorporated into the polymer structure of many plastic compounds. Broken plastic particles have an impact on the balance between soil microbial diversity and the environment. Numerous reviews have addressed how MPs and microbes interact in the soil environment (Fig. 8.6).

When microplastics are consumed by resident soil fauna, they affect the growth, physiology, reproduction and gut health. Peng et al. (2017) reported that, soil fauna are capable to digest microplastics, which will block the absorption of food while passing through the gut. Earthworms play key role in organic matter decomposition, nutrient cycling and improve structure of soil. Microplastics show negative effect on earthworms behavior, development, oxidative response, gut microbiota and gene expression. In recent years, earthworms have been utilized extensively as a biological indicator of soil contamination to assess the impact of microplastics on soil environment. Earthworms can ingest MPs due to their smaller size. Through physical

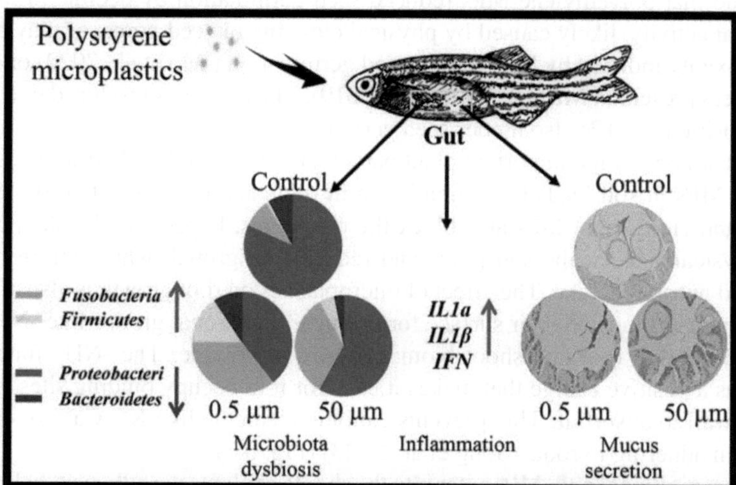

Fig. 8.6 Polystyrene microplastics induced gut microbiota dysbiosis in zebrafish (Jin et al., 2019)

adsorption, complexation, cation-π bonding interaction, and electrostatic attraction, heavy metals adhere to the surface of microplastics. Earthworm gut eventually experience an increase in heavy metal accumulation due to the desorption of heavy metals that have been absorbed by microplastics (Li et al., 2022). Histopathological analysis carried out by Jiang et al. (2020) indicated that the intestinal cells of earthworms were damaged after exposure to polystyrene microplastics. They have also reported that there was DNA damage to the earthworm (*Eisenia foetida*) after consuming polystyrene particles.

8.1.11 Effect of MPs on Plant Growth

As MPs enter into the soil, they accumulate, endure and slowly disrupt all soil processes and they also alter soil biodiversity and harm soil ecosystem. Plant growth is generally associated with physical specialization of cells, physiology and reproduction. All these processes are dependent on growth conditions. If there is any deviation from optimal growth condition, all these processes will be affected which results in reduced growth of plants and their development. Plant growth and development are inhibited by MP stress through two main mechanisms: (i) direct impacts resulting from the physical blockage and (ii) indirect effects due to a decrease in soil productivity.

Seed germination is the initial phase of plant development. Presence of MPs in soil blocks the pores in seed capsule. This leads to reduction in water uptake and imbibition. Reduction in imbibitions leads to lower rate of seed germination. De silva et al. (2022) employed biospeckle optical coherence tomography (BOCT) technique and found that polyethylene MPs reduced seed germination by decline in internal biological activity, likely caused by physical clogging of seed pores, or may be due to cytotoxicity induced by MPs during seed germination (Bao et al., 2022) or due to plasticizers present in MPs (Balestri et al., 2019). The leachates produced during the decomposition of MPs also hinder seed germination.

Roots are one of the important plant parts that are affected by MP stress. Occurrence of MPs in soil leads to accumulation in roots and this restricts root activity, growth and elongation. MPs also block the root pores. Especially the sharp-edged MPs physically block the soil pores and reduce root growth which inturn reduce water and nutrient uptake. The effect of microplastics on root growth is also affected by the charge groups on their surface for example, functional groups like -NH_2 and -SO_3H decreased root and shoot biomass and root volume. The -NH_2 functional group has a positive charge that makes it easy for it to occupy binding sites and for the cell wall to absorb it. This prevents cationic elements like K^+, Ca^{+2} or H^+ and Pb^{+2} from adhering to roots (Song et al., 2023) (Fig. 8.7).

By transpirational pull, MPs enter into the shoots of plants via xylem vessels. From there, they enter leaves. MPs have negative impact on growth of leaves and shoots. They reduce photosynthesis of leaves. MP pollution reduces plant yield by reducing growth and development. Many studies have reported that there was reduction in

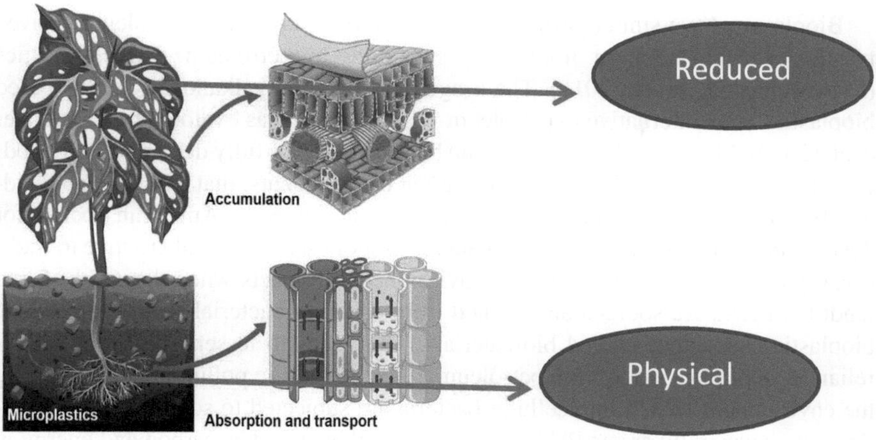

Fig. 8.7 Effect of microplastic pollution on plant roots and shoots

rice grain yield in MP contaminated soil by approximately 20–30%. A number of important physiological processes, including photosynthesis, redox regulation, ionic homeostasis and hormone regulation, are altered in conjunction with the loss in physical growth that MP causes. Due to their sensitivity to stressful situations, these processes can be altered, which will reduce crop development.

Microplastic stress also affects the plant growth indirectly by altering many soil processes. The presence of MPs leads to alteration of soil bulk density, structure. It affects the nutrient transformations, which adversely affects the nutrient uptakes by plants. The soil enzyme activity and microbes population is also reduced, which affects the nutrient availability and uptake. Reduction in nutrient uptake leads to reduction in crop yields.

8.1.12 Possible Solution

Significant amounts of microplastics are present in the soil environment. Reducing the plastic inputs should be treated as top priority at global level. Prioritizing plastic recycling through different methods can also serve as a possible solution. Use of biodegradable plastics is the best option to curb the problem. Biodegradable or bioplastic (BP) mulch films, initially synthesized in the 1970s, are now seen as a more environmentally friendly substitute for plastic polyethylene mulch. According to American Society for Testing and Materials (ASTM), biodegradable plastic is a naturally occurring substance, that, in aerobic environments, can be broken down by microorganisms and transformed into carbon dioxide (CO_2), water (H_2O) and biomass. These biodegradable plastics (bioplastics) can be easily degraded by enzyme activity of microorganisms.

Bioplastics have similar structure and properties like that of petroleum derived plastics, so they can be used as a substitute to petroleum derived plastics (Mascarenhas & Aruna, 2017). The usage of polyhydroxyalkanoates (PHA)-based bioplastics as an alternative to petroleum-based plastics was explored by Meereboer et al. (2020). PHA-based bioplastics, can be used to create fully decomposable products that break down in both soil and marine environments, that exhibit biodegradability in both anaerobic and aerobic conditions, as defined by American Society for Testing and Materials standards. This serves as more sustainable alternative to traditional plastics, which have negative environmental impacts when disposed of into landfills. PHAs are sourced and refined from different bacterial strains to produce bioplastics and other related biomaterials. This leads to lessening of the world's reliance on plastics made from petroleum, reducing plastic pollution and protecting the environment. When intracellular bacteria are subjected to some dietary restrictions, they generally create PHAs, which are further stored as carbon and energy in specialized sub-cellular organelles known as carbonosomes (Tanamool et al., 2013). It is generally recognized that the occurrence of an abundant supply of carbon (C) or limiting dietary elements such as nitrogen (N) and phosphorus (P) can induce the formation of polyhydroxybutyrate (PHB) in bacteria. Usually, living organisms frequently have higher C requirements than N. The carbon-to-nitrogen (C:N) ratio dictates how bacteria can utilize an organic material.

8.2 Conclusion

The growing prevalence of microplastics in agricultural soils poses major environmental challenges that threaten soil health and ecosystem functionality. Their sources ranging from plastic mulch to atmospheric deposition contribute to alarming concentrations, which cause major disruption to soil structure, microbial activity and nutrient cycles, ultimately hindering the crop growth. Addressing this issue requires an urgent action on a global scale, prioritizing the reduction of plastic inputs into agricultural soils. Implementing effective recycling strategies and shifting to biodegradable alternatives, such as polyhydroxyalkanoate (PHA)-based bioplastics, can mitigate the adverse impacts of traditional plastics on soil environment. These bioplastics not only serve as a sustainable substitute with similar properties to conventional plastics, but they also promote ecological balance due to their natural degrading ability in various environments. By adopting these measures, we can protect soil health, achieve agricultural sustainability and combat the growing threat of plastic pollution in our ecosystems.

References

Balestri, E., Menicagli, V., Ligorini, V., Fulignati, S., Galletti, A. M. R., & Lardicci, C. (2019). Phytotoxicity assessment of conventional and biodegradable plastic bags using seed germination test. *Ecological Indicators, 102*, 569–580.

Bao, Y., Pan, C., Li, D., Guo, A., & Dai, F. (2022). Stress response to oxytetracycline and microplastic-polyethylene in wheat (*Triticumaestivum* L.) during seed germination and seedling growth stages. *Science of the Total Environment, 806*, 150553.

Brennecke, D., Duarte, B., Paiva, F., Caçador, I., & Canning-Clode, J. (2016). Microplastics as vector for heavy metal contamination from the marine environment. *Estuarine, Coastal and Shelf Science, 178*, 189–195.

Bullard, J., Ockelford, A., McKenna Neuman, C., & O'Brien, P. (2020). Wind erosion of microplastics from soils: implications for microplastic dispersal and distribution. In: EGU General Assembly; 4–8 May 2020; Vienna (p. 7847).

Chen, M., Cao, M., Zhang, W., Chen, X., Liu, H., Ning, Z., Peng, L., Fan, C., Wu, D., Zhang, M., & Li, Q. (2024). Efect of biodegradable PBAT microplastics on the C and N accumulation of functional organic pools in tropical latosol. *Environment International, 183*, Article 108393.

Chen, Y., Gao, B., Yang, Y., Pan, Z., Liu, J., Sun, K., & Xing, B. (2022). Tracking microplastics biodegradation through CO_2 emissions: Role of photoaging and mineral addition. *Journal of Hazardous Materials, 439*, Article 129615.

Cristina, R. C., Rebeca, M. F., Marola, S. Y., & Xosé-Antón, Á. S. (2023). Leaching and bioavailability of dissolved organic matter from petrol-based and biodegradable plastics. *Marine Environmental Research, 176*, Article 105607.

De Silva, Y. S. K., Rajagopalan, U. M., Kadono, H., & Li, D. (2022). Effects of microplastics on lentil (*Lens culinaris*) seed germination and seedling growth. *Chemosphere, 303*, Article 135162.

De Tender, C., Devriese, L. I., Haegeman, A., Maes, S., Vangeyte, J., Cattrijsse, A., Dawyndt, P., & Ruttink, T. (2017). Temporal dynamics of bacterial and fungal colonization on plastic debris in the North Sea. *Environmental Science & Technology, 51*(13), 7350–7360.

Esders, E. M., Georgi, C., Babel, W., & Thomas, C. K. (2022). Quantitative detection of aerial suspension of particles with a full-frame visual camera for atmospheric wind tunnel studies. *Aerosol Science and Technology., 56*(6), 530–544.

Gasperi, J., Wright, S. L., Dris, R., Collard, F., Mandin, C., Guerrouache, M., Langlois, V., Kelly, F. J., & Tassin, B. (2018). Microplastics in air: Are we breathing it in? *Current Opinion in Environmental Science and Health, 1*, 1–5.

Geyer, R., Jambeck, J. R., & Law, K. L. (2017). Production, use, and fate of all plastics ever made. *Science Advances, 3*(7), Article e1700782.

Hu, X., Gu, H., Sun, X., Wang, Y., Liu, J., Yu, Z., Li, Y., Jin, J., & Wang, G. (2023). Distinct influence of conventional and biodegradable microplastics on microbe-driving nitrogen cycling processes in soils and plastispheres as evaluated by metagenomic analysis. *Journal of Hazardous Materials, 451*, Article 131097.

Huang, Y., Liu, Q., Jia, W., Yan, C., & Wang, J. (2020). Agricultural plastic mulching as a source of microplastics in the terrestrial environment. *Environmental Pollution, 260*, Article 114096.

Imhof, H. K., Ivleva, N. P., Schmid, J., Niessner, R., & Laforsch, C. (2013). Contamination of beach sediments of a subalpine lake with microplastic particles. *Current Biology, 23*, R867–R868.

Jiang, X., Chang, Y., Zhang, T., Qiao, Y., Klobučar, G., & Li, M. (2020). Toxicological effects of polystyrene microplastics on earthworm (*Eiseniafetida*). *Environmental Pollution, 259*, Article 113896.

Jin, Y., Xia, J., Pan, Z., Yang, J., Wang, W., & Fu, Z. (2019). Polystyrene microplastics induce microbiota dysbiosis and inflammation in the gut of adult zebrafish. *Environmental Pollution, 235*, 322–329.

Khan, T. F., & Sikder, A. H. F. (2024). Microplastic can decrease enzyme activities and microbes in soil. *Open Journal of Soil Science, 14*(01), 1–12.

Li, J., Yu, S., Yu, Y., & Xu, M. (2022). Effects of microplastics on higher plants: A review. *Bulletin of Environmental Contamination and Toxicology, 109*(2), 241–265.

Lian, Y., Shi, R., Liu, J., Zeb, A., Wang, Q., Wang, J., Yu, M., Li, J., Zheng, Z., Ali, N., & Bao, Y. (2024). Effects of polystyrene, polyethylene, and polypropylene microplastics on the soil-rhizosphere-plant system: Phytotoxicity, enzyme activity, and microbial community. *Journal of Hazardous Materials, 465*, Article 133417.

Liang, Y., Lehmann, A., Yang, G., Leifheit, E. F., & Riling, M. C. (2021). Effects of Microplastic fibers on soil aggregation and enzyme activities are organic matter dependent. *Frontiers in Environmental Science, 9*, Article 650155.

Liu, X. E., Li, X. G., Hai, L., Wang, Y. P., Fu, T. T., Turner, N. C., & Li, F. M. (2014). Film mulched ridge-furrow management increases maize productivity and sustains soil organic carbon in a dryland cropping system. *Journal of Soil Science Society of America, 78*, 1434–1441.

Liu, Y., Huang, Q., Hu, W., Qin, J., Zheng, Y., Wang, J., Wang, Q., Xu, Y., Guo, G., Hu, S., & Xu, L. (2021). Effects of plastic mulch film residues on soil-microbe-plant systems under different soil pH conditions. *Chemosphere, 267*, Article 128901.

Lozano, Y. M., Aguilar-Trigueros, C. A., Onandia, G., Maaß, S., Zhao, T., & Rillig, M. C. (2021). Effects of microplastics and drought on soil ecosystem functions and multifunctionality. *Journal of Applied Ecology, 58*(5), 988–996.

Mao, Y., Li, H., Gu, W., Yang, G., Liu, Y., & He, Q. (2020). Distribution and characteristics of microplastics in the Yulin River, China: Role of environmental and spatial factors. *Environmental Pollution, 265*, Article 115033.

Mascarenhas, J., & Aruna, K. (2017). Screening of Polyhydroxyalkonates (Pha) accumulating bacteria from diverse habitats. *Journal of Global Biosciences, 6*(3), 4835–4848.

Meereboer, K. W., Misra, M., & Mohanty, A. K. (2020). Review of recent advances in the biodegradability of polyhydroxyalkanoate (PHA) bioplastics and their composites. *Green Chemistry, 22*(17), 5519–5558.

Panno, S. V., Kelly, W. R., Scott, J., Zheng, W., McNeish, R. E., Holm, N., Hoellein, T. J., & Baranski, E. L. (2019). Microplastic contamination in karst groundwater systems. *Ground Water, 57*, 189–196.

Peng, J., Wang, J., & Cai, L. (2017). Current understanding of microplastics in the environment: Occurrence, fate, risks, and what we should do. *Intergrated Environmental Assessment and Management, 13*(3), 476–482.

Plastics Europe, Plastics–The Facts 2019. An Analysis of European Plastics Production, Demand and Waste Data, Brussels, Belgium.

Prata, J. C., da Costa, J. P., Lopes, I., Duarte, A. C., & Rocha-Santos, T. (2020). Environmental exposure to microplastics: An overview on possible human health effects. *Science of the Total Environment, 702*, Article 134455.

Qi, Y., Beriot, N., Gort, G., Lwanga, E. H., Gooren, H., Yang, X., & Geissen, V. (2020). Impact of plastic mulch film debris on soil physicochemical and hydrological properties. *Environmental Pollution, 266*, Article 115097.

Rezaei, M., Riksen, M. J. P. M., Sirjani, E., Sameni, A., & Geissen, V. (2019). Wind erosion as a driver for transport of light density microplastics. *Science of the Total Environment, 669*, 273–281.

Rillig, M. C., & Lehmann, A. (2020). Microplastic in terrestrial ecosystems. *Science, 368*(6498), 1430–1431.

Rong, L., Wang, Y., Meidl, P., Wang, L., & Sun, H. (2023). Microplasticsafect soybean rhizosphere microbial composition and function during vegetative and reproductive stages. *Ecotoxicology and Environmental Safety, 252*, Article 114577.

Ryan, P. G., Moore, C. J., van Franeker, J. A., & Moloney, C. L. (2009). Monitoring the abundance of plastic debris in the marine environment. *Philosophical Transactions of the Royal Society—B, 364*, 999–2012.

Sey, B. K., Manceur, A. M., Whalen, J. K., Gregorich, E. G., & Rochette, P. (2008). Small-scale heterogeneity in carbon dioxide, nitrous oxide and methane production from aggregates of a cultivated sandy-loam soil. *Soil Biology and Biochemistry, 40*, 2468–2473.

Sharma, V. K., Ma, X., Guo, B., & Zhang, K. (2021). Environmental factors-mediated behavior of microplastics and nanoplastics in water: A review. *Chemosphere, 271*, Article 129597.

Shi, Z., Xiong, L., Liu, T., & Wu, W. (2022). Alteration of bacterial communities and co-occurrence networks as a legacy effect upon exposure to polyethylene residues under field environment. *Journal of Hazardous Materials, 426*, Article 128126.

Song, Z., Zhao, X., Dong, Y., Bai, L., Wang, S., & Gao, M. (2023). Effects of polystyrene nanoplastics with different functional groups on the accumulation and toxicity of Pb on dandelion. *Chemosphere, 310*, Article 136874.

Tanamool, V., Imai, T., Danvirutai, P., & Kaewkannetra, P. (2013). Biopolymer generation from sweet sorghum juice: Screening, isolation, identifcation, and fermentative polyhydroxyalkanoate production by *Bacillus aryabhattai. Turkish Journal of Biology, 37*(3), 259–264.

Xiao, S., Cui, Y., Brahney, J., Mahowald, N. M., & Li, Q. (2023). Long-distance atmospheric transport of microplasticfibres influenced by their shapes. *Nature Geoscience, 16*, 863–870.

Zhang, G., Zhang, F., & Li, X. (2019). Effects of polyester microfibers on soil physical properties: Perception from a field and a pot experiment. *Science of the Total Environment, 670*, 1–7.

Zhang, G. S., & Liu, Y. F. (2018). The distribution of microplastics in soil aggregate fractions in southwestern China. *Science of the Total Environment, 642*, 12–20.

Zhang, K., Xiong, X., Hu, H., Wu, C., Bi, Y., Wu, Y., Zhou, B., Lam, P. K. S., & Liu, J. (2017). Occurrence and characteristics of microplastic pollution in Xiangxi Bay of three Gorges Reservoir, China. *Environmental Science and Technology, 51*, 3794–3801.

Zhang, Y., Kang, S., Allen, S., Allen, D., Gao, T., & Sillanpää, M. (2020). Atmospheric microplastics: A review on current status and perspectives. *Earth-Science Reviews, 203*, 103118–103132.

Zhang, Y. G., Yang, S., Fu, M. M., Cai, J. P., Zhang, Y. Y., Wang, R. Z., Xu, Z. W., Bai, Y. F., & Jiang, Y. (2015). Sheep manure application increases soil exchangeable base cations in a semi-arid steppe of Inner Mongolia. *Journal of Arid Land, 7*(3), 361–369.

Zhao, W., Zhang, J. B., Mueller, C., & Cai, Z. C. (2018). Effects of pH and mineralisation on nitrification in a subtropical acid forest soil. *Soil Research, 56*, 275–283.

Zhou, X., Xiao, C., Zhang, B., Chen, T., & Yang, X. (2024). Effects of microplastics on carbon release and microbial community in mangrove soil systems. *Journal of Hazardous Materials, 465*, Article 133152.

Chapter 9
Bioremediation Techniques of Microplastic Waste Management

Anjali Mishra, Rajak Preeti Phoolchandra, Ekta Verma, Neelam Atri, and Shivam Yadav

Abstract Plastics' unique physical and chemical properties have rendered them indispensable components of modern life and technology. Plastic waste does not easily break down in the environment, and the process of decomposition is slow. These plastics undergo fragmentation into microplastics (MPs) and nanoplastics (NPs). The slow degradation rate of MPs contributes to their pervasive contamination across various environmental ecosystems, including terrestrial and aquatic systems, posing detrimental effects on both ecosystems and human health. Among plastic pollutants, microplastics, which are plastic fragments smaller than 5 mm, have gained attention because of their potential harm to living organisms. This chapter explains different types of plastics, their properties, and factors that affect how they degrade. It discusses how organisms interact with microplastics and the harmful effects these particles have on living beings. To manage plastic waste, various disposal methods, such as incineration, landfilling, recycling, and transformation, have been globally implemented. However, due to their harmful effects, the utilization of bacterial, fungal, and microalgal strains for the elimination of environmental microplastics through eco-friendly bioremediation techniques has proven highly effective. Bioremediation and biodegradation are explicated herein. Additionally, new methods for biodegradation, like enzyme-based techniques, are being introduced. Certain types of fungi and bacteria, along with their enzymes, such as oxidoreductases and hydrolases, play a significant role in breaking down plastics. This chapter emphasizes diverse bioremediation techniques and their underlying mechanisms, as well as successful biotechnological strategies, including biostimulation, bioaugmentation, and enzymatic biodegradation, which are crucial for enhancing the efficacy of microplastics degradation in contaminated environments. The use of microbial communities and

A. Mishra · R. P. Phoolchandra · S. Yadav (✉)
Department of Botany, University of Allahabad, Prayagraj, India
e-mail: sy1510bhu@gmail.com

E. Verma
Post Graduate Department of Botany, Magadh University, Bodh-Gaya, India

N. Atri
Department of Botany, MMV, BHU, Varanasi, India

V. Kumar et al. (eds.), *Microplastics Pollution*,
https://doi.org/10.1007/978-3-031-99774-7_9

specially modified microorganisms with their enzymes are also effective ways to manage plastic pollution.

Keywords Microplastics · Bioremediation · Phytoremediation · Bioaugmentation · Biodegradation

9.1 Introduction

The twentieth century has been referred to as the "polymer age" or the "age of plastic." Plastic is a synthetic polymer made from crude oil, natural gas, and coal (nonrenewable resources) during the industrial revolution and subsequent developments in chemical and material sciences. Petroleum-based plastics are characterised by their light weight, flexibility, strength, durability, insulating capacity, cheap production expenses, and ease of transportation. The widespread usage of synthetic plastics in our toys, cars, construction materials, clothing, dishes, and packaging had a greater impact on our lives than any technological innovation made during this time period. The existence of natural polymers, such as rubber and silk, does not contribute to "white pollution" because they naturally deteriorate due to the inclusion of nitrogen and oxygen in their polymeric structure, enabling the polymer to biodegrade (La Mantia, 2002). According to the plastics trade organization, global plastic manufacturing in the 1950s was two million tonnes plastic production, which exceeded 353 million tonnes in 2019 and is predicted to double by 2040 (OECD, 2022a). By 2020, 367 million tonnes were produced, with 40.5% for packaging, 20.4% for construction, and 32.3 million tonnes for the motor vehicle industry (OECD, 2022b). Several mechanisms, including thermal, catalytic, mechanical, chemical, ozonation, and photo-oxidation, disintegrate large polymeric polymers into "micro- and nano-sized particles." Microplastics (hereafter MPs) are plastic particles and fragments measuring less than 5 mm in size (Thompson et al., 2004). According to the Swiss non-profit EA Earth Action, China is the largest microplastics polluter of aquatic bodies, releasing 787,069 tonnes, with India ranking second at 391,879 tonnes. A rapid rise in population has resulted in widespread and inefficient disposal practices, which contributed to both environmental contamination and waste management challenges. MPs reach the environment as primary particles or as byproducts of larger plastic products. Small sized MPs aid in the transportation of high concentration of organic and inorganic pollutants which persist in the environment and consequently may impact the ecosystem and poses environmental risks due to their interactions with high chemical contaminants (Sridharan et al., 2021). MPs also change the texture of the soil by affecting soil porosity and bulk density, creating cracks on soil surface (de Souza Machado et al., 2019; Wang et al., 2019). The accumulation of MPs in plankton and marine mammals causes internal blockages, cell damage, and disruptions to food webs and reproductive activity reduces leading to a decline in biodiversity (Mathalon, 2014; Barrick et al., 2021). According to Du et al. (2021), biological, thermal, and photocatalytic degradation approaches are helpful for the destruction of

MPs, and further studies are necessary to determine their effectiveness in ensuring a healthy environment (Andrady, 2011). Presently, many researches are ongoing for investigating the biodegradation of MPs by living entities like bacteria, fungi, algae, and other genetically modified organisms. Various advanced oxidation processes (AOPs) are being explored for their potential to degrade MPs through depolymerization, breaking them down into oligomers, dimers, and monomers (Haider et al., 2019). An eco-friendly approach used for MPs cleanup is bioremediation, which involves bacteria, fungi, or plants. This multidisciplinary approach includes several key steps: MPs evaluation, screening and studying microorganisms, and in-situ toxicity evaluation. Microbial remediation process converts MPs into substrate for biotransformation using chemical depolymerization. To learn more about the microbial degradation of various plastic monomers, more research is required (Wierckx et al., 2018). This chapter focuses on the study of MPs and recent advancements in various bioremediation techniques. Additionally, it analyzes further investigations and strategies to enhance knowledge of the challenges associated with MPs biodegradation.

9.2 Microplastics

The US National Oceanic and Atmospheric Administration (NOAA) suggested the size of MPs ranges from 1 nm to < 5 mm (Andrady, 2011; Arthur et al., 2009). Thompson et al. (2004) referred to these synthetic polymers forming tiny plastic particles as MPs. These polymers, including polyethylene (PE), acrylonitrile butadiene styrene (ABS), polypropylene (PP), polystyrene (PS), polyvinyl chloride (PVC), and polyethylene terephthalate (PET), are often combined with various additives such as antioxidants, adhesives like bisphenol-A (BPA), and flame retardants to enhance the properties of plastics by improving their durability and resistance to mechanical, chemical, and electrical stress. Additionally, they provide protection against environmental factors such as light radiation, mold, and humidity, making plastics more versatile for a wide range of applications (Hahladakis et al., 2018). According to the report by Boucher and Friot (2017), globally seven major sources of microplastics have been identified, including city dust, marine coatings, personal care products, plastic pellets, road markings, synthetic textiles, and tires. Based on their origins, microplastics are classified into two categories: Primary MPs and Secondary MPs. Figure 9.1 illustrates the sources and toxic effects associated with MPs.

9.2.1 Primary Microplastics

Primary microplastics are synthetic polymers of regular shapes ranging from 1 to 5 mm, intentionally and artificially generated for specific applications such as exfoliants in skincare products to remove dead cells, abrasive media for polishing, chemical formulations, and grinding pads (Cole et al., 2013) and air blasting technology

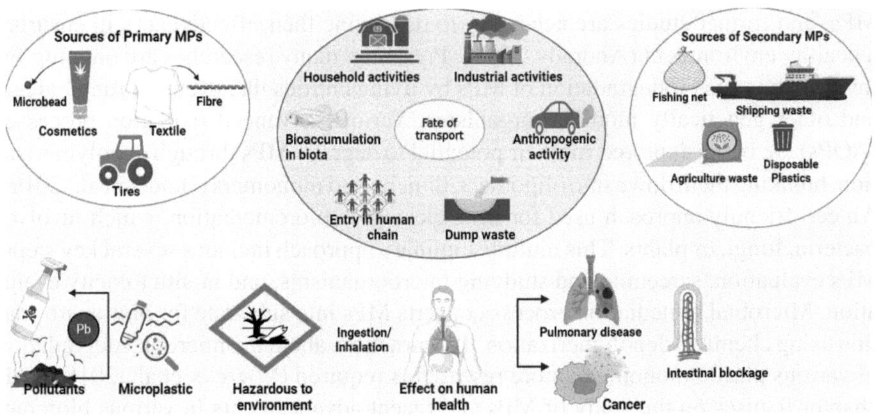

Fig. 9.1 Source and toxicological effect of microplastics

(Haider et al., 2019). Car tires, road markings with paint, artificial grass, plastic materials, and man-made fibres are the most frequent sources of MPs. The tiny fragments are released by tire wear caused by road friction and then, carried by rainwater, enter the drainage system, eventually ending up in aquatic environments. Microbeads, which are less than 2 mm in size and made of materials such as PE, PP and PS, were patented in the 1980s. Polyether sulfone (PES), polyamide (PA) and PET MPs are derived from textile-derived primary plastics (Gong & Xie, 2020). Typically, microbeads flow into sewage systems along with wastewater because of their tiny size, inability to dissolve in water, and slow breakdown, which is commonly used in agriculture. Synthetic fibres from clothing washing release over 1900 microfibres per garment into wastewater, which sewage treatment plants fail to remove, leading to environmental MP pollution.

9.2.2 *Secondary Microplastics*

The plastic particles formed by the breakdown or scattering of MPs due to different environmental factors or human activities are referred to as secondary microplastics (Andrady, 2011; Gigault et al., 2018). Secondary microplastics are the broken-down remnants of larger plastics, mainly resulting from environmental processes such as biodegradation, photodegradation, oxidation caused by heat, decomposition due to high temperatures, and chemical reactions like hydrolysis (Sharma & Chatterjee, 2017). MPs in aquatic environments originate from the disintegration of larger plastic items caused by exposure to sunlight (Benítez et al., 2013; Cooper & Corcoran, 2010), leading to partial fragmentation rather than complete mineralization. These fragments, along with toxic additives like BPA and phthalates (Crain et al., 2007; Kwon et al., 2014) persist and pose risks to marine life and ecosystems. Secondary

MPs are largely contributed by vessels, businesses involved in recreational activities like fishing, tourism, fish farming, and marine operations. Fishing trash, including plastic nets, ropes, and styrofoam, is a major source of marine MPs contamination due to its widespread use and disposal. Modern agricultural methods employ a variety of plastic polymers in things such as greenhouse covers and mulching sheets, which can breakdown into secondary microplastics owing to weathering, abrasion, or microbial action. These MPs can remain in soil, be moved by wind or water erosion, and infiltrate aquatic habitats. They get into agro-ecosystems directly through sewage sludge, irrigation water, or air deposition, and indirectly through the breakdown of plastic leftovers (Gao et al., 2020; Zhou et al., 2020b).

9.2.3 Biological and Toxicological Effects

MPs have an immense effect on aquatic organisms and ecosystems, affecting both small species and larger animals. Marine animals frequently ingest the smaller-sized MPs instead of their natural food. Smaller MPs are frequently mistaken for their natural diet by marine animals because of their similar size, shape, or appearance. These MP accumulates such as in primary producers and subsequently impact marine biodiversity and ecological processes. Furthermore, MPs promote malnutrition in marine fauna by leading to intestinal blockage, which prevents the absorption of sufficient nutritional value. Malnutrition can impair reproductive capacities and contribute to population reduction due to deficiencies in assimilated food (Romeo et al., 2015; Mkuye et al., 2022; Li et al., 2022). MPs have also been found in studies to influence marine animal physiology, including oxidative stress, inflammatory gene expression, pro-apoptotic regulation, the immune system and the central nervous system, via altered gene regulation. According to Lei et al. (2018a; 2018b), exposure to 70-μm particles of PA, PVC, PE, and PS caused gastrointestinal toxicity in zebrafish and nematodes. Microalgal populations' distribution and growth rates vary depending on the presence of MPs. Khoironi et al. (2019) reported a similar finding for *Spirulina* sp. High MP concentrations, along with low light intensity, damage the algae's cell wall and photosynthetic machinery (Khoironi et al., 2019).

The agricultural soil becomes integrated with a large number of MPs due to irrigation, sewage sludge, and atmospheric deposition (van den Berg et al., 2020) which changes the soil structure (Guo et al., 2020), water-holding capacity, and microbial activity, making the soil less suitable for microorganisms while also affecting nutrient cycling. Also, MPs can accumulate in the soil, be absorbed by plants, and transported up to the food web and into other ecological sectors (Boots et al., 2019). Assessing soil microbial activities, plant biomass, soil quality, and tissue components of spring onion can be beneficial for determining the influence of MPs on plants (de Souza Machado et al., 2019). Using surgical face masks during the global pandemic of COVID-19 contributes to high environmental plastic pollution and creates many health-related issues and other chronic effects on living organisms (Fadare & Okoffo, 2020).

The influence of contact between humans and MPs can be explained by three common pathways including skin contact, inhalation, and food intake through synthetic fibres of MPs. A recent study shows that the traces of MPs present in sugar are 0.44 MPs/gm, in salt 0.11 MPs/gm, in alcohol 0.03 MPs/gm, and in bottled water 0.09 MPs/gm (Cox et al., 2019). Primary producers have the efficacy to accumulate MPs and transport them to another trophic level at the rate of 80 g per day through biomagnification (Enyoh et al., 2019). A study conducted by Liao and Colleagues (2019), demonstrated that, the non-degradable (PVC, PE, PP, and PS) and biodegradable (PLA) microplastic, when degraded release hazardous chemicals like plasticizers, polychlorinated biphenyls (PCBs), and heavy metal including Cr (VI) and Cr (III) in surrounding (Liao et al., 2019). Wind, agricultural runoff, and discharge from wastewater facilities contribute to MPs entering the human body through inhalation, leading to health problems like respiratory irritation, cellular toxicity, and inflammation (Dong et al., 2020). According to one study, 50 nm-sized PS particles exhibit hydrophobicity, shape, and chemical composition, as well as toxicity to genetic material and lung epithelial cells and macrophages (Calu-3 and THP-1) (Paget et al., 2015). MPs entering the body via dermal absorption through products with harmful additives, such as cosmetics or detergents, may trigger immune system disruptions, oxidative stress, hormonal imbalances, genetic damage, and even cancer development (Vethaak & Legler, 2021).

9.3 Common MPs Waste Disposal Methods

Since microplastic is present in every environmental zone, it has been a contentious topic of discussion in recent years. Because of its increased usage and low degradation capacity, many strategies are in consideration to get rid of plastic pollution like breaking it down and transforming it to any other useful product by employing different techniques. The approaches like landfilling, incineration, recycling, transformation and bioremediation are the commonly used method to eliminate MPs from environment.

9.3.1 Landfilling

Landfills are designated sites or facilities for disposing of solid waste and waste residuals in controlled and monitored manner (Zabalata, 2008). Landfills are characterized as the most ancient and common waste disposal method around the world. As landfilling is a common way of disposing waste, but it comes under some criticism for not being safe and potentially responsible for health risks and numerous environmental impacts. Various researchers and regulators point objections for disposing of different types of waste in landfilling (Abdel-Shafy & Mansour, 2018). Experts of Waste management suggested that, landfills have an adverse impact on the environment. There are several online resources available which discusses, the environmental impacts of landfill activities (Eriksson & Finnveden, 2009). Because of the improper sorting at the beginning of waste collection causes the plastic waste to accumulate in landfills, where it could be exposed to deleterious conditions that harm the ecosystem and human health (He et al., 2019). Usually, all of the solid waste generated from cosmopolitan cities, along with plastic waste is being disposed in landfills. Of all plastic waste generated, 65% is disposed of in landfills (Sarker, 2011). In the landfills plastic waste takes thousands of years to break down that makes MP or plastic hazardous for soil fertility as well as environmentally unsafe for mankind (Sarker, 2011; Tansel & Yildiz, 2011). Presence of limited oxygen in landfills minimise the degradation of polythene (Tollner et al., 2010). Once waste is discarded in a landfill, it goes through a different kind of physical, chemical, and biological changes that lead to the formation of highly polluted water called leachate. Due to various reasons like kind of waste, decomposition level, weather conditions, landfill site features, economic factors and landfilling technologies, different type of leachate form in the environment (Sundt et al., 2014). Plastics of various compositions present in landfills are a significant source of generating hazardous gaseous emissions and leachate contamination (Zhang et al., 2004). Toxic compounds generated due to plastic waste includes benzene, ethylbenzene, toluene, trimethylbenzenes, xylenes, and BPA which are mentioned in studies of many researchers (Tsuchida et al., 2011; Xu et al., 2011). After waste decomposition in landfills, it produces various detrimental landfill gases and greenhouse gases which disrupts the ozone layer and responsible for global warming and climate change (Modin et al., 2012). Landfills are significant contributors of methane, accounting for 45% of the UK's methane emissions according to Osazee (2021). As per data from 2010, the USA ranked first among ten nations in methane emissions, totalling 129.7 mmtCO_2-eq (GMIL, 2011).

From different period of time, plastic pollutants in dumpsites breaks down mechanically or physically, forming MPs (Rodrigues et al., 2020; Vithanage et al., 2021) that diffuses into the surroundings, increasing the quantity of MPs (Alimi et al., 2018; Imhof et al., 2013; Sundt et al., 2014). As a result, landfills are a significant source of microplastic pollution (Sun et al., 2021; Xu et al., 2020). Although landfills are included in linear waste management, they also contribute to integrated management by handling end-of-life residues from recycling and incineration processes (Singh et al., 2022). Nevertheless, landfills should only be considered as a final option because of their greater negative effects on the environment, higher chances of pollution, resource depletion, and need for land (Liu et al., 2018). As a result, landfills should only accept trash that comes from recycling, feedstock production, or energy generation processes. Landfill mining provides a different option for bringing landfilled materials back into the circular economy, whether through recycling (such as metals) or converting waste into energy (such as plastics) (Burlakovs et al., 2019). According to Zhou et al. (2018), the number of plastics discarded in Chinese landfills has the potential to produce 38% of the country's yearly consumption of energy, if recycled. The use of waste-to-energy may be included in cleaning projects (such as recycling of landfill plastics) and comprehensive waste management, primarily by enhancing the gasification processes of that region, with calorific values determined by thermal-gravimetric analysis (Burlakovs et al., 2019). However, the capabilities of this technology are still restricted.

9.3.2 Incineration

The incineration method is the second most frequently used way to get rid of plastic waste according to Zhang et al. (2004). Sarker (2011) stated that burning 25% of plastic waste addresses landfill issues like space and soil fertility, but also releases harmful gases like carbon dioxide, heavy metals, PAHs, PCBs, toxic free radicals, furans, and dioxins into the environment (Astrup et al., 2009; Shen et al., 2021). The residents of nearby areas are affected by different illnesses due to these harmful gases released from the incineration sites. Therefore, neither landfill nor incineration methods are eco-friendly. The incineration process is a crucial element of the circular economy. After incinerating the waste, recovery of energy is vital component in both developed and developing countries (Jeswani & Azapagic, 2016; Medina Jimenez et al., 2017). Because of combustion, fly ash and bottom ash are produced and diffused in the environment easily. Recent studies indicates that plastic waste is not completely eradicated through incineration and may be present as in the form of residues and waste ash with microplastics (Shen et al., 2021; Yang et al., 2021). Assessing MPs in various waste management systems could offer important insights into the rising levels of microplastics and the potential exposure risks for organisms. Compared to landfilling, incineration technology can be beneficial for decreasing the mass and volume of municipal solid waste (MSW), needing less land and efficiently recovering electrical and thermal energy (Quina et al., 2014; Shen et al., 2020a; Xing et al., 2020). With limited oxygen availability, techniques like incineration and pyrolysis are generally used in small town and township for centralized treatment of rural waste. Because of the complex composition of plastic product and variability in the incineration techniques, the MSW is not fully eliminated and leads to generation of MPs and other toxic pollutants (Yang et al., 2021). Apart from this burning plastic in an open environment can form habitats for different microbes and release harmful chemicals that can potentially harm the ecosystem.

Like landfilling, incineration is also considered as an important disposing method of plastic waste around the globe (Geyer et al., 2017; Yang et al., 2020a; Ye et al., 2020). Not all the plastic present in incineration process burn completely, and may remain in the bottom or could be carried to the environment through recycling or improper disposal in the form of MPs (He et al., 2019). For a safe environment, it is essential to evaluate the quality of MPs originating from bottom ash. Due to the composition of different compounds in plastics of MSW, can potentially release different heavy metals after incineration of the waste (Yang et al., 2020b). Heavy metals have the potential to enhance the range of pollution by entering in the environment through surface runoff. Currently, the bottom ash form after incineration is either disposed of or kept on the incineration site, which could lead to the generation of secondary pollution or health-related issues. Some studies provide information about heavy metal contamination through MP pollution in different environments (Deng et al., 2020; Koelmans et al., 2016; Li et al., 2020). According to certain research, it is believed that there might be some link between MP pollution and heavy metal contamination in certain locations like water, soil, and sediments. There is limited research available on MPs and heavy metals found in the leachate through bottom ash, formed by pyrolysis and incineration of urban waste.

9.3.3 Recycling

Out of all the plastic waste produced, just 10% of is being recycled worldwide, according to Sarker (2011). Among all municipal waste generated in India, plastic accounts approximately 9.22% (Central Pollution Control Board, 2009), with 60% of it being able to be recycled (Singh, 2013). Still multiple factors impede the effectiveness of plastic recycling, resulting in the production of inferior, degraded materials (Awaja & Pavel, 2006). There are different contaminants present in plastic waste like the additives which make the recycling process more difficult. Plastic recycling needs the individual processing of diverse plastic varieties. For manual segregation of distinct plastic types present in solid waste, the recycling process demands a substantial allocation of human labor resources. Although recycling is a more desirable alternative to landfilling and incineration for plastic waste management, it is a relatively costly and inefficient process. Therefore, it is need to explore the novel degradation techniques to address the challenges associated with plastic waste disposal (Yamada-Onodera et al., 2001; Zhang et al., 2004). The process of recycling depends on different steps including characterization of plastic by quality, degrading capacity, transforming them into granules, and finally modification into pellets for future recycling (Brown et al., 2023). In industries recycling processes using mechanical friction or similar techniques, can increase the microplastics in wash water, which can be released into the environment as wastewater (Brown et al., 2023). The process of recycling first starts with collection followed by sorting, grinding, extrusion, and finally selling to the manufacturer (Schneider et al., 2015; Walker et al., 2018). Sorting of waste based on the type of plastic possess a challenge and can impact the overall quality of recycled plastics. When good quality plastics are obtained from manufacturers, that termed as primary recycling whereas secondary recycling gets lower quality polymers from contaminated plastics waste. Preferably, recycled plastics should be utilized in applications that are long-lasting and robust (Walker et al., 2018; Ignatyev et al., 2014). The recycling of plastics, particularly from consumer waste, is constrained by expensive recycling procedures compared to inexpensive virgin plastics, deterioration, and impact of plastics restricting their usage and cycles of recycling, and the limited recycling properties of certain plastic products. Typically, recycling plastic does not make financial sense; however, it could be profitable in closed-loop systems (Bernardo et al., 2016). Taking into account the environmental consequences, recycling is the favoured approach for handling waste (Bernardo et al., 2016; Chilton et al., 2010). One of the examples of loss in recycling is that PET and PE plastics uses half of the energy to make new polymers than manufacturing it from recycled plastic (Arena et al., 2003). Nevertheless, incineration may be preferred over recycling due to high organic contamination and limited potential to substitute virgin plastics (Lazarevic et al., 2010).

Recycling can also be limited due to many environmental effects like using non-renewable energy, transporting materials and adding fillers or additives (Burlakovs et al., 2019; Gu et al., 2017). Even with many difficulties, recycling continues to be a focus in waste management. It conserves resources and energy, cuts down on pollution, minimizes landfill usage, generates employment opportunities, boost local economies, diminishes resource imports, and overall benefits the environment. Some plastics cannot be recycled because they are mixed, contaminated, or degraded. However, they can still be used for feedstocks or energy recovery due to various factors such as thermal behaviour, heterogeneity, moisture, impurity content, and ash production (Burlakovs et al., 2019). However, recycling continues to be a vital aspect of sustainable waste management, with advancements in recycling methods and regulations improving its sustainability and efficiency.

9.3.4 Transformation (Recovery)

Wastewater treatment plants (WWTPs) are characterized by the presence of high amounts of MPs which can be valuable if isolated properly and reused for other applications by considering the data on their presence. Various methods are used to manage plastic, but specific techniques for converting microplastics remain impractical. Although there are various methods exist for transformation of plastics, one of them is plastic into fuel (Kunwar et al., 2016; Rajmohan et al., 2020). For instance, turning plastics into fuel can be appropriate for Low density polyethylene (LDPE), High density polyethylene (HDPE), PP, PET, PVC, and PS, enabling energy to be regained through co-pyrolysis. Apart from this, a different type of gas called syngas (a combination of $H_2 + CO$), can be derived from different plastics (LDPE, HDPE, PS, PET, and PP) through catalytic pyrolysis (Saad & Williams, 2016). Plastic can also be used as self-compacting concrete or clay composites in building materials. The global initiative, "Operation Clean Sweep" was carried out by few countries including North America, Europe and the United Kingdom for mitigating the plastic contamination in marine ecosystem. Various operational strategies were guided in this initiative to the plastic production industries for releasing of different polymer particles from production to recycling including resin pellets, flakes and fibres into the open environment (Operation Clean Sweep, 2018). Apart from this, converting discarded MPs or plastic into reusable form could be sustainable and utilize for managing plastic waste.

New technologies like ECONYL© are used to create textiles by recycling nylon waste collected from landfills and oceans, resulting in regeneration of nylon for different uses, completing the circular economy loop. In 2013, an open -source plastic recycling project (Hakken's Precious Plastic Project) was able to produce new plastic products from recycled plastic by employing set of techniques for crushing, melting and mouldings. However, it is important to note that the majority of these options are focused on addressing larger pieces of plastic and have not been designed specifically for managing MPs (Thiounn et al., 2020). Many companies such as Sekisui Chemical Co. in Osaka, (Japan) and Sumitomo Chemical Corp. in Tokyo, (Japan) is commercializing the chemical recycling of MPs and microplastic films via pyrolysis in lab settings (Thiounn et al., 2020). Studies have shown that mixture of three waste streams containing microplastic fibres and other plastic can be changed into char and tarry products at temperatures of 300, 425 and 550 °C by process of slow pyrolysis (Dussud et al., 2018). Due to concerning increase in microplastic production, research has been conducted on transforming MPs into reusable products and energy, in order to mitigate the volatilization of microplastic powders or resin powders into the environment. Many strategies are primarily focuses on converting polyolefins (PP, PS and PE) into production of heat energy through pyrolysis (Aguado et al., 2002). Due to contamination, mixing, composition or degradation many plastics are not recycled and can be used to produce energy by using methods like pyrolysis and incineration (Prata et al., 2019).

9.4 Bioremediation Techniques for Microplastic Waste Management

One of the best eco-friendly approaches is bioremediation, which involves using living organisms such as bacteria, fungi, and plants to remove toxins or pollutants from contaminated places (Dash et al., 2013). The microorganisms use organic compounds as their energy source by breaking down the pollutant and mitigating the toxic effects such as caused by oil spills. Polyethylene terephthalate (PET), a common plastic bottle material is degraded by the bacterium *Ideonella sakaiensis* (Emisha et al., 2024). Plastic mineralization is the process by which natural agents like fungi, bacteria, and algae establish biofilms on plastics, leading to enzymatic decomposition of their structure. This process generates oligomers, dimers, monomers, carbon dioxide, and water, although it may also release stable toxic chemicals that pose risks to the ecosystem. Aerobic environments lead to the production of water and carbon dioxide, while anaerobic conditions result in carbon dioxide and methane (Zhang et al., 2020a; Giacomucci et al., 2020). Bioremediation can lower microplastic levels in the environment (Fig. 9.2).

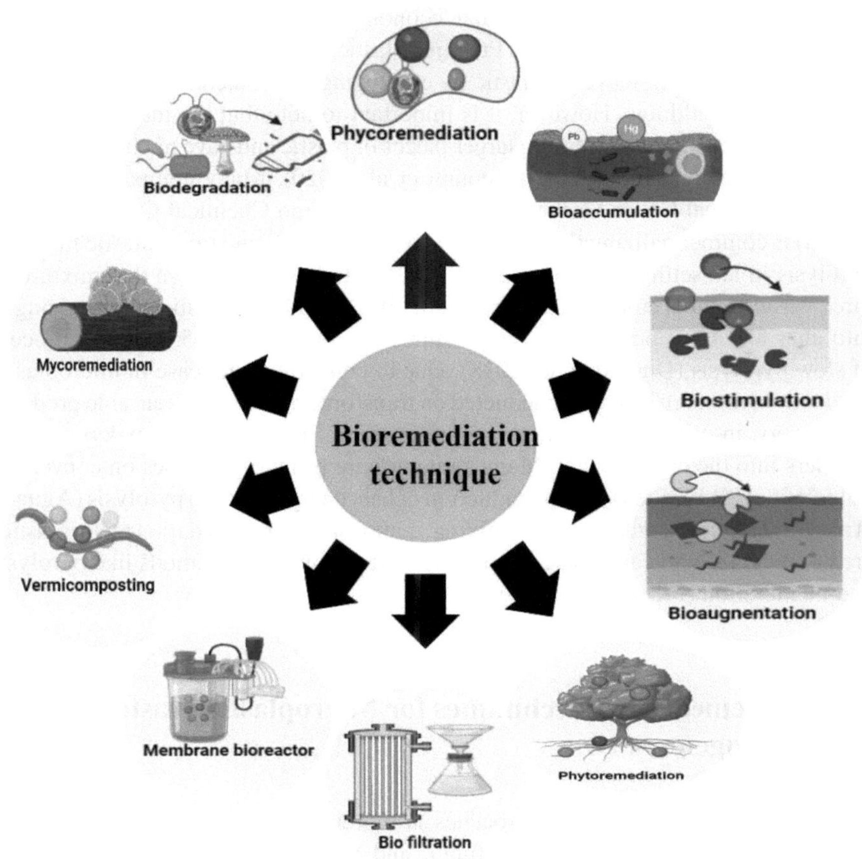

Fig. 9.2 Various bioremediation techniques employed for MPs disposal

9.4.1 Microbial Bioremediation

Hydrophobic polymers enzymatically degrade via microbial (bacteria, actinomycete and fungus) metabolic pathways into water, carbon dioxide, inorganic compounds, and methane, without generating harmful by products (Alshehrei, 2017; Bacha et al., 2021). Microbial breakdown of polymers uses molecular oxygen as a terminal electron acceptor, leading to the production of carbon dioxide and water (Müller, 2005). Anaerobic bacteria use sulphate (SO_4^{2-}), iron (Fe), and nitrate (NO_3^-) as electron acceptors to break down polythene into H_2S, methane, and C residues. When microorganisms attach to polymers, two stages of processes occur: (1) depolymerisation, in which larger organic molecules are broken down by microbes extracellular enzyme activity into small-sized hydrophilic molecules that can pass the bacterial membrane and be used as energy sources, and (2) during mineralisation, small polymers convert into monomers and oligomers of low-molecular weight inorganic molecules (Gu, 2003). Plastic biodegradation potential is influenced by functional groups, linear or branching chain length, polymer density, crystallinity, polymer nature, and toughness. Polyolefins, often known as low density polyethylene, are chemically inert. Microorganisms degrade polyethylene via the carbonyl group produced during oxidation (Awasthi et al., 2017).

Plastic biodegradation consists of three processes: biodeterioration, bio fragmentation, and mineralization.

9.4.1.1 Biodeterioration

Relies on specific type of enzymes secreted by microorganisms to disrupt the polymer surface. This process involves a variety of changes, including physical, chemical, and mechanical properties in the material. The efficacy of biodeterioration is regulated by environmental conditions (Lucas et al., 2008). The bacterial and fungal microorganisms *Phaeosphaeria spartinicola*, *P. halima*, and *Mycosphaerella* species can develop biofilms on plastic polymers, ingest microaplastic particles, and metabolise them for growth and energy which they then metabolize for energy and growth (Kawai et al., 2019).

9.4.1.2 Biofragmentation

Occurs when plastic polymers are reduced into tiny fragments by enzyme action (Kjeldsen et al., 2018). Using oxygenase, oxygen molecules can be transformed into less dangerous alcohol and peroxyl compounds after being eliminated from polymeric carbon chain molecules. The insertion of functional groups like carbonyl or hydroxyl groups leads the polymer to become polar, which can be accomplished by enzymatic activity during free radical oxidation (Ali et al., 2021; Shah et al., 2008).

9.4.1.3 Mineralization

Produces small-sized monomers by microbial enzyme's activity. Some of these may penetrate across semipermeable membranes and be used in biomass production for microbial cell growth and reproduction (Kale et al., 2015), whereas bigger monomers are impermeable to membranes and cannot be assimilated directly. Using biotransformation approach employs specific enzymes produced by microbes to convert non-assimilate monomers into intermediate (Ali et al., 2021). This process also generates CO_2 and CH_4, which can be assimilated or used by microorganisms (Shah et al., 2008).

9.4.2 Key Microorganisms in MPs Bioremediation

9.4.2.1 Fungal Degradation

Fungi are a varied group of microorganisms that are widely spread and have a wide range of reproductive abilities. They have the ability to facilitate oxidation and the transportation of chemicals. Fungi, along with bacteria, have the ability to stick to and feed on microplastics. Fungi can decrease the hydrophobic properties of microplastics by promoting the formation of different chemical bonds between different groups like carboxyl, carbonyl, and ester found in MPs (Harms et al., 2011). Nowadays, research is being conducted on the breakdown of MPs by using fungi in different habitats with some reported advancements (Hyde et al., 1998). Fungi release some enzymes that break down large organic compounds into smaller ones for reabsorption. The utilization of polymers such as plastics as a carbon and electron source is made possible by a powerful enzymatic system, adsorption capacity and natural biosurfactants (Purohit et al., 2018). Lately, there has been a growing interest in utilizing fungi for microplastic biodegradation as well as other purposes. Assessment and research will be in high demand in coming years, to check the degradation property of fungal enzymes working on MPs, which are not degraded easily being resistant to corrosion and oxidation (Yuan et al., 2020). Fungi synthesize certain enzymes like proteases, amylases and lipases which are capable of breaking down organic substances like proteins, starches, and lipids respectively. Some fungi are fully capable of breaking down any type of organic compound. Cellulose, hemicellulose, and pectin are all molecules that can serve as carbon sources for a limited number of species. Fungi effectively break down naturally existing, intricate polymers that are resistant to microbes, like lignin, keratin, and chitin (Anastasi et al., 2012). Various forms of beneficial mycoremediation approaches include: Ligninolytic Fungal Degradation, Brown Rot Fungi (BRF), Mycorrhizal Fungal Degradation (MFD), Soil Fungal Biosorption and White Rot Fungi (WRF) (Anastasi et al., 2012; Periasamy et al., 2019). Hydrolases and oxidoreductases are the main enzyme groups responsible for breaking down plastic in biodegradation processes. Extensive studies have been conducted on these enzymes due to their applications in industrial and natural processes. The enzymes are responsible for breaking down various molecules like lignocellulose and conversion of triacylglycerol (TAG) into fatty acids (FA) in both natural and industrial settings. They are utilized in bioremediation processes as well as in the food and textile industries (Hou & Majumder, 2021).

Mycota strain	Microplastics	Enzyme(s) involved	Remarks	References
A. flavus	Polyethylene (PEDX3)	Laccase and laccase-like multicopper oxidases and manganese peroxidase	There is terminal oxidation, cleavage of chains, and fatty acid metabolization	Zhang et al. (2020)
A. flavus	Polyurethae (ITCC6051)	Esterase	The enzyme degrades urethane bond and polyol segment which are major component of polyurethane. The enzymes have dual activity of endopolyurethanase and exopolyurethenases which have synergistic effect of polyurethane degradation	Mathur and Prasad (2012)
A. niger	PE	Extracellular and intracellular Depolymerases	The weight loss was 8% in one month confirmed by using SEM	Raaman et al. (2012)
A. japonicum	PE	Depolymerases	12% breakdown of plastic was majored in 4 weeks confirmed by using SEM	Raaman et al. (2012)
P. simplicissimum	Polyethylene	Laccase and manganese peroxidise	The enzymes production was high in submerged fermentation and the activity of manganese peroxidase was greater as compared to laccase enzyme	Sowmya et al. (2015)
T. viride	LPDE	–	The weight loss in 45 days was observed for LPDE films	Munnir et al. (2018)
Cladosporium cladosporioides, Xepiculopsis graminea, P. griseofulvum, Leptosphaeria sp.	Polyester polyurethane (PPU)	–	Due to efficient enzymatic activity on plastic degradation between 6 and 14 days of growth showed visible clearance of growing culture	Brunner et al. (2018)
Pestalotiopsis microspore, Curvularia senegalensis, Fusarium solani	PUR	Serine hydrolase	Trasparent visuals of culture shown the result that PUR degraded successfully	Russell et al. (2011)

(continued)

(continued)

Mycota strain	Microplastics	Enzyme(s) involved	Remarks	References
T. insolens, F. solani	PET film	Cutinase	Weight loss upto 96% was observed at 70 °C within 96 h. Cutinase converted PET into TPA	Ronkvist et al. (2009)
P. citrinum	PET pellets	Polyesterase	The enzyme obtained from fungi were able degrade PET plastic	Liebminger et al. (2009)

Adapted from Solanki et al. (2022)

9.4.2.2 Algal Degradation

Investigations have also been initiated to study the capability of algae to break down MPs. Algae are commonly employed as test subjects to investigate the harmful effects of MPs. Nevertheless, algae also play a role in breaking down microplastics. Algae are safe and simple to manage microorganisms. Through the secretion of extracellular polymeric substances, microorganisms are able to colonize the surface of microplastics, potentially enhancing their ability to degrade them. Upon attaching to plastic surfaces, algae were seen to release a variety of exopolysaccharides (EPS) and ligninolytic enzymes, potentially aiding in the degradation of plastic. Different type of algal species including *Scenedesmus dimorphous, Anabaena spiroides, Navicula pupula*, and *Oscillatoria* sp. can adhere to submerged plastics, particularly PE (Chia et al., 2020). Kumar et al. (2017) found that the blue-green alga *Anabaena spiroides* degraded 8.18% of LDPE. *Chlorella vulgaris* was recently demonstrated to break down BPA, a prevalent plastic additive, to a point lower than what can be detected (Chia et al., 2020). Compared to other bioremediation approaches, phytoremediation for MPs removal is not as widely used, raising questions about its effectiveness and whether it is only limited to algae, which are not true plants. Polymers, such as plastics, being present, can enhance the production of extracellular polymeric substances (Song et al., 2020). Moreover, various polymers elicit varying levels of production, leading to variations in both quantity and quality. This influences the range in cohesion of biofilms and in the end the biomass of the organisms colonizing (Lagarde et al., 2016). The formation of biofilm on the surface of microplastic can vary from different environments like the quality and quantity of biomass, cell surface matrix, etc. Also, the biomass of algae varies from freshwater to marine water, as stated by Chen et al. (2019). Apart from this, various other components play important roles in the biofilm formation like, attachment of microalgae to the surface of microplastic, including the size of plastic and its degradation, the duration of exposure to the environment (Fu et al., 2019), topography (Amaral-Zettler et al., 2020), temperature (Chen et al., 2019), nutrient availability and saltiness (Li et al., 2019).

Research has shown that various algae have the potential for degradation of MPs. One of the research projects shown that algal species, *Phormidium lucidum,* and *Oscillatoria subbrevis* have the great property of degrading MPs, using LDPE as a nutrient source (Sarmah & Rout, 2018). Various studies have been conducted to check the effectiveness of microalgae in the biodegradation of MPs, and results are also promising like green alga *Scenedesmus dimorphus, diatom, Navicula pupula,* blue-green alga (BGA) *Anabaena spiroides* have ability to breaking down both low- and high-density polyethylene, with slightly more effective for LDPE (Kumar et al., 2017). A type of BGA, *Spirulina* sp., have the efficacy to break down MPs of PE and PP. After breakdown, these MPs have different functional groups on their surfaces such as carboxylic acid, carbonyl, and hydroxyl groups (Hadiyanto et al., 2021). In some way microalgae have capability to breakdown MPs, but in the most of cases MPs have become the major threat to microalgae, by hindering their growth, disrupting photosynthetic apparatus and disturbing metabolic pathways. Moreover, MPs after breakdown release many toxic chemicals into environment such as polychlorinated biphenyls, dichlorodiphenyltrichloroethane, plasticizers and heavy metals (Campanale et al., 2020). The harmful chemicals emitted by MPs can damage the cell membrane or plasma membrane of microalgae, leading to impeding growth and finally cell disruption (Capolupo et al., 2020; Song et al., 2020).

Hence, more investigations are required to find the efficacy of microalage to degrade microplastic. The gaps need to be covered includes (1) identifying the types of additives emitted from microplastics, (2) Assessing how these chemicals hinder microalgae growth, and (3) finding the properties of MPs and environmental impact on microalgal growth. An assessment done by Kumar et al. (2017) demonstrated, that BGA *Anabaena spiroides*, had the 8.18% highest rate of LDPE degradation followed by *Navicula pupula* and diatom at 4.44% then *Scenedesmus dimorphus* with 3.74% rate. One of the studies conducted by Sarmah and Rout on nontoxic freshwater cyanobacteria (*Phormidium lucidum* and *Oscillatoria subbrevis*) demonstrated that, the algae can effectively colonize the surface of PE and breakdown LDPE without adding pretreatment or pro-oxidant. The algae proliferate rapidly, easy to handle and found easily. Gulnaz and Dincer (2009) conducted a study to check the breakdown of BPA by *Aeromonas hydrophila* bacteria and *Chlorella vulgaris*. The result showed that without any estrogenic effects, the algae and bacterial mixture quickly degraded the BPA. A comparable study was done by Hirooka et al. (2005), where a green algae *Chlorella fusca var. vacuolata* is utilized to degrade BPA into small molecules without any estrogenic effects. In *P. tricornutum,* Kim et al. cloned PETese enzyme, which was capable of acting on PET and PETG plastics (Kim et al., 2020). This result shows that using microalage for breaking MPs can be sustainable and eco-friendly approach.

9.4.2.3 Microbial Consortia

Numerous studies have provided information about the degradation properties of MPs by bacteria. However, not all individual microbial strains have a whole set of enzymes required for the effective breakdown of MPs, therefore using a variety of microbes with complementary enzymatic activities can efficiently degrade MPs. Microbial consortia are described as a community of different microbes that interact, coexist, and work synergistically. These consortia can degrade a variety of polymers more efficiently as it has a wider range of enzymatic properties than individual strains. Another resistant compound like pectin also degraded effectively and underwent bioremediation by using microbial consortia (Martínez-Gallardo et al., 2021). The degradation property of MPs is also influenced by the texture or composition of polymers like recycled plastics show some alterations in their structure due process of recycling that typically have a higher possibility of microbial degradation than pure plastic polymers (Tokiwa et al., 2009). A consortium of four bacterial strains belonging to the genus *Aneuinibacillus* and *Brevibacillus* have the efficacy to break PE and PP microplastics and produce low-density MPs. Another study conducted to assess the degradation efficiency of MPs by using mixture of bacterium including *Bacillus, Pseudomonas* and *Staphylococcus* have degraded 40 μm polyethylene sheets resulted in weight loss of 42.5, 7.5 and 20% under 40 days of exposure (Chen et al., 2021). A study conducted by Auta et al. (2022), demonstrated the efficacy of microbial consortia in degrading PET and PS microplastic. For assessment, they used local microbial consortia that were capable of breaking down PET and PS microplastic. The investigation was done in the in-situ condition of mangrove soil contaminated with plastics and microplastics, eight bacterial strains were examined to see PET and PS plastic degradation potential. The growth of the bacterium was tracked by counting colony-forming units at different intervals and majoring optical density at 600 nm in the presence of microplastic contamination. By using various techniques like energy-dispersive X-ray spectroscopy (EDS) and scanning electron microscopy (SEM), it has been possible to examine the surface and structural alterations in MPs that happened due to the enzymatic activities of microbes. The result showed observable changes related to peak removal and polymer elongation, indicating that microbial consortia were able to change the chemical structure of PET and PS microplastic. The study followed up to 90 days and the result showed an 18% reduction in body weight of MPs, treated with microbial consortia in mangrove soil. The researchers concluded that the microbial consortia have better efficacy in degrading microplastic rather than single microorganisms.

9.4.2.4 Degradation by Soil Fauna

Plastic production increasing rapidly as a result MPs or NPs pollution also get increased in the soil, contributing to greater interactions between MPs and soil invertebrates including slugs, grubs, mealworms, centipedes, waxworms, super worms, and earthworms. Various soil invertebrates have microbial colonies in their gut for digestion that can be utilized for the degradation of MPs suggested in many studies (Yang et al., 2014, 2015, 2020). Different polymers like PS, PE, and PVC can be degraded and metabolized by *Tenebrio molitor* (Brandon et al., 2018), *Galleria mellonella* (Lou et al., 2020), and *Zophobas atratus* (Peng et al., 2022). Another soil invertebrate *Achatina fulica* or snail has the potential to consume and break down PS polymers (Song et al., 2020). *Lumbricus terrestris* characterized by having different bacterial strains in their gut, can degrade LDPE microplastics (Lwanga et al., 2018). Around the world, earthworms are the most widespread soil animal, which is crucial for maintaining soil health. Earthworm consumes large amounts of soil and in addition, produce 2–15% of organic matter (Whalen et al., 2000). Earthworms symbiotically work with other microorganisms and play a critical role in soil structure and fertility which in turn contributes to ecosystem services like the decomposition of organic matter. Some characteristics of earthworms make them beneficial for decreasing soil pollution, consisting of MP contamination. Researchers are interested in applying earthworms in various applications because of their capacity to consume and decompose different materials including plastics and MPs (Rillig et al., 2017; Lwanga et al., 2017; Kim, 2015). By consuming LDPE, the size of *Lumbricus terrestris* shrinks without any change in plastic mass, suggesting that there was no degradation occurs by feeding of plastics (Lwanga et al., 2018).

Eisenia andrei can consume the materials that are linked with lactic acid oligomers or co-polymers, but degradation was limited due to the complexity of polymers (Alauzet et al., 2002). By this, it was suggested that earthworms have limited degradation properties but can modify the structure of consumed polymers. By using their digestive tract, earthworms can break down plastic polymers into monomers or smaller particles. The degradation will happen more quickly, as an increased surface area is available for microbial colonization and enzyme activity. Earthworms also compromise their survival, growth, and reproduction when introduced to MP contamination (Lwanga et al., 2017; Cao et al., 2017; Lithner et al., 2011). Many researches have demonstrated that *Tenebrio molitor* or meal worms use their gut microbiota to degrade PS plastics by following lignin digestive pathways. This is because the majority of soil invertebrates use plant components as their energy source. Using synthetic polymers as an energy source is linked with linkages that are similar to those in lignin and other natural polymers (Mantimin et al., 2023).

Soil invertebrates, cosmopolitan distribution including snails (*Achatina fulica*) frequently contact with soil MPs. The snails can consume MPs through various dietary sources like contaminated soil or plastic-infected leaves and release MPs in their faeces. Snail faeces having PS can form biodegradable intermediates with new oxidized functional groups, as evidenced by FTIR and 1 H NMR spectra. During biodegradation of PS, *Tenebrio genus* can also produce biodegradable intermediates (Brandon et al., 2018; Peng et al., 2019). *Tenebrio moliter* larvae can degrade PS and PVC plastics and release CO_2 as an end product because their gut microbiome consists of Streptococcaceae and Spiroplasmataceae family's microbes (Peng et al., 2020). The total amount of PVC consumed by *T. molitor* is partially mineralize, leads to change the PVC and release 2.9% of total PVC as chloride. *Tenebrio obscurus* or dark meal worms belonging to genus *Tenebrio* can potentially breakdowns PS and PE by using its gut microbiota (Peng et al., 2019). When gut microbiota uses PE and PS polymers in aerobic conditions as their carbon source, they release CO_2 as byproduct through oxidation whereas in anaerobic condition, release CO_2 and CH_4 as by product (Zhang et al., 2015). Another soil species, *Zophobas atratus* (super worm) consume styrofoam, a type of PS as their only food supply at an average daily rate of 0.58 mg. By using this metabolic functioning super worm can degrade long-chains of PS monomers into low molecular weight products (Wang et al., 2022). As laboratory results cannot define the accurate result because there is difference between controlled lab environment and real-world situations. Therefore, there is need to assess the long-term effect of MPs on earthworm population and soil ecosystem dynamics. Utilization of soil animals and insects for remediation of MPs and NPs can be very useful and economically cheaper, and doesn't produce secondary pollution, provides long-term sustainability for future generation.

9.4.2.5 Genetically Modified Organisms

Laboratory interventions for modifying microbes genetically are emerging as long-term bioremediation techniques for MP contamination. As not all microorganisms are equally capable of breaking down MPs, genetic engineering and synthetic biology techniques are used to develope synthetically modified strains for effective biodegradation of MPs. Furthermore, the use of other gene editing tools, including the most famous clustered regularly interspaced palindromic repeats (CRISPR-Cas9), zinc finger proteins, with expertise in bioinformatics and computational biology has simplified the design process of enzymes and gives the basic information of enzyme action, associated genes and biodegradation pathways (Singh et al., 2022). By using gene modifying tools, the biodegradation pathways and by-products of MP biodegradation can be predicted easily. To enhance MPs biodegradation capacity, various microorganisms are genetically modified, including bacteria, fungi, algae, and diatoms. *Ideonella sakaiensis* 201-F6 synthesize enzyme PET hydrolases that hydrolyze PET plastic into terephthalic acid and ethylene glycol (Pellis et al., 2023). The photosynthetic microalga *Phaeodactylum tricornutum,* turns into the PET degrader, by inserting a foreign gene encoding PET hydrolytic enzymes (Moog et al., 2019). The genome of *E. coli* was edited by inserting PETase enzyme and enhance the degradation of MP fibres in wastewater sludge by 17.4-fold (Zurier et al., 2023). The cloning and expressing the PETase gene in the green alga *Chlamydomonas reinhardtii* have shown the effective PET breakdown in 4 weeks under optimum condition (Kim et al., 2020). PETase gene introduced in the yeast, *Yarrowia lipolytica* has showed, it could degrade PET effectively. The degradation was also enhanced by adding olive oil to the mutant culture and release of 66% terephthalic acid. Oliveira and Almeida cloned the pudA gene of *Comamonas acidovorans* which code esterase enzyme and introduced it in the *E. coli* to construct genetically modified MPs degrading strain (Oliveira et al., 2019). Enhancing biofilm properties by genetic engineering approaches also increased the accumulation of MPs in the biofilm. To optimize the binding of PVC molecules, *Pseudomonas aeruginosa* was altered by deleting the wspF methyl esterase gene, that increased the synthesis of exopolymer compounds (Liu et al., 2021). Another study found that cloning the gene encoding the polyester hydrolase enzyme, which breaks down PET, from *Pseudomonas aestusnigri* to *E.*

coli improved PET biodegradation (Bollinger et al., 2020). There have been considerable attempts at genetic modification of microorganisms as well as enzymes to improve MP biodegradation.

9.5 Types of MPs Bioremediation Based on Location

9.5.1 Ex-Situ Bioremediation of MPs

Ex-situ MPs bioremediation systems, including biopiles, windrows, land farming, biofiltration and membrane bioreactors involve removing MPs pollutants on the basis of depth from polluted areas and transporting them to another location for treatment (Philp & Atlas, 2005).

9.5.1.1 Biopile

This full-scale technology cleans excavated toxic soil by piling it above ground. To ensure oxygen availability for aerobic microbial decomposition of MPs compounds and their subsequent conversion to their basic constituents, such as water and CO_2, it is supplemented with nutrients (urea, compost), organic matter (manure or peat), pH adjusters, and aeration systems (Dias et al., 2015). Heating systems in biopiles can help to minimize remediation time and improve microbial performance in the MPs contaminated sites (Gomez & Sartaj, 2014). Limitations include the necessity for strong engineering, potential challenges with electrical power at remote places, and excessively heated air can dry up the soil, which will block microbial activity and slow down the process of bioremediation (Sanscartier et al., 2009).

9.5.1.2 Windrow

This bioremediation technique involves turning the pile of polluted soil regularly with equipment to increase the degrading activity of indigenous and/or transitory hydrocarbonoclastic bacteria for remediation of MPs contaminated soil. Water is supplied to aerate the soil, and nutrients and pollutants are evenly spread to ensure adequate mixing with microbes for better degradation (Barr et al., 2002). Coulon et al. (2010) found that windrow treatment performed better than biopile treatment for hydrocarbon removal.

9.5.1.3 Land Farming

It is a cheap approach that requires less complexity of equipment for managing MPs polluted soil. Excavation and tilling are two treatment strategies that can be used differently at a MPs contaminated site depending on the climate, soil type, and other factors (Nikolopoulou et al., 2013). The high oxygen availability created by adding bulking agents and co-substrates, aeration, and microbial metabolism allows autochthonous microorganisms to biodegrade MPs pollutants aerobically when excavated toxic soils are spread out on the previously prepared bed and then periodically tilled (Philp & Atlas, 2005). Landfarming has a well-known structured design for water management and improved treatment efficiency. Polluted water, including dissolved MPs, is prevented from running off and contaminating surrounding areas.

9.5.1.4 Biofiltration

As in various industrial processes, plastics are used and a high quantity of MPs is generated in industrial wastewater, which is released into the environment, posing a serious threat to the ecosystem, and for this capturing MPs from wastewater and preventing MPs from entering into the environment is done by wastewater treatment plants (WWTPs). Although traditional methods of treatment are adopted, they show limited effectiveness in eliminating MPs (Eerkes-Medrano et al., 2015). For eliminating MPs and improving the removal of micropollutants from wastewater requires advanced treatment technologies like biofiltration and membrane filtration. Biofiltration is characterize by employing microbes which is fixed on filter media for degradation of organic pollutants and capturing MPs. Biofilter units works by letting wastewater flow through porous material, enhancing microorganisms to form biofilm on it (Zhang et al., 2022). The MPs provide a surface for biofilm expansion for the microbial community to grow and degrade organic pollutants, as MPs pass to the filter, they get stuck into the slimy biofilm formed by microbial colony and can be degraded in some amount (Li et al., 2020). As MPs come into the contact of biofilm, they stuck within the microbial structure or cling to the filter media surface (Li et al., 2020). For effective working of filters, periodic backwashing is carried out to trap MPs and eliminate biofilm (Li et al., 2020).

The MP removal efficiency is influenced by the interaction between MP and membrane filters, as MPs stuck in the membrane filter pores can potentially create fouling issues (Sun et al., 2019). It is important to clean the membrane filters regularly to get optimal results, avoid blockages, and uphold ideal performance. Increasing the efficiency of MP removal from wastewater is enhanced by sequentially using biofiltration and membrane filtration. The biofiltration process starts with capturing MPs and organic pollutants on biofilm and filter media, then degrade them biologically. For capturing fine MPs or NPs membrane filtration can be used effectively and release more clear effluent. This combined method can help to eliminate MPs more effectively without creating fouling in the filtration system (Zhang et al., 2020). The combination of both physical and biological processes contributes to a complete solution for MPs removal. Although biofiltration and membrane filtration show potential for removing MPs but still face some obstacles. The major issue related to the effect of MPs on the growth of microbial communities present in the filter. Study indicates that PVC can harm the structure of microbial communities, with no significant impact (Li et al., 2020). More research is required to comprehend the lasting effects of MPs on microbial dynamics and functionality. Another issue related to these techniques is membrane fouling, as MPs and other substances adhere to a membrane filter decrease the efficiency of filters. To keep the membrane performance high, usage of low-cost operational techniques and efficient cleaning methods are important (Wang et al., 2019). For better performance, there is need to investigate improved cleaning techniques and also the material for membrane that have properties to prevent fouling.

9.5.1.5 Membrane Bioreactors

MBR is a sophisticated wastewater treatment method that combines biological treatment (including aerobic and anaerobic processes) with membrane filtration. Conventional biological treatment plants use clarifiers to separate sludge from treated water whereas membrane bioreactors use advanced microfiltration or ultrafiltration techniques to remove sludge. In MBRs, microplastics are removed through microorganisms by absorbing them and accumulate in their biomass which is further filtered out by the membrane. Results have shown that in WWTPs, activated sludge is responsible for building up MPs, that are efficiently removed in secondary settling stages. Compared to advanced MBRs, a different type of bioreactor known as the A2O bioreactor system is often used in WWTPs and shows less effectiveness in eliminating MPs. The effectiveness is comprised mainly due to the reversal of sludge with 20% MPs into the water phase which was previously absorbed and another issue is related to the decomposition of MPs in A2O bioreactors that it occurs at a very low pace. Studies have demonstrated that some microorganisms can degrade MPs in bioreactors including *Rhodococcus* bacteria which have the potential to degrade 6.4% of polypropylene in 40 days, and *Ideonella sakaiensis* which can break PET plastic in 6 weeks. In WWTPs, the hydraulic retention time (HRT) ranges between 7 and 14 h, which is insufficient for bacteria to perform degradation that results in reduced the effectiveness of traditional activated sludge processes of MP removal. MBRs are emerging as potential technologies in WWTPs because of its efficient removal of microplastic at rate reaching up to 99.9% (Talvitie et al., 2017). MBR involves membrane filtration with traditional activated sludge process and the biofilm on the permeable side of the MBR system can effectively capture most of the MPs. In MBR system, effect of adsorption plays an important in removing most of the MPs. The removal efficiency of MPs also depends on the its size, as MBR systems with 0.1 μm pore sizes of membranes can theoretically eliminate microplastics (Li et al., 2020).

9.5.2 In-Situ Bioremediation of MPs

For sustainable and eco-friendly mitigation of microplastics, various bioremediation methods are classified because not a single method have the complete property of eliminating MPs from contaminated areas. In-situ bioremediation is one of them which employed to remove MPs from polluted samples or soils right at the source of contamination. In-situ bioremediation is divided into two primary categories namely, intrinsic bioremediation and improved in-situ bioremediation. Both approaches work naturally either by alone or with some other processes for removing contaminants. Intrinsic bioremediation includes biosparging and phytoremediation whereas bioventing is an example of enhanced in-situ bioremediation (Azubuike et al., 2016). These approaches are recently used to treat contaminated soil and ground water and that might be beneficial for cleaning MPs contaminated soil (Sharma, 2019).

9.5.2.1 Bioaugmentation

Bioaugmentation is a process of adding specially grown microbes to a polluted area MPs to help clean it up faster thereby reducing clean up time and expense as well (Thompson et al., 2004). Bioaugmentation uses natural or genetically modified microbes to clean up hazardous waste sites polluted with MPs (Mrozik et al., 2010). Improving the natural microbes in a polluted area helps clean up pollution more effectively and increases the variety of useful genes, making the microbial community more diverse (Dejonghe et al., 2001; Shukla et al., 2013). Factors which affect bioaugmentation are pH, temperature, moisture, organic matter content, aeration, nutrient content, and soil type (Hong et al., 2007). It involves using either one or group of microorganisms that can break down polymers. They can also be genetically altered to improve biodegradation activity (Kalogerakis et al., 2015).

Indigenous polymer-degrading cultures:

Research is going on to identify groups of microorganisms and/or pure cultures from microplastics from both land and sea ecosystems. This will help in developing location-specific bioaugmentation strategies for boosting MPs biodegradation. Microorganisms interacting with the surfaces of MPs and undergoing metabolic changes have shown that bioaugmentation can be effectively used (Kalogerakis et al., 2015).

Genetically modified microbes:

Researchers are enhancing the potentiality of bioaugmentation process by using genetic engineering techniques to non- native and native microbes. The techniques used were CRISER-Cas9, TALENS, chemical modification, electroporation, etc. to transfer plastic degrading genes into non-native bacteria (Zhang et al., 2004).

9.5.2.2 Biostimulation

In the contaminated environment where nutrients are scarcely present or unavailable for native microbes, which are associated with degrading and transforming contaminants, effectiveness can be increased by providing rate-limiting nutrients or growth-enhancing compounds, such as electron donors or acceptors or co-factors, whose aim is to encourage the metabolic processes or growth of the indigenous microbes in a process known as biostimulation. Biostimulation that promotes immobility or degradation of polylactic acid (PLA), polyhydroxybutyrate (PHB), and polyvinyl chloride (PVC) pollutants in situ (Paniagua-Michel & Rosales, 2015). Adams et al. (2015) studied that it fosters the bioremediation of hazardous chemicals by stimulating the development of natural oil degraders that are delivered to a highly polluted site. Some plastics are biodegradable, while others function as recalcitrant due to their polymeric nature. MPs degradation is dependent on its chemical and physical condition. Over the last decade, there has been a greater focus on the study of bioremediation methods for hazardous and xenobiotic substances. The decontamination process can be improved by pH, moisture content, salinity, biosolids, intensity of physical mixing, and temperature (Abdulsalam & Omale, 2009).

Numerous synthetic polymers used in MPs are based on petroleum hydrocarbons (PHCs) and polyaromatic hydrocarbons (PAHs). Sarkar et al. (2005) investigated the effect of biosolid (organic matter) addition at (PHC) sites, where degradation rates rose by up to 96%, and at diesel-contaminated sites, where inorganic ('N' and 'P' rich) fertilizers were given to stimulate degradation. According to Miya and Firestone (2001), PHC deterioration was much higher in Biostimulated (BS) -treated soil compared to Bioaugmented (BA)-treated soil. Bio stimulation's C: N: P ratio resulted in 60% PHC degradation efficiency compared to 34% attained by BA. As a carbon source alone, as a carbon source alone, *Micrococcus* sp. PHE3 is unable to degrade PAHs like anthracene, pyrene, and fluoranthene; however, it has been demonstrated that supplementing with naphthalene and phenanthrene enhances the degradation of all PAHs by more than 90% within a 10-month incubation timeframe (Zhang et al., 2013). Further research revealed that BS and BA are the most successful approaches for hydrocarbon degradation (Maneer-Lopez et al., 2008).

This method is superior to BA for the following reasons: (1) the native microorganisms are uniformly dispersed (Adams et al., 2015). (2) an economical and sustainable method) (3) There is no ecological disruption. The fundamental disadvantage of biostimulation is that it is dependent on environmental circumstances, is ineffective when firmly embedded in soil particles, and is non-biodegradable. Highly site-specific and necessitates substantial scientific research (Mohee et al., 2012).

9.5.2.3 Phytoremediation

This technique involves the interaction between plants and microbes to remove toxic chemicals from water and soil (Saxena et al., 2020). Plants operate as natural filters and are referred to as hyperaccumulators because they accumulate heavy metals from MPs contaminated areas. Decontamination process occur by diverse phytoremediation methods based on different types of contaminants such as phytoaccumulation/phytoextraction (root and shoot system or leaves tissue accumulated low toxic metals by direct uptake); phytotransformation/phytodegradation (organic toxic pollutant degraded via plant system in water); phytostabilization (leached hazardous substances are absorbed and stabilized on bioaccumulator root); rhizodegradation (rhizosphere microbes interacted with plant root to break down hazardous chemicals by metabolic activity and enzyme release); Rhizofilteration (aquatic plant roots uptake the toxic nanoplastics substances); phytovolatilization (toxic inorganic substances like Se, Hg, etc. are converted into volatile state by plants and released in the environment (Vidali, 2001; Saxena et al., 2020). The bioaugmentation or biostimulation procedure increases the potential of native plants on the site for the technique's successful execution. Parameters impacting phytoremediation techniques include pollutant depth, contaminant toxicity, hydrophobicity of organic carbon and its solubility, plant physiology, and ability to resist disease (Lee, 2013). Different methods of phytoremediation may differ according to MPs' co-exposure to various elements in the environment. In a related study, Manjate et al. (2020) investigated the effects of Cd and Cu co-exposure in PE on *Phragmites australis* and *Solanum photeinocarpum* and recently Yu et al., 2023 found that Cd was consistently removed by *Phragmites australis*, while Cd accumulation decreased and Zn uptake increased in the roots of *Solanum photeinocarpum.* Li et al., 2020 explored the MPs beads absorption pathway using a scanning electron microscope in *Triticum aestivum*, where polystyrene (PS) beads associated with fluorescent marker tracking start from the roots and translocate to the aerial parts by an apoplastic route. In a related study, the polymeric surface charges effectively impact the mobility across the root system of *Betula pendula* (silver birch) when treated with polyamide (PA) MP beads. This result suggest that the positive surface charged PS-NH_2 is less likely to be absorbed (Austen et al., 2022).

9.5.2.4 Bioventing

Bioventing emerging as one of the most common in-situ bioremediation techniques for microplastic-contaminated soil. The process starts with providing nutrients to native microorganisms via wells, the precise amount of oxygen required by microbes for biodegradation, which reduces the volatilization and release of MPs into the surrounding environment (Kensa, 2011). By injecting oxygen into the unsaturated zone of soil, it increases the movement of air and boosts the degradation activity of native microbes. Bioventing can also be enhanced by increasing microbial population, adding more nutrients and moisture to the polluted site which transformed the pollutant into innocuous molecules.

9.5.2.5 Biosparging

Like bioventing, biosparging starts by providing air beneath the microplastic-contaminated soil surface, to increase the microbial degradation activity which makes it easier to remove the pollutant from contaminated areas. In biosparging air is added into the saturated zone which is opposite to bioventing, and can enhance the moving of volatile organic chemicals upward to the unsaturated zone and support bioremediation processes. Two primary criteria for assessing the efficacy of biosparging include the degradable property of contaminants and the texture of soil particles (Sharma, 2020).

9.6 Conclusion

This chapter which is based on the bioremediation of MPs gives us a valuable knowledge of mitigating MPs waste. Because of less attention on microplastic pollution and limited data on microplastic waste from terrestrial and aquatic environments make it difficult to determine their environmental impact. High plastic production and limited quantitative and qualitative data on microplastic waste create a gap in assessing their ecotoxicological impact on the environment. The disintegration of microplastic can result in the formation of hazardous intermediate compounds which could negatively affect the ecosystem functioning. Due to the toxic effect of MPs, it is necessary to remove them from the environment to prevent long-term sustainability. Assessing the toxicity of goods after manufacturing is vital for their environmental impact and also for better management of plastic waste. As microorganisms shown the capability of microplastic degradation, the search for more novel strains is ongoing. Techniques like proteomics and genomics can identify and select particular microorganisms for MP degradation. Apart from proteomics and genomics, biotechnological approaches can also create engineered enzymes capable of degrading MP polymers.

References

Abdel-Shafy, H. I., & Mansour, M. S. (2018). Solid waste issue: Sources, composition, disposal, recycling, and valorization. *Egyptian Journal of Petroleum, 27*(4), 1275–1290.

Abdulsalam, S., & Omale, A. B. (2009). Comparison of biostimulation and bioaugmentation techniques for the remediation of used motor oil contaminated soil. *Brazilian Archives of Biology and Technology, 52*, 747–754.

Adams, G. O., Fufeyin, P. T., Okoro, S. E., & Ehinomen, I. (2015). Bioremediation, biostimulation and bioaugmention: A review. *International Journal of Environmental Bioremediation & Biodegradation, 3*(1), 28–39.

Aguado, A. S., Montiel, E., & Nixon, M. S. (2002). Invariant characterisation of the Hough transform for pose estimation of arbitrary shapes. *Pattern Recognition, 35*(5), 1083–1097.

Alauzet, N., Garreau, H., Bouché, M., & Vert, M. (2002). Earthworms and the degradation of lactic acid-based stereocopolymers. *Journal of Polymers and the Environment, 10*, 53–58.

Ali, S. S., Elsamahy, T., Koutra, E., Kornaros, M., El-Sheekh, M., Abdelkarim, E. A., Zhu, D., & Sun, J. (2021). Degradation of conventional plastic wastes in the environment: A review on current status of knowledge and future perspectives of disposal. *Science of the Total Environment, 771*, Article 144719.

Alimi, O. S., Farner Budarz, J., Hernandez, L. M., & Tufenkji, N. (2018). Microplastics and nanoplastics in aquatic environments: Aggregation, deposition, and enhanced contaminant transport. *Environmental Science & Technology, 52*(4), 1704–1724.

Alshehrei, F. (2017). Biodegradation of synthetic and natural plastic by microorganisms. *Journal of Applied & Environmental Microbiology, 5*(1), 8–19.

Amaral-Zettler, L. A., Zettler, E. R., & Mincer, T. J. (2020). Ecology of the plastisphere. *Nature Reviews Microbiology, 18*(3), 139–151.

Anastasi, A., Tigini, V., & Varese, G. C. (2012). The bioremediation potential of different ecophysiological groups of fungi. In *Fungi as bioremediators* (pp. 29–49). Springer Berlin Heidelberg.

Andrady, A. L. (2011). Microplastics in the marine environment. *Marine Pollution Bulletin, 62*(8), 15961605.

Arena, U., Mastellone, M. L., & Perugini, F. (2003). Life cycle assessment of a plastic packaging recycling system. *The International Journal of Life Cycle Assessment, 8*, 92–98.

Arthur, C., Baker, J. E., & Bamford, H. A. (2009). *Proceedings of the International Research Workshop on the Occurrence, Effects, and Fate of Microplastic Marine Debris, September 9–11, 2008*. Tacoma, WA, USA: University of Washington Tacoma.

Astrup, T., Møller, J., & Fruergaard, T. (2009). Incineration and co-combustion of waste: Accounting of greenhouse gases and global warming contributions. *Waste Management & Research, 27*(8), 789–799.

Austen, K., MacLean, J., Balanzategui, D., & Hölker, F. (2022). Microplastic inclusion in birch tree roots. *Science of the Total Environment, 808*, 152085.

Auta, H. S., Abioye, O. P., Aransiola, S. A., Bala, J. D., Chukwuemeka, V. I., Hassan, A., Aziz, A., & Fauziah, S. H. (2022). Enhanced microbial degradation of PET and PS microplastics under natural conditions in mangrove environment. *Journal of Environmental Management, 304*, Article 114273.

Awaja, F., & Pavel, D. (2006). *Design aspects of used lubricating oil re-refining*. Elsevier.

Awasthi, S., Srivastava, P., Singh, P., Tiwary, D., & Mishra, P. K. (2017). Biodegradation of thermally treated high-density polyethylene (HDPE) by Klebsiella pneumoniae CH001. *3 Biotech, 7*(5), 332.

Bacha, A. U. R., Nabi, I., & Zhang, L. (2021). Mechanisms and the engineering approaches for the degradation of microplastics. *Acs Es&t Engineering, 1*(11), 1481–1501.

Barr, D., Finnamore, J. R., Bardos, R. P., Weeks, J. M., & Nathanail, C. P. (2002). *Biological methods for assessment and remediation of contaminated land: Case studies*. CIRIA.

Barrick, A., Champeau, O., Chatel, A., Manier, N., Northcott, G., & Tremblay, L. A. (2021). Plastic additives: Challenges in ecotox hazard assessment. *PeerJ, 9*, Article e11300.

Benítez, A., Sánchez, J. J., Arnal, M. L., Müller, A. J., Rodríguez, O., & Morales, G. (2013). Abiotic degradation of LDPE and LLDPE formulated with a pro-oxidant additive. *Polymer Degradation and Stability, 98*(2), 490–501.

Bernardo, C. A., Simões, C. L., & Pinto, L. M. C. (2016). Environmental and economic life cycle analysis of plastic waste management options. A review. In *AIP Conference Proceedings* (Vol. 1779, No. 1). AIP Publishing.

Bollinger, A., Thies, S., Knieps-Grünhagen, E., Gertzen, C., Kobus, S., Höppner, A., Ferrer, M., Gohlke, H., Smits, S. H., & Jaeger, K. E. (2020). A novel polyester hydrolase from the marine bacterium Pseudomonas aestusnigri—Structural and functional insights. *Frontiers in Microbiology, 11*, 114.

Boots, B., Russell, C. W., & Green, D. S. (2019). Effects of microplastics in soil ecosystems: Above and below ground. *Environmental Science & Technology, 53*(19), 11496–11506.

Boucher, J., & Friot, D. (2017). *Primary microplastics in the oceans: A global evaluation of sources.*

Brandon, A. M., Gao, S. H., Tian, R., Ning, D., Yang, S. S., Zhou, J., Wu, W. M., & Criddle, C. S. (2018). Biodegradation of polyethylene and plastic mixtures in mealworms (larvae of Tenebrio molitor) and effects on the gut microbiome. *Environmental Science & Technology, 52*(11), 6526–6533.

Brown, E., MacDonald, A., Allen, S., & Allen, D. (2023). The potential for a plastic recycling facility to release microplastic pollution and possible filtration remediation effectiveness. *Journal of Hazardous Materials Advances, 10*, Article 100309.

Brunner, I., Fischer, M., Rüthi, J., Stierli, B., & Frey, B. (2018). Ability of fungi isolated from plastic debris floating in the shoreline of a lake to degrade plastics. *PLoS ONE, 13*(8), Article e0202047.

Burlakovs, J., Kriipsalu, M., Porshnov, D., Jani, Y., Ozols, V., Pehme, K. M., Rudovica, V., Grinfelde, I., Pilecka, J., Vincevica-Gaile, Z., & Turkadze, T. (2019). Gateway of landfilled plastic waste towards circular economy in Europe. *Separations, 6*(2), 25.

Campanale, C., Massarelli, C., Savino, I., Locaputo, V., & Uricchio, V. F. (2020). A detailed review study on potential effects of microplastics and additives of concern on human health. *International Journal of Environmental Research and Public Health, 17*(4), 1212.

Cao, D., Wang, X., Luo, X., Liu, G., & Zheng, H. (2017, April). Effects of polystyrene microplastics on 1006 the fitness of earthworms in an agricultural soil. In *IOP Conference Series: Earth and Environmental 1007 Science* (Vol. 61, No. 1, p. 012148). IOP Publishing.

Capolupo, M., Sørensen, L., Jayasena, K. D. R., Booth, A. M., & Fabbri, E. (2020). Chemical composition and ecotoxicity of plastic and car tire rubber leachates to aquatic organisms. *Water Research, 169*, Article 115270.

Chen, L., He, P., Shao, L., Zhang, H., & Lü, F. (2019). Municipal solid waste (MSW) landfill: A source of microplastics?-Evidence of microplastics in landfill leachate. *Water Research, 159*, 38–45.

Chen, X., Chen, X., Liu, Q., Zhao, Q., Xiong, X., & Wu, C. (2021). Used disposable face masks are significant sources of microplastics to environment. *Environmental Pollution, 285*, Article 117485.

Chia, W. Y., Tang, D. Y. Y., Khoo, K. S., Lup, A. N. K., & Chew, K. W. (2020). Nature's fight against plastic pollution: Algae for plastic biodegradation and bioplastics production. *Environmental Science and Ecotechnology, 4*, Article 100065.

Chilton, T., Burnley, S., & Nesaratnam, S. (2010). A life cycle assessment of the closed-loop recycling and thermal recovery of post-consumer PET. *Resources, Conservation and Recycling, 54*(12), 1241–1249.

Cole, M., Lindeque, P., Fileman, E., Halsband, C., Goodhead, R., Moger, J., & Galloway, T. S. (2013). Microplastic ingestion by zooplankton. *Environmental Science & Technology, 47*(12), 6646–6655.

Cooper, D. A., & Corcoran, P. L. (2010). Effects of mechanical and chemical processes on the degradation of plastic beach debris on the island of Kauai, Hawaii. *Marine Pollution Bulletin, 60*(5), 650–654.

Coulon, F., Al Awadi, M., Cowie, W., Mardlin, D., Pollard, S., Cunningham, C., Risdon, G., Arthur, P., Semple, K. T., & Paton, G. I. (2010). When is a soil remediated? Comparison of biopiled and windrowed soils contaminated with bunker-fuel in a full-scale trial. *Environmental Pollution, 158*(10), 3032–3040.

Cox, K. D., Covernton, G. A., Davies, H. L., Dower, J. F., Juanes, F., & Dudas, S. E. (2019). Human consumption of microplastics. *Environmental Science & Technology, 53*(12), 7068–7074.

Crain, D. A., Eriksen, M., Iguchi, T., Jobling, S., Laufer, H., LeBlanc, G. A., & Guillette, L. J., Jr. (2007). An ecological assessment of bisphenol-A: Evidence from comparative biology. *Reproductive Toxicology, 24*(2), 225–239.

Dash, H. R., Mangwani, N., Chakraborty, J., Kumari, S., & Das, S. (2013). Marine bacteria: Potential candidates for enhanced bioremediation. *Applied Microbiology and Biotechnology, 97*(2), 561–571.

de Souza Machado, A. A., Lau, C. W., Kloas, W., Bergmann, J., Bachelier, J. B., Faltin, E., Becker, R., Görlich, A. S., & Rillig, M. C. (2019). Microplastics can change soil properties and affect plant performance. *Environmental Science & Technology, 53*(10), 6044–6052.

Dejonghe, W., Boon, N., Seghers, D., Top, E. M., & Verstraete, W. (2001). Bioaugmentation of soils by increasing microbial richness: Missing links. *Environmental Microbiology, 3*(10), 649–657.

Deng, J., Guo, P., Zhang, X., Su, H., Zhang, Y., Wu, Y., & Li, Y. (2020). Microplastics and accumulated heavy metals in restored mangrove wetland surface sediments at Jinjiang Estuary (Fujian, China). *Marine Pollution Bulletin, 159*, Article 111482.

Dias, R. L., Ruberto, L., Calabró, A., Balbo, A. L., Del Panno, M. T., & Mac Cormack, W. P. (2015). Hydrocarbon removal and bacterial community structure in on-site biostimulated biopile systems designed for bioremediation of diesel-contaminated Antarctic soil. *Polar Biology, 38*, 677–687.

Dong, C. D., Chen, C. W., Chen, Y. C., Chen, H. H., Lee, J. S., & Lin, C. H. (2020). Polystyrene microplastic particles: In vitro pulmonary toxicity assessment. *Journal of Hazardous Materials, 385*, Article 121575.

Du, H., Xie, Y., & Wang, J. (2021). Microplastic degradation methods and corresponding degradation mechanism: Research status and future perspectives. *Journal of Hazardous Materials, 418*, Article 126377.

Dussud, C., Hudec, C., George, M., Fabre, P., Higgs, P., Bruzaud, S., Delort, A. M., Eyheraguibel, B., Meistertzheim, A. L., Jacquin, J., & Cheng, J. (2018). Colonization of non-biodegradable and biodegradable plastics by marine microorganisms. *Frontiers in Microbiology, 9*, 1571.

Eerkes-Medrano, D., Thompson, R. C., & Aldridge, D. C. (2015). Microplastics in freshwater systems: A review of the emerging threats, identification of knowledge gaps and prioritisation of research needs. *Water Research, 75*, 63–82.

Emisha, L., Wilfred, N., Kavitha, S., Halder, G., Haldar, D., Patel, A. K., Singhania, R. R., & Pandey, A. (2024). Biodegradation of microplastics: Advancement in the strategic approaches towards prevention of its accumulation and harmful effects. *Chemosphere, 346*, Article 140661.

Enyoh, C. E., Verla, A. W., Verla, E. N., Ibe, F. C., & Amaobi, C. E. (2019). Airborne microplastics: A review study on method for analysis, occurrence, movement and risks. *Environmental Monitoring and Assessment, 191*, 1–17.

Eriksson, O., & Finnveden, G. (2009). Plastic waste as a fuel-CO_2-neutral or not? *Energy & Environmental Science, 2*(9), 907–914.

Fadare, O. O., & Okoffo, E. D. (2020). Covid-19 face masks: A potential source of microplastic fibers in the environment. *The Science of the Total Environment, 737*, Article 140279.

Fu, D., Zhang, Q., Fan, Z., Qi, H., Wang, Z., & Peng, L. (2019). Aged microplastics polyvinyl chloride interact with copper and cause oxidative stress towards microalgae Chlorella vulgaris. *Aquatic Toxicology, 216*, Article 105319.

Gao, D., Li, X. Y., & Liu, H. T. (2020). Source, occurrence, migration and potential environmental risk of microplastics in sewage sludge and during sludge amendment to soil. *Science of the Total Environment, 742*, Article 140355.

Geyer, R., Jambeck, J. R., & Law, K. L. (2017). Production, use, and fate of all plastics ever made. *Science Advances, 3*(7), Article e1700782.

Giacomucci, L., Raddadi, N., Soccio, M., Lotti, N., & Fava, F. (2020). Biodegradation of polyvinyl chloride plastic films by enriched anaerobic marine consortia. *Marine Environmental Research, 158*, Article 104949.

Gigault, J., Ter Halle, A., Baudrimont, M., Pascal, P. Y., Gauffre, F., Phi, T. L., El Hadri, H., Grassl, B., & Reynaud, S. (2018). Current opinion: What is a nanoplastic? *Environmental Pollution, 235*, 1030–1034.

Gomez, F., & Sartaj, M. (2014). Optimization of field scale biopiles for bioremediation of petroleum hydrocarbon contaminated soil at low temperature conditions by response surface methodology (RSM). *International Biodeterioration & Biodegradation, 89*, 103–109.

Gu, F., Guo, J., Zhang, W., Summers, P. A., & Hall, P. (2017). From waste plastics to industrial raw materials: A life cycle assessment of mechanical plastic recycling practice based on a real-world case study. *Science of the Total Environment, 601*, 1192–1207.

Gu, J. D. (2003). Microbiological deterioration and degradation of synthetic polymeric materials: Recent 1104 research advances. *International Biodeterioration & Biodegradation, 52*(2), 69–91.

Gulnaz, O., & Dıncer, S. (2009). Biodegradation of bisphenol A by Chlorella vulgaris and Aeromonas hydrophilia. *Journal of Applied Biological Sciences, 3*(2), 79–84.

Guo, J. J., Huang, X. P., Xiang, L., Wang, Y. Z., Li, Y. W., Li, H., Cai, Q. Y., Mo, C. H., & Wong, M. H. (2020). Source, migration and toxicology of microplastics in soil. *Environment International, 137*, Article 105263.

Hadiyanto, H., Khoironi, A., Dianratri, I., Suherman, S., Muhammad, F., & Vaidyanathan, S. (2021). Interactions between polyethylene and polypropylene microplastics and Spirulina sp. microalgae in aquatic systems. *Heliyon, 7*(8).

Hahladakis, J. N., Velis, C. A., Weber, R., Iacovidou, E., & Purnell, P. (2018). An overview of chemical additives present in plastics: Migration, release, fate and environmental impact during their use, disposal and recycling. *Journal of Hazardous Materials, 344*, 179–199.

Haider, T. P., Völker, C., Kramm, J., Landfester, K., & Wurm, F. R. (2019). Plastics of the future? The impact of biodegradable polymers on the environment and on society. *Angewandte Chemie International Edition, 58*(1), 50–62.

Harms, H., Schlosser, D., & Wick, L. Y. (2011). Untapped potential: Exploiting fungi in bioremediation of hazardous chemicals. *Nature Reviews Microbiology, 9*(3), 177–192.

He, P., Chen, L., Shao, L., Zhang, H., & Lü, F. (2019). Municipal solid waste (MSW) landfill: A source of microplastics?—Evidence of microplastics in landfill leachate. *Water Research, 159*, 38–45.

Hirooka, T., Nagase, H., Uchida, K., Hiroshige, Y., Ehara, Y., Nishikawa, J. I., Nishihara, T., Miyamoto, K., & Hirata, Z. (2005). Biodegradation of bisphenol A and disappearance of its estrogenic activity by the green alga Chlorella fusca var. vacuolata. *Environmental Toxicology and Chemistry: An International Journal, 24*(8), 1896–1901.

Hong, Q., Zhang, Z., Hong, Y., & Li, S. (2007). A microcosm study on bioremediation of fenitrothioncontaminated soil using Burkholderia sp. FDS-1. *International Biodeterioration & Biodegradation, 59*(1), 55–61.

Hou, L., & Majumder, E. L. W. (2021). Potential for and distribution of enzymatic biodegradation of polystyrene by environmental microorganisms. *Materials, 14*(3), 503.

https://www.oecd.org/en/about/news/press-releases/2022/02/plastic-pollution-is-growing-relentlessly-as-waste-management-and-recycling-fall-short.html

Hyde, K. D., Jones, E. G., Leaño, E., Pointing, S. B., Poonyth, A. D., & Vrijmoed, L. L. (1998). Role of fungi in marine ecosystems. *Biodiversity & Conservation, 7*, 1147–1161.

Ignatyev, I. A., Thielemans, W., & Vander Beke, B. (2014). Recycling of polymers: A review. *Chemsuschem, 7*(6), 1579–1593.

Imhof, H. K., Ivleva, N. P., Schmid, J., Niessner, R., & Laforsch, C. (2013). Contamination of beach sediments of a subalpine lake with microplastic particles. *Current Biology, 23*(19), R867–R868.

Jeswani, H. K., & Azapagic, A. (2016). Assessing the environmental sustainability of energy recovery from municipal solid waste in the UK. *Waste Management, 50*, 346–363.

Kale, S. K., Deshmukh, A. G., Dudhare, M. S., & Patil, V. B. (2015). Microbial degradation of plastic: A review. *Journal of Biochemical Technology, 6*(2), 952–961.

Kalogerakis, N., Arff, J., Banat, I. M., Broch, O. J., Daffonchio, D., Edvardsen, T., Eguiraun, H., Giuliano, L., Handå, A., López-de-Ipiña, K., & Marigomez, I. (2015). The role of environmental biotechnology in exploring, exploiting, monitoring, preserving, protecting and decontaminating the marine environment. *New Biotechnology, 32*(1), 157–167.

Kawai, F., Kawabata, T., & Oda, M. (2019). Current knowledge on enzymatic PET degradation and its possible application to waste stream management and other fields. *Applied Microbiology and Biotechnology, 103*, 4253–4268.

Kensa, V. M. (2011). Bioremediation-an overview. *Journal of Industrial Pollution Control, 27*(2), Article 161168.

Khoironi, A., Anggoro, S., & Sudarno, S. (2019). Evaluation of the interaction among microalgae Spirulina sp, plastics polyethylene terephthalate and polypropylene in freshwater environment. *Journal of Ecological Engineering, 20*(6), 161–173.

Kim, H. (2015). A study on the utilization of the earthworms Eisenia fetida and Eisenia andrei for the disposal of polymers.

Kim, J. W., Park, S. B., Tran, Q. G., Cho, D. H., Choi, D. Y., Lee, Y. J., & Kim, H. S. (2020). Functional expression of polyethylene terephthalate-degrading enzyme (PETase) in green microalgae. *Microbial Cell Factories, 19*, 1–9.

La Mantia, F. P. (Ed.). (2002). *Handbook of plastics recycling*. iSmithers Rapra Publishing.

Sridharan, S., Kumar, M., Bolan, N. S., Singh, L., Kumar, S., Kumar, R., & You, S. (2021). Are microplastics destabilizing the global network of terrestrial and aquatic ecosystem services? *Environmental Research, 198*, 111243.

Thompson, R. C., Olsen, Y., Mitchell, R. P., Davis, A., Rowland, S. J., John, A. W., McGonigle, D., & Russell, A. E. (2004). Lost at sea: Where is all the plastic? *Science, 304*(5672), 838–838.

Chapter 10
Microbial Degradation of Microplastics

Giorgia Perpetuini and Rosanna Tofalo

Abstract Microplastics pollution has become a widely discussed concern in recent years as these tiny particles have infiltrated every corner of the environment. Microplastics, defined as plastic particles smaller than 5 mm, pose significant environmental and health challenges due to their persistence, widespread distribution, and potential to carry toxins. Several methods have been suggested to break down microplastics. Bioremediation is a highly effective method that shows great potential. Bioremediation of microplastics refers to the use of biological organisms, typically microbes such as bacteria, fungi, and algae, to degrade or transform microplastic particles into less harmful substances. In this chapter the roles of bacteria, fungi and their enzymes in plastic biodegradation is discussed. Moreover, the main advantages and disadvantages of this approach are analysed. The last part of the chapter explores the main challenges in microbial degradation, as well as the use of advancement such as nanotechnology and their potential role in resolving plastic pollution challenges.

10.1 Introduction

The use of plastic materials doubled between 2000 and 2019, reaching about 460 million metric tons by 2019. The global plastic usage could nearly triple by 2060, reaching 1321 million metric tons. This growth is mainly linked to the industrial growth and the consumer demand for plastic products across multiple sectors. Similarly, the plastic waste increased, with over 350 million metric tons of plastic waste produced each year worldwide. Without specific policies the plastic waste will reach about 1 billion metric tons by 2060. Only 21% of plastic has been recycled or incinerated (Law, 2017), while the rest is weathered and broken down into tiny fragments that are defined as microplastics (MP). Microplastics (plastic particles, manufactured or debris, $\leq$ 5 mm in length) are present worldwide and can accumulate in the environment (Batel et al., 2016). One of the methods proposed to decrease

G. Perpetuini (✉) · R. Tofalo
Department of Bioscience and Technology for Food, Agriculture and Environment, University of Teramo, Teramo, Italy
e-mail: gperpetuini@unite.it

V. Kumar et al. (eds.), *Microplastics Pollution*,
https://doi.org/10.1007/978-3-031-99774-7_10

MP pollution is the use of microorganisms (Krueger et al., 2015). The ability of certain microbes to degrade MP depends on their metabolic evolution due to their exposure to xenobiotic polymers released into the environment from human activities (Meng et al., 2021). The selective pressure exerted by the environment forces the adaptative evolution of microorganisms: they can adapt to environmental conditions thanks to their high genome plasticity, metabolic versatility, and capacity to exchange genetic materials and to appropriately mix and match the desired catabolic pathways (Yuan et al., 2020).

10.2 Microplastics Degrading Bacteria

Several studies highlighted the ability of bacteria to degrade MP. Bacteria can utilize polymer materials as their exclusive carbon source in minimal media and modify the polymer's structural integrity, chemical characteristics, tensile strength, and average molecular weight. These bacteria have different origins e.g. sediments, landfills, sludge, wastewater, dump sites, marine environments, insects, recycling sites, and cow dung (Wang et al., 2020).

Different species have been shown to degrade polyethylene terephthalate (PET) (Table 10.1).

Generally, the bacteria involved in PET degradation are Gram-positive and belong to the phylum *Actinobacteria* (Acero et al., 2011). The main bacteria showing this activity belong to the genera *Thermobifida* and *Thermomonospora* (Acero et al., 2011; Danso et al., 2018). During PET degradation microbes adhere to PET surface and release PET-degrading enzymes. According to Roth et al. (2014), the main enzymes involved are serine hydrolases such as cutinases (EC 3.1.1.74), lipases (EC 3.1.1.3), carboxylesterases (EC 3.1.1.1), PET hydrolase, and tannase. These enzymes hydrolyze the ester bond of PET forming bis(2-hydroxyethyl) terephthalate (BHET) and monohydroxyethyl terephthalate (MHET), which are further degraded into terephthalic acid (TPA) and ethylene glycole (EG), respectively (Yoshida et al., 2016).

Yoshida et al. (2016) isolated *Ideonella sakaiensis* 201-F6 strain which uses PET as a major carbon source. The *I. sakaiensis* genome encodes a unique enzyme, identified as a tannase, which degrades mono(2-hydroxyethyl) terephthalic acid, in addition to the hydrolase. The secreted PET hydrolase generates the intermediate MHET. MHET is likely internalized by the cell and then hydrolyzed by MHETase. The obtained monomers are destroyed in a downstream step and utilized for bacterial metabolism. Auta et al. (2018) investigated the polypropylene (PP) degradation using bacteria isolated from marine sediments. *Bacillus* sp. 27 strain and *Rhodococcus* sp. 36 strain used PP as carbon source, and induced a weight loss of 4% and 6.4%, respectively after 40 days of incubation.

Patil (2018) screened different bacteria (*Bacillus amylolyticus*, *B. subtilis*, *Pseudomonas putida,* and *Pseudomonas fluorescens*) isolated from soil for their ability to degrade polyethylene (PE). The strains were incubated in a broth medium containing

Table 10.1 Main bacterial species involved in MP degradation

Bacteria	Target MP	Reference
Rhodococcus ruber	PE	Auta et al. (2018)
Ideonella sakaiensis	PET	Yoshida et al. (2016)
Pseudomonas putida	PS	Mohan et al. (2016)
Bacillus cereus	PE, PS	Auta et al. (2017)
Bacillus gottheilii	PE, PET, PS	Auta et al. (2017)
Exiguobacterium sp.	PS	Yang et al. (2014)
Stenotrophomonas maltophilia	PLA	Jeon and Kim (2013a)
Paenibacillus sp.	PE	Yuan et al. (2020)
Bacillus subtilis	PU	Shah et al. (2013)
Chelatococcus sp.	PE	Jeon and Kim (2013b)
Enterobacter asburiae	PE	Yang et al. (2014)
Pseudomonas aeruginosa	PS, PU	Shimpi et al. (2012)
Arthrobacter sp.	LDPE	Sangeetha Devi et al. (2015)
Acinetobacter sp.	PP	Auta et al. (2018)
Lysinibacillus fusiformis	PE	Restrepo-Flórez et al. (2014)
Comamonas acidovorans	PU	Nakajima-Kambe et al. (1997)

Abbreviations: *PE* polyethylene; *PET* polyethylene terephthalate; *PS* polystyrene; *PLA* polylactic acid; *PU* polyurethane; *LDPE* low-density polyethylene; *PP* polypropylene

PE film for a month. *Bacillus amylolyticus* reduced 32% of PE film weight, followed by *P. fluorescens* with 22%, *P. putida* 18%, and *B. subtilis* 14%. A similar ability was detected for *P. aeruginosa* and *P. stutzeri* (Agrawal & Singh, 2016).

Several studies demonstrated the ability of bacteria to degrade low-density polyethylene (LDPE) which is composed of short branches of carbon chains and is therefore very stable (Delacuvellerie et al., 2019; Shresta et al., 2019). For instance, Delacuvellerie et al. (2019) studied the MP degradation ability of *Alcanivorax borkumensis* (a hydrocarbonoclastic bacteria growing on alkanes). The LDPE degradation mediated by this bacterium involves various enzymes, such as alkane hydroxylase (alkB1, alkB2), cytochrome P50, and ferredoxin. However, the degradation conditions should be optimized in terms of temperature, supplemental carbon source, molecular weight of LDPE, UV treatment in order to open new perspectives to use this bacterium alone or in consortium to recycle LDPE.

Some bacteria can degrade modified plastics. Shimpi et al. (2012) showed that *Pseudomonas aeruginosa* degraded polystyrene (PS)-poly(lactic acid) (PLA) nanocomposites, with a degradation rate of 10%. Brominated high-impact polystyrene (HIPS) was degraded by *Bacillus* spp. with a weight loss of 23% in 30 days. However, despite the potential of bacteria in MP degradation, it should be noted that the number of bacteria selected for MP-degradation is small, and the degradation rate is low (0–15%), probably because MP are poorly biodegradable. Moreover, the degradation process is long (0–3 months).

10.3 Extremophilic Bacteria and Microplastic Degradation

Extremophilic bacteria are microbes able to survive under extreme environmental conditions such as high temperature, pH, or salinity. They have been proposed as a promising tool to degrade MP, since they show unique metabolic pathways and enzymes able to degrade otherwise persistent polymers. The main extremophiles with a potential application in MP degradation are reported in Table 10.2.

For instance, *Thermobifida fusca* can degrade PET by enzymatic hydrolysis with the support of high temperatures that imply higher flexibility of the polymer chain and lower crystallinity, increasing its accessibility to bacterial enzymes. Thermophilic environments enhance the bioavailability of such polymers, besides increasing the activity of thermophilic enzymes, which generally are more stable and resistant to denaturation than those derived from mesophilic bacteria. The main enzymes involved in this process (polyester hydrolases and cutinases) are thermophilic enzymes and degrade PET into TPA and EG which are more easily assimilated by other organisms or further processed for recycling (Müller et al., 2005; Yoshida et al., 2016).

Another promising group of estremophiles are alkaliphilic bacteria, which can degrade LDPE, a common plastic in packaging. These bacteria live in high-pH environments, which facilitate some chemical changes in plastics, such as increased brittleness which increases susceptibility to microbial degradation. Certain species, like *Bacillus pseudofirmus*, have been shown to degrade PE by as much as 10% under conditions of long-term exposure. The degradation rate increases when iron oxide nanoparticles are provided. These nanoparticles enhance bacterial attachment and biofilm formation on the plastic surface, further accelerating degradation. The effect is probably due to the properties of the nanoparticles, such as magnetism and electrostatic charge (Cada et al., 2019). In particular, a polymer weight losses of 7% were observed for the pure culture, while in the presence of nanoparticles an increase of about 10% was observed (Cada et al., 2019).

Halophilic bacteria can survive in high-salt environments such as saline lakes and marine waters. These microbes could be used for the bioremediation of oceanic and coastal ecosystems since plastic debris is accumulated in these niches. They can form biofilms on plastic surfaces, which seem essential for initiating biodegradation. Examples include species of the genus *Erythrobacter* growing in marine plastic debris and playing a role in biofilm development that protects not only bacterial communities but even enhances degradation by extracellular enzymes, which are maintained in close contact with the plastic surfaces (Debroas et al., 2017).

Psychrophilic bacteria can develop in temperatures lower than 15 °C, and they play a major role in the biodegradation of plastics in cold marine environments, such as deep-sea sediments and polar waters. Such bacteria show specific adaptations strategies to counteract freezing, such as antifreeze proteins and changes in cell membrane structure, enabling their activity at low temperatures where enzymes from other organisms cannot function. Species of the genera *Shewanella* and *Moritella* can

Table 10.2 Extremophilic bacteria capable of degrading microplastics

Extremophile	Bacterium	Plastic target	Degradation characteristics	References
Thermophilic	*Thermobifida fusca*	PET	Produces polyester hydrolases and cutinases active at 55–70 °C, breaking PET into TPA and EG	Müller et al. (2005)
	Brevibacillus borstelensis	LDPE	Reduces LDPE molecular weight by 11% after 30 days at 50 °C	Hadad et al. (2005)
	Clostridium thermocellum	PET	Capable of PET degradation with over 60% weight loss after two weeks at 60 °C	Oda et al. (2018)
Alkaliphilic	*Bacillus pseudofirmus*	PE	Degrades polyethylene at pH 11; degradation enhanced by iron oxide nanoparticles that support biofilm formation	Cada et al. (2019)
	Bacillus krulwichiae	LDPE	Causes LDPE weight loss under high-pH conditions; secretes extracellular enzymes that perform hydrolysis under alkaline conditions	Atanasova et al. (2021)
Halophilic	*Erythrobacter litoralis*	PE, PP	Forms biofilms on PE and PP debris in marine environments, maintaining enzyme activity on plastic surfaces	Debroas et al. (2017)
	Halomonas titanicae	General plastics	Degrades plastics in high-salt conditions using salt-tolerant enzymes	Shivlata and Satyanarayana (2015)
Psychrophilic	*Shewanella oneidensis*	PCL	Exhibits plastic degradation activity in cold environments (below 15 °C), suitable for polar and deep-sea conditions	Tokiwa and Calabia (2009)
	Moritella marina	PCL	Degrades PCL in high-pressure, low-temperature marine environments, suggesting applicability in deep-sea plastic bioremediation	Yoshida et al. (2016)

Abbreviations: *PP* polypropylene; *PCL* polycaprolactone; *TPA* terephthalic acid; *EG* ethylene glycole degrade polycaprolactone (PCL) under high-pressure and low-temperature conditions. Therefore, they can be used for cleaning up plastic pollution in deep-sea environments where cold temperatures and high pressures generally inhibit microbial activity (Tokiwa & Calabia, 2009).

Despite the promising results, the industrial application of extremophiles for plastic biodegradation is still limited by technical challenges. In extremophilic microorganisms, growth and biomass production are usually slower compared to those of mesophilic ones; limiting their scalability in industrial processes. Furthermore, their enzymes can be substrate-specific and susceptible to inhibition by degradation by-products. These limitations could be overcome through genetic engineering. The enzymes could be modified to enhance their stability and substrate affinity. Moreover, metagenomic studies could be useful to identify, isolate, and characterize genes from environmental samples with high plastic-degrading activities. Further knowledge and technological advances in the exploitation of extremophiles and their enzymes are likely to contribute to a sustainable biological solution to mitigate the problem of plastic pollution.

10.4 Fungal Degradation of Microplastics

The biodegradation of plastics by fungi relies on the secretion of various extracellular enzymes that break down complex synthetic polymers into simpler, absorbable compounds (Table 10.3). Among these, oxidases, peroxidase, laccase, monooxygenase, esterase, cutinase, lipase, protease, urease, and hydrolases play crucial roles. Oxidative enzymes, such as laccases, peroxidases, and manganese peroxidases (MnP), initiate the degradation of nonpolar, stable plastics by incorporating oxygen-containing functional groups into the polymer chains. These reactions, often initiated by reactive oxygen species (ROS), make the polymers more hydrophilic and accessible to enzymatic action, thus promoting further degradation. For instance, *Phanerochaete chrysosporium* secretes MnP and laccase to oxidize PE and PP, creating carbonyl and hydroxyl groups that weaken the polymer structure, leading to fragmentation (Okal et al., 2023).

Hydrolases, including cutinases, lipases, and esterases, specifically target plastics with ester linkages, such as PET and PU. These enzymes lack the lid structure found in lipases, giving cutinases, like those from *Aspergillus tubingensis*, a wide substrate-binding groove that improves their accessibility to the plastic surface. They catalyze the hydrolysis of ester bonds, breaking down polyester-based plastics into intermediates like TPA and EG, which fungi can metabolize via the tricarboxylic acid (TCA) cycle. Although lipases show some activity against PET, their efficiency is often limited by a hydrophobic lid over the active site. Additions, such as surfactants, can boost lipase activity by improving its interaction with the polymer surface, especially for PET (Taniguchi et al., 2019). Oxidase and hydrolase initiate MP degradation forming small fragments which are further assimilated and mineralized in the cell (Urbanek et al., 2020). However, some polymers, such as polyolefins (PE,

Table 10.3 Main fungal species able to degrade microplastics

Target plastic	Fungi	References
PE	*Alternaria alternata, Zalerion maritimum, Trametes versicolor*	Gao et al. (2022), Paco et al. (2017), Iiyoshi et al. (1998)
PET	*Moniliophthora roreri, Theromyces lanuginosus, Thermobifida fusca, Fusarium solani*	Eberl et al. (2009), Vazquez-Alcantara et al. (2021)
PU	*Cladosporium* spp., *Aspergillus flavus, Aspergillus tubingensis, Aspergillus oryzae, Aspergillus niger*	Alvarez-Barragan et al. (2016), Khan et al. (2017), Khruengsai et al. (2022), Garrido et al. (2012), Ohtaki et al. (2006)
PS	*Penicillium variabile*	Tian et al. (2017)

PS, PP, PVC) show a compact structure which make them difficult to be degraded. Some fungal species *Fusarium falciforme, Purpureocillum lilacinum, Zelererion maritimum, Psychotria flavida, Humboldtia brunonis, Paecilomyces lilacinum* and *Lasiodiplodia theobromae* release oxidoreductase such as peroxidase, laccase, and oxygenase which facilitate their degradation (Sivan, 2011). In presence of synthetic polymers, microbes are exposed to carbon starvation and oxidative stress which induce the release of oxidative enzymes, H_2O_2, and ROS. The oxidases modify the polyolefin surface altering the morphology, the molecular weight, and the terminal group composition. These modifications result in plastic degradation (Saravanan et al., 2021).

Laccases and peroxidases act similarly. They promote redox reactions which form ROS, that in turn oxidise the polymer favoring the degradation (Okal et al., 2023). They are produced by various fungi, including *Trametes versicolor* and *Pleurotus ostreatus*, and exhibit broad substrate specificity. Microplastics, particularly those containing aromatic or synthetic polymeric backbones, can be attacked by laccases, leading to partial depolymerization or surface modifications that enhance biodegradability (Okal et al., 2023).

Esterases can cleave ester bonds of plastic polymers forming acid- and alcohol-based molecules with polar groups which are further metabolized in not harmful compounds (Han et al., 2017). Esterases are subjected to carbon catabolite repression; therefore, their expression is inhibited in presence of easily metabolizable sugars e.g. glucose and fructose (Tanaka & Gomi, 2021). The main transcriptional regulators are Mig1/CreA/Cre1, CreB and CreC which bind the promoters of genes encoding the esterases suppressing their expression when the glucose is present in high concentration. When simple sugars are limited, CreA, Mig1, and CreB are degraded and the esterases are produced (Tanaka & Gomi, 2021).

Moreover, several basidiomycetes and ascomycetes produce proteases, amidases, and ureases involved in the degradation of PU and polyamides (PA) (Okal et al., 2023). Very few studies have been carried out to explore the molecular mechanisms underlying fungal MP degradation. Fungal carbon metabolism is regulated by the type and composition of carbon source. Generally, microbial degradation of plastics is accelerated when simple sugars are present in low concentration, and the genes involved in biofilm formation are up-regulated (Okal et al., 2023). In the presence of synthetic polymers, the microbes use the limited carbon sugars available to colonize the plastic surface, and degradation enzymes are produced. According to Vrsanska et al. (2015) the expression of fungal laccases is induced by plastic polymers. In general, these enzymes are repressed in the presence of high glucose concentration, while they are activated by nitrogen deficiency and starvation. Moreover, the occurrence of heavy metals ions like Cu^{2+}, Mn^{2+}, and Zn^{2+} induces laccases expression during the degradation of aromatic compounds (Galhaup et al., 2002).

Fungi also use mechanical and chemical means to accelerate degradation. As fungal mycelia colonize plastics, they exert pressure on the polymer surface, and combined with enzymatic activity can form surface pits and cracks, increasing the area available for degradation. Studies indicate that *Zalerion maritimum* degrades PE by forming oligomers with carbonyl and hydroxyl groups, making the polymer more susceptible to enzymatic attack and further breakdown. Environmental pretreatments, like UV radiation or thermal oxidation, can also create functional groups on plastics that fungal enzymes more readily target, enhancing overall degradation rates (Okal et al., 2023).

These enzymatic processes highlight fungi's potential in bioremediation applications for managing plastic waste. Challenges persist, especially in optimizing enzyme efficiency and stability across varied environmental conditions. Ongoing research focuses on regulatory pathways and gene expression patterns that enhance enzyme production in fungi, aiming to maximize plastic degradation for industrial use.

10.5 Role of Microbial Consortia and Biofilm in Microplastic Degradation

Microbial consortia are made up of different microbial species and represent a valuable tool to improve MP degradation, since they establish synergistic interactions, where metabolites produced by one species serve as substrates for others, thus facilitating a sequential breakdown of complex polymers. This interspecies cooperation minimizes the accumulation of toxic byproducts that could inhibit microbial growth and activity (Dobretsov et al., 2013). In addition to their degradative capacities, the biofilm-forming capacity of microbial consortia accelerates plastic degradation by creating persistent, surface-attached communities that concentrate metabolic activity on plastic surfaces. This biofilm layer facilitates enhanced interactions between enzymes and substrates while safeguarding microbial cells from diverse environmental stresses, hence maintaining degrading capacity throughout time (Yuan et al., 2020).

Microbial biofilms are aggregates of microorganisms adhered to a biotic or an abiotic surface embedded in a self-produced matrix made up of exopolysaccharides which provides structural integrity and facilitates cell-to-cell communication.

The microbes present in the biofilm are able to better face environmental stresses since they are protected by the matrix. Biofilm architecture includes the presence of channels for water and nutrient transportation, creating a unique microenvironment within the biofilm. It should be noted that sessile cells show different gene expression profiles compared to those in planktonic states. This altered expression affects their metabolic activities, stress responses, and virulence factors, making them more resilient in hostile environments. The main steps involved in biofilm formation are reported in Fig. 10.1.

Biofilms can impair the structure and function of MP through several mechanisms, such as altering surface features, degrading additives, secreting enzymes that change or degrade MP, and releasing metabolic byproducts (Miao et al., 2019). The biofilm-mediated breakdown of MP can be divided into 4 steps: (i) the microorganisms adhere on the MP surface and modify its characteristics (e.g., hydrophilicity); (ii) the leaching of additives and monomers from the MP intensifies; (iii) microbial enzymes degrade MP, resulting in embrittlement and diminished mechanical stability; (iv) the accumulation of water and microbial filaments infiltrates the MP, resulting in its decomposition and subsequent utilization by microbes (Yuan et al., 2020). It seems that biofilms utilize MP as adhesion substrates, carbon sources, and energy sources, targeting and destroying them through the activity of specific enzymes. Lobelle and Cunliffe (2011) examined the biofilm development on the surfaces of

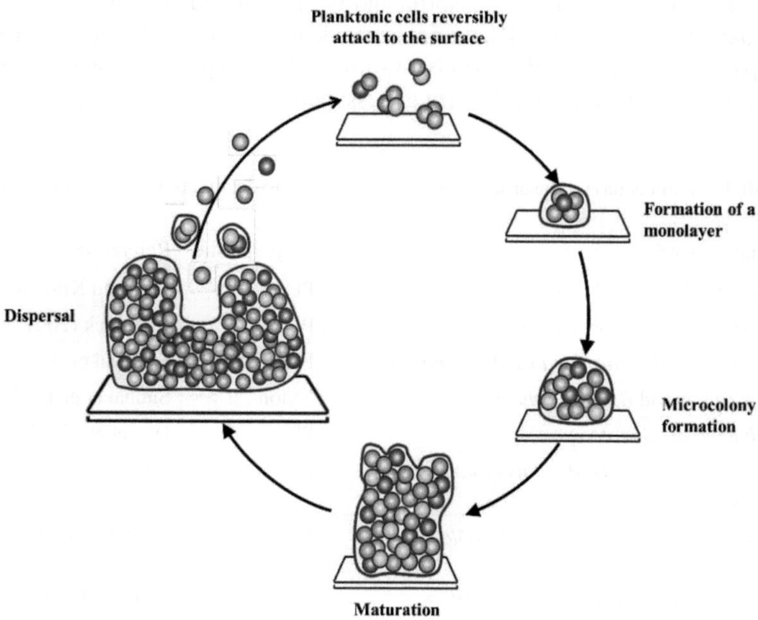

Fig. 10.1 Mechanism of biofilm formation. At the beginning the planktonic cells attach reversibly to biotic or abiotic surfaces; then the cells produce an extracellular exopolysaccharide matrix and form microcolonies. With cell division, a mature biofilm is gradually formed. Finally, the dispersion of bacterial cells from the biofilm is observed

PE microplastics for 3 weeks. Microbial biofilms developed rapidly on the MP and coincided with significant changes in the physicochemical properties of the plastic (the hydrophilicity markedly increased over the 3 weeks). Hossain et al. (2019) tested a biofilm of *Acinetobacter calcoaceticus*, *Burkholderia cepacia*, and *Escherichia coli* against PP and noted that degraded PP discs exhibited the greatest bacterial abundance, highlighting the significance of surface roughness and physicochemical characteristics of MP in bacterial colonisation.

Some examples of microbial consortia used for MP degradation are reported below and in Table 10.4.

Park and Kim (2019) showed that a consortia of *Bacillus sp.* and *Paenibacillus sp.* separated from landfill sediments, caused a reduction in the weight and particle size of PE, with 14.7% dry weight loss and a 22.8% decrease in diameter within 60 days. A consortium made up of *Rhodococcus sp.* and *Bacillus sp.* was effective against PP in vitro, this is evidenced by surface roughening, cracking, and weight loss (Singh & Wahid, 2015).

Roberts et al. (2020) obtained an 8% loss in PET weight after 8 weeks using a consortium of five strains of *Pseudomonas* and *Bacillus* species. Gao and Sun (2021) developed a synthetic consortium made up of *Ochrobactrum, Halomonas*, and *Exiguobacterium* species and their associated enzymes that decomposed 30 mm × 20 mm PET and PE fragments after two weeks of incubation on a minimum medium.

The degradation of MP by consortia and biofilm should be better studied since the breakdown products of MP have not been adequately characterized. In particular, the impact of biofilms on MP depends on various physical properties e.g. MP surface chemistry and structure, including roughness, particle size, and hydrophobicity.

Table 10.4 Main bacterial consortia involved in microplastic degradation, highlighting the target plastics

Bacterial consortia	Target plastic	References
Bacillus sp., *Paenibacillus* sp.	PE	Park and Kim (2019)
Rhodococcus sp. *Actinobacter* sp.	PE	Charnock (2021)
Bacillus cereus, Bacillus pumilus, Arthrobacter spp.	HDPE, LDPE	Satlewal et al. (2008)
Bacillus cereus and *Bacillus sphaericus*	Nylon	Sudhakar et al. (2007)
Arthrobacter sp., *Streptomyces* sp.	PE	Han et al. (2020)
Lysinibacillus fusiformis, Bacillus aryabhattai, Microbacterium oxydans	LDPE	Montazer et al. (2020)
Aspergillus niger, Aspergillus flavus, Aspergillus oryzae	PE	De Souza et al. (2021)
Bacillus siamensis, Bacillus wiedmannii	PE	Maroof et al. (2021)
Aneurinibacillus sp., *Brevibacillus* sp.	LDPE, PE	Skariyachan et al. (2018)
Bacillus vallismortis, Pseudomonas protegens, Stenotrophomonas sp., *Paenibacillus* sp.	LDPE	Skariyachan et al. (2017)

10.6 Microbial Enzymes Involved in Microplastic Degradation

The degradation is mediated by several enzymes and the main are hydrolases and oxidoreductases (Table 10.5). The biodegradation comprises 4 steps: biodeterioration, biofragmentation or depolymerization, assimilation, and mineralization (Mohanan et al., 2020) (Fig. 10.2).

Table 10.5 Main microbial enzymes involved in MP degradation

Enzyme	Microorganism	Polymer type	Reference
Manganese peroxidase	*Phanerochaete chrysosporium*, *Trametes versicolor*	PE	Iiyoshi et al. (1998)
Alkane hydroxylase	*Pseudomonas* sp.		Yoon et al. (2012)
Laccase	*Rhodococcus ruber*, *Aspergillus flavus*, *Pleurotus ostreatus*		Santo et al. (2013), Zhang et al. (2020), Gomez-Mendez et al. (2018)
Esterase	*Lentinus tigrinus*	PS	Tahir et al. (2013)
PETase and MHETase	*Ideonella sakaiensis*	PET	Yoshida et al. (2016)
Suberinase	*Streptomyces scabies*		Jabloune et al. (2020)
Cutinase	Various microbial sources	PU, PET	Danso et al. (2019), Jabloune et al. (2020)
6-Aminohexanoate-dimer hydrolase	*Flavobacterium*, *Pseudomonas*	Nylon	Negoro (2000)

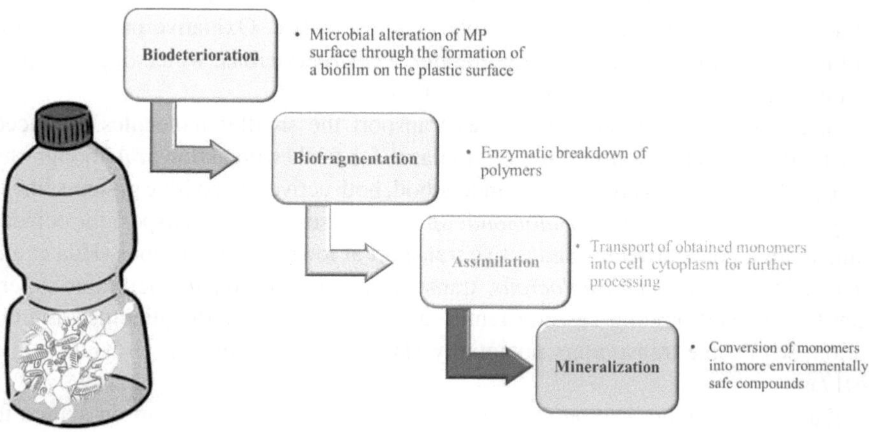

Fig. 10.2 Microplastic degradation by microorganisms involves colonization of plastic surfaces, enzymatic breakdown of polymers into smaller molecules, and further mineralization into harmless end products

During the first step, microbes colonize plastic surfaces and form biofilms, leading to cracks and holes in the material. They also reduce the pH, which accelerates chemical degradation. Microbial attachment begins with adherence to the hydrophobic plastic surface, often made easier by the introduction of hydrophilic groups. For hydrophobic plastics like PE, biofilms enhance interactions between bacteria and plastics favoring plastic degradation. Fungi also attach to plastics using hyphae, causing localized swelling and weakening the polymer structure. Microbes use polymers as a carbon source, with additives like plasticizers promoting their growth. Exopolysaccharides stabilize biofilms and improve adhesion, allowing better exchanges between hydrophilic and hydrophobic phases, which boosts microbial penetration into the plastic architecture.

During the first step, microbes colonize plastic surfaces and form biofilms, leading to cracks and holes in the material. They also reduce the pH, which accelerates chemical degradation. Microbial attachment begins with adherence to the hydrophobic plastic surface, often made easier by the introduction of hydrophilic groups. For hydrophobic plastics like PE, biofilms enhance interactions between bacteria and plastics favoring plastic degradation. Fungi also attach to plastics using hyphae, causing localized swelling and weakening the polymer structure. Microbes use polymers as a carbon source, with additives like plasticizers promoting their growth. Exopolysaccharides stabilize biofilms and improve adhesion, allowing better exchanges between hydrophilic and hydrophobic phases, which boosts microbial penetration into the plastic architecture.

During the depolymerization phase, microbes secrete extracellular enzymes such as peroxidases, lipases, depolymerases, and esterases. These enzymes break plastics into smaller oligomers and monomers that can pass through microbial membranes during assimilation. This biofragmentation involves two key reactions: reducing the polymer's molecular weight and oxidizing smaller molecules to prepare them for further enzymatic action. Hydrolytic enzymes cleave polymer bonds via nucleophilic attacks, creating fragments that microbes can assimilate. Oxidative processes also aid in this fragmentation, as seen in *Rhodococcus rhodocrous*, which degrades pre-oxidized PE oligomers (Gravouil et al., 2017).

In the assimilation stage, microbes transport the smaller molecules produced during depolymerization into their cytoplasm for further breakdown. Although the transport mechanisms are not fully understood, both active and passive systems likely play a role. For example, *Pseudomonas sp.* DG17 uses passive transport for octadecane at high concentrations and active transport at lower concentrations (Hua et al., 2013). In *Rhodococcus rhodocrous*, transporters from the major facilitator superfamily and ATP-binding cassette family are upregulated for PE oligomer assimilation, with some transporters potentially also aiding in oxidation (Gravouil et al., 2017).

During the mineralization step, the microbe can use MP as a carbon source to produce more environmentally safe compounds. In particular, microbial biomass, CO_2, and H_2O are produced during aerobic degradation, while CO_2, H_2O, and CH_4 or H_2S are the main products of anaerobic degradation (Chigwada & Tekere, 2023). Studies using isotopic tracing and CO_2 quantification have confirmed complete plastic mineralization (Yang et al., 2020a, 2020b). Intermediates from plastic breakdown can also enter other metabolic pathways. For instance, PE degradation produces acetic acid, which enters the Krebs cycle as acetyl-CoA or contributes to lipid formation. Similarly, polyethersulfone degradation by *Pseudomonas sp.* AKS2 produces succinate, a Krebs cycle intermediate. Polystyrene breaks down into phenylacetate, which is metabolized as phenylacetyl-CoA. In *Ideonella sakaiensis*, TPA degrades into protocatechuic acid (PCA), which further breaks down into pyruvate and oxaloacetate before complete mineralization. Key enzymes in these processes include esterases, lipases, cutinases, peroxidases, and laccases (Alshehrei, 2017).

Hydrolases and oxidoreductases are the primary enzymes involved in MP degradation. Hydrolases like cutinases, esterases, keratinases, and lipases degrade plastics by incorporating water molecules into polymers, breaking them down via nucleophilic substitution (Edwards et al., 2022). Oxidoreductases, including hydroxylases, laccases, lignin peroxidases, and manganese peroxidases, degrade plastics by adding oxygen, hydrogen, or electrons to the polymers (Carmen, 2021).

PETase are the most studied enzymes. Yoshida et al. (2016) described the PET degradation process. It is a 2-step reaction mediated by PETase and MHTEase. PETase converts PET to mono(2-hydroxyethyl) terephthalic acid (MHET), and TPA and bis(2-hydroxyethyl)-TPA, MHETase hydrolyzes MHET into TPA and EG. According to Li et al. (2019) EG can be further oxidized in glycolaldehyde, glyoxal, glycolate, and glyoxylate which are used for the production of agro-, aroma- and polymer chemicals, or pharmaceuticals. The EG metabolic pathway has been described in *Pseudomonas putida* which can use EG as a sole source of carbon and energy producing glycolic acid and glyoxylic acid (Wehrmann et al., 2017). Muller et al. (2005) investigated the mechanism of PET-degrading cutinase, and suggested 2 mechanisms. The first one is based on the modification of polyester fibers and involves surface-modifying enzymes or hydrolases, including lipases, cutinases, carboxylesterases and proteases. These enzymes do not hydrolyze the PET polymer chains, but degrade the surface of the polymer. The second one is based on the hydrolysis of the polymer. However, only few cutinases have been recognized as PET hydrolases that can degrade the inner block of PET.

The degradation of polyurethane (PU) is mediated by PU-esterases (PudA). Akutsu et al. (1998) proposed a 2-step process. At the beginning a surface-binding domain favors PudA adsorption to the surface, and then the hydrolysis of ester bonds occurs.

Islam et al. (2019) demonstrated that fusing the cutinase enzyme from *Thermomonospora curvata* (Tcur1278) with the anchor peptide Tachystatin A2 (TA2) significantly enhanced the binding and degradation of polyester-polyurethane nanoparticles. Anchor peptides, which are metal-binding adhesion molecules, provide an effective method for targeted degradation of MP and polymers such as PET, PP, and PU, even under ambient conditions.

Magnin et al. (2019) recently highlighted the synergistic effects of esterase and amidase enzymes in a proposed stepwise mechanism for degrading various PU derivatives. Using proximity ligation-based metagenomic analysis, researchers identified multiple enzymes within a microbial consortium that contribute to PU breakdown. These enzymes, which metabolize PU intermediates, include dioxygenases, decarboxylases, dehydrogenases, transferases, ligases, hydrolases, isomerases, and peroxidases (Gaytán et al., 2020). However, most PU-active enzymes studied to date predominantly target ester-linked PU rather than polyurethane ethers. This discrepancy is likely due to the carbonyl group present in ester-linked PUR, which makes them more susceptible to enzymatic hydrolysis, while ether-linked PU are more resistant.

Polystyrene (PS) is one of the more challenging plastics to biodegrade due to its high hydrophobicity, aromatic structure, and large molecular weight, which make it particularly resistant to microbial breakdown. However, recent studies have identified certain bacterial strains capable of initiating PS degradation under specific conditions. For instance, *Exiguobacterium sp.*, *Pseudomonas putida*, and *Rhodococcus ruber* have shown potential in PS biodegradation through pathways involving oxidation of the styrene monomer side chains. This process often begins with the enzymatic conversion of styrene into styrene oxide by styrene monooxygenase. Styrene oxide is then converted into phenylacetaldehyde, which subsequently degrades to phenylacetic acid. This compound is further transformed into phenylacetyl-CoA, which can then enter the tricarboxylic acid (TCA) cycle for energy production and biomass generation (Jacquin et al., 2019).

A unique aspect of polystyrene degradation is observed in *Pseudomonas putida* strain CA-3, which can utilize phenylacetic acid as a substrate to accumulate polyhydroxyalkanoates (PHAs) as byproducts during PS metabolism. This pathway, known as the PACoA catabolon, represents an alternate fate for PS degradation intermediates, where instead of full mineralization, the intermediates are redirected into biosynthetic pathways for storage polymers like PHAs. The PACoA pathway involves enzymes that oxidize and convert the phenylacetic acid intermediates through β-oxidation-like processes, eventually forming medium-chain-length PHAs (Jacquin et al., 2019).

10.7 Factors Influencing Microbial Degradation of Microplastics

Microbial degradation of plastics is influenced by several factors including the properties of the MP, environmental conditions, and microbial community dynamics. The main chemical characteristics of MP which influence the microbial degradation ability are: the molecular weight, crystallinity, and the presence of functional groups or additives. In particular, lower crystallinity is generally associated to a higher biodegradability, since the increased amorphous regions make the material more accessible to microbial enzymes (Yuan et al., 2020). In addition, the occurrence of additives such as plasticizers, stabilizers, and flame retardants can either inhibit or facilitate microbial colonization depending on their effects on the MP's physical structure and the types of microbial communities involved.

Concerning the environmental factors it seems that elevated temperatures generally increase microbial metabolic rates, thereby accelerating plastic degradation processes. UV exposure can initiate photo-oxidation, leading to the formation of surface cracks and reactive functional groups, which in turn increase the plastic's bioavailability for microbial degradation. In aquatic environments, pH and salinity influence microbial diversity and enzyme activity, further impacting the efficiency of plastic breakdown (Yuan et al., 2020). The presence of HNO_3, H_2SO_4, O_3, Cu^{2+}, Fe^{3+}, Mn^{3+}, Co^{2+} favors MP degradation, since they stimulate the release of extracellular enzymes (Yuan et al., 2020).

Microbial community composition is another critical determinant of plastic degradation. Studies show that bacteria, fungi, and biofilms formed on plastic surfaces contribute to the breakdown process. Biofilms, particularly, allow for the prolonged adherence of microorganisms to the plastic substrate, facilitating the continuous release of enzymes that can degrade plastic polymers. Some bacterial consortia are capable of synergistic degradation, where one species breaks down initial compounds into intermediates that another species can further metabolize. For instance, consortia involving *Bacillus* and *Paenibacillus* have demonstrated the ability to degrade PE by forming dense biofilms, which promote the breakdown of complex polymeric structures (Yuan et al., 2020). Moreover, it should be noted that microbial growth requires optimal conditions in terms of temperature, pH, oxygen availability, and activity water. For instance, during the mineralization process, CO_2 is released and its accumulation can impair the degradation process especially in the case of fungi, since it slows down fungal metabolism. Similarly, the high occurrence of oxygen and glucose harms the activity of respiratory enzymes (Okal et al., 2023).

Research suggests that the use of tailored microbial consortia or biofilm-forming microbes could enhance plastic degradation efficiency, offering a promising strategy for addressing plastic pollution in various ecosystems (Yuan et al., 2020).

10.8 Case Study: Microbial Degradation of Marine Microplastics

The accumulation of MP in marine environment represents an ecological problem of particular interest because of its negative effects on biodiversity and ecosystem health. The development of novel massive DNA sequencing techniques allowed to study the microbial community inhabiting the plastic surface in marine environment, and the term "plastisphere" was introduced. MP can represent an ecological niche which help microbes to face harsh conditions providing physical support, and nutrient supply to microbes present on it. The plastisphere is made up of bacteria fungi, viruses and algae (Wright et al., 2020). The composition of MP biofilms in marine environment fluctuates over time (Jacquin et al., 2019). At the beginning the pioneer microorganisms form assemblages and start covering the plastic surface. During the subsequent days the microbes attach to the plastic surface irreversibly, finally, the continuous recruitment, loss, and replacement of species transform the structure into a hotspot for diverse interacting communities (Zhai et al., 2023), resulting in a mature plastisphere (Zhai et al., 2023).

Some studies revealed that the first colonizing bacteria are diatoms, cyanobacteria, green algae, Gammaproteobacteria and Alphaproteobacteria (Bryant et al., 2016). In particular, the dominant early colonizers are members of the families *Saprospiraceae*, *Oscillatoriaceae*, *Oceanospirillaceae*, *Alteromonadaceae*, *Vibrio* spp., *Alcanivorax* spp., *Aestuariicella* spp., and *Marinobacter* spp. (Zhai et al., 2023). Mature plastisphere is made up of Bacteroidetes, *Alphaproteobacteria*, *Rhodobacteraceae*, *Rhodospirillaceae*, flavobacteria, *Planctomycetaceae*, and *Phyllobacteriaceae* (Pinto et al., 2019).

The composition of the plastisphere is influenced by several factors including geography, environmental factors, and MP characteristics (Yang et al., 2020a, 2020b). Some studies demonstrated that plastisphere communities reflect the biogeographic origin (Zhai et al., 2023). However, other factors can shape these microbial communities. Wright et al. (2021) demonstrated that the light availability played a key role in the definition of biofilm composition. Similarly, Oberbeckmann et al. (2018) and Jacquin et al. (2019) revealed a strong effect of salinity gradients on the presence of Gammaproteobacteria on plastic surface.

The MP degradation in marine environment is slower than in the pelagic zone, probably because of the unfavorable ratio C/N which impair hydrocarbon-based products degradation (Ishigaki et al., 2004). Actually, the number of studies concerning marine plastics degradation are limited, and include the enrichment and isolation of plastic degrading microorganisms in the offshore environment. The main species/genera isolated from marine environment with plastic-degradation ability are: *Kocuria palustris*, *Bacillus pumilus*, *Bacillus subtilis*, *Bacillus sphericus*, *Bacillus cereus*, *Arthrobacter* spp., *Pseudomonas* spp., *Lysinibacillus* spp., *Halomonas* spp., *Ochrobactrum* spp., *Alcanivorax* spp., *Hyphomonas* spp., *Cycloclasticus* spp., *Stanieria* spp., *Pseudomorphium* spp. (Zhai et al., 2023). Sudhakar et al. (2008) demonstrated that a strain of *Pseudomonas* spp., isolated from plastics immersed for several days in Bay of Bengal, degraded PP, LDPE and HDPE.

Bollinger et al. (2020) identified a new polyester hydrolase named PE-H in *Pseudomonas aestusnigri*. A site-directed mutagenesis approach was also used to obtain a variant with significantly improved activity and a binding model was proposed and important evidences about the structural features required for efficient polyester degradation were obtained and highlighted that *P. aestusnigri* may prove as a prolific source for such enzymes. Gao and Sun (2021) characterized a marine community (*Exiguobacterium* sp., *Halomonas* sp. and *Ochrobactrum* sp.) able to degrade PET and PE with a high degradation rate (24 h for PE). Further studies are necessary to better investigate the microbial community structure in the plastisphere and biofilms in order to better exploit the metabolism and MP degradation ability of marine bacteria.

10.9 Conclusion and Future Perspectives

The microbial degradation of MP presents a significant environmental challenge. In fact, the efficacy of microorganisms in breaking down various synthetic polymers remains limited. The biodeterioration is often limited by the recalcitrant nature of MP because of their crystallinity and hydrophobic properties. Moreover, only a few species can degrade MP, and often the degradation rate is low. Moreover, environmental conditions such as temperature, pH, and nutrient availability can impair this process and the optimization of these conditions is still a challenge. To overcome these limitations, further studies are necessary and should be focused on the identification of new microbial consortia which may enhance degradation efficiency compared to single species approaches. This could leverage synergistic interactions among microbes to improve overall MP breakdown. Another interesting perspective could be the combination of microbial degradation with physicochemical methods (e.g., UV treatment) which could enhance the overall efficiency of MP waste management. Moreover, enzymes obtained from various microbial communities, in particular extreme environments, should be functionally characterized. This includes the engineering of enzymes such as cutinases, peroxidases, and PETases through synthetic biological approaches to enhance their catalytic efficiency, stability, and substrate specificity for the different types of MP in various contaminated environments. Moreover, new species with degradation ability should be characterized. The isolation of microbes inhabiting the gut of insects known for their plastic-degrading abilities could uncover new microbial candidates and enzymes for biotechnological applications. The development of microbial consortia could allow to exploit the synergistic action of microbial consortia to improve the rate and pathways of degradation. Degradation by consortia mimics natural ecosystems where many organisms are involved in such breakdown of complex compounds. It is envisaged that metabolically engineered consortia involving organisms with complementary pathways of degradation will prove highly effective for multistep degradation of various types of MP. Integration of microbial degradation into current approaches to waste management, such as composting and wastewater treatment plants, thus provides exciting potential in the mitigation strategies for microplastic pollution. For example, composting plants provide optimum controlled environments for microbial growth and enzymatic activities. In wastewater treatment, microbial communities could be enhanced to degrade microplastics, reducing their release into aquatic ecosystems.

More studies based on omics approaches are necessary to identify the metabolic networks, regulation, and functional genes in microbial-driven breakdown process for MP. These will constitute the basis for specific manipulations necessary to enhance degradation performance.

However, any risks to local ecosystems have to be assessed to ensure the microbial species introduced or engineered will not affect the native biodiversity or disrupt preexisting microbial communities. Biosafety protocols and regulatory frameworks will become really important in the days to come as microbial degradation technologies are coming closer to practical realities.

These future outlooks represent a growing emphasis on interdisciplinary research that bridges microbiology, environmental science, materials engineering, and synthetic biology. As all these lines of research continue to improve, microbial degradation of MP may offer future prospects for mitigating the environmental burden of microplastics, with possible transformation of plastic waste management toward a more circular economy.

References

Acero, E. H., Ribitsch, D., Steinkellner, G., Gruber, K., Greimel, K., Eiteljoerg, I., Trotscha, E., Wei, R., Zimmermann, W., Zinn, M., Cavaco-Paulo, A., Freddi, G., Schwab, H., & Guebitz, G. (2011). Enzymatic surface hydrolysis of PET: Effect of structural diversity on kinetic properties of cutinases from Thermobifida. *Macromolecules, 44*, 4632–4640.

Agrawal, P., & Singh, R. K. (2016). Breaking Down of Polyethylene by *Pseudomonas* Species. *International Journal of Scientific and Engineering Research, 7*, 124–127.

Akutsu, Y., Nakajima-Kambe, T., Nomura, N., & Nakahara, T. (1998). Purification and properties of a polyester polyurethane-degrading enzyme from *Comamonas acidovorans* TB-35. *Applied and Environmental Microbiology, 64*, 62–67.

Alshehrei, F. (2017). Biodegradation of synthetic and natural plastic by microorganisms. *Journal of Applied and Environmental Microbiology, 5*, 8–19.

Alvarez-Barragan, J., Domínguez-Malfavon, L., Vargas-Suarez, M., Gonzalez- Hernandez, R., Aguilar-Osorio, G., & Loza-Tavera, H. (2016). Biodegradative activities of selected environmental fungi on a polyester polyurethane varnish and polyether polyurethane foams. *Applied and Environmental Microbiology, 82*, 5225–5235.

Atanasova, N., Stoitsova, S., Paunova-Krasteva, T., & Kambourova, M. (2021). Plastic degradation by extremophilic bacteria. *International Journal of Molecular Sciences, 22*, 5610.

Auta, H. S., Emenike, C. U., & Fauziah, S. H. (2017). Screening of *Bacillus* strains isolated from mangrove ecosystems in Peninsular Malaysia for microplastic degradation. *Environmental Pollution, 231*, 1552–1559.

Auta, H. S., Emenike, C. U., Jayanthi, B., & Fauziah, S. H. (2018). Growth kinetics and biodeterioration of polypropylene microplastics by *Bacillus* sp. and *Rhodococcus* sp. isolated from mangrove sediment. *Marine Pollution Bulletin, 127*, 15–21.

Batel, A., Linti, F., Scherer, M., Erdinger, L., & Braunbeck, T. (2016). Transfer of benzo[a]pyrene from microplastics to Artemia nauplii and further to zebrafish via a trophic food web experiment: CYP1Ainduction and visual tracking of persistent organic pollutants. *Environmental Toxicology and Chemistry, 35*, 1656–1666.

Bollinger, A., Thies, S., Knieps-Grünhagen, E., Gertzen, C., Kobus, S., Höppner, A., Ferrer, M., Gohlke, H., Smits, S. H. J., & Jaeger, K.-E. (2020). A novel polyester hydrolase from the marine bacterium *Pseudomonas aestusnigri*—structural and functional insights. *Frontiers in Microbiology, 11*, 114.

Bryant, J. A., Clemente, T. M., Viviani, D. A., Fong, A. A., Thomas, K. A., Kemp, P., Karl, D. M., White, A. E., & De Long, E. F. (2016). Diversity and activity of communities inhabiting plastic debris in the North Pacific Gyre. *mSystems 1*(3), e00024–16.

Cada, E. J. G., Muyot, M. L. C., Sison, J. M. C., & Baculi, R. Q. (2019). Enhanced *in vitro* biodegradation of low-density polyethylene using alkaliphilic bacterial consortium supplemented with iron oxide nanoparticles. *Philippine Science Letters, 12*, 55–69.

Carmen, S. (2021). Microbial capability for the degradation of chemical additives present in petroleum-based plastic products: A review on current status and perspectives. *Journal of Hazardous Materials, 402*, Article 123534.

Charnock, C. (2021). Norwegian soils and waters contain mesophilic, plastic-degrading bacteria. *Microorganisms, 9*, 94.
Chigwada, A. D., & Tekere, M. (2023). The plastic and microplastic waste menace and bacterial biodegradation for sustainable environmental clean-up a review. *Environmental Research, 231*, Article 116110.
Danso, D., Schmeisser, C., Chow, J., Zimmermann, W., Wei, R., Leggewie, C., Li, X., Hazen, T., & Streit, W. R. (2018). New insights into the function and global distribution of polyethylene terephthalate PET-degrading bacteria and enzymes in marine and terrestrial metagenomes. *Applied Environmental Microbiology, 84*, e02773-e2817.
Danso, D., Chow, J., & Streit, W. R. (2019). Plastics: Environmental and biotechnological perspectives on microbial degradation. *Applied and Environmental Microbiology, 85*, e01095-e1119.
De Souza, G. C., Sheriff, R. S., Ullanat, V., Shrikrishna, A., Joshi, A. V., Hiremath, L., & Entoori, K. (2021). Fungal biodegradation of low-density polyethylene using consortium of *Aspergillus* species under controlled conditions. *Heliyon, 7*, Article e07008.
Debroas, D., Mone, A., & Ter Halle, A. (2017). Plastics in the North Atlantic garbage patch: A boat-microbe for hitchhikers and plastic degraders. *Science of the Total Environment, 599*, 1222–1232.
Delacuvellerie, A., Cyriaque, V., Gobert, S., Benali, S., & Wattiez, R. (2019). The Plastisphere in marine ecosystem hosts potential specific microbial degraders including *Alcanivorax borkumensis* as a key player for the low density polyethylene degradation. *Journal of Hazardous Materials, 380*, 1–11.
Dobretsov, S., Abed, R. M., & Teplitski, M. (2013). Mini-review: Inhibition of biofouling by marine microorganisms. *Biofouling, 29*, 423–441.
Eberl, A., Heumann, S., Brückner, T., Araujo, R., Cavaco-Paulo, A., Kaufmann, F., Kroutil, W., & Guebitz, G. M. (2009). Enzymatic surface hydrolysis of poly (ethylene terephthalate) and bis (benzoyloxyethyl) terephthalate by lipase and cutinase in the presence of surface active molecules. *Journal of Biotechnology, 143*, 207–212.
Edwards, S., León-Zayas, R., Ditter, R., Laster, H., Sheehan, G., Anderson, O., Beattie, T., & Mellies, J. L. (2022). Microbial consortia and mixed plastic waste: Pangenomic analysis reveals potential for degradation of multiple plastic types via previously identified PET degrading bacteria. *International Journal of Molecular Sciences, 23*, 5612.
Galhaup, C., Goller, S., Peterbauer, C. K., Strauss, J., & Haltrich, D. (2002). Characterization of the major laccase isoenzyme from *Trametes pubescens* and regulation of its synthesis by metal ions. *Microbiology, 148*, 2159–2169.
Gao, R., Liu, R., & Sun, C. (2022). A marine fungus *Alternaria alternata* FB1 efficiently degrades polyethylene. *Journal of Hazardous Materials, 431*, Article 128617.
Gao, R., & Sun, C. (2021). A marine bacterial community capable of degrading poly (ethylene terephthalate) and polyethylene. *Journal of Hazardous Materials, 416*, Article 125928.
Garrido, S. M., Kitamoto, N., Watanabe, A., Shintani, T., & Gomi, K. (2012). Functional analysis of FarA transcription factor in the regulation of the genes encoding lipolytic enzymes and hydrophobic surface binding protein for the degradation of biodegradable plastics in *Aspergillus oryzae*. *Journal of Bioscience and Bioengineering, 113*, 549–555.
Gaytán, I., Sánchez-Reyes, A., Burelo, M., Vargas-Suárez, M., Liachko, I., Press, M., Sullivan, S., Cruz-Gómez, M. J., & Loza-Tavera, H. (2020). Degradation of recalcitrant polyurethane and xenobiotic additives by a selected landfill microbial community and its biodegradative potential revealed by proximity ligation-basedmetagenomic analysis. *Frontiers in Microbiology, 10*, 2986.
Gomez-Mendez, L. D., Moreno-Bayona, D. A., Poutou-Pinales, R. A., Salcedo-Reyes, J. C., Pedroza-Rodriguez, A. M., Vargas, A., & Bogoya, J. M. (2018). Biodeterioration of plasma pretreated LDPE sheets by *Pleurotus ostreatus*. *PLoS ONE, 13*, Article e0203786.
Gravouil, K., Ferru-Clément, R., Colas, S., Helye, R., Kadri, L., Bourdeau, L., Moumen, B., Mercier, A., & Ferreira, T. (2017). Transcriptomics and lipidomics of the environmental strain *Rhodococcus ruber* point out consumption pathways and potential metabolic bottlenecks for polyethylene degradation. *Environmental Science and Technology, 51*, 5172–5181.

Hadad, D., Geresh, S., & Sivan, A. (2005). Biodegradation of polyethylene by the thermophilic bacterium *Brevibacillus borstelensis*. *Journal of Applied Microbiology, 98*(5), 1093–1100.

Han, X., Liu, W., Huang, J.-W., Ma, J., Zheng, Y., Ko, T.-P., Xu, L., Cheng, Y.-S., Chen, C.-C., & Guo, R.-T. (2017). Structural insight into catalytic mechanism of PET hydrolase. *Nature Communication 8*, 1–6.

Han, Y. N., Wei, M., Han, F., Fang, C., Wang, D., Zhong, Y. J., Guo, C. L., Shi, X. Y., Xie, Z. K., & Li, F. M. (2020). Greater biofilm formation and increased biodegradation of polyethylene film by a microbial consortium of *Arthrobacter* sp. and *Streptomyces* sp. *Microorganisms 8*, 1979.

Hossain, M. R., Jiang, M., Wei, Q., & Leff, L. G. (2019). Microplastic surface properties affect bacterial colonization in freshwater. *Journal of Basic Microbiology, 59*, 54–61.

Hua, F., Wang, H. Q., Li, Y., & Zhao, Y. C. (2013). Trans-membrane transport of n-octadecane by *Pseudomonas* sp. DG17. *Journal of Microbiology, 51*, 791–799.

Iiyoshi, Y., Tsutsumi, Y., & Nishida, T. (1998). Polyethylene degradation by lignin-degrading fungi and manganese peroxidase. *Journal of Wood Science, 44*, 222–229.

Ishigaki, T., Sugano, W., Nakanishi, A., Tateda, M., Ike, M., & Fujita, M. (2004). The degradability of biodegradable plastics in aerobic and anaerobic waste landfill model reactors. *Chemosphere, 54*, 225–233.

Islam, S., Apitius, L., Jakob, F., & Schwaneberg, U. (2019). Targeting microplastic particles in the void of diluted suspensions. *Environmental International, 123*, 428–435.

Jabloune, R., Khalil, M., Moussa, I. E. B., Simao-Beaunoir, A. M., Lerat, S., Brzezinski, R., & Beaulieu, C. (2020). Enzymatic degradation of p-Nitrophenyl esters, polyethylene terephthalate, cutin, and suberin by Sub1, a suberinase encoded by the plant pathogen *Streptomyces scabies*. *Microbes and Environments 35*, ME19086.

Jacquin, J., Cheng, J., Odobel, C., Pandin, C., Conan, P., Pujo-Pay, M., Barbe, V., Meistertzheim, A. L., & Ghiglione, J.-F. (2019). Microbial ecotoxicology of marine plastic debris: A review on colonization and biodegradation by the "plastisphere." *Frontiers in Microbiology, 10*, 865.

Jeon, H. J., & Kim, M. N. (2013a). Biodegradation of poly(l-lactide) (PLA) exposed to UV irradiation by a mesophilic bacterium. *International Biodeterioration and Biodegradation, 85*, 289–293.

Jeon, H. J., & Kim, M. N. (2013b). Isolation of a thermophilic bacterium capable of low molecular-weight polyethylene degradation. *Biodegradation, 24*, 89–98.

Khan, S., Nadir, S., Shah, Z. U., Shah, A. A., Karunarathna, S. C., Xu, J., Khan, A., Munir, S., & Hasan, F. (2017). Biodegradation of polyester polyurethane by *Aspergillus tubingensis*. *Environmental Pollution, 225*, 469–480.

Khruengsai, S., Sripahco, T., & Pripdeevech, P. (2022). Biodegradation of polyester polyurethane by embarria clematidis. *Frontiers in Microbiology, 13*, Article 874842.

Krueger, M. C., Harms, H., & Schlosser, D. (2015). Prospects for microbiological solutions to environmental pollution with plastics. *Applied Microbiology and Biotechnology, 99*, 8857–8874.

Law, K. L. (2017). Plastics in the marine environment. *Annual Review of Marine Science, 9*, 205–229.

Li, W. J., Jayakody, L. N., Franden, M. A., Wehrmann, M., Daun, T., Hauer, B., Blank, L. M., Beckham, G. T., Klebensberger, J., & Wierckx, N. (2019). Laboratory evolution reveals the metabolic and regulatory basis of ethylene glycol metabolism by *Pseudomonas putida* KT2440. *Environmental Microbiology, 21*, 3669–3682.

Lobelle, D., & Cunliffe, M. (2011). Early microbial biofilm formation on marine plastic debris. *Marine Pollution Bulletin, 62*, 197–200.

Magnin, A., Pollet, E., Perrin, R., Ullmann, C., Persillon, C., Phalip, V., & Avérous, L. (2019). Enzymatic recycling of thermoplastic polyurethanes: Synergistic effect of an esterase and an amidase and recovery of building blocks. *Waste Management, 85*, 141–150.

Maroof, L., Khan, I., Yoo, H. S., Kim, S., Park, H., Ahmad, B., & Azam, S. (2021). Identification and characterization of low density polyethylene-degrading bacteria isolated from soils of waste disposal sites. *Environmental Engineering Research, 26*, Article 200167.

Meng, X., Yang, L., Liu, H., Li, Q., Xu, G., Zhang, Y., Guan, F., Zhang, Y., Zhang, W., & Wu, N. (2021). Protein engineering of stable IsPETase for PET plastic degradation by Premuse. *International Journal of Biological Macromolecules, 180*, 667–676.

Miao, L., Wang, P., Hou, J., Yao, Y., Liu, Z., Liu, S., & Li, T. (2019). Distinct community structure and microbial functions of biofilms colonizing microplastics. *Science of the Total Environment, 650*, 2395–2402.

Mohan, A. J., Sekhar, V. C., Bhaskar, T., & Nampoothiri, K. M. (2016). Microbial assisted High Impact Polystyrene (HIPS) degradation. *Bioresource Technology, 213*, 204–207.

Mohanan, N., Sharma, P. K., & Levin, D. B. (2020). Characterization of an intracellular poly(3-hydroxyalkanoate) depolymerase from the soil bacterium, *Pseudomonas putida* LS46. *Polymer Degradation and Stability, 175*, Article 109127.

Montazer, Z., Habibi Najafi, M. B., & Montazer, D. B. (2020). Challenges with verifying microbial degradation of polyethylene. *Polymers, 12*, 123.

Müller, R.-J., Schrader, H., Profe, J., Dresler, K., & Deckwer, W.-D. (2005). Enzymatic degradation of poly(ethylene terephthalate): Rapid hydrolyse using a hydrolase from *Thermobifida fusca. Macromolecular Rapid Communications, 26*, 1400–1405.

Nakajima-Kambe, T., Onuma, F., Akutsu, Y., & Nakahara, T. (1997). Determination of the polyester polyurethane breakdown products and distribution of the polyurethane degrading enzyme of *Comamonas Acidovorans* strain TB-35. *Journal of Fermentation and Bioengineering 83*, 456–460.

Negoro, S. (2000). Biodegradation of nylon oligomers. *Applied Microbiology and Biotechnology, 54*, 461–466.

Oberbeckmann, S., Kreikemeyer, B., & Labrenz, M. (2018). Environmental factors support the formation of specific bacterial assemblages on microplastics. *Frontiers in Microbiology, 8*, 2709.

Oda, Y., Urakami, T., & Tonomura, K. (2018). Polycaprolactone depolymerase produced by *Alcaligenes faecalis. FEMS Microbiology Letters 152*(339–343).

Ohtaki, S., Maeda, H., Takahashi, T., Yamagata, Y., Hasegawa, F., Gomi, K., Nakajima, T., & Abe, K. (2006). Novel hydrophobic surface binding protein, HsbA, produced by *Aspergillus oryzae. Applied and Environmental Microbiology, 72*, 2407–2413.

Okal, E. J., Heng, G., Magige, E. A., Khan, S., Wu, S., Ge, Z., Zhang, T., Mortimer, P. E., & Xu, J. (2023). Insights into the mechanisms involved in the fungal degradation of plastics. *Ecotoxicology and Environmental Safety, 262*, Article 115202.

Paço, A., Duarte, K., da Costa, J. P., Santos, P. S., Pereira, R., Pereira, M., Freitas, A. C., Duarte, A. C., & Rocha-Santos, T. A. (2017). Biodegradation of polyethylene microplastics by the marine fungus *Zalerion maritimum. Science of the Total Environment, 586*, 10–15.

Park, S. Y., & Kim, C. G. (2019). Biodegradation of micro-polyethylene particles by bacterial colonization of a mixed microbial consortium isolated from a landfill site. *Chemosphere, 222*, 527–533.

Patil, R. C. (2018). Screening and characterization of plastic degrading bacteria from Garbage soil. *British Journal of Environmental Sciences, 6*, 33–40.

Pinto, M., Langer, T. M., Hüffer, T., Hofmann, T., & Herndl, G. J. (2019). The composition of bacterial communities associated with plastic biofilms differs between different polymers and stages of biofilm succession. *PLoS ONE, 14*, Article e0217165.

Restrepo-Flórez, J.-M., Bassi, A., & Thompson, M. R. (2014). Microbial degradation and deterioration of polyethylene—A review. *International Biodeterioration and Biodegradation, 88*, 83–90.

Roberts, C., Edwards, S., Vague, M., Leon-Zayas, R., Scheffer, H., Chan, G., Swartz, N. A., & Mellies, J. L. (2020). Environmental consortium containing *Pseudomonas* and *Bacillus* species synergistically degrades polyethylene terephthalate plastic. *mSphere, 5*, e01151–20.

Roth, C., Wei, R., Oeser, T., Then, J., Follner, C., Zimmermann, W., & Strater, N. (2014). Structural and functional studies on a thermostable polyethylene terephthalate degrading hydrolase from *Thermobifida fusca. Applied Microbiology and Biotechnology, 98*, 7815–7823.

Sangeetha Devi, R., Rajesh Kannan, V., Nivas, D., Kannan, K., Chandru, S., & Robert Antony, A. (2015). Biodegradation of HDPE by *Aspergillus* spp. from marine ecosystem of Gulf of Mannar, India. *Marine Pollution Bulletin, 96*, 32–40.

Santo, M., Weitsman, R., & Sivan, A. (2013). The role of the copper-binding enzyme–laccase–in the biodegradation of polyethylene by the actinomycete *Rhodococcus ruber*. *International Biodeterioration and Biodegradation, 84*, 204–210.

Saravanan, A., Kumar, P. S., Vo, D.-V.N., Jeevanantham, S., Karishma, S., & Yaashikaa, P. (2021). A review on catalytic-enzyme degradation of toxic environmental pollutants: Microbial enzymes. *Journal of Hazardous Materials, 419*, Article 126451.

Satlewal, A., Soni, R., Zaidi, M., Shouche, Y., & Goel, R. (2008). Comparative biodegradation of HDPE and LDPE using an indigenously developed microbial consortium. *Journal of Microbiology and Biotechnology, 18*, 477–482.

Shah, Z., Krumholz, L., Aktas, D. F., Hasan, F., Khattak, M., & Shah, A. A. (2013). Degradation of polyester polyurethane by a newly isolated soil bacterium, *Bacillus subtilis* strain MZA-75. *Biodegradation, 24*, 865–877.

Shimpi, N., Borane, M., Mishra, S., & Kadam, M. (2012). Biodegradation of polystyrene (PS)-poly(lactic acid) (PLA) nanocomposites using Pseudomonas aeruginosa. *Macromolecules Research, 20*, 181–187.

Shivlata, L., & Satyanarayana, T. (2015). Thermophilic and alkaliphilic Actinobacteria: Biology and potential applications. *Frontiers in Microbiology, 6*, 1014.

Shresta, J. K., Joshi, D. R., Regmi, P., & Badahit, G. (2019). Isolation and Identification of Low Density Polyethylene (LDPE) degrading *Bacillus* spp. from a soil of landfill site. *Acta Scientific Microbiology, 2*, 30–34.

Singh, L., & Wahid, Z. A. (2015). Methods for enhancing bio-hydrogen production from biological process: A review. *Journal of Industrial and Engineering Chemistry, 21*, 70–80.

Sivan, A. (2011). New perspectives in plastic biodegradation. *Current Opinion in Biotechnology, 22*, 422–426.

Skariyachan, S., Patil, A. A., Shankar, A., Manjunath, M., Bachappanavar, N., & Kiran, S. (2018). Enhanced polymer degradation of polyethylene and polypropylene by novel thermophilic consortia of *Brevibacillus* sps. and *Aneurinibacillus* sp. screened from waste management landfills and sewage treatment plants. *Polymer Degradation and Stability, 149*, 52–68.

Skariyachan, S., Setlur, A. S., Naik, S. Y., Naik, A. A., Usharani, M., & Vaisist, K. S. (2017). Enhanced biodegradation of low- and high-density polyethylene by novel bacterial consortia formulated from plastic contaminated cow dung under thermophilic conditions. *Environmental Science and Pollution Research, 24*, 8443–8457.

Sudhakar, M., Doble, M., Murthy, P. S., & Venkatesan, R. (2008). Marine microbe-mediated biodegradation of low- and high-density polyethylenes. *International Biodeterioration and Biodegradation, 61*, 203–213.

Sudhakar, M., Priyadarshini, C., Doble, M., Sriyutha Murthy, P., & Venkatesan, R. (2007). Marine bacteria mediated degradation of nylon 66 and 6. *International Biodeterioration and Biodegradation, 60*, 144–151.

Tahir, L., Ali, M. I., Zia, M., Atiq, N., Hasan, F., & Ahmed, S. (2013). Production and characterization of esterase in Lantinus tigrinus for degradation of polystyrene. *Polish Journal of Microbiology, 62*, 101–108.

Tanaka, M., & Gomi, K. (2021). Induction and repression of hydrolase genes in *Aspergillus oryzae*. *Frontiers in Microbiology, 12*, Article 677603.

Taniguchi, I., Yoshida, S., Hiraga, K., Miyamoto, K., Kimura, Y., & Oda, K. (2019). Biodegradation of PET: Current status and application aspects. *ACS Catalysis, 9*, 4089–4105.

Tian, L., Kolvenbach, B., Corvini, N., Wang, S., Tavanaie, N., Wang, L., Ma, Y., Scheu, S., Corvini, P.F.-X., & Ji, R. (2017). Mineralisation of 14C-labelled polystyrene plastics by *Penicillium variabile* after ozonation pre-treatment. *New Biotechnology, 38*, 101–105.

Tokiwa, Y., & Calabia, B. P. (2009). Biodegradability of plastics. *International Journal of Molecular Sciences, 10*, 3722–3742.

Urbanek, A. K., Mironczuk, A. M., García-Martín, A., Saborido, A., de la Mata, I., & Arroyo, M. (2020). Biochemical properties and biotechnological applications of microbial enzymes involved in the degradation of polyester-type plastics. *Biochimica et Biophysica Acta—Proteins and Proteomics, 1868*, Article 140315.

Vazquez-Alcantara, L., Oliart-Ros, R. M., García-Borquez, A., & Pena-Montes, C. (2021). Expression of a Cutinase of *Moniliophthora roreri* with Polyester and PET-plastic residues degradation activity. *Microbiology Spectrum, 9*, e00976-e01021.

Vrsanska, M., Buresova, A., Damborsky, P., & Adam, V. (2015). Influence of different inducers on ligninolytic enzyme activities. *Journal of Metallomics and Nanotechnologies, 3*, 64–70.

Wang, Z., Xin, X., Shi, X., & Zhang, Y. (2020). A polystyrene-degrading *Acinetobacter* bacterium isolated from the larvae of Tribolium castaneum. *Science of the Total Environment, 726*, Article 138564.

Wehrmann, M., Billard, P., Martin-Meriadec, A., Zegeye, A., & Klebensberger, J. (2017). Functional role of lanthanides in enzymatic activity and transcriptional regulation of pyrroloquinoline quinone-dependent alcohol dehydrogenases in *Pseudomonas putida* KT2440. *mBio, 8*, e00570–e00617.

Wright, R. J., Erni-Cassola, G., Zadjelovic, V., Latva, M., & Christie-Oleza, J. A. (2020). Marine plastic debris: A new surface for microbial colonization. *Environmental Science and Technology, 54*, 11657–11672.

Wright, R. J., Langille, M. G. I., & Walker, T. R. (2021). Food or just a free ride? A meta-analysis reveals the global diversity of the Plastisphere. *ISME Journal, 15*, 789–806.

Yang, J., Yang, Y., Wu, W. M., Zhao, J., & Jiang, L. (2014). Evidence of polyethylene biodegradation by bacterial strains from the guts of plastic-eating waxworms. *Environmental Science and Technology, 48*, 13776–13784.

Yang, Y., Liu, W., Zhang, Z., Grossart, H. P., & Gadd, G. M. (2020a). Microplastics provide new microbial niches in aquatic environments. *Applied Microbiology and Biotechnology, 104*, 6501–6511.

Yang, Y., Wang, J., & Xia, M. (2020b). Biodegradation and mineralisation of polystyrene by plastic-eating superworms Zophobas atratus. *Science of the Total Environment, 708*, Article 135233.

Yoon, M. G., Jeon, H. J., & Kim, M. N. (2012). Biodegradation of polyethylene by a soil bacterium and AlkB cloned recombinant cell. *Journal of Bioremediation and Biodegradation, 3*, 1–8.

Yoshida, S., Hiraga, K., Takehana, T., Taniguchi, I., Yamaji, H., Maeda, Y., Toyohara, K., Miyamoto, K., Kimura, Y., & Oda, K. (2016). A bacterium that degrades and assimilates poly(ethylene terephthalate). *Science, 351*, 1196–1199.

Yuan, J., Ma, J., Sun, Y., Zhou, T., Zhao, Y., & Yu, F. (2020). Microbial degradation and other environmental aspects of microplastics/plastics. *Science of the Total Environment, 715*, Article 136968.

Zhai, X., Zhang, X.-H., & Yu, M. (2023). Microbial colonization and degradation of marine microplastics in the plastisphere: A review. *Frontiers in Microbiology, 14*, 1127308.

Zhang, J., Gao, D., Li, Q., Zhao, Y., Li, L., Lin, H., Bi, Q., & Zhao, Y. (2020). Biodegradation of polyethylene microplastic particles by the fungus Aspergillus flavus from the guts of wax moth Galleria mellonella. *Science of the Total Environment, 704*, Article 135931.

Chapter 11
Technological Advancements in Microplastic Waste Management

Anuradha Kumari, Arjun Kumar Mehara, Pranjal Singh, Prachi Marwaha, Ankit Kumar Singh, and Mayank Kumar Singh

Abstract Microplastics (MPs), originating from the breakdown of larger plastic materials (with a size of less than 5 mm) or as primary synthetic particles, represent a pressing global environmental concern due to their persistence, ubiquity, and complex interactions with ecosystems and human health. This chapter provides a comprehensive overview of MPs waste, tracing its origins, environmental distribution, and toxicological effects across terrestrial, aquatic, and atmospheric systems. Various physical, chemical, biological, and integrated treatment technologies are discussed including dendrimer technology, alongside advanced detection tools such as SEM–EDS, NMR, FTIR, Raman spectroscopy, and GC/MS. Special emphasis is placed on emerging nanotechnological interventions and AI-based real-time monitoring systems that show promise for scalable and sustainable microplastic removal. Furthermore, the chapter highlights current policy frameworks, global collaborative efforts, and future directions essential for effective microplastic governance. The integration of scientific innovation, industrial responsibility, and governmental action is presented as critical to mitigating the long-term impacts of microplastic pollution and achieving a circular and resilient environmental economy.

Keywords Microplastics · Waste management · Membrane bioreactors · Ultrafiltration · Nanofiltration · Dendrimer technology · Recycling technologies

A. Kumari · A. K. Mehara · A. K. Singh
University Department of Botany, Lalit Narayan Mithila University, Darbhanga, Bihar, India

P. Marwaha
University Department of Home Science, Lalit Narayan Mithila University, Darbhanga, Bihar, India

P. Singh
Department of Plastic Technology, Harcourt Butler Technical University, Kanpur, Uttar Pradesh, India

M. K. Singh (✉)
Central Michigan University-College of Medicine, Mount Pleasant, MI, USA
e-mail: singh2mk@cmich.edu

The National Dendrimer and Nanotechnology Center, Mount Pleasant, MI, USA

V. Kumar et al. (eds.), *Microplastics Pollution*,
https://doi.org/10.1007/978-3-031-99774-7_11

11.1 Background

Plastics are anthropogenic, high-molecular-mass synthetic organic polymers, mainly derived from hydrocarbons, such as crude oil and natural gas. They are strong, pliable, lightweight, and easily shaped. Its versatility and durability have led to its extensive adoption in everyday products, ranging from packaging materials to automotive components and medical devices (Kurniawan et al., 2023). The origins of plastic can be traced back to the early nineteenth century, when the first synthetic polymers were developed. This marked the beginning of a transformative era in material science, as innovations in plastic production paved the way for a range of applications that have fundamentally altered consumer behavior and industrial practices (Science History Institute, 2024).

Over the decades, the production and consumption of plastic have escalated dramatically. According to recent data from the International Union for Conservation of Nature (IUCN), the production of plastic exceeds 460 million metric tons each year. The majority of plastic pollution comes from single-use products, such as bottles, caps, cigarettes, shopping bags, cups, and straws. Due to its slow and non-degradable nature, an estimated 20 million metric tons of plastic litter accumulate in the environment each year and expected to increase significantly by 2040. This led to a pervasive presence of plastics in modern life. As a result, plastic waste has become a significant contributor to global pollution, affecting both terrestrial and marine ecosystems (IUCN, 2024).

Environmental key factors, such as sunlight (especially ultraviolet rays), water, wind, or biological processes involving bacteria and enzymes, help to breakdown plastics. These plastic litters are then further converted into inorganic chemicals, methane, CO_2, biomass, and water; depending on their molecular weight, chemical composition, polymer structure, climate conditions and the type of organisms involved (Lokesh et al., 2023). When plastics degrade through various environmental processes, primarily due to mechanical abrasion, photodegradation, and chemical breakdown, and this leading to formation of microplastics. Microplastics are water-insoluble, synthetic, solid, small plastic particles with a diameter ranging from 1 μm to 5 mm. Thus, the polymer matrix can be of primary or secondary origin, with a regular or irregular shape and size (Bermúdez & Swarzenski, 2021). Every year, an estimated 600,000 tons of fishing gear enter the ocean, primarily from commercial and military vessels. The majority of microplastics (plastic bottles, bags, toiletries, building materials, and clothing) found in water bodies, almost 80 to 90%, originate from land-based sources. Additionally, plastic incinerators generate bottom ash, which is a terrestrial source of microplastic pollution. Approximately 10–20% of microplastics are released into the marine environment, primarily through activities such as seashore tourism, industrial fishing, and maritime operations (Dayal et al., 2024).

11.2 Introduction

Based on their sources, MPs can be classified into primary and secondary. Primary MPs are used to manufacture plastic products and plastic microbeads, which are added to personal care products, such as hand and facial cleansers, toothpaste, and cosmetics. They are directly released into the environment as pre-production microplastics pellets. Secondary MPs are fragments of plastics that derive from the degradation of larger plastic products (e.g., rope, packaging, and clothing). Researchers have also classified microplastics according to their shapes as microbeads, nurdles, fibers, foam, and fragments. Moreover, MPs have been further classified into prominent MPs (1–5 mm), small MPs (0.3–1 mm), and nano plastics (< 0.3 mm), based on the characteristics of continuous breakdown (Wu et al., 2019).

MPs persist in the environment for an extended period due to their high polymer content, slow degradation rate, and resistance to natural breakdown. It affects all the core components of the ecosystem, whether they are terrestrial, atmospheric, marine, or freshwater environments. MPs accumulation in soil, air, and water sources can inhibit plant growth, disrupting shoot and root development, and impact crop health. It can also alter soil structure and nutrient uptake, as well as the health of soil microorganisms (Hasan & Tarannum, 2024; Sajjad et al., 2022). Their presence can also be observed in food chains. Microplastics can upgrade oxidative stress and inflammation in intestinal cell lines. This could lead to increased cytotoxicity and disruption of gut barrier integrity (Mattioda et al., 2023; Wen et al., 2022). It has been observed that the co-exposure of harmful substances, such as nonylphenol, might intensify the toxic effects of MPs. This led to damage of cell membranes and impaired intestinal function (Ding et al., 2024).

Altered gut microbiota and increased intestinal permeability may also facilitate the entry of other toxins into the bloodstream after long-term exposure to MPs. This can result in significant health risks (Wang et al., 2024; Wen et al., 2022). Microplastic ingestion has been found to cause severe health complications in aquatic animals, like kidney damage, gill inflammation, oxidative stress, and reproductive organ abnormalities (Jeong et al., 2024). Humans, as top consumers in food chains, are particularly susceptible to the effects of MPs through the consumption of contaminated aquatic food (Mahu et al., 2022). Plastic food containers, which contain polypropylene (PP), and polyethylene Terephthalate (PET) have been linked to the formation of MPs. Research shows that thermal treatments, such as hot water and microwaving, significantly increase the release of MPs from these containers. This can induce potential cytotoxic effects on human intestinal health upon ingestion (Jin et al., 2024).

MPs are a major part of landfills due to improper practice of waste management. Treatment methods like utilization of membrane bioreactors, ultrafiltration, nanofiltration and nanotechnology, show promising and varying effectiveness in microplastic removal. Biological treatments may not fully degrade MPs within short time frames due to their tendency to adhere to sediments. Also, the presence of microplastics in landfill leachate can lead to soil and water contamination, requiring effective treatment strategies (Kabir et al., 2023).

11.3 Innovative Approach

MPs can adsorb pathogenic microorganisms, viruses, and biomolecules (such as antibiotics, allergens, and lipopolysaccharides (LPS), from the environment depending on their high specific surface area, chemical composition, surface charge and hydrophobicity (Yang et al., 2023). This can impact human health and marine life. Therefore, it is crucial to adopt and develop efficient, scalable, and environmentally friendly technological methods for removing and degrading microplastics from various environments. Technical Approaches in Microplastic waste management involve physical, chemical, biological and integrative methods. Physical methods use mechanical processes to separate microplastics from the ecosystem and are often integrated with other approaches to enhance their efficiency. Membrane bioreactors (MBRs) and ultrafiltration (UF) systems have shown high removal efficiencies, particularly effective in wastewater treatment plants (WWTPs) due to their ability to capture microplastics as small as 10 μm, with MBRs achieving up to 99.9% removal rates (Chand et al., 2024; Ramos et al., 2023). Sedimentation and flotation processes rely on the density differences between microplastics and water. For instance, dissolved air flotation (DAF) has been reported to remove up to 95% of microplastics (Bodzek & Pohl, 2023). Sieving and screening techniques are often used as a pretreatment step in WWTPs to remove debris and microplastics before advanced treatment processes, especially effective methods for removing larger microplastics (Bodzek et al., 2024). Magnetic separation uses magnetic materials to attract and remove microplastics that have magnetic properties or are coated with magnetic materials (Li et al., 2024).

Chemical methods use chemical agents to degrade or aggregate microplastics. Coagulation and flocculation achieve removal efficiencies of up to 90% and are widely used in wastewater treatment to aggregate microplastics into larger particles, which can then be removed through sedimentation or filtration (Dayal et al., 2024). Photocatalytic oxidation, Fenton oxidation, and ozonation are Advanced Oxidation Processes (AOPs), which break down microplastics into smaller, less harmful fragments and are effective in degrading microplastics. For example, photocatalytic degradation has been shown to break down microplastics into environmentally harmless molecules (Li et al., 2024). Electrocoagulation achieves removal efficiency up to 90%. This technique uses an electric current to remove microplastics by destabilizing suspended particles (Bodzek & Pohl, 2023).

Biological methods offer a sustainable and eco-friendly alternative to physical and chemical processes, utilizing microorganisms and enzymes to break down microplastics. Species of *Pseudomonas* and *Aspergillus* have shown promise in breaking down polyethylene and polypropylene (Ali et al., 2024; Park & Park, 2021). Enzymes such as lipase and esterase break down the polymer chains and reduce the size and toxicity of microplastics (Li et al., 2024). Biofilms absorb and degrade microplastics and are particularly effective in wastewater treatment plants. These biofilms are composed of microorganisms and extracellular polymeric substances (Ali et al., 2024).

Integrated approaches achieve higher removal efficiency and address the limitations of individual techniques by combining physical, chemical, and biological methods. A combination of membrane bioreactors (MBRs) and activated sludge processes has been shown to achieve near-complete removal of microplastics. For example, MBR systems have been reported to remove up to 99.9% of microplastics (Ramos et al., 2023) (Bodzek & Pohl, 2023). Filtration systems remove the degraded fragments, while Advanced oxidation processes (AOPs) degrade microplastics. AOPs integrated with filtration systems have been proposed as a highly effective strategy for microplastic removal. (Li et al., 2024). The use of microorganisms can enhance the efficiency of subsequent physical and chemical removal methods to degrade microplastics and reduce their size and toxicity as well (Ali et al., 2024; Park & Park, 2021). Thus, the chapter will further highlight technological advancements in microplastic waste management.

11.4 Traditional Approach

High levels of MPs in wastewater present a serious environmental concern, not only because of the particles themselves but also due to their ability to attract and bind various contaminants on their hydrophobic surfaces. These include heavy metals such as copper, zinc, arsenic, lead, cadmium, chromium, and iron, as well as persistent pollutants and substances from pharmaceuticals and personal care products (Verla et al., 2019; Yuan et al., 2022). These microparticles often contain toxic additive contaminants, such as polybrominated diphenyl ethers, polychlorinated biphenyls, and phthalate esters, which are byproducts for their chemical production, highlighting the challenges involved in removing them from aquatic ecosystems (Elkhatib & Oyanedel-Craver, 2020; Yuan et al., 2022; Zhang & Chen, 2019). Recent studies suggest that MPs adsorb toxic heavy metals and emerging contaminants possessing a growing threat to human health, particularly as they can contaminate food and water supplies (Khalid et al., 2021; Singh et al., 2022; Sun et al., 2021). Traces of specific polymers have already been identified in human blood and tissue (Ragusa et al., 2020). As a result, removing microplastics from wastewater is vital. A range of treatment methods is currently being investigated for removing microplastics from the environment. These methods are generally characterized into three main types: physical, chemical and biological approaches.

For treating microplastics in wastewater, various physical separation techniques are widely used. These are pressure-driven membranes, including microfiltration (MF), ultrafiltration (UF), nanofiltration (NF), and reverse osmosis (RO) membranes, as well as sedimentation and density separation. Studies indicate that ultrafiltration can achieve removal efficiencies of approximately 53–96% for microplastics of different sizes. Advanced reverse osmosis systems have shown up to 71.88% removal efficiency in specific leachate treatment scenarios (Sun et al., 2019). Membrane processes, such as microfiltration, ultrafiltration, and reverse osmosis, are effective in removing microplastics from leachate. Integrating advanced oxidation processes

(AOPs) with biodegradation and membrane filtration can improve overall treatment performance and help to tackle emerging contaminants in leachate (Liu et al., 2019). These methods are discussed further.

Microfiltration membranes have pore sizes ranging from 0.1 to 10 μm and operate at low pressures, typically between 0 and 2 bar (Moradihamedani, 2021). It is effective for removing large microplastics (MPs) and suspended solids. Studies show MF can remove 81.5–100% of MPs, depending on the membrane material, pore size, and operating conditions. Microplastics with a size larger than the pore size of the MF membrane are separated from the wastewater by being retained on the membrane surface or pores. The size of MPs in treated waters of conventional DWTPs is generally 100 μm, and studies show that conventional DWTPs are not very effective in removing small-sized MPs (Pivokonsky et al., 2018). Similarly, the dominance of MPs smaller than 500 μm in the effluent of WWTPs indicates that small MPs are not effectively removed in WWTPs, and especially small MPs are discharged into the receiving aquatic environments (Long et al., 2019). Pizzichetti et al. (2020) found 96–99.6% removal using different MF membranes; nevertheless, sharp-edged MPs have the potential to damage and seep through the membranes. MF is typically used as a pre-treatment before UF and NF. However, small MPs (< 1 μm) may still pass through or damage the membrane, and precautions must be taken to prevent membrane abrasion.

Ultrafiltration (UF) membranes function at pressures between 1 and 10 bar and contain pores that range in size from 1 to 100 nm. Many experiments show that UF membranes have remarkable MP removal efficiency, frequently above 96%. For example, Ma et al. (2018) used a commercial PVDF UF membrane to remove 100% of polyethylene MPs. According to Tadsuwan and Babel (2022), the addition of UF to a traditional WWTP increased the elimination of total MP from 86.14% to 96.97%.

Luogo et al. (2022) demonstrated that a ZRO2 UF membrane performed better than a SiC MF membrane in removing MPs from laundry wastewater. It is also found that UF membranes are vulnerable to fouling, particularly by tiny, hydrophobic MPs, which can lower overall effectiveness. Removal performance is further influenced by surface charge, hydrophilicity, and membrane composition. When incorporated into multi-stage treatment systems, UF membranes offer a dependable way to remove MPs from effluents. High-molecular-weight organic molecules, bacteria, viruses, and suspended particles are frequently eliminated from water and wastewater. Smaller MPs (1–5000 μm) are effectively retained by UF membranes because of their finer pore size when compared to microfiltration.

Nanofiltration (NF) membrane is a relatively new technology in wastewater treatment systems. Nanofiltration membranes have pore sizes ranging from 1 to 10 nm and operate at pressures of 5–15 bar. The size of pores in NF membranes is such that even small uncharged solutes are highly rejected. At the same time, the surface's electrostatic properties allow monovalent ions to be transmitted reasonably well, with multivalent ions being mostly retained. These characteristics have made NF membranes extremely useful. However, due to the tiny pore sizes of NF membranes, the high pressure required for water to pass through the membranes limits their widespread use. It has been reported that the use of NF membranes in water treatment increases the energy requirement for treatment by 60–150% (Abdel-Fatah, 2018).

Studies on NF for MP removal are limited. Kara et al. (2022) showed 99% MP removal from landfill leachate using NF after UF. Barbier et al. (2022) reported very low MP levels (0.002–0.018 MP/L) post-NF in drinking water treatment. Since the NF membrane exhibits properties between those of ultrafiltration (UF) and reverse osmosis (RO), both charge and size of the particle play an essential role in the NF rejection mechanism. Simpson et al. (1987) have described NF as a charged UF system. However, NF has the advantage of lower operating pressure compared to RO, and higher organic rejection compared to UF. For colloids and large molecules, physical sieving is a better rejection mechanism, whereas for ions and lower-molecular-weight substances, solution diffusion and the charge effect of the membrane aid in the separation process.

Reverse Osmosis (RO) membranes are dense membranes (pore size: 1 nm) (Acarer, 2023). These membranes are operated at high pressures (> 20 bar) to ensure water passes through them, but their energy requirements are quite high, which is a significant disadvantage. The separation process in RO membranes is based on the sorption–diffusion mechanism (Licona et al., 2018). RO membranes are used in the desalination of seawater, as well as in the treatment of drinking water and wastewater. They can remove even monovalent ions, in addition to all contaminants that MF, UF, and NF membranes can reject from water (Ansari et al., 2021; Indika et al., 2021).

Studies report up to 99.8% MP removal using reverse osmosis (RO) membranes (Sun et al., 2021). However, fibers and small fragments can still appear in permeates. For example, Dalmau-Soler et al. (2021) found 0.06 MP/L in RO-treated effluent, with ~ 60% fibers, likely due to their ability to pass lengthwise through membranes.

Ziajahromi et al. (2017) reported that the MP abundance in the effluent of the RO system, which has a treatment capacity of 48 million liters per day, is 0.21 MP/L, and approximately 10 million MPs are released from the RO effluent to the aquatic environment in 1 day. However, it is interesting that fibers are present in the permeate even after wastewater treatment with the RO membrane, which is classified as non-porous and can even retain ions. RO systems use a lot of energy but provide very high efficiency.

Membrane filtration technology is usually coupled with the application of biological catalysts and referred to as Membrane Bioreactors (MBRs) (Xiao et al., 2018). MBR was demonstrated to be more effective than single biodegradation treatment or dynamic membrane technology for the removal of MPs. MBR includes multiple procedures, such as pretreatment, bioreactor, and membrane filtration. In a recent study at three WWTPs, Michielssen et al. evaluated the capability of different unit processes in removing MPs, including secondary activated sludge treatment, tertiary treatment by granular sand filtration and MBR. According to their results, the highest removal efficiency of MPs occurred in MBR, i.e., 99.4% (Michielssen et al., 2016). The abundance of MP fibers in the effluent and raw water in the secondary treatment was similar, and the removal proportion of fibers was about 30% after sand filtration in the MBR process. Fibers accounted for > 50% of MPs in WWTP influent;

however, they were more challenging to remove in comparison to other MPs shapes, including particles, bead, and foam (Lares et al., 2018). According to the total MPs abundance, MBR caused relatively high efficiency for the removal of MPs, even up to 99.9% (Talvitie et al., 2017).

The sedimentation technique is based on gravitational settling and helps in removing suspended MP contamination from any liquid. This technique is used for both primary and secondary treatment. This process is used before other treatment techniques because it does not entirely remove pollutants at this stage (Gies et al., 2018; Hidayaturrahman & Lee, 2019).

According to studies which have been carried out at WWTP of some countries, the efficiency of this technology is different and relatively high (57–64% (Bui et al., 2020) and 91.7% (Ngo et al., 2019) and affected by two factors: density and shape (Bui et al., 2020; Ngo et al., 2019). Sedimentation is effective for denser MPs, but it is not as effective with lighter MPs. Pivokonský et al. achieved 88% MP removal by sedimentation and coagulation-flocculation (Pivokonský et al., 2020). Another finding from Zhang et al. found that sedimentation followed by filtration and the GAC (granular activated carbon) process is more efficient (Zhang et al., 2020).

Density separation method is widely used for removing MPs. The basic principle of this method is to isolate MPs by making the surrounding particles denser. By achieving this, MPs would have a lighter density, as we know that low-density particles float and high-density particles settle down. This is usually done by adding a brine solution.

Samples may be undergone two separation steps, including a reduction process that provides reduce sample volume (e.g., via using nets during collection), separation step commonly by filtration and/or density separation by employing of NaCl (1.2 g cm^{-3}) which explained by Thompson et al. due to being low cost, easy availability and eco-friendly (Thompson et al., 2004).

Most common methods to extract MPs from sediments are elutriation column, pressurized fluid extraction, sediment-MPs isolation (SMI) and density separation (Claessens et al., 2013; Coppock et al., 2017; Fuller & Gautam, 2016) Brine solutions are very effective for separation, but they are costly and harmful to the environment (Fok et al., 2019; Fuller & Gautam, 2016). However, the National Oceanic and Atmospheric Administration (NOAA) has suggested using lithium meta tungstate (LMT) (1.62 g/cm^3) for density separation technology (Herring, 2015).

Preliminary tests suggest that adding a centrifugation step after density separation in saline solutions enhances the ability to separate MP (Stolte et al., 2015). Centrifugation is used to separate low-density particles from higher-density sediments. Centrifugation has been suggested by Claessens et al. as a following step after floatation to improve the extraction capacity of MP fibers and granules from sediments (Claessens et al., 2013; Stolte et al., 2015).

11.5 Chemical and Biological Degradation Methods

In the water purification process, chemical treatment can be used, which involves applying different chemicals in a series of reactions. In such methods, microplastics are forcibly removed by adding specifically targeted chemicals (Ahmed et al., 2021). Chemical treatments are often beneficial where microplastics cannot be removed from wastewater by physical and biological methods. Such methods convert the harmful microplastics into less harmful chemicals. Coagulation, flocculation, electrochemical oxidation, photolytic degradation and ozonation are some examples of chemical treatment processes.

The flocculation/coagulation procedure is used to separate the pre-existing colloidal particles into the solution by neutralizing their charge, forming floccules, and subsequently removing them through sedimentation or filtration. Aluminum sulfate ($Al_2(SO_4)_3$), Ferric sulfate ($Fe_2(SO_4)_3$) and Ferric chloride ($FeCl_3$) are some of the most common coagulants (Akinnawo et al., 2023). Combining two coagulants, polyacrylamides and alum, can effectively remove low-density polyethylene, high-density polyethylene and polypropylene from wastewater up to 92%, 84% and 96%, respectively (Monira et al., 2021). A combination of both coagulation and flocculation methods is often considered an effective way to remove microplastics. The coagulation process has a major drawback, including the production of a significant amount of sludge, which exacerbates environmental issues (Osman et al., 2023).

Electrochemical oxidation method is used to break various organic pollutants, including microplastics, into non-harmful byproducts such as carbon dioxide and water vapor (Du et al., 2021). This method is a cost-effective way of treating wastewater via indirect cathode and anode oxidation, one of the sustainable and affordable methods utilized in the electrochemical oxidation process. Ning et al. (2023) used a Ceo2-doped PbO2 composite anode, fabricated for electrochemical oxidation of polyvinyl chloride (PVC), and it was observed that the size of the polyvinyl chloride is significantly reduced, and many cracks appear on the surface of the microplastics. Falco et al. (2024) used boron-doped diamond (BDD) electrodes to remove polystyrene from wastewater and observed a removal efficiency of more than 65%. Electrochemical oxidation efficiency depends on the surface area, material, current strength, type, and concentration of the electrolyte used. (Du et al., 2021; Falco et al., 2024).

Ozonation is used as a tertiary treatment process in wastewater treatment plants to remove any residual coagulation byproducts and purify the sewage. Furthermore, the microplastic polymer can be split into functional groups that contain oxygen through ozonation (Bui et al., 2020). It was found that most (89.9%) of the remaining microplastics were eliminated by the ozonation process (Bui et al., 2020). Ozonation in wastewater treatment plants is known for changing the physicochemical properties of water, migration of plasticizer from polymer matrix into water, and structural changes in microplastics present in water (Ziembowicz & Kida, 2024).

For improving the efficiency of the ozonation process, α-MnO_2 and α-FeOOH are used as a catalyst, and polyethylene could be efficiently broken down under the attack of O_3/OH and decomposition of the polymer is increased under the above-mentioned setup (Hu & Hu, 2023). The major drawback of the ozonation process is that an insufficient ozonation process can produce intermediary compounds due to the production of reactive oxygen species, causing damage to public health and the environment.

Photocatalytic degradation is considered a very successful and promising technology for treating hazardous organic pollutants, such as microplastics from wastewater (Liu et al., 2019). During this process, a semiconductor material absorbs visible or ultraviolet light, which produces free radicals such as superoxide and oxygen radicals that oxidize the MPs (Zhu et al., 2019). The material absorbs light above the photocatalytic energy of a semiconductor's band gap. A transition from the valence band to the conduction band of an electron band results in positive holes in the valence band. Ultimately, MPs are degraded by the radicals produced by this process, specifically superoxide and hydrogen. Iron-zinc oxide nanocomposite is a semiconductor material used in the photocatalytic breakdown of polyethylene (Lam et al., 2020). Ariza-Tarazona et al. (2023) showed that the photocatalytic degradation of polyethylene terephthalate can be achieved using C, N-TiO_2/SiO_2 photocatalysts, and the loss of mass in this process ranged between 9.35 and 16.22%. The factors affecting the rate of photocatalytic degradation are the solution's pH, the ease of generating free radicals necessary for extracting hydrogen atoms from polymer chains, the termination rate, the size of microplastic particles and time (Xie et al., 2023).

The biological approach utilizes various plant and animal species to address the issue of microplastic contamination in aquatic systems and wastewater. Biodegradation involves microbial processes that can break down microplastics into smaller, safer molecules, but the efficiency varies depending on the microbial species and environmental conditions. Scientists are investigating possible treatment methods to break down microplastics through decomposition using different biological species. It has been investigated how organisms can contribute to degrading microplastics present in water and wastewater.

However, in case of fungal degradation most organisms can flourish in a wide range of terrestrial and aquatic ecosystems and many environmental settings. They can withstand various hazardous chemicals and metals and produce enzymes from outside the cell, including organic biosurfactants like hydrophobics, which help break down complex polymers into simpler monomers (Olicón-Hernández et al., 2017). Fungi that break down polyethylene mainly include *Aspergillus* and *Cladosporium*. Their primary source of carbon is polyethylene, which subsequently decomposes using an enzyme outside of cells. By decreasing the hydrophobicity of plastic particles, these fungi promote the formation of various chemical bonds. The development and degeneration of multiple microplastics are significantly influenced by the maximum number of these fungi, which have a wide distribution and efficient reproduction (Dey et al., 2022). According to a study by Das et al. (2018), low-density polyethylene can be biodegraded by *Aspergillus* and *Fusarium* species. *Aspergillus flavus*, a type of fungus strain that was discovered among piles of plastic debris near the coast, showed the ability to degrade high-density polyethylene, whereas *Aspergillus tubingensis* efficiently degrade polyethylene (Devi et al., 2015).

Further, bacteria are an omnipresent and diverse class of microorganisms that are present in soil, water, and the atmosphere, among other settings (Yuan et al., 2020). They have a well-known ability to degrade contaminants like microplastics. Several studies have focused on using pure bacterial cultures in carefully monitored lab settings to assess the degradation. Most of these bacterial cultures can be found in culture collections or obtained through enrichment culture from sludge, sediment, and wastewater. Researchers can study metabolic processes and use pure lines to assess their effects, unlike ecological factors that influence the breakdown of MPs.

Furthermore, this method allows for tracking the entire microplastic process through changes in active bacteria. The ability of bacterial isolates to degrade microplastics was examined in studies. Researchers examined a mangrove silt bacterial culture's ability to degrade polypropylene after it had been purified. Research indicates that two strains of *Rhodococcus* sp. Type 36 and *Bacillus* sp. Type 27 may support polypropylene polymers with microplastic losses of 4.0% and 6.4%, respectively (Yuan et al., 2020). Bacteria were evident on the surface of the treated polypropylene due to the numerous pores and other irregularities.

11.6 Microplastic Recycling Technologies

Between 1950 and 2015, global plastic production increased significantly, from 2 million tons to 380 million tons per year. However, this also generated a vast amount of plastic waste, accumulating 6300 million tons. Out of all that, only about 9% was recycled, 12% was burned, and the remaining 79% ended up in landfills or were released into the environment (Geyer et al., 2017). These days, it is believed that most plastic products, around 90%, are used only once and then thrown away. According to the UNEP (United Nations Environment Program), approximately 3 million tons of microplastics and 5.3 million tons of MPs end up in the environment every year, with a significant portion likely ending up in the ocean (Division et al., 2018). Borrelle and Lau, along with their teams, found that tackling plastic pollution needs countries to work together quickly. They also pointed out that recycling plastic waste could significantly help reduce the amount of plastic ending up in the environment, especially in the ocean (Borrelle et al., 2020; Lau et al., 2020).

In 2009, IUPAC noted that recycling plastic waste was becoming one of the key emerging technologies. They saw it as crucial for supporting the circular economy and began addressing the pollution caused by plastic waste (Chen et al., 2021a; Gomollón-Bel., 2019), aiming to create an eco-friendlier world. Currently, plastic recycling is divided into four types: primary recycling, secondary recycling (also known as mechanical recycling), tertiary recycling, which involves chemical processes, and quaternary recycling, which focuses on energy recovery (Chen et al., 2021a).

Primary and Secondary Recycling are often grouped under mechanical recycling, which involves steps such as sorting, cleaning, crushing, melting, and reshaping the plastic materials. Mechanical recycling starts with the collection of plastic waste from different places like homes, factories etc. Plastics are then sorted by color, type, and other features using manual or automated methods like near-infrared sensors and air classifiers. Proper sorting is crucial for maintaining the purity of recycled materials and ensuring the quality of the final products (Choi et al., 2023). Once sorted, plastics are cleaned to remove dirt, glue, labels, and other residues. This is typically done utilizing detergents and water in specialized washing equipment. This step is essential for improving the quality of recycled plastic and ensuring smooth processing in later stages (Geng et al., 2022). After cleaning, the plastics are shredded or granulated into uniform flakes or pellets, making them easier to handle and process further. In the final stage, flakes or pellets of plastic are melted and reformed using processes such as extrusion or injection molding, resulting in new plastic material or intermediate products (Feil & Pretz, 2020).

The key difference between primary and secondary mechanical recycling lies in the types of plastic processed. Primary recycling deals exclusively with clean, uncontaminated plastic waste, such as post-industrial scrap or post-consumer plastic waste with a known source, allowing the recycled product to retain properties close to those of the original material (Rahimi & García, 2017). And this is why primary recycling is often called "closed-loop recycling." Production of plastic bottles using a blend of recycled PET and new, virgin PET is a typical example of this. Secondary recycling involves thermal and mechanical processing techniques, where the polymer's chemical structure remains unchanged. However, the material reprocessed often results in end products with lower performance or quality, like discoloration, deterioration in mechanical strength and rheological properties, compared to the original plastic. This restricts their application to products with lower performance demands, a process commonly referred to as downcycling (Helbig et al., 2022).

Mechanical recycling faces economic challenges, particularly when declining oil prices reduce the cost of virgin plastic, making recycling less financially viable. Still, it plays a vital role in plastic waste management by recovering material for reuse. It complements other recycling approaches and makes a significant contribution to the circular economy by reducing resource consumption and lowering environmental pollution.

Tertiary recycling, also known as chemical recycling, is an increasingly flexible method being developed to address the global microplastic waste problem. Unlike conventional recycling methods that often reduce plastic through mechanical processing, chemical recycling works at the molecular level. It breaks plastic waste down into monomers or other beneficial compounds, making it possible to produce new materials that match the quality of virgin plastic. This approach helps close the loop in plastic use and supports the goal of the circular economy (Dogu et al., 2021). Chemical recycling methods, such as glycolysis and pyrolysis, have the potential to increase the range of plastics that can be recycled by breaking down polymers into their original oligomers, monomers, or other valuable chemical feedstocks.

Glycolysis is considered one of the most promising chemical recycling methods for breaking down PET on an industrial scale due to its gentle reaction conditions and the use of a less volatile reagent, ethylene glycol. Some companies, such as Shell Polyester, Dow Chemical, and DuPont, have already built commercial or pilot-scale glycolysis units (Xin et al., 2021). This process breaks down polyester plastics through transesterification, converting them back into their monomer forms, such as bis(2-hydroxyethyl) terephthalate (BHET). López-Fonseca et al. observed that using optimized conditions can result in high monomer recovery, demonstrating the efficiency of the process and its potential for use in recycling systems (López-Fonseca et al., 2010). Glycolysis products can be separated and purified using methods like adsorption, cooling crystallization, and hot water extraction. The resulting oligomers can be reused to produce recycled polyethene terephthalate (PET) (Xin et al., 2021). Despite these notable benefits, challenges such as long reaction times and reduced yields of BHET hinder its commercial production.

Pyrolysis, a thermal decomposition method, breaks down plastic such as polystyrene (PS), polypropylene (PP), and polyethylene (PE) utilizing heat in the absence of oxygen, and produces gas, oil, and char (Dai et al., 2022). Pyrolysis enables efficient recovery of hydrocarbons from plastic waste, offering potential applications in energy production or feedstock of crude petrochemical. The study suggests that pyrolysis of waste polystyrene (PS) and polyethylene (PE) results in varying product distribution, with PS generating a higher proportion of liquid products. At the same time, PE tends to produce more gaseous outputs (Kiran et al., 2000). Thermogravimetric analysis was used to determine the optimal temperature for efficient pyrolysis, with a low heating rate employed to maximize liquid yields. These findings highlight the potential in plastic waste management process by converting difficult-to-recycle plastic waste into valuable products, helping to reduce pollution and conserve resources (Klaimy et al., 2021; Padmanabhan et al., 2022). However, these waste management approaches require high energy input and produce complex mixtures that must undergo additional processing. Recently, the concept of upcycling has gained momentum. This approach builds on chemical recycling and the valorization of plastic waste, aiming to produce materials with improved properties compared to the original, by applying catalytic techniques that require less energy and utilizing depolymerization processes. Upcycling offers an efficient way to refine common plastics into higher-value products. Emerging methods, such as hydrogenolysis, photo reforming, and microwave-assisted processes, are being increasingly explored in industry to produce fine chemicals and value-added functional materials (Chen et al., 2021a). These products often retain properties and structural characteristics similar to those of the original plastics (Chen et al., 2021b; Lim & Thian, 2021). Solvolysis using supercritical or subcritical fluids, such as water or alcohol, has become an effective method for recycling plastic waste. These fluids, which exhibit properties between those of gases and liquids, offer high solubility and reactivity, enabling the breakdown of polymers and partial oxidation for material recovery (Wei & Hadigheh, 2023). Hydrogenolysis is a sustainable and waste-free depolymerization technique that cleaves chemical bonds, especially C–C bonds, using hydrogen, which can be sourced renewably, making it both economical and environmentally friendly for transforming organic compounds (Kumar et al., 2020). Photo reforming under ambient conditions typically yields low selectivity and efficiency. Nevertheless, studies have shown that polymers such as polyurethanes, polyethylene terephthalate (PET), and polylactic acid (PLA) can be photo-reformed at room temperature using an economical CdS/CdOx catalyst in alkaline solution, although the toxicity of the catalyst remains a concern (Yadav et al., 2021; Zhao et al., 2021).

Quaternary recycling, commonly known as incineration, is another widely used method in the processing industry for managing mixed plastic waste. Waste particles are often incinerated to produce heat for power generation, as they have a high calorific value ranging from 20 to 40 MJ/kg, which is similar to crude oil, at approximately 45 MJ/kg (Martin et al., 2021). Many countries utilize energy recovery from plastic waste incineration as a strategy to reduce landfill usage and partially substitute fossil fuel-based energy sources, offering potential environmental advantages. Incineration emits toxic gases, including greenhouse gases, furans, and dioxins, as well as microplastic-laden ash that contaminates soil and water, thereby increasing ecological risks and remediation costs (Chen et al., 2021a).

11.7 Emerging Nanotechnology in Microplastic Removal and Management

A man-made, spheroidal nanomaterial known as a "Dendrimer" has gained attention due to its multifunctional, three-dimensional structure, typically ranging from 1 to 10 nm in size. These dendrimers can be precisely designed and synthesized. Structurally, they consist of a central core, an interior cavity, and surface functional groups. The interior cavity is capable of entrapping guest molecules, while the surface groups offer customizable sites for various physical or chemical interactions, allowing for tailored applications across fields such as medicine, biotechnology, materials science, and agriculture (Tomalia, 2005). Historically, Tomalia-type Poly(amidoamine) (PAMAM) dendrimers (Fig. 11.1) are recognized as the first dendrimer family to be synthesized and become commercially available (https://www.sigmaaldrich.com).

In the context of microplastic waste management, dendrimers present a novel and powerful solution as an ideal platform for capturing or degrading microplastic particles and associated toxins. Microplastics, due to their small size and chemical persistence, are difficult to detect, remove, or degrade using conventional methods. Dendrimers offer multiple mechanisms that can help tackle this challenge effectively: (i) Selective Binding and Adsorption: The highly branched, tunable surface of dendrimers can be chemically modified to target specific microplastic particles or associated pollutants (e.g., Bisphenol A and phthalates). The peripheral functional groups interact electrostatically or via hydrophobic interactions, allowing dendrimers to capture these particles from aquatic environments. (ii) Encapsulation and Complexation: The interior cavity of dendrimers allows them to act as molecular sponges. Microplastic-derived toxins or nanoplastics can be trapped within the dendritic voids, preventing them from entering food chains. (iii) Catalytic Degradation: Metal-dendrimer complexes (e.g., Fe^{3+} or TiO_2-PAMAM hybrids) can generate reactive oxygen species (ROS) capable of breaking down polymer

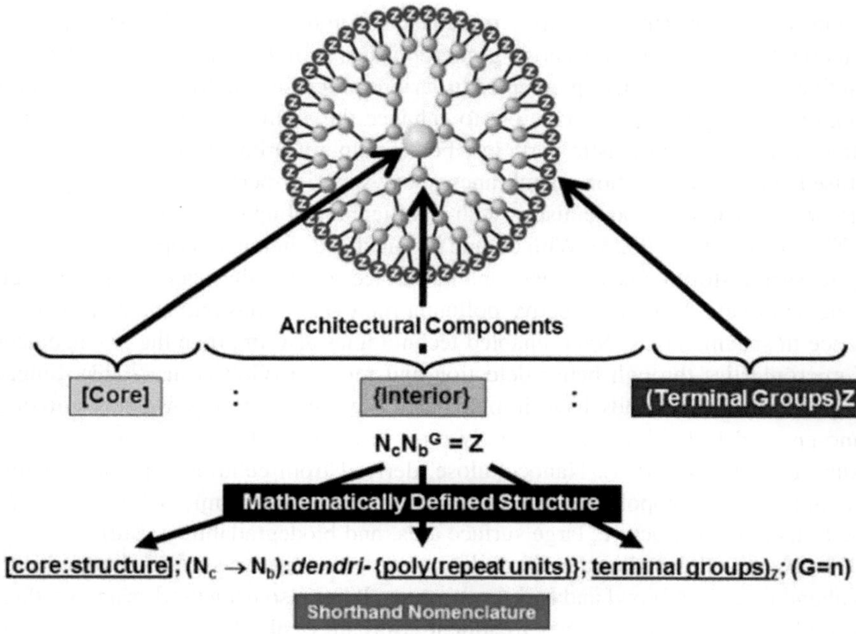

Fig. 11.1 Architectural view (3 dimensional components) of PAMAM dendrimer. Graphic provided by NanoSynthons (i.e., Dendrimer Nanotechnology Company located in Mt. Pleasant, Michigan, USA)

chains. This leads to photo-oxidative or catalytic degradation of microplastics in the environment, especially in sun-drenched water bodies. (iv) Integration into Smart Membranes: Dendrimers can be grafted onto filtration membranes to enhance selectivity, antifouling properties, and binding capacity. These membranes are efficient in filtering micro- and nanoplastics from municipal and industrial wastewater streams. (v) Low Toxicity and Green Chemistry Potential: Unlike many other organic nanoparticles, dendrimers are especially biodegradable which can be even re-engineered for fast degradability in environmental conditions. (vi) Detection and Sensing of Microplastics: Dendrimer-based fluorescent sensors selectively bind with plastic additives like phthalates or Bisphenol A. Their high sensitivity enables early detection of low concentration microplastics in water bodies (i.e. PAMAM dendrimers with fluorophores can bind plastic additives and emit measurable signals under UV light). This all makes them a sustainable choice for environmental remediation applications (Abdelhamid & Mathew, 2021; Iordachescu et al., 2024; Khosravi et al., 2022; Pan et al., 2022). In conclusion, dendrimer nanotechnology has represented an innovative frontier in the fight against microplastic pollution and purifying ground water (https://www.nanotech-now.com). Its unique nanoscale architecture, chemical versatility, and proven effectiveness in capturing and degrading pollutants render it a valuable tool for future circular and resilient environmental systems.

Further, use of nanotechnology in advanced filtration systems has shown great potential for detecting and removing microplastics from the environment. Over the last five years, it has been implemented in various processes, such as membrane filtration, photocatalysis, and adsorption, to enhance the ability of filtration systems to attract and trap microplastic particles. For example, Carbon nanoparticles (CNP) infused into nano-filtration membranes show superior performance in capturing microplastics and microorganisms, with an efficiency of up to 95% (Bano & Gupta, 2023; Kamaraj et al., 2023). With unique physical and chemical properties, such as a higher surface-to-volume ratio and tunable surface charges, they can be implemented at the nanoscale. It offers effective pollutant removal options and enhances performance in specific areas. Nano-enabled technologies also improve the management of microplastics through better detection and removal (Goh et al., 2022). Singapore and the Netherlands have implemented advanced filtering systems utilizing nano-enabled technology to efficiently remove microplastics from water systems (Alam & Rahman, 2024). Nanocellulose, derived from cellulose, the most abundant and versatile biopolymer on Earth can effectively capture microplastic particles due to its porous structure, large surface area, and biodegradability (Burhani et al., 2024). Nano cellulose has a hydrophilic nature with negative surface charge due to its abundance of carboxyl and hydroxyl groups. It has also been used as an excellent antifouling agent in wastewater treatment (Burhani et al., 2024). Nano filters and membranes are emerging as crucial technologies for removing microplastics from water and wastewater.

Micro/nanorobots (MNRs) are self-propelled robots that can be tailored for specific functionalities, such as magnetic guidance or enzymatic degradation, and represent another cutting-edge technology for microplastic removal. They can be deployed for large-scale applications and operate efficiently in aqueous environments (Li et al., 2023; Urso & Pumera, 2022). These magnetic nanorobots are rapid and efficient, with the ability to capture and remove over 90% of nanoplastics from wastewater within 120 min, highlighting their potential in filtration processes (Velikov et al., 2024).

Recent studies show that magnetic nanoparticles (MNPs), like Fe_3O_4, can selectively bind to microplastics and nanoplastics from water by adsorbing them to the nanoparticle surface, enabling separation through magnetic fields for microplastic removal (Aragón et al., 2024). Their effectiveness is attributed to their high surface-to-volume ratio and tunable surface functionalization, which can be optimized for specific microplastic types (Gaß et al., 2023). Furthermore, advancements in nanobubble-assisted flotation can effectively remove microplastics smaller than 10 μm better than traditional methods by up to 17% (Jia et al., 2024).

The environmental impact of nanotechnology for microplastic removal is a critical consideration, as it is an emerging technology. Nanocellulose is biodegradable and environmentally friendly, with minimal ecological risk (Burhani et al., 2024). Similarly, magnetic nanoparticles and micro/nanorobots do not introduce harmful byproducts into the environment. They are designed for safe operation (Velikov et al., 2024). However, nano-based technology for microplastic removal needs further investigation. Nanoparticles can accumulate in ecosystems, and their long-term interactions with biota need to be thoroughly assessed for ecological safety (Rizvi et al., 2024). Research has also shown that specific nanomaterials, such as those used in photocatalytic degradation, can effectively break down microplastics without generating harmful byproducts (Xiao et al., 2024). Additionally, using biodegradable nanomaterials for microplastic removal is a key factor in minimizing their environmental footprint.

11.8 Innovative Technologies for Microplastic Detection and Monitoring

Rising awareness of microplastic contamination has driven notable progress in the development of detection methods. Some advanced analytical techniques commonly used for detecting microplastics are SEM–EDS, NMR, FTIR, GC/MS, Raman spectroscopy, and NIR (Wright et al., 2013). SEM–EDS is a highly effective microscopy technique used to examine the surface characteristics of microplastics and detect any additives associated with them. FTIR and Raman spectroscopy are among the most effective methods for chemical characterization, offering accurate, non-destructive analysis by generating spectra based on the interaction of light with molecular structures. Near-Infrared (NIR) and Nuclear Magnetic Resonance (NMR) have recently emerged as valuable tools for identifying and analyzing microplastics. NIR allows for swift and non-invasive detection, with no need for complex sample preparation, and is capable of recognizing particles as small as 1 mm in environmental samples. On the other hand, NMR offers accurate quantification by assessing the unique magnetic behavior of microplastic polymers (Tirkey & Upadhyay, 2021).

Scanning Electron Microscopy (SEM) utilizes a focused beam of high-energy electrons to scan the surface of a sample. The interaction between the electrons and the sample generates secondary electrons, which are collected to provide detailed insights into the sample's surface structure and morphology. The interaction between the electron beam and the sample also leads to the emission of characteristic X-ray photons, which are specific to individual elements, as electrons transition to lower energy levels after being excited. An energy dispersive spectroscopy (EDS) detector identifies and distinguishes the characteristics X-rays emitted by various elements present in the sample (Schwaferts et al., 2019). SEM produces detailed micrographs of the sample's surface, while EDS offers quantitative data on its elemental composition. This integrated approach is commonly known as SEM/EDX or SEM/EDS analysis. Pan et al. (2019) reported that EDS elemental analysis of microplastics showed prominent nitrogen peaks on the surface of polystyrene, polypropylene, and polyethylene, suggesting the presence of biomass. This strong nitrogen signal indicates bioaccumulation, highlighting significant interactions between microplastic and biological organisms.

SEM–EDS enables rapid differentiation between plastic and non-plastic pellets and is capable of identifying small particles that may be overlooked during visual inspection. This technique utilizes backscattered electron imaging and elemental spectra to analyze the composition of pellets, where plastic samples exhibit a prominent carbon peak along with minor signals from other elements, whereas non-plastic pellets lack a significant carbon signature (Blair et al., 2019). Cooper et al. demonstrated that SEM offers high-resolution imaging of microplastic surface features, revealing that chemical and physical weathering cause structural damage such as fractures, grooves, pits, horizontal notches, and vermiculite textures, which contribute to the progressive breakdown of plastic particles into smaller fragments (Cooper & Corcoran, 2010).

Environmental scanning electron microscopy (ESEM), a specialized form of SEM, allows the analysis of moist or hydrated samples by maintaining them in a low-pressure nitrogen atmosphere, preventing degradation that would typically occur under the high vacuum conditions of conventional SEM (Schwaferts et al., 2019). SEM–EDS has become a vital technique for analyzing microplastic particles within samples, enabling researchers to track the degradation of plastics into micro-sized pollutants that pose a significant risk to aquatic and terrestrial life (Scimeca et al., 2018). A key limitation of quantitative SEM–EDS is that it requires samples to be either flat and polished or extremely thin, with minimal surface irregularities, to ensure accurate analysis. While SEM–EDS is generally regarded as a costly technique due to the high price of the equipment and the detailed nature of the analysis, the sample preparation and examination process can be relatively quick, straightforward, and low-cost once the necessary infrastructure is in place (Wagner et al., 2016).

In the past decade, quantitative nuclear magnetic resonance (qNMR) has been increasingly recognized as a reliable technique for the quantitative detection and analysis of a wide range of substances, including microplastics (Papini et al., 2023; Peez & Imhof, 2020; Peez et al., 2019). Peez et al. (2019) were among the first to apply 1H NMR spectroscopy for the identification and quantification of polyethylene, polyethylene terephthalate, and polystyrene particles in model samples. In their analysis, polyethylene was deuterated using toluene and deuteron, polystyrene in chloroform, and poly-ethylene terephthalate utilizing a mixture of chloroform and trifluoroacetic acid. The acquired spectra were processed using the NMR software MestReNova.

In a separate study, Peez et al. (2018) utilized quantitative NMR through both the integration method and calibration curve approach (INT-qNMR). To address matrix signal interference following chemical digestion, they applied the peak fitting technique (PF-qNMR) to enhance spectral clarity. This mathematical approach enables the separation of overlapping signals, allowing accurate integration of the isolated signal areas. This offers a key advantage compared to FTIR and Raman spectroscopy, as FTIR is prone to matrix interference, while Raman signals can be affected by fluorescence.

In another study, one-dimensional 1H NMR spectroscopy was employed to accurately identify resonance signals corresponding to polybutylene adipate-co-terephthalate (PBAT) in soil samples that had been intentionally spiked (Nelson et al., 2019). The versatility, accuracy, and precision of NMR (1H/13C) spectroscopy for microplastic detection have been effectively demonstrated in studies conducted by

Günther et al. and Peez et al. (Günther & Imhof, 2023; Peez et al., 2018, 2019). Papini et al. in their study of sediment samples collected from the Mediterranean Sea near the mouth of the Tiber River, focused on detecting polystyrene and acrylonitrile butadiene styrene (ABS). NMR spectroscopy revealed the presence of a polystyrene while ABS was not detected in the samples (Papini et al., 2023). High-resolution NMR spectroscopy enables accurate identification and quantification of PET microplastic at atomic resolution, regardless of particle size. In samples from Lake Constance, PET exhibited varied chain lengths, as shown through diffusion and relaxation behavior. The developed workflow achieved a detection limit of 19.2 ng, with an 88% $\pm$ 25% recovery rate for spiked samples, and an estimated PET concentration of 335 $\pm$ 200 ng/m^3. This method not only quantifies PET mass effectively but also reveals polymer characteristics that are inaccessible through conventional microplastic detection techniques (Dukek et al., 2024).

The high cost associated with qNMR methods remains a significant barrier to their routine use in analytical applications. This issue was previously highlighted by Primpke et al. who identified the expense of deuterated solvents and the substantial acquisition and operational cost of high-field NMR instruments (ranging from 300 to 1000 MHz) as key challenges (Primpke et al., 2020a). A major economic challenge could be addressed by adapting the method to low-field NMR instruments (30–100 MHz), which have already proven effective for characterizing polymers such as polyisoprene and various styrene-based copolymers (Minkler et al., 2020). A key benefit of low-field NMR spectrophotometers is their affordability and lower operating costs, as they do not require liquid helium and nitrogen for magnet cooling (Chakrapani et al., 2019). Cost saving can also be achieved by substituting commonly used deuterated solvents with more affordable non-deuterated alternatives.

Fourier Transformation Infrared (FT-IR) spectroscopy enables precise identification of microplastic polymer types by analyzing their unique infrared spectral signatures. FTIR spectroscopy identifies microplastics by detecting molecular vibrations triggered by infrared radiation, which produce wavelength-specific spectra based on the sample's composition and structure. In these analyses, specific wavelengths of infrared (IR) radiation are absorbed to excite molecular vibrations, enabling quantification through characteristic IR spectra (Veerasingam et al., 2020). FTIR operates through three key enhancement techniques: Focal Plane Array (FPA), Attenuated total reflection (ATR), and micro-FTIR. ATR-FTIR is ideal for analyzing large or opaque microplastic samples without extensive preparation, as it involves direct surface contact between the sample and the crystal (Shim et al., 2016). Focal Plane Array (FPA)-FTIR enables high-throughput chemical imaging by capturing detailed spectral data over a broad sample area, making it efficient for identifying multiple microplastic particles simultaneously. Micro-FTIR combines infrared spectroscopy with microscopy, allowing precise identification and mapping of microplastics down to 10 μm in size within complex environmental matrices.

Song et al. used ATR-FTIR to detect microplastics in sea-surface microlayer and beach sand samples, identifying particles smaller than 50 μm in size. They observed that smaller particle sizes improved detection efficiency due to higher concentrations of the particles. FTIR also revealed the types of polymers and offered insights into their origins and characteristics (Song et al., 2015). Another study employed ATR-FTIR with a single-element MCT detector to identify various microplastics in beach debris. Spectral data were collected over the 4000–650 cm^{-1} range with 128 scans at a resolution of 8 cm^{-1} (Andrades et al., 2018). Tagg et al. used Fenton's reagent and FPA-based reflectance micro-FTIR imaging to detect microplastics (MPs) at various stages and depths of a wastewater treatment plant (WWTP). This method enabled automated identification of MPs in samples with high organic content, without visual sorting or special sample preparation. Results showed a significant drop in MP fragments after primary settlement, which accounted for 76.9% of total MPs detected. Nonetheless, the final effluent still contained an average of 1.5 million particles per litre (MPL-1) (Tagg et al., 2020).

Tadsuwan and Babel analyzed water and sludge samples to identify the types, shapes, and sources of microplastics (MPs) using sieving and FTIR spectroscopy. The influent contained 26.6 ± 11.8 MPs/L, with significant removal after both grit trapping and secondary treatment. Despite this, up to 2.32×10^9 MPs may be discharged daily at peak flow. Sludge samples contained $8.12 \pm 0.28 \times 10^3$ MPs/kg (dry weight), indicating a potential source of soil contamination. Most MPs were polyester fibres, along with polypropylene, polyethylene, silicone and polystyrene (Tadsuwan & Babel, 2021). ATR-FTIR has certain limitations, such as the risk of damaging fragile or heavily weathered microplastics due to the pressure exerted by the probe. Additionally, microscopic microplastic particles may adhere to the probe tip due to electrostatic attraction (Shim et al., 2016). FTIR is limited to detecting microplastics down to 20 μm, whereas micro-FTIR extends this capability to particles smaller than 10 μm (Löder & Gerdts, 2015).

Raman and Fourier-transform infrared (FTIR) spectroscopy are among the most widely used techniques for identifying microplastics. FTIR and Raman spectroscopy are practical for identifying microplastic types, sizes and quantities, but are limited by lengthy analysis times, complex sample preparation, and reduced sensitivity to the smallest particles. While Raman can detect particles as small as 50 nm in spiked samples, the practical detection limits in environmental samples are approximately 10–20 μm for FTIR and 1 μm for Raman spectroscopy (Hu et al., 2022; Xu et al., 2019). In Raman spectroscopy, a monochromatic laser, typically with a wavelength of 500–800 nm, irradiates the sample contaminated with microplastics. The laser interacts with molecular vibrations, causing frequency shifts in the light that is backscattered. These shifts, known as Raman shifts, produce distinct spectra that help to identify the material composition of the sample (Mikac et al., 2022). Since every polymer exhibits a unique Raman spectral fingerprint, this method provides a reliable and rapid way to identify plastic types within minutes.

In Raman spectroscopy, autofluorescence from organic residues on microplastics can obscure characteristic polymer peaks, hindering identification. To address this, Ghosal et al. implemented an automated algorithm that removes fluorescence interference, enhancing visibility of the underlying Raman signals. These clarified spectra can then be matched with reference libraries to determine the size, composition, and morphology of microplastic particles (Ghosal et al., 2017). Raman spectroscopy is a non-destructive method well-suited for analyzing visually distinct microplastic particles. When paired with microscopy (micro-Raman), it enables identification of particles as small as one μm. Recent advancements also integrate Raman spectral imaging to produce spatially resolved chemical maps of samples. Micro-Raman mapping provides high-resolution spectral analysis (< 1 μm) across entire membrane filters, facilitating the detection of even the tiniest particles (Chen, 2022). However, its practical application to real-world environmental samples remains limited.

Initially, analyzing single microplastic particles in natural samples was challenging. However, Sobhani et al. used confocal Raman microscopy to detect and map microplastics directly in soil and sand without sample preparation or dyes. This non-destructive method showed minimal interference from organic matter and identified five polymer types (PS, PP, PE, PET, and PVC) (Sobhani et al., 2019). Each particle produced a unique Raman spectrum, and by combining maps of characteristic peaks, the spatial distribution of microplastics was visualized.

Pyrolysis–Gas Chromatography/Mass Spectrometry (Py-GC/MS) is a technique that allows direct analysis of the chemical composition of potential microplastic samples. This method analyzes the thermal degradation products of microplastics during pyrolysis, producing distinctive program to identify polymer types (Picó & Barceló, 2020). After determining the composition of MPs, the polymer source of microparticles is accessed by matching their distinct thermal degradation products with reference program from known polymer standards. Including a thermal desorption step before pyrolysis allows for the simultaneous analysis of microplastic additives during the Py-GC/MS process (Primpke et al., 2020b). According to researchers, Py-GC/MS is generally suitable for microplastic particles larger than 500 μm, although the smallest size successfully analyzed to date is 100 μm (Li et al., 2017).

Hermabessiere et al. aimed to optimize and validate a Py-GC/MS method, establish detection limits for eight common polymers, and apply it to environmental microplastics. Method optimization focused on various GC parameters using polyethylene and polystyrene microspheres (Hermabessiere et al., 2018). The optimized Py-GC/MS method operates at a pyrolysis temperature of 700 °C, a split ratio of 5, and an injector temperature of 300 °C. Its performance was evaluated through repeatability, precision tests, and detection limit calculation, all below 1 μg. Despite a reduction in signal overtime, identification remained reliable. The method was compared with Raman micro spectroscopy and successfully applied to the environmental samples from seawater, marine organisms, and beach sediments.

Okoffo et al. conducted a quantitative analysis using pressurized liquid extraction in combination with pyrolysis gas chromatography-mass spectrometry (Py-GC/MS). They observed that the total plastic content in raw influent samples from WWTPs ranges from 840 to 3116 μg/L, translating to daily inputs of approximately 2.1 tp 196.4 kg. Over 99% of these plastics were removed during pre-treatment, likely ending up in sewage sludge, which resulted in minimal discharge via treated effluents. The plastic levels in treated effluents and reference water samples were below the reporting limits. Among the five plastics identified, PE (76.4%) and PVC (21%) were most prevalent by mass, PP, PMMA, and PP contributing less than 3% combined (Okoffo et al., 2022).

Inductively Coupled Plasma-Mass Spectrometry (ICP-MS) is increasingly used in microplastic waste management to detect trace metals associated with plastic particles. While it doesn't identify polymers directly, ICP-MS helps analyze metallic additives and contaminants adsorbed on microplastics, providing insight into their toxicity and environmental impact. Combined with other techniques, ICP-MS supports better monitoring and treatment strategies for microplastic pollution (Bolea-Fernandez et al., 2019). The concept behind single-particle (SP) ICP-MS was initially introduced by Degueldre et al. as a method for analyzing colloidal particles suspended in water. SP-ICP-MS offers simultaneous analysis of a particle's elemental makeup, estimated size and size distribution (as spherical equivalent diameter in nanometers), mass concentration (in mg/L) and particle count per milliliter (Degueldre et al., 2005; Laborda et al., 2013). ICP-MS detection of carbon is challenged by low sensitivity and high background noise. This is due to carbon's low ionization efficiency (~ 5%) from its high ionization energy (11.26 eV) and poor transmission caused by space charge effects. Additionally, carbon's widespread presence in air and water contributes to elevated background levels, resulting in relatively high detection limits, typically in the mg/L range (Riisom et al., 2018).

Trujillo et al. developed a SP-ICP-MS approach suitable for quickly screening river water for microplastics, which can subsequently be examined using more selective, though slower, methods such as Raman microscopy. In SP-ICP-MS, carbon-based particles are challenging due to interference from carbonate, organic matter, and microorganisms. They performed a 24-h pre-treatment with 10% nitric acid, which helped reduce dissolved carbon to minimal levels and digest microorganisms, enabling up to an 80% recovery of polystyrene microparticles. Microplastics were found in most samples from Spanish and French Pyrenean rivers, with Raman microscopy confirming their presence and electron microscopy with EDX revealing their morphology (Trujillo et al., 2023).

Bolea-Fernandez et al. demonstrated the detection of 1 and 2.5 μm polystyrene microspheres, mimicking microplastics, using ICP-MS in single-event mode. By monitoring transient signal spikes at m/z 13 ($^{13}C^{+}$) with a 100 ms dwell time, particle number densities and sizes were accurately determined. Results were validated using 165 Ho^{+} for lanthanide-doped beads and compared to electron microscopy measurements. The method effectively provides both size distribution and mass concentration of microplastics (Bolea-Fernandez et al., 2019).

11.9 Use of AI, Machine Learning and Real Time Monitoring Technologies

Generating extensive experimental data on microplastics for numerical modelling is difficult due to limited long-term studies and the labor-intensive nature of monitoring the methods. Key challenges include manual collection, identification, counting, and classification of microplastics. While visual identification helps categorize particles by shape and size, chemical verification, typically using FTIR or Raman Spectroscopy, is essential for confirming polymer types. However, these techniques are time-consuming and require expert input, making high throughput and long-term monitoring currently impractical. Machine learning and computer vision are increasingly integrated, with machine learning enhancing visual analysis and computer vision extracting meaningful data from images. Applied to microplastic monitoring, these technologies enabled automated detection and data collection. Computer vision objectively identifies features and improves microplastic characterization in environmental studies (Phan & Luscombe, 2023).

Lorenzo-Navarro et al. created a computer-based design to automatically detect, count, and categorize microplastics ranging from 1 to 5 mm in size (Lorenzo-Navarro et al., 2020). The system used high-resolution scans of particles on a white background to segment microplastics from RGB images. Various machine learning models were tested for classification, with the cascade classifier, leveraging shape, color, and texture, achieving the highest accuracy of 91%. Integrating machine learning with nanomaterials in wastewater treatment offers a novel way to improve contaminant removal. Techniques like supervised, unsupervised, and reinforcement learning help build predictive models that optimize nanomaterial behavior, treatment efficiency, and dosage (Oruganti et al., 2023).

AI-based microplastic imaging is advancing rapidly with the support of emerging technologies. These innovations include new algorithms, tools, and applications that play a significant role in enhancing the detection and analysis of microplastics. Compared to larger plastics, analyzing microplastic is more complex due to their small size and abundance. As a result, researchers have explored various AI-driven imaging technologies, particularly deep learning models, to improve classification, identification, and quantification in this field (Amaral-Zettler et al., 2020; Tian et al., 2020).

Core technologies driving AI-based microplastic imaging include smart unmanned aerial vehicles (UAVs), advanced artificial intelligence systems, and intelligent robotics. These tools enhance the detection, monitoring, and analysis of microplastics in various environments (Arise & Kurazume, 2021; Martin et al., 2018). Robotics technologies like drones, smart vehicles, and bio-inspired robotic fish have been developed to collect microplastics (MPs) from aquatic environments (Zhang et al., 2023; Zolich et al., 2022). One such example is Gillbert, a 3D printed, salmon sized robot with a built-in filtration system. Its gill-like filters capture MPs up to 2 mm, storing them in an internal container for later recycling and disposal (Sanfilippo et al., 2021).

Although optical, laser-based, and scanning electron microscopy techniques have significantly advanced microplastic detection, they are often intrusive, large in size, and rely on costly components (Song et al., 2015). Moreover, these techniques are confined to laboratory settings and require transporting water samples to a controlled environment, making them unsuitable for on-site detection of microplastics. There is a growing need for fast, convenient, and non-destructive in situ detection methods that can enable real-time monitoring and accurate quantification of microplastics, potentially enhanced by artificial intelligence technologies (Liu et al., 2021; Yin et al., 2021). Sarker et al. proposed a novel method combining computer vision and AI for microplastic detection. They tested three camera setups (fixed focus 2D and autofocus 2D/3D) and used a YOLOv5 model for object detection, followed by DeepSORT for tracking particles. In real-time lab tests, the system achieved 97% accuracy in counting MPs, and 96% in field tests on a local river (Sarker et al., 2024).

11.10 Challenges and Limitations in Current Technologies

Current microplastic (MP) waste technologies- including physical, chemical, and biological methods face significant challenges that limit their effectiveness. Traditional physical processes, such as filtration and sedimentation, are effective for removing larger particles but struggle to remove micro- and nanoplastics, and they require significant infrastructure and maintenance costs. Chemical treatment, including coagulation and advanced oxidation processes, may degrade microplastics but often produces toxic byproducts and is sensitive to variation in water chemistry. Biological methods, such as microbial degradation, are sustainable but suffer from slow processing times and a limited specificity to certain polymer types, which limits their large-scale applicability (Amparán et al., 2025; Sridharan et al., 2021).

Conventional detection methods, such as Fourier-transform infrared spectroscopy (FTIR), Raman spectroscopy, and scanning electron microscopy (SEM), are highly accurate but come with practical constraints. These techniques are labor-intensive, require expensive and bulky equipment and are typically restricted to laboratory settings. As a result, they are unsuitable for real-time or in situ monitoring of MPs in natural environments (Prata et al., 2018).

Emerging AI-integrated technologies, such as machine learning and computer vision, offer real-time monitoring and classification of MPs. Yet, their accuracy heavily depends on high-quality training datasets and optimal imaging conditions, which are difficult to ensure in dynamic aquatic environments. Recently, Sarker et al. have demonstrated the feasibility of using computer vision and deep learning (YOLOv5 and DeepSORT) to monitor MPs in real time. However, these technologies are still in their early stages and require further validation in various environmental conditions before they can be deployed on a broad scale (Sarker et al., 2024). While robotics solutions like UAVs and bio-inspired fish (Gillbert) are promising for remote collection, they remain in the prototype phase with limited real-world scalability. Ultimately, to overcome these limitations, there is a critical need for standardized methodologies, improved cost-effective sensing tools, and scalable real-time monitoring systems that can be used across diverse environments.

11.11 Policy and Future Aspects

The widespread use of plastic products in everyday life cannot be easily replaced, as they are well-integrated into almost every industry. There is no doubt that after the discovery of plastic, everyday life has become so convenient and easy. The plastic has also helped alleviate the socioeconomic status and livelihood of marginalized sections, either directly or indirectly (Nøklebye et al., 2022). Microplastics, which are degraded plastic particles (usually smaller than 5 mm), have different physical and chemical properties due to their diverse sources (Ashrafy et al., 2022). They can infiltrate marine, freshwater, and terrestrial environments alike and are a substantial environmental hazard (Jolaosho et al., 2025). Moreover, advancements in transportation systems have made the whole world into a global village, making it challenging to develop effective solutions. Therefore, in the coming year, microplastic pollution is likely to continue being a major environmental issue.

However, there are several possible strategies and future perspectives that might help tackle these issues. For example, reducing the release of microplastics into the environment can be achieved by adopting advanced waste management technologies and recycling methods. Searching for and creating new alternative materials that may effectively replace plastic is also important. Implementing policies such as extended producer responsibility (EPR), regulating single-use plastics, banning microbeads, improving solid waste collection, and encouraging public awareness campaigns and education programs for responsible consumption of plastic products and their disposal habits will help limit the demand for plastic products for long-term success. Therefore, we need global coordination on microplastic waste management policies and strategies that can lead us to consistent and effective approaches.

11.12 Global Policies and Efforts for Microplastic Waste Management

The Australian government introduced a ban on the use of single-use plastic, a major source of microplastics, in the Capital Territory in 2011. Within seven years, regular polythene bag consumption had reduced by around 2600 tons (Macintosh et al., 2019). A similar policy measure has been introduced in about 90 countries as well (Nielsen et al., 2019).

The Global Partnership on Plastic Pollution and Marine Litter (GPML) was launched at the United Nations Conference on Sustainable Development (Rio + 20) in 2012. GPML is a multi-stakeholder partnership collaboration among governments, researchers, agencies, civil society, and the commercial sector, aimed at protecting the global marine ecosystem by reducing marine litter and plastic pollution, especially microplastics (Alam & Rahman, 2024).

In 2015, the United States passed the Microbead-Free Waters Act, which aims to ban the marketing, production, packaging, and sale of personal care items containing plastic microbeads. This law aims to reduce plastic microbeads in waterways, which is a significant source of microplastics (Borriello, 2023). From 2015 to 2020, the Plastic Microbeads Free Waters Act was introduced in many countries in Asia, Europe and South America (UNEP, 2017; Alam & Rahman, 2024).

A global awareness campaign, Clean Seas, started in 2017 to address marine litter and microplastics. With over 69 countries joining, it became the biggest global campaign pledging to improve laws and encourage change. The campaign involved governments, the public, and the commercial sector to reduce plastic use by 2022 (UN Environment Annual Report, 2017). The Basel Convention, a global environmental treaty which was initially focused on controlling and managing the transboundary movement of hazardous wastes and their disposal, has been extended to include plastic waste, including microplastics. This amendment provides member states with a legal framework to implement advanced waste management and recycling technologies, reducing plastic waste movement between nations and controlling its environmental impact (Basel Convention, n.d.).

APEC focuses on sustainable marine ecosystems in the Asia–Pacific, addressing climate change and managing marine litter, especially microplastics, through regional cooperation among member economies. The EU's 2018 Strategy for Plastics in a Circular Economy aims to reduce and eliminate plastic waste through better design, novel plastic recycling technologies, and a reduction in single-use plastics. It sets goals for recyclable packaging by 2030 and supports recycled material use in new product investments.

11.13 Global Collaboration Between Academia, Industry, and the Government

Academic researchers may play a crucial role in effectively addressing and managing microplastic issues by scientifically studying the complex interaction of microplastics with various ecosystems and other pollutants. They may standardize the methodologies and develop new, novel, innovative remediation strategies and policies to tackle microplastics. Researchers may help in the identification and isolation of microorganisms that can effectively metabolize synthetic polymers and degrade plastics (Kurniawan et al., 2023). However, industries may help in developing biodegradable and reusable plastics while being environmentally friendly. They may collaborate with research institutions and NGOs to promote and advocate for international agreements and regulations, as well as invest in research and development for innovative microplastic recycling, removal and remediation technologies. Further governments bodies are responsible for implementing laws to address microplastic pollution by banning the use of microplastics in personal care products, cosmetics, and cleaning products. They can require labelling for products that contain microplastics, helping consumers make informed choices. Regulations can be implemented to limit the release of microplastics into waterways from industrial processes. Governments can set targets to reduce microplastic pollution and establish rules to ensure these targets are met. They can create programs to research and monitor microplastic pollution, better understanding the issue and informing policy. Financial incentives can be offered to encourage businesses and researchers to develop microplastic-free technologies. Education campaigns can inform the public about the risks of microplastic pollution. International cooperation can lead to the establishment of common standards and regulations to reduce microplastic pollution, and fines can be imposed on those who violate these regulations.

11.14 Conclusions

The widespread infiltration of microplastics into environmental systems underscores the urgent need for multidisciplinary action. Current treatment technologies, including dendrimer technology, filtration membranes, chemical degradation, and microbial remediation offer significant potential but face limitations in scalability, specificity, and economic viability. Advanced analytical tools have improved our ability to detect and characterize microplastics, while recent innovations in nanotechnology and artificial intelligence are opening new frontiers in monitoring and removal strategies. However, technological advancement alone cannot resolve this complex issue. A coordinated global response involving researchers, industries, and policymakers is essential. Regulations such as bans on microbeads, improvements in waste infrastructure, extended producer responsibility, and international treaties must work in tandem with innovations in biodegradable alternatives and closed-loop recycling systems. Public awareness and behavioral change will further support the reduction of microplastic release. Moving forward, fostering collaborative research, promoting circular economy practices, and enacting stringent environmental policies will be central to effectively combating microplastic pollution and safeguarding planetary health.

Acknowledgements Authors are grateful to Ms. Linda S. Nixon, Vice President of Operations at NanoSynthons (Mt. Pleasant, Michigan, USA), for providing the dendrimer graphics and editing this manuscript.

References

Abdel-Fatah, M. A. (2018). Nanofiltration systems and applications in wastewater treatment: Review article. *Ain Shams Engineering Journal, 9*(4), 3077–3092. https://doi.org/10.1016/j.asej.2018.08.001

Abdelhamid, H. N., & Mathew, A. P. (2021). Cellulose-based materials for water remediation: Adsorption, catalysis, and antifouling. *Frontiers in Chemical Engineering, 3*.https://doi.org/10.3389/fceng.2021.790314

Acarer, S. (2023). A review of microplastic removal from water and wastewater by membrane technologies. *Water Science & Technology, 88*(1), 199–219. https://doi.org/10.2166/wst.2023.186

Ahmed, S., Mofijur, M., Nuzhat, S., Chowdhury, A. T., Rafa, N., Uddin, M. A., Inayat, A., Mahlia, T., Ong, H. C., Chia, W. Y., & Show, P. L. (2021). Recent developments in physical, biological, chemical, and hybrid treatment techniques for removing emerging contaminants from wastewater. *Journal of Hazardous Materials, 416*, Article 125912. https://doi.org/10.1016/j.jhazmat.2021.125912

Akinnawo, S. O., Ayadi, P. O., & Oluwalope, M. T. (2023). Chemical coagulation and biological techniques for wastewater treatment. *Analele Universităţii "Ovidius" Constanţa. Seria Chimie/ "Ovidius" University Annals of Chemistry, 34*(1), 14–21. https://doi.org/10.2478/auoc-2023-0003

Alam, M. J., & Rahman, M. M. (2024). Various conventional and advanced management techniques and policies adopted at the global level for microplastics. In *Elsevier eBooks* (pp. 429–440). https://doi.org/10.1016/b978-0-443-29804-2.00018-4

Ali, S. S., Elsamahy, T., Al-Tohamy, R., & Sun, J. (2024). A critical review of microplastics in aquatic ecosystems: Degradation mechanisms and removing strategies. *Environmental Science and Ecotechnology, 21*, Article 100427. https://doi.org/10.1016/j.ese.2024.100427

Amaral-Zettler, L. A., Zettler, E. R., & Mincer, T. J. (2020). Ecology of the plastisphere. *Nature Reviews Microbiology, 18*(3), 139–151. https://doi.org/10.1038/s41579-019-0308-0

Amparán, M. A. A., Palacios, A., Flores, G. M., & Olivera, P. M. C. (2025). Review and future outlook for the removal of microplastics by physical, biological and chemical methods in water bodies and wastewaters. *Environmental Monitoring and Assessment, 197*(4). https://doi.org/10.1007/s10661-025-13883-0

Andrades, R., Santos, R. G., Joyeux, J., Chelazzi, D., Cincinelli, A., & Giarrizzo, T. (2018). Marine debris in Trindade Island, a remote island of the South Atlantic. *Marine Pollution Bulletin, 137*, 180–184. https://doi.org/10.1016/j.marpolbul.2018.10.003

Ansari, M., Obaidi, M. A. A., Hadadian, Z., Moradi, M., Haghighi, A., & Mujtaba, I. M. (2021). Performance evaluation of a brackish water reverse osmosis pilot-plant desalination process under different operating conditions: Experimental study. *Cleaner Engineering and Technology, 4*, 100134. https://doi.org/10.1016/j.clet.2021.100134

Aragón, D., García-Merino, B., Barquín, C., Bringas, E., Rivero, M. J., & Ortiz, I. (2024). Advanced green capture of microplastics from different water matrices by surface-modified magnetic nanoparticles. *Separation and Purification Technology, 354*, Article 128813. https://doi.org/10.1016/j.seppur.2024.128813

Arise, M., & Kurazume, R. (2021). Development of garbage collecting robot for marine microplastics. *The Proceedings of JSME Annual Conference on Robotics and Mechatronics (Robomec), 2021*(0), 1P2–G09. https://doi.org/10.1299/jsmermd.2021.1p2-g09

Ariza-Tarazona, M. C., Siligardi, C., Carreón-López, H. A., Valdéz-Cerda, J. E., Pozzi, P., Kaushik, G., Villarreal-Chiu, J. F., & Cedillo-González, E. I. (2023). Low environmental impact remediation of microplastics: Visible-light photocatalytic degradation of PET microplastics using bio-inspired C, N-TiO_2/SiO_2 photocatalysts. *Marine Pollution Bulletin, 193*, Article 115206. https://doi.org/10.1016/j.marpolbul.2023.115206

Ashrafy, A., Liza, A. A., Islam, M. N., Billah, M. M., Arafat, S. T., Rahman, M. M., & Rahman, S. M. (2022). Microplastics pollution: A brief review of its source and abundance in different aquatic ecosystems. *Journal of Hazardous Materials Advances, 9*, Article 100215. https://doi.org/10.1016/j.hazadv.2022.100215

Bano, S., & Gupta, P. (2023). Advanced carbon nanoparticle-based filtration systems for water disinfection and microplastics removal. *Migration Letters, 20*(S13), 151–160. https://doi.org/10.59670/ml.v20is13.6279

Barbier, J., Dris, R., Lecarpentier, C., Raymond, V., Delabre, K., Thibert, S., Tassin, B., & Gasperi, J. (2022). Microplastic occurrence after conventional and nanofiltration processes at drinking water treatment plants: Preliminary results. *Frontiers in Water, 4*.https://doi.org/10.3389/frwa.2022.886703

Basel Convention. (n.d.). *Overview*. Copyright 2011 by Basel Convention. https://www.basel.int/Implementation/Plasticwaste/Overview/tabid/8347/Default.aspx

Bermúdez, J., & Swarzenski, P. (2021). A microplastic size classification scheme aligned with universal plankton survey methods. *MethodsX, 8*, Article 101516. https://doi.org/10.1016/j.mex.2021.101516

Blair, R. M., Waldron, S., Phoenix, V. R., & Gauchotte-Lindsay, C. (2019). Microscopy and elemental analysis characterisation of microplastics in sediment of a freshwater urban river in Scotland, UK. *Environmental Science and Pollution Research, 26*(12), 12491–12504. https://doi.org/10.1007/s11356-019-04678-1

Bodzek, M., & Pohl, A. (2023). Removal of microplastics in unit processes used in water and wastewater treatment: A review. *Archives of Environmental Protection*. https://doi.org/10.24425/aep.2022.143713

Bodzek, M., Pohl, A., & Rosik-Dulewska, C. (2024). Microplastics in wastewater treatment plants: Characteristics, occurrence and removal technologies. *Water, 16*(24), 3574. https://doi.org/10.3390/w16243574

Bolea-Fernandez, E., Rua-Ibarz, A., Velimirovic, M., Tirez, K., & Vanhaecke, F. (2019). Detection of microplastics using inductively coupled plasma-mass spectrometry (ICP-MS) operated in single-event mode. *Journal of Analytical Atomic Spectrometry, 35*(3), 455–460. https://doi.org/10.1039/C9JA00379G

Borrelle, S. B., Ringma, J., Law, K. L., Monnahan, C. C., Lebreton, L., McGivern, A., Murphy, E., Jambeck, J., Leonard, G. H., Hilleary, M. A., Eriksen, M., Possingham, H. P., De Frond, H., Gerber, L. R., Polidoro, B., Tahir, A., Bernard, M., Mallos, N., Barnes, M., & Rochman, C. M. (2020). Predicted growth in plastic waste exceeds efforts to mitigate plastic pollution. *Science, 369*(6510), 1515–1518. https://doi.org/10.1126/science.aba3656

Borriello, A. (2023). Preferences for microplastic marine pollution management strategies: An analysis of barriers and enablers for more sustainable choices. *Journal of Environmental Management, 344*, Article 118382. https://doi.org/10.1016/j.jenvman.2023.118382

Bui, X., Vo, T., Nguyen, P., Nguyen, V., Dao, T., & Nguyen, P. (2020). Microplastics pollution in wastewater: Characteristics, occurrence and removal technologies. *Environmental Technology & Innovation, 19*, Article 101013. https://doi.org/10.1016/j.eti.2020.101013

Burhani, D., Voet, V. S., Folkersma, R., Maniar, D., & Loos, K. (2024). Potential of nanocellulose for microplastic removal: Perspective and challenges. *Tetrahedron Green Chem, 3*, Article 100045. https://doi.org/10.1016/j.tgchem.2024.100045

Chakrapani, S. B., Minkler, M. J., & Beckingham, B. S. (2019). Low-field1H-NMR spectroscopy for compositional analysis of multicomponent polymer systems. *The Analyst, 144*(5), 1679–1686. https://doi.org/10.1039/c8an01810c

Chand, R., Iordachescu, L., Bäckbom, F., Andreasson, A., Bertholds, C., Pollack, E., Molazadeh, M., Lorenz, C., Nielsen, A. H., & Vollertsen, J. (2024). Treating wastewater for microplastics to a level on par with nearby marine waters. *Water Research, 256*, Article 121647. https://doi.org/10.1016/j.watres.2024.121647

Chen, B. (2022). Current status and trends of research on microplastic fugacity characteristics and pollution levels in mangrove wetlands. *Frontiers in Environmental Science, 10*.https://doi.org/10.3389/fenvs.2022.1021274

Chen, H., Wan, K., Zhang, Y., & Wang, Y. (2021a). Waste to wealth: Chemical recycling and chemical upcycling of waste plastics for a great future. *Chemsuschem, 14*(19), 4123–4136. https://doi.org/10.1002/cssc.202100652

Chen, X., Wang, Y., & Zhang, L. (2021b). Recent progress in the chemical upcycling of plastic wastes. *Chemsuschem, 14*(19), 4137–4151. https://doi.org/10.1002/cssc.202100868

Choi, J., Lim, B., & Yoo, Y. (2023). Advancing plastic waste classification and recycling efficiency: Integrating image sensors and deep learning algorithms. *Applied Sciences, 13*(18), 10224. https://doi.org/10.3390/app131810224

Claessens, M., Van Cauwenberghe, L., Vandegehuchte, M. B., & Janssen, C. R. (2013). New techniques for the detection of microplastics in sediments and field collected organisms. *Marine Pollution Bulletin, 70*(1–2), 227–233. https://doi.org/10.1016/j.marpolbul.2013.03.009

Cooper, D. A., & Corcoran, P. L. (2010). Effects of mechanical and chemical processes on the degradation of plastic beach debris on the island of Kauai, Hawaii. *Marine Pollution Bulletin, 60*(5), 650–654. https://doi.org/10.1016/j.marpolbul.2009.12.026

Coppock, R. L., Cole, M., Lindeque, P. K., Queirós, A. M., & Galloway, T. S. (2017). A small-scale, portable method for extracting microplastics from marine sediments. *Environmental Pollution, 230*, 829–837. https://doi.org/10.1016/j.envpol.2017.07.017

Dai, L., Zhou, N., Lv, Y., Cheng, Y., Wang, Y., Liu, Y., Cobb, K., Chen, P., Lei, H., & Ruan, R. (2022). Pyrolysis technology for plastic waste recycling: A state-of-the-art review. *Progress in Energy and Combustion Science, 93*, Article 101021. https://doi.org/10.1016/j.pecs.2022.101021

Dalmau-Soler, J., Ballesteros-Cano, R., Boleda, M. R., Paraira, M., Ferrer, N., & Lacorte, S. (2021). Microplastics from headwaters to tap water: Occurrence and removal in a drinking water treatment plant in Barcelona Metropolitan area (Catalonia, NE Spain). *Environmental Science and Pollution Research, 28*(42), 59462–59472. https://doi.org/10.1007/s11356-021-13220-1

Das, M. P., Kumar, S., & Das, J. (2018). Fungal-mediated deterioration and biodegradation study of low-density polyethylene (LDPE) isolated from municipal dump yard in Chennai, India. *Energy Ecology and Environment, 3*(4), 229–236. https://doi.org/10.1007/s40974-018-0085-z

Dayal, L., Yadav, K., Dey, U., Das, K., Kumari, P., Raj, D., & Mandal, R. R. (2024). Recent advancement in microplastic removal process from wastewater—A critical review. *Journal of Hazardous Materials Advances, 16*, Article 100460. https://doi.org/10.1016/j.hazadv.2024.100460

Degueldre, C., Favarger, P., & Wold, S. (2005). Gold colloid analysis by inductively coupled plasma-mass spectrometry in a single particle mode. *Analytica Chimica Acta, 555*(2), 263–268. https://doi.org/10.1016/j.aca.2005.09.021

Devi, R. S., Kannan, V. R., Nivas, D., Kannan, K., Chandru, S., & Antony, A. R. (2015). Biodegradation of HDPE by Aspergillus spp. from marine ecosystem of Gulf of Mannar, India. *Marine Pollution Bulletin, 96*(1–2), 32–40. https://doi.org/10.1016/j.marpolbul.2015.05.050

Dey, S., Anand, U., Kumar, V., Kumar, S., Ghorai, M., Ghosh, A., Kant, N., Suresh, S., Bhattacharya, S., Bontempi, E., Bhat, S. A., & Dey, A. (2022). Microbial strategies for degradation of microplastics generated from COVID-19 healthcare waste. *Environmental Research, 216*, Article 114438. https://doi.org/10.1016/j.envres.2022.114438

Ding, F., Wang, H., Li, Y., Leng, X., Gao, J., & Huang, D. (2024). Polystyrene microplastics with absorbed nonylphenol induce intestinal dysfunction in human Caco-2 cells. *Environmental Toxicology and Pharmacology, 107*, Article 104426. https://doi.org/10.1016/j.etap.2024.104426

Division, E., Ryberg, M. W., Laurent, A., Hauschild, M., & Facility, G. E. (2018). *Mapping of global plastics value chain and plastics losses to the environment: with a particular focus on marine environment*. https://wedocs.unep.org/20.500.11822/26745

Dogu, O., Pelucchi, M., Van De Vijver, R., Van Steenberge, P. H., D'hooge, D. R., Cuoci, A., Mehl, M., Frassoldati, A., Faravelli, T., & Van Geem, K. M. (2021). The chemistry of chemical recycling of solid plastic waste via pyrolysis and gasification: State-of-the-art, challenges, and future directions. *Progress in Energy and Combustion Science, 84*, 100901. https://doi.org/10.1016/j.pecs.2020.100901

Du, H., Xie, Y., & Wang, J. (2021). Microplastic degradation methods and corresponding degradation mechanism: Research status and future perspectives. *Journal of Hazardous Materials, 418*, Article 126377. https://doi.org/10.1016/j.jhazmat.2021.126377

Dukek, P., Schleheck, D., & Kovermann, M. (2024). High-resolution NMR spectroscopic approaches to quantify PET microplastics pollution in environmental freshwater samples. *Chemosphere, 367*, Article 143657. https://doi.org/10.1016/j.chemosphere.2024.143657

Elkhatib, D., & Oyanedel-Craver, V. (2020). A critical review of extraction and identification methods of microplastics in wastewater and drinking water. *Environmental Science & Technology, 54*(12), 7037–7049. https://doi.org/10.1021/acs.est.9b06672

Falco, G., Fenti, A., Galoppo, S., Chianese, S., Musmarra, D., Cocca, M., Mallardo, S., & Iovino, P. (2024). Promoting removal of polystyrene microplastics from wastewater by electrochemical treatment. *Journal of Water Process Engineering, 68*, Article 106418. https://doi.org/10.1016/j.jwpe.2024.106418

Feil, A., & Pretz, T. (2020). Mechanical recycling of packaging waste. In *Elsevier eBooks* (pp. 283–319). https://doi.org/10.1016/b978-0-12-817880-5.00011-6

Fok, L., Lam, T. W. L., Li, H., & Xu, X. (2019). A meta-analysis of methodologies adopted by microplastic studies in China. *The Science of the Total Environment, 718*, Article 135371. https://doi.org/10.1016/j.scitotenv.2019.135371

Fuller, S., & Gautam, A. (2016). A procedure for measuring microplastics using pressurized fluid extraction. *Environmental Science & Technology, 50*(11), 5774–5780. https://doi.org/10.1021/acs.est.6b00816

Gaß, H., Kloos, T. M., Höfling, A., Müller, L., Rockmann, L., Schubert, D. W., & Halik, M. (2023). Magnetic removal of micro- and nanoplastics from water—from 100 nm to 100 μm Debris Size. *Small, 20*(10). https://doi.org/10.1002/smll.202305467

Geng, X., Song, N., Zhao, Y., & Zhou, T. (2022). Waste plastic resource recovery from landfilled refuse: A novel waterless cleaning method and its cost-benefit analysis. *Journal of Environmental Management, 306*, Article 114462. https://doi.org/10.1016/j.jenvman.2022.114462

Geyer, R., Jambeck, J. R., & Law, K. L. (2017). Production, use, and fate of all plastics ever made. *Science Advances, 3*(7). https://doi.org/10.1126/sciadv.1700782

Ghosal, S., Chen, M., Wagner, J., Wang, Z., & Wall, S. (2017). Molecular identification of polymers and anthropogenic particles extracted from oceanic water and fish stomach—A Raman micro-spectroscopy study. *Environmental Pollution, 233*, 1113–1124. https://doi.org/10.1016/j.envpol.2017.10.014

Gies, E. A., LeNoble, J. L., Noël, M., Etemadifar, A., Bishay, F., Hall, E. R., & Ross, P. S. (2018). Retention of microplastics in a major secondary wastewater treatment plant in Vancouver, Canada. *Marine Pollution Bulletin, 133*, 553–561. https://doi.org/10.1016/j.marpolbul.2018.06.006

Goh, P., Kang, H., Ismail, A., Khor, W., Quen, L., & Higgins, D. (2022). Nanomaterials for microplastic remediation from aquatic environment: Why nano matters? *Chemosphere, 299*, Article 134418. https://doi.org/10.1016/j.chemosphere.2022.134418

Gomollón-Bel, F. (2019). Ten chemical innovations that will change our world: IUPAC identifies emerging technologies in Chemistry with potential to make our planet more sustainable. *Chemistry International, 41*(2), 12–17. https://doi.org/10.1515/ci-2019-0203

Günther, M., & Imhof, W. (2023). Simultaneous quantification of microplastic particles by non-deuterated (NoD) 1H-qNMR from samples comprising different polymer types. *The Analyst, 148*(5), 1151–1161. https://doi.org/10.1039/d2an01751b

Hasan, M. M., & Tarannum, M. N. (2024). Adverse impacts of microplastics on soil physicochemical properties and crop health in agricultural systems. *Journal of Hazardous Materials Advances, 17*, Article 100528. https://doi.org/10.1016/j.hazadv.2024.100528

Helbig, C., Huether, J., Joachimsthaler, C., Lehmann, C., Raatz, S., Thorenz, A., Faulstich, M., & Tuma, A. (2022). A terminology for downcycling. *Journal of Industrial Ecology, 26*(4), 1164–1174. https://doi.org/10.1111/jiec.13289

Hermabessiere, L., Himber, C., Boricaud, B., Kazour, M., Amara, R., Cassone, A., Laurentie, M., Paul-Pont, I., Soudant, P., Dehaut, A., & Duflos, G. (2018). Optimization, performance, and application of a pyrolysis-GC/MS method for the identification of microplastics. *Analytical and Bioanalytical Chemistry, 410*(25), 6663–6676. https://doi.org/10.1007/s00216-018-1279-0

Herring, C. (2015). *Laboratory methods for the analysis of microplastics in the marine environment: Recommendations for quantifying synthetic particles in waters and sediments*. http://hdl.handle.net/11329/1076

Hidayaturrahman, H., & Lee, T. (2019). A study on characteristics of microplastic in wastewater of South Korea: Identification, quantification, and fate of microplastics during treatment process. *Marine Pollution Bulletin, 146*, 696–702. https://doi.org/10.1016/j.marpolbul.2019.06.071

Hu, J., & Hu, J. (2023). Mineralization characteristics and behavior of polyethylene microplastics through ozone-based treatment. *Chemosphere, 349*, Article 140839. https://doi.org/10.1016/j.chemosphere.2023.140839

Hu, R., Zhang, K., Wang, W., Wei, L., & Lai, Y. (2022). Quantitative and sensitive analysis of polystyrene nanoplastics down to 50 nm by surface-enhanced Raman spectroscopy in water. *Journal of Hazardous Materials, 429*, Article 128388. https://doi.org/10.1016/j.jhazmat.2022.128388

Indika, S., Wei, Y., Hu, D., Ketharani, J., Ritigala, T., Cooray, T., Hansima, M. A. C. K., Makehelwala, M., Jinadasa, K. B. S. N., Weragoda, S. K., & Weerasooriya, R. (2021). Evaluation of performance of existing RO drinking water stations in the North Central Province, Sri Lanka. *Membranes, 11*(6), 383.https://doi.org/10.3390/membranes11060383

Iordachescu, L., Papacharalampos, K., Barritaud, L., Denieul, M., Plessis, E., Baratto, G., Julien, V., & Vollertsen, J. (2024). Microplastics in an advanced wastewater treatment plant: Sustained and robust removal rates unfazed by seasonal variations. *Microplastics and Nanoplastics, 4*(1). https://doi.org/10.1186/s43591-024-00097-3

Jeong, E., Lee, J., & Redwan, M. (2024). Animal exposure to microplastics and health effects: A review. *Emerging Contaminants, 10*(4), Article 100369. https://doi.org/10.1016/j.emcon.2024.100369

Jia, M., Farid, M. U., Ho, Y., Ma, X., Wong, P. W., Nah, T., He, Y., Boey, M. W., Lu, G., Fang, J. K., Fan, J., & An, A. K. (2024). Advanced nanobubble flotation for enhanced removal of sub-10 μm microplastics from wastewater. *Nature Communications, 15*(1). https://doi.org/10.1038/s41467-024-53304-3

Jin, T., Liu, Y., Lyu, H., He, Y., Sun, H., Tang, J., & Xing, B. (2024). Plastic takeaway food containers may cause human intestinal damage in routine life usage: Microplastics formation and cytotoxic effect. *Journal of Hazardous Materials, 475*, Article 134866. https://doi.org/10.1016/j.jhazmat.2024.134866

Jolaosho, T. L., Rasaq, M. F., Omotoye, E. V., Araomo, O. V., Adekoya, O. S., Abolaji, O. Y., & Hungbo, J. J. (2025). Microplastics in freshwater and marine ecosystems: Occurrence, characterization, sources, distribution dynamics, fate, transport processes, potential mitigation strategies, and policy interventions. *Ecotoxicology and Environmental Safety, 294*, Article 118036. https://doi.org/10.1016/j.ecoenv.2025.118036

Kabir, M. S., Wang, H., Luster-Teasley, S., Zhang, L., & Zhao, R. (2023). Microplastics in landfill leachate: Sources, detection, occurrence, and removal. *Environmental Science and Ecotechnology, 16*, Article 100256. https://doi.org/10.1016/j.ese.2023.100256

Kamaraj, P., Sridhar, V. V., Tharumasivam, S. V., Parthasarathy, S., Bupesh, G., Raju, N. K., Sahoo, U. K., Nanda, A., & Saravanan, K. M. (2023). Carbon nanoparticles fabricated microfilm: A potent filter for microplastics debased water. *Environmental Pollution, 336*, Article 122502. https://doi.org/10.1016/j.envpol.2023.122502

Kara, N., Erkan, H. S., & Engin, G. O. (2022). Characterization and removal of microplastics in landfill leachate treatment plants in Istanbul, Turkey. *Analytical Letters, 56*(9), 1535–1548. https://doi.org/10.1080/00032719.2022.2137808

Khalid, N., Aqeel, M., Noman, A., Khan, S. M., & Akhter, N. (2021). Interactions and effects of microplastics with heavy metals in aquatic and terrestrial environments. *Environmental Pollution, 290*, Article 118104. https://doi.org/10.1016/j.envpol.2021.118104

Khosravi, M. J., Hosseini, S. M., & Vatanpour, V. (2022). Polyamidoamine dendrimers-Mil-125(Ti) MOF embedded polyethersulfone membrane for enhanced removal of heavy metal, antibiotic and dye from water. *Journal of Environmental Chemical Engineering, 10*(6), Article 108644. https://doi.org/10.1016/j.jece.2022.108644

Kiran, N., Ekinci, E., & Snape, C. (2000). Recyling of plastic wastes via pyrolysis. *Resources Conservation and Recycling, 29*(4), 273–283. https://doi.org/10.1016/s0921-3449(00)00052-5

Klaimy, S., Lamonier, J., Casetta, M., Heymans, S., & Duquesne, S. (2021). Recycling of plastic waste using flash pyrolysis—Effect of mixture composition. *Polymer Degradation and Stability, 187*, Article 109540. https://doi.org/10.1016/j.polymdegradstab.2021.109540

Kumar, A., Von Wolff, N., Rauch, M., Zou, Y., Shmul, G., Ben-David, Y., Leitus, G., Avram, L., & Milstein, D. (2020). Hydrogenative depolymerization of nylons. *Journal of the American Chemical Society, 142*(33), 14267–14275. https://doi.org/10.1021/jacs.0c05675

Kurniawan, T. A., Haider, A., Mohyuddin, A., Fatima, R., Salman, M., Shaheen, A., Ahmad, H. M., Al-Hazmi, H. E., Othman, M. H. D., Aziz, F., Anouzla, A., & Ali, I. (2023). Tackling microplastics pollution in global environment through integration of applied technology, policy instruments, and legislation. *Journal of Environmental Management, 346*, Article 118971. https://doi.org/10.1016/j.jenvman.2023.118971

Laborda, F., Bolea, E., & Jiménez-Lamana, J. (2013). Single particle inductively coupled plasma mass spectrometry: A powerful tool for nanoanalysis. *Analytical Chemistry, 86*(5), 2270–2278. https://doi.org/10.1021/ac402980q

Lam, S., Sin, J., Zeng, H., Lin, H., Li, H., Chai, Y., Choong, M., & Mohamed, A. R. (2020). Green synthesis of Fe-ZnO nanoparticles with improved sunlight photocatalytic performance for polyethylene film deterioration and bacterial inactivation. *Materials Science in Semiconductor Processing, 123*, Article 105574. https://doi.org/10.1016/j.mssp.2020.105574

Lares, M., Ncibi, M. C., Sillanpää, M., & Sillanpää, M. (2018). Occurrence, identification and removal of microplastic particles and fibers in conventional activated sludge process and advanced MBR technology. *Water Research, 133*, 236–246. https://doi.org/10.1016/j.watres.2018.01.049

Lau, W. W. Y., Shiran, Y., Bailey, R. M., Cook, E., Stuchtey, M. R., Koskella, J., Velis, C. A., Godfrey, L., Boucher, J., Murphy, M. B., Thompson, R. C., Jankowska, E., Castillo, A. C., Pilditch, T. D., Dixon, B., Koerselman, L., Kosior, E., Favoino, E., Gutberlet, J., . . . Palardy, J. E. (2020). Evaluating scenarios toward zero plastic pollution. *Science, 369*(6510), 1455–1461.https://doi.org/10.1126/science.aba9475

Li, J., Liu, H., & Chen, J. P. (2017). Microplastics in freshwater systems: A review on occurrence, environmental effects, and methods for microplastics detection. *Water Research, 137*, 362–374. https://doi.org/10.1016/j.watres.2017.12.056

Li, W., Wang, J., Xiong, Z., & Li, D. (2023). Micro/nanorobots for efficient removal and degradation of micro/nanoplastics. *Cell Reports Physical Science, 4*(11), Article 101639. https://doi.org/10.1016/j.xcrp.2023.101639

Li, Y., Chen, P., Tang, Y., Yang, Y., Zhou, C., Bu, J., & Zhong, S. (2024). Microplastics in water: A review of characterization and removal methods. *Sustainability, 16*(10), 4033. https://doi.org/10.3390/su16104033

Licona, K., De O Geaquinto, L., Nicolini, J., Figueiredo, N., Chiapetta, S., Habert, A., & Yokoyama, L. (2018). Assessing potential of nanofiltration and reverse osmosis for removal of toxic pharmaceuticals from water. *Journal of Water Process Engineering, 25*, 195–204.https://doi.org/10.1016/j.jwpe.2018.08.002

Lim, B. K. H., & Thian, E. S. (2021). Biodegradation of polymers in managing plastic waste—A review. *The Science of the Total Environment, 813*, Article 151880. https://doi.org/10.1016/j.scitotenv.2021.151880

Liu, P., Qian, L., Wang, H., Zhan, X., Lu, K., Gu, C., & Gao, S. (2019). New insights into the aging behavior of microplastics accelerated by advanced oxidation processes. *Environmental Science & Technology, 53*(7), 3579–3588. https://doi.org/10.1021/acs.est.9b00493

Liu, T., Yu, S., Zhu, X., Liao, R., Zhuo, Z., He, Y., & Ma, H. (2021). In-situ detection method for microplastics in water by polarized light scattering. *Frontiers in Marine Science, 8*.https://doi.org/10.3389/fmars.2021.739683

Löder, M. G. J., & Gerdts, G. (2015). Methodology used for the Detection and Identification of Microplastics—A critical appraisal. In *Springer eBooks* (pp. 201–227). https://doi.org/10.1007/978-3-319-16510-3_8

Lokesh, P., Shobika, R., Omer, S., Reddy, M., Saravanan, P., Rajeshkannan, R., Saravanan, V., & Venkatkumar, S. (2023). Bioremediation of plastics by the help of microbial tool: A way for control of plastic pollution. *Sustainable Chemistry for the Environment, 3*, Article 100027. https://doi.org/10.1016/j.scenv.2023.100027

Long, Z., Pan, Z., Wang, W., Ren, J., Yu, X., Lin, L., Lin, H., Chen, H., & Jin, X. (2019). Microplastic abundance, characteristics, and removal in wastewater treatment plants in a coastal city of China. *Water Research, 155*, 255–265. https://doi.org/10.1016/j.watres.2019.02.028

López-Fonseca, R., Duque-Ingunza, I., De Rivas, B., Arnaiz, S., & Gutiérrez-Ortiz, J. (2010). Chemical recycling of post-consumer PET wastes by glycolysis in the presence of metal salts. *Polymer Degradation and Stability, 95*(6), 1022–1028. https://doi.org/10.1016/j.polymdegradstab.2010.03.007

Lorenzo-Navarro, J., Castrillon-Santana, M., Santesarti, E., De Marsico, M., Martinez, I., Raymond, E., Gomez, M., & Herrera, A. (2020). SMACC: A system for microplastics automatic counting and classification. *IEEE Access, 8*, 25249–25261. https://doi.org/10.1109/access.2020.2970498

Luogo, B. D. P., Salim, T., Zhang, W., Hartmann, N. B., Malpei, F., & Candelario, V. M. (2022). Reuse of water in laundry applications with micro- and ultrafiltration ceramic membrane. *Membranes, 12*(2), 223. https://doi.org/10.3390/membranes12020223

Ma, B., Xue, W., Hu, C., Liu, H., Qu, J., & Li, L. (2018). Characteristics of microplastic removal via coagulation and ultrafiltration during drinking water treatment. *Chemical Engineering Journal, 359*, 159–167. https://doi.org/10.1016/j.cej.2018.11.155

Macintosh, A., Simpson, A., Neeman, T., & Dickson, K. (2019). Plastic bag bans: Lessons from the Australian Capital Territory. *Resources Conservation and Recycling, 154*, Article 104638. https://doi.org/10.1016/j.resconrec.2019.104638

Mahu, E., Datsomor, W. G., Folorunsho, R., Fisayo, J., Crane, R., Marchant, R., Montford, J., Boateng, M. C., Oti, M. E., Oguguah, M. N., & Gordon, C. (2022). Human health risk and food safety implications of microplastic consumption by fish from coastal waters of the eastern equatorial Atlantic Ocean. *Food Control, 145*, Article 109503. https://doi.org/10.1016/j.foodcont.2022.109503

Martín, A. J., Mondelli, C., Jaydev, S. D., & Pérez-Ramírez, J. (2021). Catalytic processing of plastic waste on the rise. *Chem, 7*(6), 1487–1533. https://doi.org/10.1016/j.chempr.2020.12.006

Martin, C., Parkes, S., Zhang, Q., Zhang, X., McCabe, M. F., & Duarte, C. M. (2018). Use of unmanned aerial vehicles for efficient beach litter monitoring. *Marine Pollution Bulletin, 131*, 662–673. https://doi.org/10.1016/j.marpolbul.2018.04.045

Mattioda, V., Benedetti, V., Tessarolo, C., Oberto, F., Favole, A., Gallo, M., Martelli, W., Crescio, M. I., Berio, E., Masoero, L., Benedetto, A., Pezzolato, M., Bozzetta, E., Grattarola, C., Casalone, C., Corona, C., & Giorda, F. (2023). Pro-Inflammatory and cytotoxic effects of polystyrene microplastics on human and murine intestinal cell lines. *Biomolecules, 13*(1), 140. https://doi.org/10.3390/biom13010140

Michielssen, M. R., Michielssen, E. R., Ni, J., & Duhaime, M. B. (2016). Fate of microplastics and other small anthropogenic litter (SAL) in wastewater treatment plants depends on unit processes employed. *Environmental Science Water Research & Technology, 2*(6), 1064–1073. https://doi.org/10.1039/c6ew00207b

Mikac, L., Rigó, I., Himics, L., Tolić, A., Ivanda, M., & Veres, M. (2022). Surface-enhanced Raman spectroscopy for the detection of microplastics. *Applied Surface Science, 608*, Article 155239. https://doi.org/10.1016/j.apsusc.2022.155239

Minkler, M. J., Kim, J. M., Shinde, V. V., & Beckingham, B. S. (2020). Low-field 1 H NMR spectroscopy: Factors impacting signal-to-noise ratio and experimental time in the context of mixed microstructure polyisoprenes. *Magnetic Resonance in Chemistry, 58*(12), 1168–1176. https://doi.org/10.1002/mrc.5022

Monira, S., Bhuiyan, M. A., Haque, N., & Pramanik, B. K. (2021). Assess the performance of chemical coagulation process for microplastics removal from stormwater. *Process Safety and Environmental Protection, 155*, 11–16. https://doi.org/10.1016/j.psep.2021.09.002

Moradihamedani, P. (2021). Recent advances in dye removal from wastewater by membrane technology: A review. *Polymer Bulletin, 79*(4), 2603–2631. https://doi.org/10.1007/s00289-021-03603-2

Nelson, T. F., Remke, S. C., Kohler, H. E., McNeill, K., & Sander, M. (2019). Quantification of synthetic polyesters from biodegradable Mulch films in soils. *Environmental Science & Technology, 54*(1), 266–275. https://doi.org/10.1021/acs.est.9b05863

Ngo, P. L., Pramanik, B. K., Shah, K., & Roychand, R. (2019). Pathway, classification and removal efficiency of microplastics in wastewater treatment plants. *Environmental Pollution, 255*, Article 113326. https://doi.org/10.1016/j.envpol.2019.113326

Nielsen, T. D., Holmberg, K., & Stripple, J. (2019). Need a bag? A review of public policies on plastic carrier bags—Where, how and to what effect? *Waste Management, 87*, 428–440. https://doi.org/10.1016/j.wasman.2019.02.025

Ning, Z., Duan, X., Li, Y., Zhao, X., & Chang, L. (2023). Degradation of polyvinyl chloride microplastics via electrochemical oxidation with a CeO_2–PbO_2 anode. *Journal of Cleaner Production, 432*, 139668. https://doi.org/10.1016/j.jclepro.2023.139668

Nøklebye, E., Adam, H. N., Roy-Basu, A., Bharat, G. K., & Steindal, E. H. (2022). Plastic bans in India—Addressing the socio-economic and environmental complexities. *Environmental Science & Policy, 139*, 219–227. https://doi.org/10.1016/j.envsci.2022.11.005

Okoffo, E. D., Rauert, C., & Thomas, K. V. (2022). Mass quantification of microplastic at wastewater treatment plants by pyrolysis-gas chromatography–mass spectrometry. *The Science of the Total Environment, 856*, Article 159251. https://doi.org/10.1016/j.scitotenv.2022.159251

Olicón-Hernández, D. R., González-López, J., & Aranda, E. (2017). Overview on the biochemical potential of filamentous fungi to degrade pharmaceutical compounds. *Frontiers in Microbiology, 8*.https://doi.org/10.3389/fmicb.2017.01792

Oruganti, R. K., Biji, A. P., Lanuyanger, T., Show, P. L., Sriariyanun, M., Upadhyayula, V. K., Gadhamshetty, V., & Bhattacharyya, D. (2023). Artificial intelligence and machine learning tools for high-performance microalgal wastewater treatment and algal biorefinery: A critical review. *The Science of the Total Environment, 876*, Article 162797. https://doi.org/10.1016/j.scitotenv.2023.162797

Osman, A. I., Hosny, M., Eltaweil, A. S., Omar, S., Elgarahy, A. M., Farghali, M., Yap, P., Wu, Y., Nagandran, S., Batumalaie, K., Gopinath, S. C. B., John, O. D., Sekar, M., Saikia, T., Karunanithi, P., Hatta, M. H. M., & Akinyede, K. A. (2023). Microplastic sources, formation, toxicity and remediation: A review. *Environmental Chemistry Letters, 21*(4), 2129–2169. https://doi.org/10.1007/s10311-023-01593-3

Padmanabhan, S., Giridharan, K., Stalin, B., Kumaran, S., Kavimani, V., Nagaprasad, N., Jule, L. T., & Krishnaraj, R. (2022). Energy recovery of waste plastics into diesel fuel with ethanol and ethoxy ethyl acetate additives on circular economy strategy. *Scientific Reports, 12*(1). https://doi.org/10.1038/s41598-022-09148-2

Pan, Z., Guo, H., Chen, H., Wang, S., Sun, X., Zou, Q., Zhang, Y., Lin, H., Cai, S., & Huang, J. (2019). Microplastics in the Northwestern Pacific: Abundance, distribution, and characteristics. *The Science of the Total Environment, 650*, 1913–1922. https://doi.org/10.1016/j.scitotenv.2018.09.244

Pan, Y., Gao, S., Ge, C., Gao, Q., Huang, S., Kang, Y., Luo, G., Zhang, Z., Fan, L., Zhu, Y., & Wang, A. (2022). Removing microplastics from aquatic environments: A critical review. *Environmental Science and Ecotechnology, 13*, Article 100222. https://doi.org/10.1016/j.ese.2022.100222

Papini, G., Petrella, G., Cicero, D. O., Boglione, C., & Rakaj, A. (2023). Identification and quantification of polystyrene microplastics in marine sediments facing a river mouth through NMR spectroscopy. *Marine Pollution Bulletin, 198*, Article 115784. https://doi.org/10.1016/j.marpolbul.2023.115784

Park, H., & Park, B. (2021). Review of microplastic distribution, toxicity, analysis methods, and removal technologies. *Water, 13*(19), 2736. https://doi.org/10.3390/w13192736

Peez, N., Becker, J., Ehlers, S. M., Fritz, M., Fischer, C. B., Koop, J. H. E., Winkelmann, C., & Imhof, W. (2019). Quantitative analysis of PET microplastics in environmental model samples using quantitative 1H-NMR spectroscopy: Validation of an optimized and consistent sample clean-up method. *Analytical and Bioanalytical Chemistry, 411*(28), 7409–7418. https://doi.org/10.1007/s00216-019-02089-2

Peez, N., & Imhof, W. (2020). Quantitative 1H-NMR spectroscopy as an efficient method for identification and quantification of PVC, ABS and PA Microparticles. *The Analyst, 145*(15), 5363–5371. https://doi.org/10.1039/d0an00879f

Peez, N., Janiska, M., & Imhof, W. (2018). The first application of quantitative 1H NMR spectroscopy as a simple and fast method of identification and quantification of microplastic particles (PE, PET, and PS). *Analytical and Bioanalytical Chemistry, 411*(4), 823–833. https://doi.org/10.1007/s00216-018-1510-z

Phan, S., & Luscombe, C. K. (2023). Recent trends in marine microplastic modeling and machine learning tools: Potential for long-term microplastic monitoring. *Journal of Applied Physics, 133*(2). https://doi.org/10.1063/5.0126358

Picó, Y., & Barceló, D. (2020). Pyrolysis gas chromatography-mass spectrometry in environmental analysis: Focus on organic matter and microplastics. *TrAC Trends in Analytical Chemistry, 130*, Article 115964. https://doi.org/10.1016/j.trac.2020.115964

Pivokonsky, M., Cermakova, L., Novotna, K., Peer, P., Cajthaml, T., & Janda, V. (2018). Occurrence of microplastics in raw and treated drinking water. *The Science of the Total Environment, 643*, 1644–1651. https://doi.org/10.1016/j.scitotenv.2018.08.102

Pivokonský, M., Pivokonská, L., Novotná, K., Čermáková, L., & Klimtová, M. (2020). Occurrence and fate of microplastics at two different drinking water treatment plants within a river catchment. *The Science of the Total Environment, 741*, Article 140236. https://doi.org/10.1016/j.scitotenv.2020.140236

Pizzichetti, A. R. P., Pablos, C., Álvarez-Fernández, C., Reynolds, K., Stanley, S., & Marugán, J. (2020). Evaluation of membranes performance for microplastic removal in a simple and low-cost filtration system. *Case Studies in Chemical and Environmental Engineering, 3*, Article 100075. https://doi.org/10.1016/j.cscee.2020.100075

Plastic pollution. (n.d.). IUCN (2024). https://iucn.org/resources/issues-brief/plastic-pollution

Plastics strategy. (n.d.). Environment. https://environment.ec.europa.eu/strategy/plastics-strategy_en

Prata, J. C., Da Costa, J. P., Duarte, A. C., & Rocha-Santos, T. (2018). Methods for sampling and detection of microplastics in water and sediment: A critical review. *TrAC Trends in Analytical Chemistry, 110*, 150–159. https://doi.org/10.1016/j.trac.2018.10.029

Primpke, S., Christiansen, S. H., Cowger, W., De Frond, H., Deshpande, A., Fischer, M., Holland, E. B., Meyns, M., O'Donnell, B. A., Ossmann, B. E., Pittroff, M., Sarau, G., Scholz-Böttcher, B. M., & Wiggin, K. J. (2020a). Critical Assessment of analytical methods for the harmonized and cost-efficient analysis of microplastics. *Applied Spectroscopy, 74*(9), 1012–1047. https://doi.org/10.1177/0003702820921465

Primpke, S., Fischer, M., Lorenz, C., Gerdts, G., & Scholz-Böttcher, B. M. (2020b). Comparison of pyrolysis gas chromatography/mass spectrometry and hyperspectral FTIR imaging spectroscopy for the analysis of microplastics. *Analytical and Bioanalytical Chemistry, 412*(30), 8283–8298. https://doi.org/10.1007/s00216-020-02979-w

Ragusa, A., Svelato, A., Santacroce, C., Catalano, P., Notarstefano, V., Carnevali, O., Papa, F., Rongioletti, M. C. A., Baiocco, F., Draghi, S., D'Amore, E., Rinaldo, D., Matta, M., & Giorgini, E. (2020). Plasticenta: First evidence of microplastics in human placenta. *Environment International, 146*, Article 106274. https://doi.org/10.1016/j.envint.2020.106274

Rahimi, A., & García, J. M. (2017). Chemical recycling of waste plastics for new materials production. *Nature Reviews Chemistry, 1*(6). https://doi.org/10.1038/s41570-017-0046

Ramos, R. L., Santos, C. R. D., Drumond, G. P., De Souza Santos, L. V., & Amaral, M. C. S. (2023). Critical review of microplastic in membrane treatment plant: Removal efficiency, environmental risk assessment, membrane fouling, and MP release. *Chemical Engineering Journal, 480*, Article 148052. https://doi.org/10.1016/j.cej.2023.148052

Riisom, M., Gammelgaard, B., Lambert, I. H., & Stürup, S. (2018). Development and validation of an ICP-MS method for quantification of total carbon and platinum in cell samples and comparison of open-vessel and microwave-assisted acid digestion methods. *Journal of Pharmaceutical and Biomedical Analysis, 158*, 144–150. https://doi.org/10.1016/j.jpba.2018.05.038

Rizvi, N. B., Sarwar, A., Waheed, S., Iqbal, Z. F., Imran, M., Javaid, A., Kim, T. H., & Khan, M. S. (2024). Nano-based remediation strategies for micro and nanoplastic pollution: A review. *Journal of Contaminant Hydrology, 265*, Article 104380. https://doi.org/10.1016/j.jconhyd.2024.104380

Sajjad, M., Huang, Q., Khan, S., Khan, M. A., Liu, Y., Wang, J., Lian, F., Wang, Q., & Guo, G. (2022). Microplastics in the soil environment: A critical review. *Environmental Technology & Innovation, 27*, Article 102408. https://doi.org/10.1016/j.eti.2022.102408

Sanfilippo, F., Tang, M., & Steyaert, S. (2021). The Aquatic Surface Robot (ANSWER), a lightweight, low cost, multipurpose unmanned research vessel. In *Communications in computer and information science* (pp. 251–265). https://doi.org/10.1007/978-3-030-71711-7_21

Sarker, M. A. B., Imtiaz, M. H., Holsen, T. M., & Baki, A. B. M. (2024). Real-time detection of microplastics using an AI camera. *Sensors, 24*(13), 4394.https://doi.org/10.3390/s24134394

Schwaferts, C., Niessner, R., Elsner, M., & Ivleva, N. P. (2019). Methods for the analysis of submicrometer- and nanoplastic particles in the environment. *TrAC Trends in Analytical Chemistry, 112*, 52–65. https://doi.org/10.1016/j.trac.2018.12.014

Science History Institute. (2024, October 11). *History and future of plastics|Science History Institute*. https://www.sciencehistory.org/education/classroom-activities/role-playing-games/case-of-plastics/history-and-future-of-plastics/

Scimeca, M., Bischetti, S., Lamsira, H. K., Bonfiglio, R., & Bonanno, E. (2018). Energy dispersive X-ray (EDX) microanalysis: A powerful tool in biomedical research and diagnosis. *European Journal of Histochemistry*. https://doi.org/10.4081/ejh.2018.2841

Shim, W. J., Hong, S. H., & Eo, S. E. (2016). Identification methods in microplastic analysis: A review. *Analytical Methods, 9*(9), 1384–1391. https://doi.org/10.1039/c6ay02558g

Simpson, A., Kerr, C., & Buckley, C. (1987). The effect of pH on the nanofiltration of the carbonate system in solution. *Desalination, 64*, 305–319. https://doi.org/10.1016/0011-9164(87)90104-4

Singh, S., Trushna, T., Kalyanasundaram, M., Tamhankar, A. J., & Diwan, V. (2022). Microplastics in drinking water: A macro issue. *Water Science & Technology Water Supply, 22*(5), 5650–5674. https://doi.org/10.2166/ws.2022.189

Sobhani, Z., Amin, M. A., Naidu, R., Megharaj, M., & Fang, C. (2019). Identification and visualisation of microplastics by Raman mapping. *Analytica Chimica Acta, 1077*, 191–199. https://doi.org/10.1016/j.aca.2019.05.021

Song, Y. K., Hong, S. H., Jang, M., Han, G. M., Rani, M., Lee, J., & Shim, W. J. (2015). A comparison of microscopic and spectroscopic identification methods for analysis of microplastics in environmental samples. *Marine Pollution Bulletin, 93*(1–2), 202–209. https://doi.org/10.1016/j.marpolbul.2015.01.015

Sridharan, S., Kumar, M., Singh, L., Bolan, N. S., & Saha, M. (2021). Microplastics as an emerging source of particulate air pollution: A critical review. *Journal of Hazardous Materials, 418*, Article 126245. https://doi.org/10.1016/j.jhazmat.2021.126245

Stolte, A., Forster, S., Gerdts, G., & Schubert, H. (2015). Microplastic concentrations in beach sediments along the German Baltic coast. *Marine Pollution Bulletin, 99*(1–2), 216–229. https://doi.org/10.1016/j.marpolbul.2015.07.022

Sun, J., Dai, X., Wang, Q., Van Loosdrecht, M. C., & Ni, B. (2019). Microplastics in wastewater treatment plants: Detection, occurrence and removal. *Water Research, 152*, 21–37. https://doi.org/10.1016/j.watres.2018.12.050

Sun, T., Wang, S., Ji, C., Li, F., & Wu, H. (2021). Microplastics aggravate the bioaccumulation and toxicity of coexisting contaminants in aquatic organisms: A synergistic health hazard. *Journal of Hazardous Materials, 424*, Article 127533. https://doi.org/10.1016/j.jhazmat.2021.127533

Tadsuwan, K., & Babel, S. (2021). Microplastic contamination in a conventional wastewater treatment plant in Thailand. *Waste Management & Research the Journal for a Sustainable Circular Economy, 39*(5), 754–761. https://doi.org/10.1177/0734242x20982055

Tadsuwan, K., & Babel, S. (2022). Microplastic abundance and removal via an ultrafiltration system coupled to a conventional municipal wastewater treatment plant in Thailand. *Journal of Environmental Chemical Engineering, 10*(2), Article 107142. https://doi.org/10.1016/j.jece.2022.107142

Tagg, A. S., Sapp, M., Harrison, J. P., Sinclair, C. J., Bradley, E., Ju-Nam, Y., & Ojeda, J. J. (2020). Microplastic monitoring at different stages in a wastewater treatment plant using reflectance Micro-FTIR Imaging. *Frontiers in Environmental Science, 8*.https://doi.org/10.3389/fenvs.2020.00145

Talvitie, J., Mikola, A., Koistinen, A., & Setälä, O. (2017). Solutions to microplastic pollution—Removal of microplastics from wastewater effluent with advanced wastewater treatment technologies. *Water Research, 123*, 401–407. https://doi.org/10.1016/j.watres.2017.07.005

Thompson, R. C., Olsen, Y., Mitchell, R. P., Davis, A., Rowland, S. J., John, A. W. G., McGonigle, D., & Russell, A. E. (2004). Lost at sea: Where is all the plastic? *Science, 304*(5672), 838. https://doi.org/10.1126/science.1094559

Tian, C., Fei, L., Zheng, W., Xu, Y., Zuo, W., & Lin, C. (2020). Deep learning on image denoising: An overview. *Neural Networks, 131*, 251–275. https://doi.org/10.1016/j.neunet.2020.07.025

Tirkey, A., & Upadhyay, L. S. B. (2021). Microplastics: An overview on separation, identification and characterization of microplastics. *Marine Pollution Bulletin, 170*, Article 112604. https://doi.org/10.1016/j.marpolbul.2021.112604

Tomalia, D. A. (2005). Birth of a new macromolecular architecture: Dendrimers as quantized building blocks for nanoscale synthetic polymer chemistry. *Progress in Polymer Science, 30*(3–4), 294–324. https://doi.org/10.1016/j.progpolymsci.2005.01.007

Trujillo, C., Pérez-Arantegui, J., Lobinski, R., & Laborda, F. (2023). Improving the detectability of microplastics in river waters by single particle inductively coupled plasma mass spectrometry. *Nanomaterials, 13*(10), 1582. https://doi.org/10.3390/nano13101582

UN Environment Annual Report 2017. (n.d.). UNEP—UN environment programme. https://www.unep.org/resources/un-environment-annual-report-2017

Urso, M., & Pumera, M. (2022). Nano/microplastics capture and degradation by autonomous nano/microrobots: A perspective. *Advanced Functional Materials, 32*(20). https://doi.org/10.1002/adfm.202112120

Veerasingam, S., Ranjani, M., Venkatachalapathy, R., Bagaev, A., Mukhanov, V., Litvinyuk, D., Mugilarasan, M., Gurumoorthi, K., Guganathan, L., Aboobacker, V. M., & Vethamony, P. (2020). Contributions of Fourier transform infrared spectroscopy in microplastic pollution research: A review. *Critical Reviews in Environmental Science and Technology, 51*(22), 2681–2743. https://doi.org/10.1080/10643389.2020.1807450

Velikov, D. I., Jancik-Prochazkova, A., & Pumera, M. (2024). On-the-fly monitoring of the capture and removal of nanoplastics with nanorobots. *ACS Nanoscience Au, 4*(4), 243–249. https://doi.org/10.1021/acsnanoscienceau.4c00002

Verla, A. W., Enyoh, C. E., Verla, E. N., & Nwarnorh, K. O. (2019). Microplastic–toxic chemical interaction: A review study on quantified levels, mechanism and implication. *SN Applied Sciences, 1*(11). https://doi.org/10.1007/s42452-019-1352-0

Wagner, J., Wang, Z., Ghosal, S., Rochman, C., Gassel, M., & Wall, S. (2016). Novel method for the extraction and identification of microplastics in ocean trawl and fish gut matrices. *Analytical Methods, 9*(9), 1479–1490. https://doi.org/10.1039/c6ay02396g

Wang, J., Xie, L., Wu, X., Zhao, Z., Lu, Y., & Sun, H. (2024). Impact of micro-nano plastics in daily life on human health: toxicological evaluation from the perspective of normal tissue cells and organoids. *Toxicology Research, 13*(6). https://doi.org/10.1093/toxres/tfae205

Wei, Y., & Hadigheh, S. (2023). Development of an innovative hybrid thermo-chemical recycling method for CFRP waste recovery. *Composites Part B Engineering, 260*, Article 110786. https://doi.org/10.1016/j.compositesb.2023.110786

Wen, S., Zhao, Y., Wang, M., Yuan, H., & Xu, H. (2022). Micro(nano)plastics in food system: Potential health impacts on human intestinal system. *Critical Reviews in Food Science and Nutrition, 64*(5), 1429–1447. https://doi.org/10.1080/10408398.2022.2116559

Wright, S. L., Thompson, R. C., & Galloway, T. S. (2013). The physical impacts of microplastics on marine organisms: A review. *Environmental Pollution, 178*, 483–492. https://doi.org/10.1016/j.envpol.2013.02.031

Wu, P., Huang, J., Zheng, Y., Yang, Y., Zhang, Y., He, F., Chen, H., Quan, G., Yan, J., Li, T., & Gao, B. (2019). Environmental occurrences, fate, and impacts of microplastics. *Ecotoxicology and Environmental Safety, 184*, Article 109612. https://doi.org/10.1016/j.ecoenv.2019.109612

Xiao, K., Liang, S., Wang, X., Chen, C., & Huang, X. (2018). Current state and challenges of full-scale membrane bioreactor applications: A critical review. *Bioresource Technology, 271*, 473–481. https://doi.org/10.1016/j.biortech.2018.09.061

Xiao, Y., Tian, Y., Xu, W., & Zhu, J. (2024). Photodegradation of microplastics through nanomaterials: Insights into photocatalysts modification and detailed mechanisms. *Materials, 17*(11), 2755. https://doi.org/10.3390/ma17112755

Xie, A., Jin, M., Zhu, J., Zhou, Q., Fu, L., & Wu, W. (2023). Photocatalytic technologies for transformation and degradation of microplastics in the environment: Current achievements and future prospects. *Catalysts, 13*(5), 846. https://doi.org/10.3390/catal13050846

Xin, J., Zhang, Q., Huang, J., Huang, R., Jaffery, Q. Z., Yan, D., Zhou, Q., Xu, J., & Lu, X. (2021). Progress in the catalytic glycolysis of polyethylene terephthalate. *Journal of Environmental Management, 296*, Article 113267. https://doi.org/10.1016/j.jenvman.2021.113267

Xu, J., Thomas, K. V., Luo, Z., & Gowen, A. A. (2019). FTIR and Raman imaging for microplastics analysis: State of the art, challenges and prospects. *TrAC Trends in Analytical Chemistry, 119*, Article 115629. https://doi.org/10.1016/j.trac.2019.115629

Yadav, S., Singh, D., Mohanty, P., & Sarangi, P. K. (2021). Biochemical and thermochemical routes of H2 production from food waste: A comparative review. *Chemical Engineering & Technology, 46*(2), 191–203. https://doi.org/10.1002/ceat.202000526

Yang, W., Li, Y., & Boraschi, D. (2023). Association between microorganisms and microplastics: How does it change the host-pathogen interaction and subsequent immune response? *International Journal of Molecular Sciences, 24*(4), 4065. https://doi.org/10.3390/ijms24044065

Yin, K., Wang, D., Zhao, H., Wang, Y., Guo, M., Liu, Y., Li, B., & Xing, M. (2021). Microplastics pollution and risk assessment in water bodies of two nature reserves in Jilin Province: Correlation analysis with the degree of human activity. *The Science of the Total Environment, 799*, Article 149390. https://doi.org/10.1016/j.scitotenv.2021.149390

Yuan, J., Ma, J., Sun, Y., Zhou, T., Zhao, Y., & Yu, F. (2020). Microbial degradation and other environmental aspects of microplastics/plastics. *The Science of the Total Environment, 715*, Article 136968. https://doi.org/10.1016/j.scitotenv.2020.136968

Yuan, Z., Nag, R., & Cummins, E. (2022). Human health concerns regarding microplastics in the aquatic environment—From marine to food systems. *The Science of the Total Environment, 823*, Article 153730. https://doi.org/10.1016/j.scitotenv.2022.153730

Zhang, Y., Diehl, A., Lewandowski, A., Gopalakrishnan, K., & Baker, T. (2020). Removal efficiency of micro- and nanoplastics (180 nm–125 μm) during drinking water treatment. *The Science of the Total Environment, 720*, Article 137383. https://doi.org/10.1016/j.scitotenv.2020.137383

Zhang, Y., Duan, J., Liu, R., Petropoulos, E., Feng, Y., Xue, L., Yang, L., & He, S. (2023). Efficient magnetic capture of PE microplastic from water by PEG modified Fe_3O_4 nanoparticles: Performance, kinetics, isotherms and influence factors. *Journal of Environmental Sciences, 147*, 677–687. https://doi.org/10.1016/j.jes.2023.07.025

Zhang, Z., & Chen, Y. (2019). Effects of microplastics on wastewater and sewage sludge treatment and their removal: A review. *Chemical Engineering Journal, 382*, Article 122955. https://doi.org/10.1016/j.cej.2019.122955

Zhao, X., Korey, M., Li, K., Copenhaver, K., Tekinalp, H., Celik, S., Kalaitzidou, K., Ruan, R., Ragauskas, A. J., & Ozcan, S. (2021). Plastic waste upcycling toward a circular economy. *Chemical Engineering Journal, 428*, Article 131928. https://doi.org/10.1016/j.cej.2021.131928

Zhu, K., Jia, H., Zhao, S., Xia, T., Guo, X., Wang, T., & Zhu, L. (2019). Formation of environmentally persistent free radicals on microplastics under light irradiation. *Environmental Science & Technology, 53*(14), 8177–8186. https://doi.org/10.1021/acs.est.9b01474

Ziajahromi, S., Neale, P. A., Rintoul, L., & Leusch, F. D. (2017). Wastewater treatment plants as a pathway for microplastics: Development of a new approach to sample wastewater-based microplastics. *Water Research, 112*, 93–99. https://doi.org/10.1016/j.watres.2017.01.042

Ziembowicz, S., & Kida, M. (2024). The effect of water ozonation in the presence of microplastics on water quality and microplastics degradation. *The Science of the Total Environment, 929*, Article 172595. https://doi.org/10.1016/j.scitotenv.2024.172595

Zolich, A., Faltynkova, A., Johnsen, G., & Johansen, T. A. (2022). Portable catamaran drone—An uncrewed sampling vehicle for micro-plastics and aquaculture research. *OCEANS 2022, Hampton Roads*, 1–6. https://doi.org/10.1109/oceans47191.2022.9977294

Chapter 12
Green Solution: A Sustainable Solutions Microplastics Management

Vikas Kumar Joshiya

Abstract Microplastics (MPs) pollution is a growing global challenge, with tiny plastic particles (< 5 mm) from sources like tire wear, synthetic textiles, and packaging infiltrating air, water, and soil. Each year, millions of tons plastics such as 2.9 million tons from tires alone enter the environment, posing risks to ecosystems and human health through bioaccumulation and toxicity. Conventional methods like landfilling, incineration, and mechanical recycling fail to manage these persistent pollutants effectively, often creating secondary pollution and requiring high energy. This chapter explores green solutions as a sustainable alternative to tackle MPs waste. It highlights the urgent need for eco-friendly approaches that reduce environmental harm and promote a circular economy. The discussion begins with the scale and limitations of current MPs challenges, followed by the principles of sustainability, biodegradability, and minimal impact that guide green strategies. Innovative technologies—such as biodegradable polymers (e.g., PLA), bio-based filtration, microbial degradation, and green nanotechnology—are examined for their potential to prevent and remove MPs. Practical waste management strategies, including source reduction and upcycling, are presented alongside the critical role of policies, economic incentives, and public awareness in scaling these solutions. Future research needs, like testing biodegradable materials in real conditions, are also addressed. By integrating science-based, low-impact methods, this chapter offer a roadmap for researchers, policymakers, and communities to combat MPs pollution, aiming for a cleaner, healthier planet through sustainable innovation.

Keywords MPs · Green technology · Waste management · Sustainability · Biodegradable polymers · Circular economy

V. K. Joshiya (✉)
Sri Aurobindo College (Evening), University of Delhi, New Delhi, India
e-mail: vkjoshiya@yahoo.co.in

V. Kumar et al. (eds.), *Microplastics Pollution*,
https://doi.org/10.1007/978-3-031-99774-7_12

12.1 Introduction

The improper disposal and mismanagement of plastic waste expose it to a range of physical, chemical, and biological processes such as mechanical wear, hydrolysis, photo-degradation, microbial degradation, and chemical degradation (Andrady, 2011). When the plastic is exposed to these processes, it breaks down, split apart, or dispersed into tiny particles. These tiny particles smaller than 5 mm are termed as MPs, and smaller than 100 nm are called as nanoplastics (NPs) (Cole et al., 2011). The MPs and NPs generated by fragmentation or degradation of larger pieces of plastics are known as secondary MPs. However, the MPs and NPs which are deliberately prepared for use in several industrial products such as paints, cosmetics, fertilizers, pesticides, detergents, and medicines are termed as primary MPs (Boucher & Friot, 2017). These particles are non-biodegradable which make them suitable to persist in the environment decades as these are made of polymers (Barnes et al., 2009). The persistent nature, smaller size and widespread distribution of MPs in air, water, and soil make them suitable candidate to enter into the food chain (Wright & Kelly, 2017). After entering into food chain, MPs and NPs get bio-accumulated at various trophic level and thus biomagnified too (Carbery et al., 2018). Since, these MPs contains several toxic chemical as additives during plastic manufacturing to increase their durability and safety from hazards, poses significant ecological and human health risks (Rochman et al., 2013). The accelerating production and unintentional release of MPs in the various segments of the environment is increasing the ecological toxicity (Geyer et al., 2017).

The conventional strategies of MPs mitigations such as sedimentation, mechanical filtration, and chemical treatments are not efficient (Talvitie et al., 2017). In addition, these strategies are energy intensive, costly, and limited in scope (Padervand et al., 2020). These approaches don't focuses on the root causes of MPs proliferation rather only on symptoms. Moreover, these conventional methods also fail to tackle the persistent nature of MPs which resist degradation and continuously keep accumulation in different trophic levels in various ecosystems (Law & Thompson, 2014). Therefore, there is an urgent need of finding sustainable solutions to MPs pollution.

To overcome the inefficiency of traditional approaches, at present, the green solutions are attracting the ecologist, scientists, and researchers worldwide as a promising strategy to mitigate MPs pollution. The green solutions are equipped with microorganism like bacterial, and fungi, algae and plants as eco-friendly technologies (Jacquin et al., 2019). The green solution is driven by the natural processes to biodegrade, absorb, adsorb or transform to less toxic form which provides a sustainable and cost-effective approach (Zettler et al., 2013). For example, microbial enzymes can breakdown polymeric chains while the periphytic biofilm and macrophytes can trap and stabilize the MPs in aquatic environment (McCormick et al., 2014). Similarly, the genetically engineered microorganism produces plastic degrading enzymes which present a novel strategy to overcome MPs (Martin-Gonzalez et al., 2024). The green approaches follow the principles of ecological balance which minimizes the environmental degradation and focuses on the polymeric breakdown of MPs (Krueger et al., 2015).

The current chapter focuses on the potential of the current green approaches available in present time and its scalability. Here a comprehensive assessment is carried out from microbial degradation to phytoremediation to present green solution as a sustainable strategy to address MPs. Furthermore, this chapter also explores to identify the gas in the current knowledge, propose directions for future research, and advocate for the integration of these solutions into broader environmental management framework.

12.2 Challenges in MPs Management

MPs pollution has emerged as a severe environmental crisis which is hard to manage in a sustainable manner due to its ubiquitous and persistent nature. Its effective management faces significant challenges due to its vast scale, smaller size, toxicity, limitation of conventional strategies, etc. The major challenges of MPs pollution are shown in Fig. 1.1.

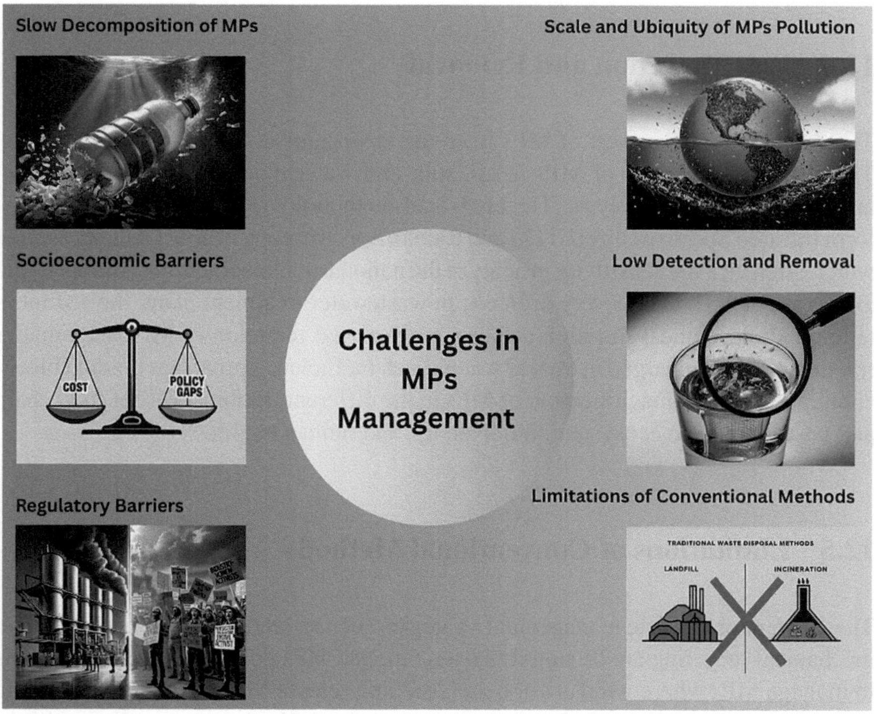

Fig. 1.1 Different challenges in MPs waste management

12.3 Scale and Ubiquity of MPs Pollution

The diverse source and large quantity of MPs has made it ubiquitous which is the greatest challenge to manage. It releases from tire wear, textiles, weathering and biological degradation of larger plastics, paints, cosmetics, detergents, medicines, fertilizers, etc. (Boucher & Friot, 2017). Globally, an estimated 2.9 million tonnes of MPs is generated from tire wear per year out of which 146,000 becomes airborne and pollutes the atmosphere (Evangeliou et al., 2020). Similarly, the synthetic fibers from clothing and household textiles release millions of microfibers (MFs) into air and water during washing due to which the indoor concentration of MPs often exceed its outdoor concentrations due to lack of proper ventilation (Dris et al., 2017). The MFs are extremely fine which get dispersed in air, water, and soil making it difficult to its complete removal.

The ubiquity of MPs has exacerbated this challenge. The researchers have found the presence of MPs even in remote areas such as Arctic ice and high altitude atmospheric samples (Allen et al., 2019). The smaller size and lighter weight enable its the long-range transport while their chemical stability is due to polymers like polyethylene (PE) and polystyrene (PS) which ensures it persistence in the environment for decades or centuries (Andrady, 2011).

12.4 Low Detection and Removal

The detection and removal of MPs from the environment is the biggest challenge. The pervasive dispersal of MPs in air, soil, and water make it difficult in systematic and quantitative analysis. The analytical methodologies such as Fourier Transform Infrared Spectroscopy (FTIR) and Raman Spectroscopy need significant sample which often lack the sensitivity to analyze the nanoparticulate fractions. Similarly, the extraction of MPs is also very difficult. In wastewater treatment plant, the sedimentation and mechanical filtration achieves only partial retention of MPs and smaller fraction of it is discharged in treated water. Such inefficiency promotes gradual bioaccumulation and biomagnification of MPs in the different trophic levels of food chain and poses threats to ecosystem, biodiversity, and human health.

12.5 Limitations of Conventional Methods

The conventional waste management strategies such as land filling, incineration and mechanical recycling are ill-suited to overcome the MPs pollution. The landfill fails to manage MPs where wind erosion and leaching release MPs in to the surrounding environment (Farid et al., 2024). For example, a landfill releases an estimated 0.16 MPs per tonnes of waste per hour due to abrasion, and weathering. In addition, the

MPs in landfill resist degradation and accumulated over time and pose long-term risks.

The incineration while reducing the volume of the waste generates the secondary MPs in its ash residue and emits toxic gas like dioxin (Verma et al., 2016). It is also very energy intensive which doesn't match with the sustainability goals and produces greenhouse gases that further exacerbate the climate change. Similarly, the mechanical recycling is also limited to MPs management as the recycling systems are designed to remove only larger particles and MPs often contaminate the recycling stream and reduces the quality of the material (Ragaert et al., 2017). In addition, it also generates MPs dust during shredding and washing and reintroduced it into the environment (Zhang et al., 2020).

12.6 Slow Decomposition of MPs

MPs exhibit very slow degradation. The predominant polymers present in MPs such as PE and polypropylene (PP) contains strong covalent bonds which make them difficult to degrade. The physical agents like sunlight, wind, and water waves can break MPs into NPs which remain in the atmosphere for more than 100 years. For instance, a plastic bottle of PE may take 450 years to disappear completely. Moreover, during their prolonged presence, MPs adsorb several toxic chemicals including heavy metals, organic pollutants, and metals, etc. which make them even more toxic.

12.7 Socioeconomic and Regulatory Barriers

The socioeconomic and regulatory dynamic acts as a significant barrier in MPs management. The substantial financial investment required for the development and implementation of specialized technologies such as filtration system of degradable polymers act as a barrier for large scale adoption for particularly economically weaker region (Prata et al., 2020). For example, the modification of wastewater treatment facility for effective removal of MPs requires expenditure in millions which is compounded by ongoing operational cost (Talvitie et al., 2017). The industries such as textiles and automotive manufacturing are the primary sources of MPs pollution but they prioritize economic profitability over environmental sustainability. This clearly exhibits the reluctance to upgrade the production methodology in the absence of fiscal incentives.

The regulatory frameworks also lag behind the scientific advancement. While European Union (EU) has restricted the use of microbeads in cosmetics, the global norms are not uniform to control MPs discharge. The tire wear and synthetic fibers which are key significant sources of MPs and MFs, they are not under the strict regulation due to the economic importance and complexity of enforcement. In poor and developing countries the poor waste management infrastructure increases the MPs pollution problem due to open landfill and unregulated incineration (Jambeck et al., 2015). Although the public awareness in increasing but it is not sufficient to drive the systematic change as many consumers are not aware of MPs sources like indoor textiles or tire dust (Gasperi et al., 2018). The gap between the scientific insight, regulatory actions and societal involvement emphasizes the demand for comprehensive, eco-friendly solutions that tackle both the technical and human dimension.

12.8 Principles of Green Solutions

The transition to green solutions for MPs waste management requires a paradigm shift from reactive, resource-intensive methods to proactive, sustainable approaches. These solutions prioritize environmental integrity, human health, and long-term ecological balance over short-term fixes. This section outlines three core principles—sustainability and circular economy, biodegradability and eco-friendly design, and minimal environmental impact—that underpin effective green strategies. Together, they provide a roadmap for reducing MPs pollution while aligning with global sustainability goals.

12.9 Sustainability and Circular Economy

Sustainability in MPs management hinges on the principles of the circular economy, which seeks to minimize waste by keeping materials in use for as long as possible (EllenMacArthur Foundation, 2013). Unlike the linear "take-make-dispose" model, a circular approach emphasizes reducing primary plastic production, reusing materials, and recycling waste into new products. For MPs, this means tackling sources like synthetic textiles and tire wear at the design stage for example, developing durable, recyclable fibers or rubber blends (Mihai et al., 2021). By closing material loops, such as upcycling MPs into construction aggregates, emissions can be curtailed significantly (Singh et al., 2017). This principle not only reduces the volume of MPs entering the environment but also conserves resources, aligning with the United Nations Sustainable Development Goals (SDG 12: Responsible Consumption and Production).

12.10 Biodegradability and Eco-Friendly Design

A cornerstone of green solutions is replacing persistent synthetic polymers with biodegradable alternatives that break down naturally without leaving harmful residues. Materials like polylactic acid (PLA) and polyhydroxyalkanoates (PHA), derived from renewable sources such as corn starch or microbial fermentation, offer viable substitutes for conventional plastics like polyethylene (Narancic & O'Connor, 2017). These biopolymers degrade into water, carbon dioxide, and biomass under composting conditions, minimizing long-term environmental accumulation. Eco-friendly design extends beyond material choice to encompass product lifecycles, for example, designing textiles that shed fewer microfibers or tires with natural rubber to reduce airborne particles (De Falco et al., 2018). By prioritizing biodegradability, green solutions address the root cause of MPs persistence, offering a preventive rather than remedial approach.

12.11 Minimal Environmental Impact

Green solutions must minimize ecological harm across their implementation. This principle rejects energy-intensive processes like incineration, which release greenhouse gases and secondary pollutants, in favor of low-impact alternatives (Verma et al., 2016). For instance, bio-based filtration systems using natural adsorbents like biochar capture MPs from water without chemical additives, preserving ecosystem health (Mohanty et al., 2018). Similarly, microbial degradation leverages naturally occurring bacteria and fungi to break down plastics, avoiding the toxicity of chemical catalysts (Yoshida et al., 2016). By reducing energy use, avoiding hazardous byproducts, and protecting biodiversity, this approach ensures that microplastic management does not exacerbate other environmental challenges like climate change or habitat loss.

These principles such as sustainability, biodegradability, and minimal impactform the basis of green solutions, guiding the development of technologies and policies that address microplastic pollution holistically. They shift the focus from end-of-pipe fixes to systemic change, setting the stage for the innovative applications explored in subsequent sections.

12.12 Green Technologies for MPs Management

The increasing crisis of MPs, urgently needed innovative technologies based on sustainability, biodegradability, and economic efficiency. Biodegradable polymers, bio-filtrations, microbial degradation are some of green technologies for MPs waste management which are shown in Fig. 1.2.

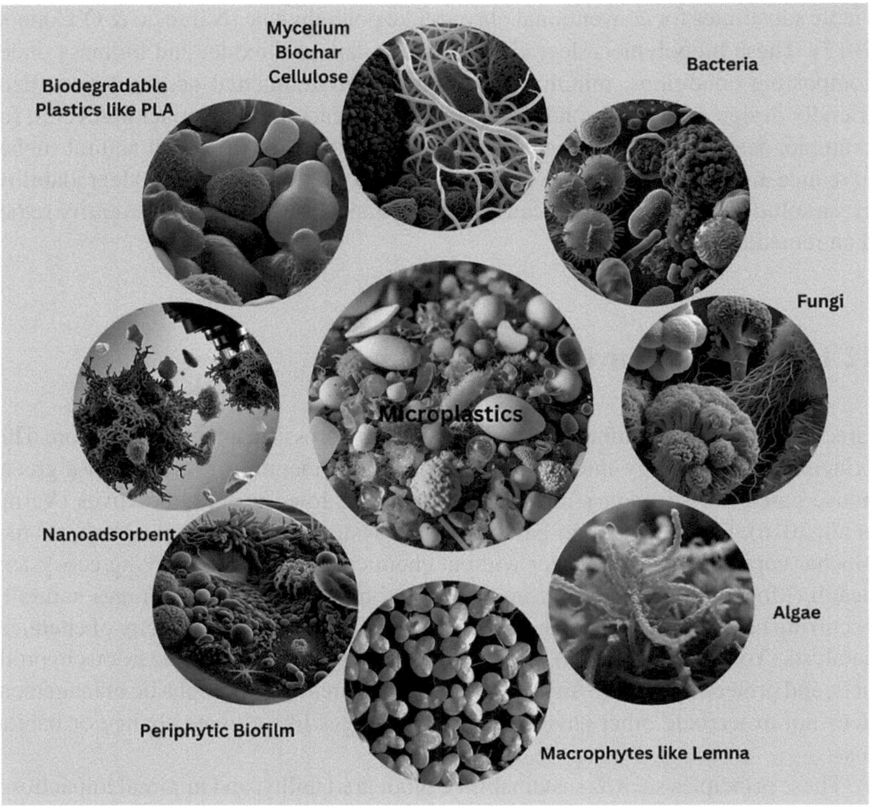

Fig. 1.2 Different green solutions to MPs pollution

12.13 Biodegradable Polymers

Since MPs are generated from made up of conventional polymers like PE, PP, PS, etc. which persist in the environment for centuries. The biodegradable polymers offer a potential green solution to reduce the MPs pollution at source. The use of the bio-based plastics such as polylactic acid, (PLA), polyhydroxyalkanoates, polybutylene succinate (PBS) polybutylene succinate adipate (PBSA), thermoplastic starch (TPS), and polycaprolactone can reduce the MPs pollution and mitigate long-term environmental accumulation as these polymers get degraded more readily under specific environmental conditions.

PLA is biodegradable polyester which is derived from the fermentation of starch rich crops such as sugarcane or corn (Narancic & O'Connor, 2017). It is the most common bio-plastics used in manufacturing of packaging materials, disposable tableware and medical implants. PLA has goof mechanical strength and elasticity required for making any durable plastic products. It degradation requires industrial composting facility which ultimately degrades into lactic acid (Jambeck et al., 2015).

PHA is also one of biodegradable polymer which is produced from bacterial fermentation. It has greater flexibility and biodegradability. When the PHA is blended with starch and combined with plasticizers, it provides a low-cost biodegradable plastic which gets degraded under specific moisture and temperature conditions. A study conducted by Narancic et al., revealed that 80% PHA can be degraded under marine condition within 90 days compared to negligible degradation of PE in the same environmental condition (Narancic et al., 2020).

PBS, and PBSA are also biodegradable aliphatic polyesters with enough mechanical strength and biodegradability. These are traditionally synthesized from petroleum resources but the recent developments have also enabled their production from renewable resources. However, the limited production and higher cost of production have restricted its widespread use as packaging materials. Its cost is being reduced by blending it with other polymers.

Similarly, plasticizing starch with water produces TPS which is a biodegradable polymer. It is often blended with other biodegradable polymers to improve its mechanical properties and water resistance. TPS is generally used in packaging agriculture, and disposable items. However, the TPS-based materials are brittle in nature and highly sensitive to moisture which is needed to be modified for its broader use. A focused research and development is needed to improve its properties.

12.14 Scalability and Practical Implementation

The scalability of biodegradable plastics depends on the capacity, production cost and infrastructure. The manufacturing of PLA has reached at industrial scale with total global production exceeding 300,000 tons per year by 2023. However, due to the high industrial fermentation cost, the production of PHA is limited till date. The lesser durability of starch-based polymers has restricted their wide use, although these are widely accessible. The development of such biodegradable polymers require huge investment in composting infrastructure as these polymers degrade under controlled environmental conditions at 50–60 °C, and high humidity.

12.15 Bio-based Filtration System

The bio-based filtration system is an eco-friendly filtration system which uses natural materials like biochar, mycelium, and cellulose that act as sustainable alternatives to synthetic fibers. Biochar is a carbon-rich material which is obtained from pyrolysis of biomass. It has a porous structure which traps the MPs even of the size of 10 μm. Mohanty et al. (2018), have showed its adsorption capacity of MPs was more than 80% in wastewater trial study. The mycelium of fungi is network of root-like structure which has been engineered into a biodegradable filter. Such filters are efficient in removing microfibers released from laundry effluent and reduce them up to 70%

(Zhang et al., 2020). Such filtration system doesn't use any chemical additives and get decomposed naturally after use. Dris et al., have found that the cellulose-based air filters have captured the synthetic fibers from the indoor air at the rate similar to HEFA filters. However, there is limited use of these filters in large scale application such as in municipal wastewater treatment where it needs further optimization to be used.

12.16 Microbial Degradation

The natural breakdown of plastic wastes into tiny pieces using enzymatic activity of microorganism is known as biodegradation. This process of biodegradation completed into following steps: (i) formation of preparatory layer, (ii) microbial settlement, (iii) biological degradation, (iv) absorption, and (v) conversion to basic compounds.

When the MPs get exposed to water, a thin layer of microbial coating starts to develop on its surface which depends on the chemical and physical properties. Further, the microbes begin to develop into the surface imperfections such as cracks and grooves of MPs. The bacterial get adhered itself on the surface of MPs with the help of extracellular polymeric substance secreted by it. This enzyme plays an important role in the degradation of MPs. The enzyme accelerates the fragmentation of carbon framework of plastic, and split it into smaller unit such as oligomers, dimers, and monomers. This enzymatic action is increases the hydrolysis and produces smaller fragments which can be absorbed by bacteria.Since, the absorption occurs only when the size of monomer is too small that can enter through the cell wall of the bacterial where the absorb it and use as source of carbon and energy. This is followed by a final phase of mineralization where these fragments are transformed into simple substances like carbon dioxide, water, or methane depending upon the availability and unavailability of oxygen.

12.17 Bacterial Degradation

As the MPs act as a surface for the development of bacterial film which uses the plastic polymers as carbon source under nutrient deficient environment. This process reduces the biomass of the polymer, molecular size as well as structural uniformity. Auta et al. (2022), have showed that *Bacillus cereus* and *Bacillus gottheilii* degraded pre-exposed MPs. These bacterial strains responded differently to different plastics. *B. cereus* was efficient in breaking the PS which reduces its weight significantly, whereas *B. gottheilii* degrade a broader range of polymers composition. Yang et al. (2015) have studied the degradation potential of *Bacillus* sp. *YP1*. The result revealed that the surface of the PE got damaged by *Bacillus* sp. *YP1* by forming pits and holes and adding carbonyl groups and reduced 10.7% weight in 60 days (Yang et al., 2015).

In a similar study performed by Shimpi et al. (2012) corroborated the degradation of PLA up to 9.9% within a month using *Pseudomonas aeruginosa.*

12.18 Fungal Degradation

Fungi are well-suited for the breakdown of MPs due to their robust metabolic systems which produces different types of extracellular enzymes. These enzymes play an important role I degradation of plastic structures. Fungi are better in degrading polymers as compared to bacteria. The studies carried out by researchers showed that fungi are capable of degrading plastics effectively. Paço et al. (2017) revealed that the fungus *Maritimum* exhibited 43% degradation rate for PE when grown in nutrient deficient environment. The *Maritimum* used plastics as carbon sole carbon source. This fungus significantly damaged the film of the PE polymer, and reduced the mass while the biomass of the fungus found to be increased.

Ultraviolet (UV) light plays an important role in the initiation of oxidation of PE by forming carbonyl groups which facilitate the attack of microbes on the shorter chains and accelerates the biodegradation. Sowmya et al. (2014) used *Trichoderma harzianum* for the degradation of UV-treated PE. The UV-treated PE led to the development of new cavities, erosion surfaces and development of new chemical functional groups which were also confirmed through FTIR and NMR studies. Yamada-Onodera et al. (2001) used *Penicillium simplicissimum* to degrade PE and achieved notable reduction in molecular weight after prolonged incubation due to enzymatic activity which breakdown the polymer into smaller fragment.

The fungus also degrade polyurethane potentially even under less ideal condition or landfill conditions. Khan et al. (2017) found that *Aspergillus tubingensis* degraded polyurethane by causing the development of visible surface cracks, pores, and loss of tensile strength. Similarly, Sánchez-Duque et al. (2020) exhibited significant degradation of polyurethane using *Fusarium solani* and found that the weight of the polyurethane was reduced by 30% in three months. The degradation of PVC was also demonstrated by Ali et al. (2014) using *Phanerochaete chrysosporium* which colonized on the surface of PVC films. This fungus used the polymers as its main source of carbon and degraded it.

12.19 Removal of MPs by Algae

The algae can settle on the surface of plastic and extracts the carbon of polymer as an energy sources for its growth. Simultaneously, it also releases enzymes which degraded the polymeric structure. The microalgae are considered as a promising option for biologically degrading MPs due to their ability to breakdown plastic compounds using their own enzymatic mechanisms. The studies have observed algae forming colonies on the synthetic surfaces of PE in wastewater, with these communities providing to be nontoxic and environmentally safe. The process of degradation begins when these algae get attached to the surface of plastic. Further they produce ligninolytic and exopolysaccharide enzyme for breaking it down. These enzymes interact with the plastic macromolecules in liquid environment, initiating decomposition. The algae utilize the polymeric carbon of plastics which resulted in species with higher cellular protein and carbohydrate levels and faster growth rates on the surface of PE.

There are five important stages in algal plastic degradation: surface fouling, corrosion, hydrolysis and penetration, leaching of components and pigment diffusion into polymers structure. Kumar et al. (2017) have found that the cyanobacterium, *Anabaena spiroides* achieved the highest degradation rate of low density polyethylene (LDPE) at 8.18% which was followed by diatom, *Navicula pupula* at 4.44%, and green alga *Scenedesmus dimorphus* at 3.74%. Sarmah and Rout demonstrated that freshwater cyanobacteria like *Phormidium lucidum* and *Oscillatoria subbrevis*, nontoxic, fast-growing and easily isolated algae can colonize and degrade LDPE without any pretreatment or additives. Gulnaz and Dıncer (2009) have investigated the degradation of bisphenol A (BPA) using *Aeromonas hydrophilia* bacteria and *Chlorella vulgaris* microalgae. They observed that the level of BPA decreased below detection limit within 168 h, showing no estrogenic effects. Similarly, Hirooka et al. (2005) reported that *Chorellafusca* var. vacuolate transformed BPA into non-estrogenic compounds.

The genetic engineering has also played an important role in improving the potential of microalgae as a microbial factories for degradation of plastics and MPs. Kim et al. (2020) used *Phaeodactylum tricornutum* to produce PETase which effectively degraded both PET and polyethylene glycol (PETG) and paved the way for sustainable method of degradation of MPs.

12.20 Removal of MPs by Macrophytes

Microphytes are widely used in biomonitoring as they act as bioindicators and bioaccumulators. The contaminated tissue of macrophytes acts as a reservoir of pollutants which transfer the MPs to the higher trophic level and prolonging their presence in aquatic systems. Sfriso et al. (2024) found MPs in 94% macrophytes samples ranging from 0.16 to 330 items per gram of fresh weight, averaging 14 items across species and sites. Rozman et al. (2023) examined long-term interactions between LDPE MPs and the floating macrophytes Lemna minor, developing a phytoremediation approach. The research revealed that Lemna minor tolerated high MPs level with 25% of introduced MPs adhering to its biomass within seven days which supported phytoremediation as a viable strategy for MPs stabilization and removal.

12.21 Degradation of MPs by Periphytic Biofilms

The periphyton, including epiphyton and epixylon, form the foundation of a widely recognized biological method for MPs degradation which is known as periphytic biofilm degradation. These biofilms play an important role in aquatic ecosystems such as stabilizing carbon and cycling nutrients. Due to their sensitivity to environmental changes, they have been utilized as bioindicators in ecotoxicological studies assessing the effects of pollution on water bodies. In freshwater ecosystems, the perihytic biofilms consist of diverse microorganisms such as cyanobacteria, algae, diatoms, and protozoans along with organic debris, either attached to submerged surfaces or suspended in water. These biofilms are classified into five types based on their substrate: epiphyton (on pants), epilithon (on rocks), epipelon (on sediments), epixylon (on woods), and epipsammon (on sand). The photoautotrophic benthic biofilms serve as primary producers in these aquatic environments.

The periphytic biofilms can significantly alter the properties and structure of MPs. Through microbial enzymes, they modify the surface features, breakdown additives, and generate metabolic byproducts. The enzymatic degradation occurs via two mechanisms: surface alteration or complete breakdown. The enzymes like oxidases, amidases, laccases, hydrolases, and peroxidases directly degrade the polymers, while others such as hydrolases focus on the change in surface structure. The microorganism breakdown the structure of large polymeric unit using these enzymes extracellularly and then absorb the resulting fragments into their cells for energy and growth through metabolic processes. Although the biofilm formation and MPs degradation are natural in aquatic environment, the rate of degradation is typically slow. Syranidou et al. (2017) tested native and enhanced microbial consortia in a simulated marine environment and found that the bioaugmentation of consortia reduced the polystyrene (PS) particles weight by 4.7% over six months as compared to just 0.19% by native consortia which suggest that the biotechnology can enhance the process. Shabbir et al. (2020) introduced a novel method of degradation of PET, PE and PP using periphytic biofilms with varying carbon source. They observed that after 60 days, the natural biofilm degradation rate was 13.24% for PET, 5.95% for PE, and 9.52% for PP which increased to 19.72%, 14.02%, and 18.02% respectively when glucose was added as a supplement.

Gao and Sun (2021) have developed an innovative strategy by reconstructing bacterial communities within biofilms. After screening several plastic waste samples, they identifies three plastic degrading bacteria and demonstrated their reconstituted community's ability to breakdown PE and PE using advanced techniques to analyze the resulting byproducts. This approach highlighted the potential of tailored marine bacterial biofilms for MPs decomposition. Liu et al. (2021) showed other methods using biofilms sticky extracellular polysaccharides to trap and aggregate MPs. They engineered a biofilm system to capture MPs for isolation and later release them via an inducible dispersion mechanism, validating this "capture-release" approach and demonstrating its effectiveness in reducing MPs levels in ocean water near sewage outfalls.

Efforts continue to explore the potential of biofilms for marine ecosystem cleanup amid rising MPs pollution. However, gaps remain in understanding microbial populations on biofilms-coated MPs, the factors driving their colonization, and their interactions with plastic substrate. Further research is essential to unravel the ecological roles of these microbial communities and optimize their use in mitigating MPs contamination in aquatic environments.

12.22 Nanotechnology

Nanotechnology provides a potential tool for MPs removal hile adhering to green principles through sustainable design. Nanoadsorbents such as grapheme oxide and cellulose nanofibre selectively bind with MPs in water and air and removal efficiently above 95% MPs even smaller than 100 nm (Wang et al., 2018). These nanoadsorbents are bio-derived; the cellulose nanofibres are derived from wood. The photocatalytic nanoparticles like titanium dioxide (TiO_2) modified with plant extracts, degrade MPs under sunlight into carbon dioxide and water. The experiments showed 70% ass reduction of PS in 24 h (Ariza-Tarazona et al., 2019). Unlike traditional catalyst, these green nanomaterials minimize the toxicity and energy use, with renewable synthesis cutting production emission by 30% (Wang et al., 2018). Such technologies are used in wastewater treatment or in air purifier which bridges the high efficiency and ecological safety.

12.23 Conclusion

MPs pollution presents a formidable challenge to global ecosystems, infiltrating air, water, and soil with persistent particles that defy conventional management. This chapter has explored green solutions as a transformative response, offering sustainable pathways to mitigate this crisis. From the vast scale of tire wear emissions to the insidious spread of indoor microfibers, the challenges are clear: traditional methods like landfilling and incineration fall short, while socioeconomic barriers hinder progress. Yet, the principles of sustainability, biodegradability, and minimal environmental impact provide a robust foundation for change. Innovative technologies such as biodegradable polymers, bio-based filtration, microbial degradation, and green nanotechnology demonstrate the potential to prevent, capture, and eliminate MPs without ecological trade-offs. These are complemented by waste management strategies that reduce sources, enhance collection, and repurpose waste, supported by real-world initiatives from community recycling to industrial biorefineries. Policy frameworks and economic incentives, paired with heightened public awareness, create the enabling environment needed to scale these solutions, turning scientific promise into practical impact. Looking ahead, the future of MPs management lies in refining these green approaches. Long-term studies on biodegradable materials, scalable microbial and nanoscale systems, and interdisciplinary collaboration will bridge current gaps, ensuring solutions are both effective and equitable. The stakes are high because MPs threaten biodiversity, human health, and climate stability but so too is the opportunity. Green solutions offer more than mitigation; they represent a shift toward a circular, resilient world where waste becomes a resource, not a burden. This chapter underscores that no single technology or policy can solve the MPs crisis alone. Success demands an integrated effort—scientists innovating, industries adapting, policymakers regulating, and societies engaging. As the evidence mounts and solutions evolve, the call is clear: act now, with purpose and unity, to reclaim our environment from the pervasive grip of MPs. Green solutions are not just a possibility, they are the imperative for a sustainable future.

References

Ali, M. I., Ahmed, S., Javed, I., Ali, N., Atiq, N., Hameed, A., & Robson, G. (2014). Biodegradation of starch blended polyvinyl chloride films by isolated *Phanerochaete chrysosporium* PV1. *International Journal of Environmental Science and Technology, 11*, 339–348.

Allen, S., Allen, D., Phoenix, V. R., Le Roux, G., Durántez Jiménez, P., Simonneau, A., et al. (2019). Atmospheric transport and deposition of microplastics in a remote mountain catchment. *Nature Geoscience, 12*(5), 339–344.

Andrady, A. L. (2011). Microplastics in the marine environment. *Marine Pollution Bulletin, 62*(8), 1596–1605.

Ariza-Tarazona, M. C., Villarreal-Chiu, J. F., Barbieri, V., Siligardi, C., & Cedillo-González, E. I. (2019). New strategy for microplastic degradation: Green photocatalysis using a protein-based porous N-TiO2 semiconductor. *Ceramics International, 45*(7), 9618–9624.

Auta, H. S., Abioye, O. P., Aransiola, S. A., Bala, J. D., Chukwuemeka, V. I., Hassan, A., et al. (2022). Enhanced microbial degradation of PET and PS microplastics under natural conditions in mangrove environment. *Journal of Environmental Management, 304*, Article 114273.

Barnes, D. K., Galgani, F., Thompson, R. C., & Barlaz, M. (2009). Accumulation and fragmentation of plastic debris in global environments. *Philosophical Transactions of the Royal Society b: Biological Sciences, 364*(1526), 1985–1998.

Boucher, J., & Friot, D. (2017). Primary microplastics in the oceans: A global evaluation of sources. IUCN.

Carbery, M., O'Connor, W., & Palanisami, T. (2018). Trophic transfer of microplastics and mixed contaminants in the marine food web and implications for human health. *Environment International, 115*, 400–409.

Cole, M., Lindeque, P., Halsband, C., & Galloway, T. S. (2011). Microplastics as contaminants in the marine environment: A review. *Marine Pollution Bulletin, 62*(12), 2588–2597.

De Falco, F., Gullo, M. P., Gentile, G., Di Pace, E., Cocca, M., Gelabert, L., et al. (2018). Evaluation of microplastic release caused by textile washing processes of synthetic fabrics. *Environmental Pollution, 236*, 916–925.

Dris, R., Gasperi, J., Mirande, C., Mandin, C., Guerrouache, M., Langlois, V., & Tassin, B. (2017). A first overview of textile fibers, including microplastics, in indoor and outdoor environments. *Environmental Pollution, 221*, 453–458.

Evangeliou, N., Grythe, H., Klimont, Z., Heyes, C., Eckhardt, S., Lopez-Aparicio, S., & Stohl, A. (2020). Atmospheric transport is a major pathway of microplastics to remote regions. *Nature Communications, 11*(1), 3381.

Farid, M. A. A., Norrahim, M. N. F., Lease, J., & Andou, Y. (2024). Microplastic reduction in municipal solid waste (MSW) landfill. In: Technical landfills and waste management: Volume 2: Municipal solid waste management (pp. 57–86). Springer Nature Switzerland.

Gao, R., & Sun, C. (2021). A marine bacterial community capable of degrading poly (ethylene terephthalate) and polyethylene. *Journal of Hazardous Materials, 416*, Article 125928.

Gasperi, J., Wright, S. L., Dris, R., Collard, F., Mandin, C., Guerrouache, M., et al. (2018). Microplastics in air: Are we breathing it in? *Current Opinion in Environmental Science & Health, 1*, 1–5.

Geyer, R., Jambeck, J. R., & Law, K. L. (2017). Production, use, and fate of all plastics ever made. *Science Advances, 3*(7), Article e1700782.

Gulnaz, O., & Dıncer, S. (2009). Biodegradation of bisphenol A by *Chlorella vulgaris* and *Aeromonas hydrophilia. Journal of Applied Biological Sciences, 3*(2), 79–84.

Hirooka, T., Nagase, H., Uchida, K., Hiroshige, Y., Ehara, Y., Nishikawa, J. I., et al. (2005). Biodegradation of bisphenol A and disappearance of its estrogenic activity by the green alga *Chlorella fusca* var. vacuolata. *Environmental Toxicology and Chemistry, 24*(8), 1896–1901.

Jacquin, J., Cheng, J., Odobel, C., Pandin, C., Conan, P., Pujo-Pay, M., et al. (2019). Microbial ecotoxicology of marine plastic debris: A review on colonization and biodegradation by the "Plastisphere." *Frontiers in Microbiology, 10*, 865.

Jambeck, J. R., Geyer, R., Wilcox, C., Siegler, T. R., Perryman, M., Andrady, A., et al. (2015). Plastic waste inputs from land into the ocean. *Science, 347*(6223), 768–771.

Khan, S., Nadir, S., Shah, Z. U., Shah, A. A., Karunarathna, S. C., Xu, J., et al. (2017). Biodegradation of polyester polyurethane by *Aspergillus tubingensis. Environmental Pollution, 225*, 469–480.

Kim, J. W., Park, S. B., Tran, Q. G., Cho, D. H., Choi, D. Y., Lee, Y. J., & Kim, H. S. (2020). Functional expression of polyethylene terephthalate-degrading enzyme (PETase) in green microalgae. *Microbial Cell Factories, 19*, 1–9.

Krueger, M. C., Harms, H., & Schlosser, D. (2015). Prospects for microbiological solutions to environmental pollution with plastics. *Applied Microbiology and Biotechnology, 99*, 8857–8874.

Kumar, R. V., Kanna, G. R., & Elumalai, S. (2017). Biodegradation of polyethylene by green photosynthetic microalgae. *Journal of Bioremediation Biodegradation, 8*(381), 2.

Law, K. L., & Thompson, R. C. (2014). Microplastics in the seas. *Science, 345*(6193), 144–145.

Liu, S. Y., Leung, M. M. L., Fang, J. K. H., & Chua, S. L. (2021). Engineering a microbial 'trap and release' mechanism for microplastics removal. *Chemical Engineering Journal, 404*, Article 127079.

MacArthur, E. (2013). Towards the circular economy. *Journal of Industrial Ecology, 2*(1), 23–44.

Martín-González, D., de la FuenteTagarro, C., De Lucas, A., Bordel, S., & Santos-Beneit, F. (2024). Genetic modifications in bacteria for the degradation of synthetic polymers: A review. *International Journal of Molecular Sciences, 25*(10), 5536.

McCormick, A., Hoellein, T. J., Mason, S. A., Schluep, J., & Kelly, J. J. (2014). Microplastic is an abundant and distinct microbial habitat in an urban river. *Environmental Science & Technology, 48*(20), 11863–11871.

Mihai, F. C., Gündoğdu, S., Markley, L. A., Olivelli, A., Khan, F. R., Gwinnett, C., et al. (2021). Plastic pollution, waste management issues, and circular economy opportunities in rural communities. *Sustainability, 14*(1), 20.

Mohanty, A. K., Vivekanandhan, S., Pin, J. M., & Misra, M. (2018). Composites from renewable and sustainable resources: Challenges and innovations. *Science, 362*(6414), 536–542.

Narancic, T., & O'Connor, K. E. (2017). Microbial biotechnology addressing the plastic waste disaster. *Microbial Biotechnology, 10*(5), 1232.

Narancic, T., Cerrone, F., Beagan, N., & O'Connor, K. E. (2020). Recent advances in bioplastics: Application and biodegradation. *Polymers, 12*(4), 920.

Paço, A., Duarte, K., da Costa, J. P., Santos, P. S., Pereira, R., Pereira, M. E., et al. (2017). Biodegradation of polyethylene microplastics by the marine fungus *Zalerion maritimum. Science of the Total Environment, 586*, 10–15.

Padervand, M., Lichtfouse, E., Robert, D., & Wang, C. (2020). Removal of microplastics from the environment. A review. *Environmental Chemistry Letters, 18*(3), 807–828.
Prata, J. C., Da Costa, J. P., Lopes, I., Duarte, A. C., & Rocha-Santos, T. (2020). Environmental exposure to microplastics: An overview on possible human health effects. *Science of the Total Environment, 702*, Article 134455.
Ragaert, K., Delva, L., & Van Geem, K. (2017). Mechanical and chemical recycling of solid plastic waste. *Waste Management, 69*, 24–58.
Rochman, C. M., Hoh, E., Kurobe, T., & Teh, S. J. (2013). Ingested plastic transfers hazardous chemicals to fish and induces hepatic stress. *Scientific Reports, 3*(1), 3263.
Rozman, U., Blažič, A., & Kalčíková, G. (2023). Phytoremediation: A promising approach to remove microplastics from the aquatic environment. *Environmental Pollution, 338*, Article 122690.
Sánchez-Duque, J. A., Arce-Villalobos, L. R., & Rodríguez-Morales, A. J. (2020). Enfermedadpor coronavirus 2019 (COVID-19) en América Latina: Papel de la atenciónprimaria en la preparación y respuesta. *Atenciónprimaria, 52*(6), 369.
Sfriso, A. A., Juhmani, A. S., Tomio, Y., Sfriso, A., Rizzolio, F., Adeel, M., et al. (2024). Microplastic accumulation and ecological impacts on benthic invertebrates: Insights from a microcosm experiment. *Marine Pollution Bulletin, 202*, Article 116231.
Shabbir, S., Faheem, M., Ali, N., Kerr, P. G., Wang, L. F., Kuppusamy, S., & Li, Y. (2020). Periphytic biofilm: An innovative approach for biodegradation of microplastics. *Science of the Total Environment, 717*, Article 137064.
Shimpi, N., Borane, M., Mishra, S., & Kadam, M. (2012). Biodegradation of polystyrene (PS)-poly (lactic acid) (PLA) nanocomposites using *Pseudomonas aeruginosa. Macromolecular Research, 20*, 181–187.
Singh, N., Hui, D., Singh, R., Ahuja, I. P. S., Feo, L., & Fraternali, F. (2017). Recycling of plastic solid waste: A state of art review and future applications. *Composites Part B: Engineering, 115*, 409–422.
Sowmya, H. V., Ramalingappa, Krishnappa, M., & Thippeswamy, B. (2014). Degradation of polyethylene by *Trichoderma harzianum*—SEM, FTIR, and NMR analyses. *Environmental monitoring and assessment, 186*, 6577–6586.
Syranidou, E., Karkanorachaki, K., Amorotti, F., Franchini, M., Repouskou, E., Kaliva, M., et al. (2017). Biodegradation of weathered polystyrene films in seawater microcosms. *Scientific Reports, 7*(1), 17991.
Talvitie, J., Mikola, A., Koistinen, A., & Setälä, O. (2017). Solutions to microplastic pollution—Removal of microplastics from wastewater effluent with advanced wastewater treatment technologies. *Water Research, 123*, 401–407.
Verma, R., Vinoda, K. S., Papireddy, M., & Gowda, A. N. S. (2016). Toxic pollutants from plastic waste-a review. *Procedia Environmental Sciences, 35*, 701–708.
Wang, Z., Li, H., Tang, F., Ma, J., & Zhou, X. (2018). A facile approach for the preparation of nano-size zinc oxide in water/glycerol with extremely concentrated zinc sources. *Nanoscale Research Letters, 13*, 1–9.
Wright, S. L., & Kelly, F. J. (2017). Plastic and human health: A micro issue? *Environmental Science & Technology, 51*(12), 6634–6647.
Yamada-Onodera, K., Mukumoto, H., Katsuyaya, Y., Saiganji, A., & Tani, Y. (2001). Degradation of polyethylene by a fungus *Penicillium simplicissimum* YK. *Polymer Degradation and Stability, 72*(2), 323–327.
Yang, Y., Chen, J., Wu, W. M., Zhao, J., & Yang, J. (2015). Complete genome sequence of *Bacillus* sp. YP1, a polyethylene-degrading bacterium from waxworm's gut. *Journal of Biotechnology, 200*, 77–78.
Yoshida, S., Hiraga, K., Takehana, T., Taniguchi, I., Yamaji, H., Maeda, Y., et al. (2016). A bacterium that degrades and assimilates poly (ethylene terephthalate). *Science, 351*(6278), 1196–1199.

Zettler, E. R., Mincer, T. J., & Amaral-Zettler, L. A. (2013). Life in the "plastisphere": Microbial communities on plastic marine debris. *Environmental Science & Technology, 47*(13), 7137–7146.

Zhang, Q., Xu, E. G., Li, J., Chen, Q., Ma, L., Zeng, E. Y., & Shi, H. (2020). A review of microplastics in table salt, drinking water, and air: Direct human exposure. *Environmental Science & Technology, 54*(7), 3740–3751.

MIX
Papier aus verantwortungsvollen Quellen
Paper from responsible sources
FSC® C105338

If you have any concerns about our products,
you can contact us on
ProductSafety@springernature.com

In case Publisher is established outside the EU,
the EU authorized representative is:
Springer Nature Customer Service Center GmbH
Europaplatz 3, 69115 Heidelberg, Germany

Printed by Libri Plureos GmbH
in Hamburg, Germany